DATE			

BAKER & TAYLOR

ORGANIC MOLECULAR CRYSTALS

Interaction, Localization, and
Transport Phenomena

ORGANIC MOLECULAR CRYSTALS

Interaction, Localization, and Transport Phenomena

Edgar A. Silinsh
Institute of Physical Energetics
Latvian Academy of Sciences, Riga

Vladislav Čápek
Institute of Physics of
Charles University, Prague

American Institute of Physics New York

AIP Press
American Institute of Physics
500 Sunnyside Boulevard
Woodbury, NY 11797-2999

Library of Congress Cataloging-in-Publication Data
Silinsh, E.
    Organic molecular crystals : interaction, localization, and transport phenomena / Edgar A. Silinsh, Vladislav Čápek.
        p. cm.
    ISBN 1-56396-069-9
    1. Molecular crystals. 2. Organic compounds--Electric properties.
I. Čápek, Vladislav. II. Title.
QD921.S53935 1994
530 .4'13--dc20   94-17192
                                                              CIP

10 9 8 7 6 5 4 3 2 1

This book is dedicated to
our wives
Dace and Jana
and children
Baiba, Sillvie, and Toms

Contents

Part 1A
Characteristic Features of Organic
Molecular Crystals

Symbols and Abbreviations

SYMBOLS

$a^\dagger$, $A^\dagger$	general particle creation operators
a, A	general particle annihilation operators
$A^\dagger$, A	electron creation and annihilation operators
$\bar{a}^\dagger$, $\bar{a}$	exciton creation and annihilation operators
$B^\dagger$, B	hole creation and annihilation operators
$b^\dagger$, $B^\dagger$	general phonon creation operators
b, B	general phonon annihilation operators
$b^\dagger$, b	creation and annihilation operators of vibronic (intramolecular) quanta
b_i	diagonal components of molecular polarizability tensor
$c^\dagger$, c	creation and annihilation operators of lattice phonons
D	diffusion coefficient; dispersion (London) interaction term; projector (projection superoperator)
$\mathbf{d}$	electric dipole moment
$\mathbf{d}_i$, $\mathbf{d}_f$	transition dipole moments
$\mathcal{E}$	electric field strength
E	energy, total energy
E_a	activation energy of thermal dissociation of charge pair (CP) state; activation energy of dark conductivity
E_a^{ph}	activation energy of photogeneration
$(E_b)_{eff}$; $(E_b)_{MP}$	formation energy of molecular polaron
$(E_b)_i$	formation energy of molecular ion
E_{CP}	energy of a charge pair (CP) state
E_{CT}	optical charge transfer energy
E_G^{Ad}	adiabatic energy gap
E_G^{Opt}	optical energy gap
E_k	kinetic energy
E_{LR}	lattice relaxation energy
E_p^+, E_p^-	self-energy levels of electronic polaron
E_{ST}	exciton self-trapping depth
E_{th}	threshold value of intrinsic photoconductivity
e	elementary electron charge ($e < 0$)
F	force
f	oscillation strength; linear coupling constant

xiii

G	charge carrier transit coefficient
g, g_0	linear coupling constants
H, h, $\mathcal{H}$	Hamiltonians, their matrix elements
$\hbar = h/2\pi$	Planck constant divided by 2π
I	resonance interaction term; initial condition term
J_{mn}, J_{ij}	resonance (hopping, transfer) integrals
$\mathbf{k}$	wave vector
k_{AI}	rate constant of direct autoionization
k'_{AI}	rate constant of autoionization after intramolecular relaxation
k_B, k	Boltzmann constant
k_i	rate constant of intramolecular relaxation
k_n	rate constant of vibrational relaxation
k_m	rate constant of internal conversion $S_n \rightarrow S_m$
L, $\mathcal{L}$	Liouville superoperators
L	Lorentz-factor tensor; thickness of a sample
L_s	mean diffusion length of S exciton
$\bar{l}$, l_o	mean free path
m_{eff}	effective mass of the quasi-particle
m_e	electron mass
m^*	relative effective mass (in m_e units)
M_P^+, M_P^-	conduction levels of molecular polarons
n	refraction coefficient (index)
n_B	Bose-Einstein distribution
$\mathcal{N}$	symbol of normal product
N	number of elementary cells in the basic region
N_t	density of trapping states
q, Q	charge
P	electronic polarization energy; probability
P_{eff}	effective electronic polarization energy
P_{id}	self-consistent energy of charge-induced dipole interaction
$\mathbf{r}$	electron coordinates
$\mathbf{R}$, R	nuclear coordinates
r_0	charge carrier separation length
r_{CT}	optical charge transfer distance
r_c	critical Coulombic radius of charge pair separation
r_{th}	thermalization length
S	exciton-phonon coupling constant; overlap factor (integral); charge carrier transit range
T	temperature; relaxation times
t	time
t_{tr}	transit time (time-of-flight) of charge carrier
u_L	longitudinal sound velocity
U, U_0	potential energy; Van der Waals interaction energy
U_{CP}	bond energy of charge pair state

$\langle v \rangle$	mean velocity
v_{th}	mean thermal velocity of charge carrier
v_d	mean drift velocity of charge carrier
$(v_d)_s$	saturation value of drift velocity
v_{max}	maximum velocity of charge carrier
W	interaction energy with phonons; hopping (transfer) rate; distribution function in Fokker-Planck equation
W_C	Coulombic interaction energy
W_{d-d}	term of dipole-dipole interaction
W_{eff}	total effective (electronic+vibrational) polarization energy
W_{q-d}	term of charge-induced dipole interaction
W_{q-Q_o}, W_{Q_o}	term of charge-quadrupole interaction
w	memory functions
z	number of molecules in a unit cell
Z	charge in $|e|$ units
α	mean molecular polarizability; absorption coefficient (cm^{-1})
α_{ij}	molecular polarizability tensor
β	relaxation rate constant of thermalization; reciprocal temperature in energy units
β_m	friction coefficient
γ, Γ	Haken-Strobl-Reineker parameters; Grover-Silbey parameters
δf	quadratic coupling constant
δg	quadratic coupling constant
$\Delta E, \quad \delta E$	bandwidth
ΔE_D	Davydov splitting
ϵ	single-particle energies
ϵ, ϵ_0	dielectric permeability (permittivity)
$\bar{\epsilon}$	mean (isotropic) dielectric permeability (permittivity)
ϵ_i	diagonal component of dielectric permeability tensor
η	quantum efficiency of photoconductivity
Θ	electric quadrupole moment
Θ_{AB}	quadrupole moment tensor
μ	phonon occupation numbers; mobility; transition dipole moment
μ_0	microscopic mobility of charge carrier
μ_{eff}	effective mobility of charge carrier
μ_{ij}	mobility tensor
ν	frequency
$\nu^{(0)}$	electron-electron interaction energy
ρ	density matrix
σ	dispersion parameter of Gaussian distribution; steepness coefficient of the Urbach rule
σ_0	interaction constant of the Urbach rule
τ	quasiparticle lifetime
τ_r	characteristic relaxation time

τ_D	decay time
τ_{tr}	transit time (time-of-flight) of charge carrier
τ_{th}	mean thermalization time
τ_h	mean hopping (residence) time of charge carrier
τ_e	characteristic electronic polarization time
τ_v	characteristic vibronic (molecular) polarization time
τ_l	characteristic lattice polarization time
τ_t	mean trapping time of charge carrier
φ	wave function, potential energy
$\Phi_0 (h\nu)$	photogenerated geminate charge pair (CP) quantum yield
Φ_0^{AI}	autoionization quantum yield
Φ_0^{CT}	optical charge transfer (CT) transition quantum yield
χ	wave function
ψ, Ψ	wave functions
ψ	field operator
ω	circular frequency
Ω	charge pair (CP) state dissociation efficiency (probability)

ABBREVIATIONS

Ac	anthracene
AI	autoionization
CP	charge pair (state)
CT	charge transfer (state, transition)
CTRW	continuous-time-random-walk (method)
GME	generalized master equations
GSLEM	generalized stochastic Liouville equation models
LB	Langmuir-Blodgett (layers)
MP	molecular polaron
Nph	naphthalene
OMC	organic molecular crystals
Pc	pentacene
Phc	phthalocyanine
Pl	perylene
Py	pyrene
S	singlet exciton
SM	Sano-Mozunder's (model)
T	triplet exciton
Tc	tetracene
TOF	time-of-flight (method)
VdW	Van der Waals (interaction)

Foreword

There is heightened scientific interest at present in the electronic properties of organic materials. These materials respond in a distinctive way to light, temperature, and an electric field. In addition, the discovery of electroluminescence and metallic conductivity in polymers, and of superconductivity in crystals, has caught the attention of the industrial world. To understand the physics underlying these phenomena, it is helpful to learn the language and the basic processes that characterize this subject. This book contains reliable experimental data and a plausible phenomenological and theoretical treatment, and it will satisfy that requirement.

This work is a tour de force of scholarly focus on the diminishing number of problems that bar the way to a satisfactory quantum mechanical description of the energetics and dynamics of charge carrier generation, transport, and recombination in single-component organic molecular crystals (OMC).

The central theme of the book is that the excited electronic state (either the exciton or charge carrier, but more appropriately the carrier) within the crystal polarizes the molecule on which it resides and also the electrons and nuclei of the molecules that are its nearest neighbors. This crystal state is referred to as a nearly small molecular polaron, to distinguish it from the well-known small polaron and large polaron. Using this theme, the authors compose a scientific symphony, clarifying and rationalizing all the known experimental results.

The organization of the book is novel in the sense that the contributions of the experimentalist (Silinsh) and the theorist (Čápek) are essentially separate. This has the pedagogical advantage of keeping the phenomenological story line more compact and in focus. As counterpoint to the phenomenological melody is the underlying theoretical supporting refrain that nevertheless emphasizes the fact that the symphony is as yet unfinished.

This book has the great advantage of representing the considered opinion of just two authors, so the presentation is laudably coherent. Its style is distinctly personal, reflecting the cultural and pedagogical inclinations of the authors. Each chapter is headed by quotations from famous scientists, philosophers, or poets, to remind the reader of the human context of science. The pedagogy is classical; each chapter begins with a statement of intent, and ends with a summary. The figures are copious and outstanding in clarity. The tables collect the most reliable data dealing with energy levels, transport, and trapping of the most intensely studied polyacenes: naphthalene, anthracene, tetracene, and pentacene.

The reader of this book will not only come away with a feeling for the mechanism of charge transport and recombination based on the nearly small molecular polaron, but also with an insight into the mode of thinking of two highly talented scientists.

The thoroughness and realism of the treatment is exemplary. At one place in the text, I could imagine myself as an observer on an electron piloted by the authors on a trip from outside the crystal to a landing within the crystal as a molecular polaron, all the while accompanied by a running commentary on what was happening to the electron and the crystal. At other places in the text, where transport and recombination were discussed, I was guided along scenic voyages within the crystal.

Over the years, many attempts have been made to produce a comprehensive electronic energy level scheme for OMC, without success. One major reason for this is that different methods (dark conductivity, photoconductivity, electroabsorption) have generated discordant results. The authors have made a major contribution in rationalizing all of the previous results, based on their polarization model. In two figures, every process that contributes to the establishment of the energy level structure is illustrated, and the relationship established among the values obtained by different measuring techniques.

The authors also discuss the controversial subject of the energy and generation mechanism of charge transfer states. This subject is also presented with a characteristic precision and attention to detail, bringing into concordance the major views on this subject.

The transport of carriers through OMC over a complete temperature range has not been accounted for on a quantum mechanical basis. At very low temperatures, there is evidence of band mobility. At higher temperatures, where the mean free path of the carrier is equal to the lattice constant, band theory should not apply. Nevertheless, the temperature dependence of the mobility is bandlike. The authors use their nearly small polaron concept, and a physical model to set up a simulation procedure that essentially ties all the loose ends together. This simulation is physically satisfying, pictorial, and should serve to direct future theorists into fruitful channels.

This book will have a stimulating effect on the advancement of theory and experiment in this field. It is an excellent choice for professionals who wish to be brought up to date on the latest and most comprehensive discussion of charge carrier generation, trapping, transport, and recombination in OMC. To a lesser extent, excitons and Langmuir-Blodgett films are also discussed. In view of its pedagogical soundness, this book would be an excellent choice as a text for an advanced undergraduate or graduate course.

It has been a pleasure for me to read this monograph.

MARTIN POPE
New York University

Preface

At present we are witnessing the birth and development of a new contemporary branch of solid-state science, *organic-solid-state physics*, with new paradigms, new rules, and interdisciplinary relationships.

In this promising field of research organic crystals of aromatic and heterocyclic molecules have become the main objects of scientific interest. This is due to the fact that in the case of aromatic and a number of heterocyclic molecular crystals, the investigator has detailed information at his disposal about their molecular and crystal structure, as well as an abundant bank of reported data on optical, electronic, and related properties. Such knowledge is an indispensable precondition for successful theoretical studies opening new promising avenues for deeper insight into the dynamics of electronic and excitonic processes in a crystal.

The development of the physics of organic molecular crystals (OMC) confirms the general tendency of modern solid-state physics to describe more and more complicated materials.

From the point of view of structural organization and the corresponding complexity of electronic processes, organic crystals occupy an intermediate state between more simple inorganic materials, such as atomic and ionic crystals, and still more complex objects, such as molecular biosystems.

In organic crystals the domination of molecular properties over the crystalline ones, caused by the weakness of Van der Waals interaction, leads to a marked tendency of localization at room temperature of charge carriers and excitons on individual molecules of the crystal. As a result, a number of qualitatively new physical properties emerge in OMC as compared with inorganic covalent or ionic crystals. Hence, for organic crystals a number of traditional concepts of solid-state physics are not valid (at least at higher temperatures), e.g., band theory of single-electron approximation, band-type charge carrier transport, etc. As a result, the researcher is compelled to find new approaches, often phenomenological ones.

As in any new branch of science, also in this field one may distinguish three main stages of development. First, collection of new experimental facts and empirical rules. Second, development of phenomenological approaches, models, and working hypotheses. And, finally, development of a comprehensive and self-consistent theory.

The present monograph, in our opinion, is an attempt to fill the gap between the above-mentioned second and third stages of development.

The principal aim of the authors was to present a comprehensive treatment of the physical principles underlying the dynamics of excitonic and electronic interaction, localization, and transport processes in OMC, in an attempt to summarize the present state of experimental and theoretical research in this field.

From the conceptual point of view this monograph extends the "static" many-electron interaction approach, as presented in *Organic Molecular Crystals. Their Electronic States* by E. A. Silinsh (Springer Verlag, Heidelberg, 1980) to the realm of electron and exciton interaction dynamics.

The basic emphasis in the monograph is focused on one of the cardinal problems of modern organic solid-state physics, namely, on interaction processes of excitons and charge carriers with the local environment, resulting in polarization and local-ization phenomena and formation of polaronic quasi-particles. The polaron theory approach is essentially the leitmotif of the whole book.

The other main problem under discussion—transport phenomena—is also directly correlated with interaction dynamics, which actually determines the time- and temperature-dependent transition of charge carriers and excitons from coherent (bandlike) to diffusive (hopping) mode of motion.

All three main topics of the book are thus mutually interrelated and are only separated into different chapters for the sake of convenience.

The origin of this book was stimulated by the desire of the authors to formulate a scientific basis for experimental investigations of electronic and excitonic processes in organic crystals. Since this new field is still in the stage of develop-ment, we discuss both recognized physical models and theories, as well as prob-lematic, controversial questions that may have, in our opinion, considerable heuristic value in stimulating further experimental and theoretical research.

In the exposition of the material under discussion we have tried to maintain a delicate balance between a more or less strict rigorous theoretical treatment and a phenomenological, empirical one throughout the whole book to satisfy both theo-rists and experimental physicists, as well as people from interdisciplinary areas. To fulfill this aim the material of the book is presented on two levels of complexity of discussion.

Thus, in the first part, Section 1A of the introductory Chapter 1, the most impor-tant characteristic features of OMC (the role of the properties of molecules, typical crystal structures of model compounds, molecular interaction forces, etc.) are briefly discussed. In our opinion, this introduction may serve as a starting point for a deeper understanding of the peculiarities of electronic processes in OMC.

The second part, Section 1B of Chapter 1, in its turn provides the theoretical basis for further extended analysis. It deals, in a generalized Hamiltonian representation, with all kinds of excitonic and electronic interactions in a molecular crystal.

Chapter 2 is devoted to the dynamics of exciton interaction with a local lattice environment in OMC, including localization and self-trapping phenomena.

Chapter 3 deals with charge carrier interaction dynamics with a local lattice environment. The formation and properties of electronic, molecular, and lattice polarons are discussed in terms of phenomenological and model Hamiltonian approaches.

Chapter 4 discusses the energy structure of electronic and molecular polaron states in perfect and real (defective) OMC. The properties of charge transfer and charge pair states are considered.

Chapter 5 presents an analytical review of contemporary theories of exciton and charge carrier transport mechanisms in molecular crystals.

Chapter 6 considers charge carrier photogeneration mechanisms in OMC, based on experimental and computer simulated studies.

Finally, Chapter 7 is devoted to experimental and computer simulated charge carrier transport studies in OMC and to the interpretation of the results in the framework of the molecular polaron model.

Section 1A of Chapter 1 was written by E. A. Silinsh, Section 1B by V. Čápek; Chapters 2 and 3 jointly by both authors; Chapter 5 by V. Čápek; and Chapters 4, 6, and 7 by E. A. Silinsh. The whole book was edited by E. A. Silinsh.

The physics of organic solids occupies a borderline between a number of related disciplines—molecular physics, solid-state physics and electronics, organic and physical chemistry, and even molecular biology. At present, organic solid-state physics also provides the material and conceptual framework for the development of a new interdisciplinary branch of science and technology known as molecular electronics.

Hence, this book is intended for specialists working in the field of physics and electronics of solids, chemical physics, organic and physical chemistry, and molecular biophysics, as well as for undergraduate and postgraduate students of these specialized disciplines.

EDGAR SILINSH
VLADISLAV ČÁPEK

Acknowledgments

The present work could hardly have been possible without assistance and cooperation on the part of a large number of colleagues, near and far. First, the authors greatly appreciate the help of their coworkers at the Institute of Physical Energetics of the Latvian Academy of Sciences in Riga and at the Institute of Physics of Charles University in Prague. The sacrifices and encouragement of our families are also recognized with deep appreciation.

Many of the ideas and concepts of this work have been presented and discussed at a number of international conferences, symposia, and seminar talks. All of those colleagues who have helped with valuable comments in private discussions and correspondence are gratefully acknowledged.

For stimulating exchange of ideas and helpful comments, the authors are especially grateful to H. Baessler, C. L. Braun, A. S. Davydov, J. L. Frankevich, H. Inokuchi, N. Karl, T. Kobayashi, K. Kojima, M. Kotani, M. V. Kurik, A. Matsui, A. Mozumder, R. W. Munn, S. Nešpurek, M. Pope, P. Reineker, W. Siebrand, J. Sworakowski, and M. Wagner. We are also very much indebted to J. Eiduss for translation of a considerable part of Chapter 4 and part of Chapter 3 from our book in Russian on *Electronic Processes in Organic Molecular Crystals*[351] and for reading the manuscript of the present book.

The authors are grateful to I. Rozīte and R. Vilīte for computerized printing of the manuscript, S. Gulbe for meticulous graphic work, and I. Muzikante, A. Klimkans, G. Shlihta, L. Taure, and U. Zandbergs for technical assistance.

We are also very grateful to the reviewers of the manuscript, Profs. M. Pope, R. W. Munn, and A. Mozumder, for highly relevant comments, suggestions, proposed corrections, and editorial amendments which helped us to improve the final version of the book.

This does not, however, free the authors from assuming full responsibility for any shortcomings or possible errors in the text.

We also wish to thank the following authors for the kind consent to reproduce their illustrative material: T. Kobayashi for presenting original electron micrographs and diagrams of molecular arrangement (Figs. 1.1, 1.4, 1.10, and 1.11); K. Kojima for presenting computer simulation data of dislocation defects (Figs. 4.20 and 4.21) (see Refs. 130,191a); R. W. Munn for the diagram in Fig. 1.2; Fig. 3.6 and Tables 3.2–3.4; H. Baessler for Figs. 2.1, 4.2, 4.4–4.6; N. Karl for Figs. 4.24, 7.1–7.11; A. Matsui for Figs. 2.4, 2.7, and 2.8.

The authors are indebted to the following publishers for permission to reproduce their copyrighted illustrations:

Elsevier Science Publishers (North Holland) for Figs. 1.7, 1.12, 2.1, 4.2–4.6, 4.8, 4.13, 4.18, 4.25, 6.11–6.13, 6.19, 6.20, 6.22–6.31, 7.12–7.23, 7.25–7.34, and Tables 1.9, 3.5, 3.6, and 3.8.

Springer-Verlag for Figs. 1.5, 2.3, 2.5, 2.6, 3.4, 3.5, 3.9, 3.11, and 6.1.

Zinātne Publishing House, Riga for Figs. 2.9, 2.10, 3.4, 3.5, 3.10, 4.8–4.11, 4.15, 4.17, 4.19, 4.23, 6.2, 6.18, and Tables 2.1, 3.7, and 4.2.

International Union of Crystallography for Figs. 1.3 and 1.6.

Gordon and Breach Science Publishers for Table 1.8.

John Wiley and Sons Ltd. for Figs. 4.26 and 7.35.

Akademic Verlag, Berlin for Figs. 2.2, 4.1, 6.4, 6.9, 6.15, 6.17, and Table 6.1.

McGraw-Hill Book Co. for Fig. 3.1.

American Institute of Physics for Figs. 1.9, 3.2, 3.3, 3.12, 5.1, 6.5, 6.8, and 6.13c.

CHAPTER 1

Basic Properties of Organic Molecular Crystals

It seems that the human mind has first to construct forms independently before we can find them in things. Kepler's marvelous achievement is a particularly fine example of the truth that knowledge cannot spring from experience alone, but only from the comparison of inventions of the mind with observed fact.

ALBERT EINSTEIN

Part 1A
Characteristic Features of Organic Molecular Crystals

1A.1. INTRODUCTION

Organic molecules in condensed solid phase form molecular crystals which differ considerably in their optical, electronic, and mechanical properties from such conventional solids as covalent (atomic) or ionic crystals. This is mainly due to weak intermolecular interaction forces of the Van der Waals type with bonding energy considerably lower than that of covalent or ionic bonds in atomic crystals.[184,288,332,351]

In a more general sense, *molecular crystals* may be defined as a specific kind of solids formed by electrically *neutral* molecules or atoms which interact with nonbonding, relatively weak interaction forces. In addition to several types of nonbonding Van der Waals interaction forces, organic molecules with specific functional groups may form intermolecular *hydrogen* bonds (see Ref. 351). Finally, in some binary molecular compounds, consisting of electron donor and acceptor molecules [so-called charge transfer (CT) complexes] the molecular interaction manifests itself via partial electron transfer from donor to acceptor molecules.[288,351]

As an example of the most simple molecular crystals one should mention the noble-gas ones in which the role of interacting "molecules" is performed by neutral helium (He), krypton (Kr), or xenon (Xe) atoms. More typical are the crystals of N_2 and O_2 molecules. These crystals, due to very weak interaction forces are stable only at low temperatures and therefore are called the cryogenic crystals.[29] They often serve as ideal models of highly symmetrical and the most elementary molecular crystals.

As a more "stable" molecular crystal, possessing a number of exotic properties, one can mention ordinary ice, the polar molecules of which interact via a network of hydrogen bonds.

However, this book is mainly devoted to *organic molecular crystals* (OMC) of aromatic and heterocyclic molecules. Actually aromatic, especially polyacene crystals, as well as some heterocyclic crystals (e.g., phthalocyanines) have served for the last two decades as popular model compounds of OMC.[288,332,351,357] This is due to the fact that, in case of aromatic and a number of heterocyclic organic crystals, investigators have at their disposal detailed information about their molecular and crystal structure, as well as abundant reported data on optical, electronic, and related properties.[288,332,351] Such knowledge is an indispensable precondition for successful studies of electronic and excitonic processes in the crystal.

In these OMC the total Van der Waals interaction forces, due to complementary Van der Waals surfaces, are considerably greater than those in noble-gas crystals. Thus, the heat of sublimation H_s of polyacene crystals ($H_s = 73-126 \text{ kJ} \cdot \text{mol}^{-1}$) lies between that of the noble-gas crystals ($H_s = 3-16 \text{ kJ} \cdot \text{mol}^{-1}$) and lattice energies of typical ionic crystals such as alkali halides ($600-1000$ $\text{kJ} \cdot \text{mol}^{-1}$).[184,332,351]

The energy range of typical Van der Waals *intermolecular* bonds ($E_{VdW} = 10^{-3}-10^{-2}$ eV) is by several orders of magnitude smaller than the intramolecular covalent bond energies ($E_{cov} = 2-4$ eV) of aromatic molecules. [Only hydrogen bonds and donor acceptor interaction bonds in CT-complexes may become competitive with covalent bonds (see Ref. 351, p. 30).]

Weak intermolecular interaction forces produce only slight changes in the electronic structure of molecules on formation of the solid phase, and, as a result, molecules in the lattice retain their identity. Thus, x-ray analysis of electron density distribution in aromatic hydrocarbon crystals shows that electronic configuration of the molecules remains practically unchanged in the crystal. Maximum electron density in this case concentrates mainly around carbon atoms and covalent intramolecular C—C and C—H bonds, dropping to practically zero value in the intermolecular space. The aromatic molecules form in the crystal something like an archipelago of molecular "islands" with delocalized π-electrons, separated by "channels" of electronless space (see Ref. 332, p. 25).

From this aspect, organic molecular crystals are closer to an "oriented molecular gas" of isolated molecules than to traditional solids such as covalent or ionic crystals with strongly bonded rigid atomic or ionic lattice characterized by practically complete loss of individual properties of the constituent particles in the crystal.

The hypothetical picture of a molecular crystal has now been confirmed by a superb electron micrograph, obtained by Uyeda, Kobayashi, *et al.*[187,397] The authors, using a highly sophisticated technique of high resolution (~ 1.5 Å) elec-

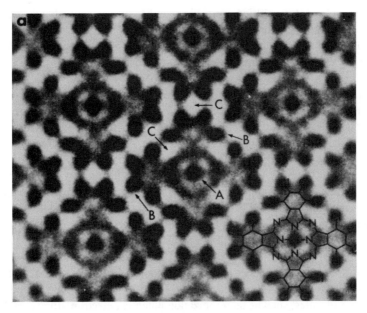

FIGURE 1.1. Electron micrograph of Cu-hexa-decachlorphthalocyanine molecules in the **ab**-plane of crystal according to Uyeda *et al.*[397]

tron microscopy, have produced an image of copper (Cu)-hexadecachlorphthalo-cyanine molecules in the **ab** plane of the crystal (see Fig. 1.1). As can be seen from Fig. 1.1 the micrograph allows one to discriminate the image of constituent atoms of every single molecule: the copper (Cu) atom at the center, as well as 16 chlorine atoms at the periphery of the molecule. The benzene and pyrrole rings are also clearly distinguishable and one can write the conventional structural formula directly on the real image of the molecule. (One may observe even a slight overlap of the electron orbitals of the chlorine atoms of neighboring molecules.)

However, the most important aspect of this micrograph is that it demonstrates that the individual molecules are actually separated by "channels" with negligible electron density. These practically electronless intermolecular "channels" actually give the visual picture of the weak, long-distance Van der Waals interaction forces.

Such electronic structure determines the basic specific features of optical and electronic properties of a molecular crystal, as well as peculiarities of their energy spectra of neutral and ionized states.[332]

The optical spectra of an isolated molecule and a molecular crystal are similar. The crystal spectrum completely retains the spectral features of that of individual molecules, including their electronic-vibronic structure.

On the other hand, certain new optical and electronic properties emerge in crystal spectra caused by collective molecular interaction. Thus, due to dispersion (polar-ization) and resonance interaction of the excited molecule with its local crystalline surroundings spectral bands show a batochrom (long-wave) shift and such specific phenomena as Davydov splitting, some change of shape of exciton bands caused by interaction with lattice phonons, etc.[288,332,351]

The domination of molecular properties over the crystalline ones, caused by the weakness of Van der Waals interaction, leads to a marked tendency of *localization* of charge carriers and excitons on individual molecules of the crystal. Due to this, a number of qualitatively new properties emerge in organic crystals, as compared with inorganic covalent or ionic crystals. As the most important, one should mention the phenomenon of electronic polarization of the electronic subsystem of surrounding lattice by charge carriers and excitons. As a result, polaron-type quasiparticles: electronic and excitonic polarons are created. We should emphasize that electronic polarization is an essentially many-electron interaction phenomenon and therefore single-electron approximation, widely used in traditional solid-state physics, is not valid and inapplicable for description of electronic states in organic molecular crystals. The self-energy of charge carriers and excitons in OMC is predominantly determined by electronic polarization, and that is why the polarization effects emerge as the leading concept of the book.

In the formation of the energy structure of conductivity levels and local trapping states in molecular crystal an equal part is played by the individual properties of constituent molecules, such as their ionization potential, electron affinity, molecular polarizability, permanent dipole and quadrupole moments, etc. Such "dualism" is a specific feature of molecular crystals which does not allow to apply traditional concepts of solid-state physics in their studies and requires a search of new approaches, often phenomenological ones (cf., e.g., Refs. 332 and 351). Therefore this book is devoted to search and development of new approaches both in terms of phenomenological models and more refined theories.

Localization effects manifest themselves in *transport* properties of polaronic quasiparticles in OMC, especially in the higher-temperature region. Thus, the charge carrier mobilities μ are low at room temperature ($\mu \leqslant 1$ cm^2/Vs), and the mean free path $\bar{l}$ of the carrier is practically equal to the lattice constant ($\bar{l} = a_o$) due to strong scattering. Consequently, one should use some kind of hopping model approach instead of conventional band-type transport model.

The weak Van der Waals interaction forces also predetermine the mechanical and elastic properties of the OMC. The lattice energy is accordingly low in molecular crystals which means low melting and sublimation temperatures, low mechanical strength, and high compressibility.[184,332,351]

Table 1.1 gives an illustrative comparison of the main characteristic features of molecular and covalent (atomic) crystals.

It should be stressed that organic molecular crystals possess complicated *structural* organization, as compared to inorganic, namely, covalent or ionic ones. As a matter of fact, in OMC there emerge several levels of structural organization. First, we should mention the intramolecular structure of covalently bound atoms of individual molecules, which practically does not change after the formation of the crystal (see Fig. 1.1).

Second, the intermolecular "infrastructure" of the atoms in the lattice should be

TABLE 1.1. *Comparison of Characteristic Features of Organic Molecular and Covalent (Atomic) Crystals*

Molecular crystal (e.g., anthracene-type crystal)	Covalent (atomic) crystal (e.g., silicon-type crystal)
Weak Van der Waals type of interaction (characteristic interaction energies $E_{VdW} = 10^{-3}$–10^{-2} eV)	Strong covalent-type interaction (characteristic interaction energies $E_{cov} = 2$–4 eV)
Marked tendency of charge carrier and exciton localization	Pronounced charge carrier delocalization
Self-energy of charge carriers and excitons determined by many-electron interaction (polarization) effects	Single-electron approximation valid
Charge carriers and excitons as polaron-type quasiparticles	Charge carriers as free holes and electrons
Low charge carrier mobilities ($\mu \approx 1$ cm²/Vs) and small mean free path ($l \approx a_0$ = lattice constant) at room temperatures	High charge carrier mobilities and large mean free path $[l = (100-1000)a_0]$
Large effective mass of charge carriers $m_{eff} = (10^{2}-10^{3})m_e$	Small effective mass of charge carriers $m_{eff} \lesssim m_e$
Hopping-type charge carrier transport dominant	Band-type charge carrier transport dominant
Excitons as molecular Frenkel-type quasiparticles	Excitons as Wannier-type quasiparticles
Low melting and sublimation temperatures, low mechanic strength, high compressibility	High melting and sublimation temperatures, high mechanical strength, low compressibility

mentioned. In this case the nuclear skeleton of the molecules may be regarded as rigid, thus the positions of nuclei in the lattice are determined by the distances of the centers of molecules and their orientation relative to the crystallographic axes (see Fig. 1.6).

These two structural levels of the nuclear subsystem clearly emerge in the vibrational spectra of OMC, namely, as high-frequency intramolecular infrared (IR) and Raman bands and low-frequency intermolecular vibrations of the lattice (the acoustic and optical phonons) (see Section 1A.4).

Finally, the *electronic* structure in OMC may be divided into three sublevels: (i) the electrons bonded to separate atoms and forming the atomic core, (ii) the valence (molecular) σ-electrons, *localized* in pairs on interatomic covalent bonds, and (iii) in the case of polyconjugated organic molecules, the valence (molecular) π-electrons *delocalized* over the whole molecule, or on a part of it.

It should be emphasized that the *electronic* structure of molecules is mainly the origin of intermolecular interaction, i.e., the formation of Van der Waals interaction forces (see Section 1A.5); they are also the source of all electronic polarization phenomena, including the formation of electronic and excitonic polarons.

Concerning the structural complexity of organic crystals one may notice that they actually occupy the intermediate region between simpler inorganic crystals and complex molecular biosystems. At present some isomorphic organic crystals serve

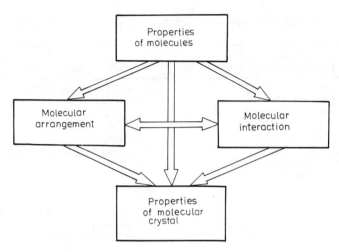

FIGURE 1.2. Schematic diagram of principal interrelations between "molecular" and "solid state" approaches in organic crystals. (After Munn.[254])

as excellent model systems for studying electronic processes in more complex molecular biosystems.

The complexity of structural organization of molecular crystals determines the nature of elementary electronic processes in these organic solids.

Thus, the three structural levels of OMC — *electronic*, nuclear *intramolecular*, and nuclear *intermolecular* (lattice) structures — determine the time domains of relaxation (polarization) processes in the crystal. Accordingly, in the time scale of relaxation processes one can conditionally separate three characteristic time domains, corresponding to the three structural levels of OMC: very fast electronic polarization ($\tau_e = 10^{-16}-10^{-15}$ s), slightly slower intramolecular (vibronic) polarization ($\tau_v = 10^{-15}-10^{-14}$ s), and slower intermolecular (lattice) polarization processes ($\tau_1 \geq 10^{-13}$ s) (see Fig. 3.9).

This illustrates how closely interrelated the structures and processes in OMC are. Let us remind the reader in this connection of the ingenious conclusion of Ludwig von Bertalanffy: "What are called structures are slow processes of long duration, functions are quick processes of short duration."[30]

As we will see later, this is one of the crucial problems of some theoretical approximations of electronic processes in OMC: whether to regard the lattice as rigid (so-called crude representation) or "moving" (the adiabatic approximation). The reader will find a detailed discussion of this problem of "structure versus process" in Section 1B.

Some basic principles in the physics of OMC, as discussed above, may be illustrated by the schematic diagram (Fig. 1.2) suggested in a review lecture by Munn.[254] This diagram, in our opinion, visualizes the interrelations between "molecular" and "solid-state" approaches, so important for understanding the essence of the problem.

Let us now briefly consider the most significant features and general implications of the individual blocks of the diagram in Fig. 1.2.

1A.2. ROLE OF THE PROPERTIES OF MOLECULES. MOLECULAR ARRANGEMENT

First, the electronic structure and the main electronic parameters of the *molecule* (ionization potential I_G, electron affinity A_G, molecular polarizability tensor β_i, presence of permanent dipole or quadrupole moments, etc.) directly influence the nature and magnitude of the molecular interaction forces.[332,351]

Second, these parameters determine the electronic polarization energy of the lattice by quasiparticles — excitons and charge carriers. It will be shown in the following chapters (Section 1B of this chapter, Chapters 2–4), that the electronic polarization term is the most important factor that influences the self-energy of the excitons and charge carriers and thus determines the energy structure of neutral and ionized states in OMC.

The second most important property of the molecules, which influences the molecular arrangement, is the *shape* of the molecule, namely, the relief of its Van der Waals interaction surface.

This surface, in case of a polyatomic organic molecule, may exhibit considerable asymmetry and geometrical complexity. Even such a relatively symmetrical molecule as anthracene (Ac), belonging to the D_{2h} symmetry group, forms a complex Van der Waals cross section in the molecular plane (see Fig. 1.3) — a relief of alternative convex and concave surfaces.

The Van der Waals surfaces of the molecule determine the mean intermolecular distance of adjacent molecules as well as their orientation in the lattice. The maximum density of molecular packing may be reached if the Van der Waals surface reliefs of the adjacent molecules are complementary to each other. In such a case, atom-atom interaction contacts are maximized giving the greatest possible additive sum of the Van der Waals interaction energies (see Section 1A.5). This results in a minimum of potential energy of the lattice. The complementary principle is highly important not only in the formation of stable molecular crystals, but also as basic structural principle in the case of more complex molecular biosystems; e.g., in the formation of substrate-enzyme, antigen-antibody complexes, etc. (By the way, the most ancient and pictorially generalized symbol of complementarity is the well-known Taoist symbol of yin–yang. In modern times this sign was used by Niels Bohr to symbolize complementarity as one of the most important principle in natural and life sciences, and in philosophy in general.[408]).

Figure 1.4 presents a schematic picture of the complementarity principle in action. As a result of inelastic slip along a partial (1/2) dislocation line in a phtalocyanine crystal (Fig. 1.4a), molecules are, before relaxation, in an energetically unstable state due to the overlap of the Van der Waals surfaces of the molecules

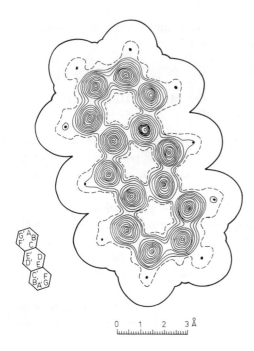

FIGURE 1.3. Anthracene. A cross-section of three-dimensional electron density map, coinciding with the molecular plane[358] and the corresponding cross section of the Van der Waals surface.[351]

(Fig. 1.4b). A new equilibrium arrangement of molecules is reached in a relaxation process in accordance with the complementarity principle (Fig. 1.4c).

Figure 1.7d demonstrates how this principle is realized in the packing of pyrene dimers in a B-type lattice (see Section 1A.3.2).

The asymmetry of polyatomic organic molecules determines low lattice symmetry of organic crystals. Even relatively symmetrical (D_{2h}) linear polyacene

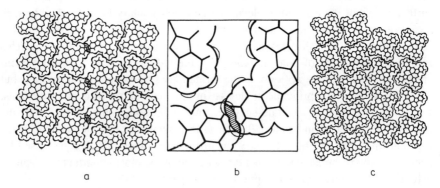

FIGURE 1.4. Schematic representation of the configuration of phthalocyanine (Phc) molecules along the partial (1/2) dislocation line bounding a stacking fault in a Phc crystal according to Ref. 187(a) unstable arrangement before relaxation, (b) enlarged fragment of (a) showing the overlaps of the Van der Waals surfaces of the molecules, (c) arrangement of the molecules according the complementarity principle after relaxation.

molecules form the lowest lattice symmetry types, namely, monoclinic naphthalene (Nph), Ac, or triclinic tetracene (Tc), pentacene (Pc) crystals. Unlike atomic (Ge, Si) or ionic (Na^+Cl^-) crystals of high symmetry, similar ones are rare among organic crystals.[184]

The situation is more dramatic for a great number of organic molecules of very low symmetry, containing heteroatoms, asymmetrical side groups, or radicals. Such molecules, due to strong steric hindrances, are not able to form regular, crystalline structures and behave like disordered solids.

The low symmetry of organic molecules is the principal cause of pronounced anisotropy of the lattice, which is a characteristic feature of many organic crystals.

The anisotropy may be regarded as one of the most important features of organic crystals. Due to the lattice anisotropy there may be corresponding anisotropy in the optical, electrical, magnetic, mechanic, and other physical properties of the crystal. This anisotropy may be still more pronounced in case of CT-complexes and ion-radical salts. The latter, is known to form quasi–one-dimensional anisotropic systems. Very sophisticated anisotropic multilayer molecular systems can be created artificially by Langmuir-Blodgett or molecular beam techniques. Such highly aniso-tropic molecular systems may have a great impact on future development of molecular electronics.

The complementary principle of molecular arrangement gives rise to another important feature of OMC. In the case of asymmetrical molecules the most compact packing, with minimum lattice energy U_{lat}, may be reached via different molecular arrangements. This means that the search for the minimum lattice energy U_{lat} may be equivocal.

Actually, in case of asymmetrical molecules there may exist several modifica-tions of lattice arrangement with very close U_{lat} values. This phenomenon gives rise to polymorphism, characteristic of a great number of low-symmetry OMC. Thus, e.g., Ac has only one stable crystallographic structure. Its rather symmetric hetero-cyclic derivate — phenazine with two N atoms in the aromatic cycle — has two modifications, but its still less symmetrical derivate — acridine, with only one N atom in the cycle—even has four modifications.[184,332] Polymorphism is typical of a great number of heterocyclic molecules, or molecules with asymmetrical side groups. Prevalence of one or other polymorphic modification depends on crystal growth technique, temperature, differences in lattice energies, and other factors. Under some conditions one may obtain a mixture of different structures, or inclu-sions of one crystalline phase in the matrix of another.[332] Mechanically induced phase transitions of the polymorphic crystals can also occur causing a rearrange-ment of the molecules in the lattice. Of course, different modifications may exhibit different optical and electrophysical properties. Thus, the change and diversity of molecular arrangement in polymorphs can directly influence the properties of the crystal. The crystal anisotropy and polymorphism of OMC will be discussed in the next section.

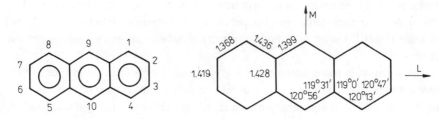

FIGURE 1.5. Anthracene. (a) structure formula; (b) structural indices of the molecule: mean bond lengths (in Å) and mean bond angles. (After Ref. 332.)

1A.3. CRYSTAL STRUCTURE OF SOME MODEL COMPOUNDS

To illustrate the crystal structure of typical organic molecular crystals we shall briefly discuss three main groups of the most popular and the most thoroughly investigated model compounds:

1. *Polyacenes* (or Ac-type crystals) with two nonparallel near-neighbor molecules in a unit cell, belonging, according to Stevens' classification[367] to lattice type *A* of aromatic crystals. Polyacenes have been for decades the most popular objects for detailed studies of electronic and excitonic processes in OMC.[288,332,351]

2. *Pyrene* and *α-perylene* which belong to lattice type *B* of aromatic crystals with two pairs of parallel adjacent molecules in a unit cell. Owing to such dimer structure of closely spaced molecules in the lattice pyrene and α-perylene crystals exhibit strong eximer fluorescence and other unusual optical properties. They have therefore been the most frequently used crystals for optical and exciton self-trapping studies.[31,32,288,332,351,367]

3. *Phthalocyanine*, possessing a cyclic polyconjugated molecular structure, often serves as a model prototype of chemically similar, biologically important porphins, such as chlorophyll and protoheme. Metal-free phthalocyanine serves as ligand for a series of metal phthalocyanines exhibiting a number of interesting photophysical properties and showing promise for practical applications.[288,351,357]

1A.3.1. Linear Polyacenes. A-type Lattice

Molecular structure: Polyacenes. Nph, Ac, tetracene Tc, and Pc belong to the D_{2h} molecular symmetry. The planar condensed ring structure of the Ac molecule (Fig. 1.5a) has been conclusively established.[301] This planarity of the molecule is due to sp^2 hybridization of three of the valence electrons of the carbon atom, the three hybridized electron orbitals being localized in the xy plane of the molecule and forming a planar trigonal configuration. The fourth of the carbon valence electrons retains $2p_z$ symmetry, namely, its orbital is directed at right angles to the xy plane of the molecule. The aromatic nature of Ac is due to 14 coplanar $2p_z$ electron orbitals of C atoms, forming, as the result of conjugation, *delocalized*

TABLE 1.2. *Crystallographic Data on Naphthalene (Nph), Anthracene (Ac), Tetracene (Tc), and Pentacene (Pc) crystals[a]*

Parameters	Nph $C_{10}H_8{}^b$	Ac $C_{14}H_{10}{}^c$	Tc $C_{18}H_{12}{}^d$	Pc $C_{22}H_{14}{}^d$
Crystal structure	Monoclinic	Monoclinic	Triclinic	Triclinic
Space group	$P2_1/a$	$P2_1/a$	$P\bar{1}$	$P\bar{1}$
a, Å	8.24	8.56	7.90	7.90
b, Å	6.00	6.04	6.03	6.06
c, Å	8.66	11.16	13.53	16.01
α, °	90.0	90.0	100.3	101.9
β, °	122.9	124.7	113.2	112.6
γ, °	90.0	90.0	86.3	85.8
V, Å^3	360	474	583	692
z	2	2	2	2
d_{calc}, g/cm^3	1.17	1.24	1.29	1.33
d_{exp}, g/cm^3	1.15	1.25	1.29	1.32
M, daltons	128.19	178.24	228.3	278.36

[a]*Notations*: V, unit cell volume (in Å^3); z, number of molecules in unit cell; d, density of the crystal (g/cm^3); M, molecular weight (in daltons). For the orientation of molecules in the lattice see Table 1.5 in Ref. 332.
[b]Taken from Ref. 6.
[c]Taken from Ref. 225.
[d]Taken from Ref. 393.

π-electron orbitals (Fig. 3.4). Thus, the presence of delocalized, relatively weakly bonded π-electrons is the origin of the main optical and electrical properties of polyacenes, including photoconductivity.[288,332,351]

The mean C — C bond length and bond angles between σ-bonds of the Ac molecule have been experimentally determined by Mathieson, Robertson, and Sinclair[225] and Sinclair, Robertson, and Mathieson[358] and later refined by Cruickshank.[76,77,79–81] Corrected mean bond lengths and bond angles of Ac are presented, according to Ref. 81 in Fig. 1.5b. X-ray analysis confirms that the Ac molecule is practically planar. Small deviations of individual carbon atom coordinates from the mean molecular plane do not exceed 0.012 Å and are caused by close intermolecular approaches of some atoms in the lattice.[76] As can be seen from Fig. 1.5b, the bond angles differ only slightly from the 120° value of ideal trigonal sp^2 hybrid orbitals.

The mean C — H bond length is 1.084 ± 0.006 Å. Effective Van der Waals thickness of the Ac molecule, i.e., the thickness of the π-electron cloud, is ca. 3.4 Å.

A cross section of the three-dimensional electron density map of Ac along the plane of the molecule[358] as well as the corresponding cross-section of the Van der Waals surface,[351] are shown in Fig. 1.3. Every contour line corresponds to an increase of electron density by ca. 0.5 electrons/Å^3. The first contour shows marked bulges due to hydrogen atoms.

Crystal Structure. The crystallographic data on Ac have been studied extensively by Robertson *et al.*[225,301,358] and later refined by Cruickshank.[76,77,78] Crystal-

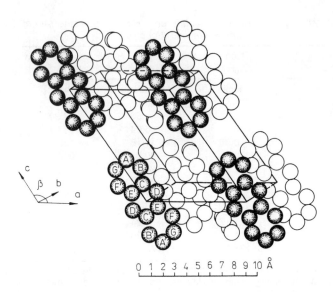

0 1 2 3 4 5 6 7 8 9 10 Å

FIGURE 1.6. Schematic structure of unit cell of an anthracene crystal (according to Ref. 358). Both unequivalent molecules of the unit cell form equal but opposite angles of 64° relative to the **ab** plane. The crystallographic data of anthracene are presented in Table 1.2.

lographic data of Nph are given in Refs. 6 and 78, those of Tc and Pc in Refs. 303 and 393 (see also Ref. 80). The reported crystallographic data of polyacenes are compiled in Table 1.2.

The Ac and Nph crystals are monoclinic ($a \neq b \neq c$, $\alpha = \gamma = 90°$, $\beta \neq 90°$) and belong to space group $P2_{1/a}$ with two molecules in a unit cell ($z = 2$).

For illustration, the positions of Ac molecules in the unit cell are shown in Fig. 1.6 (the hydrogen atoms omitted for clarity). The structure is practically closely packed, due to the complementarity principle, each molecule has 12 nearest neighbors. Molecule I of the unit cell which is situated at the center of symmetry (0, 0, 0) is transformed by the glide plane operation into molecule II at (1/2, 1/2, 0). Tetracene and Pc crystals are triclinic ($a \neq b \neq c$, $\alpha \neq \beta \neq \gamma \neq 90°$) and belong to space group $P\bar{1}$ with two molecules in a unit cell ($z = 2$) (see Table 1.2). The orientation of the two nonequivalent (I and II) molecules in the triclinic unit cell of Tc and Pc are not directly related by symmetry elements. However, the symmetry centers of both molecules are situated at sites (0, 0, 0) and (1/2, 1/2, 0) respectively, forming an arrangement in the unit cell, which bears a close resemblance to the Ac structure (cf. Fig. 1.6). This structure belongs to space groups $P\bar{1}$ with pseudo-$P2_{1/a}$ symmetry.[303]

Slight deviations from the monoclinic symmetry in Tc and Pc crystals (see Table 1.5 in Ref. 332) result in closer packing of molecules in the lattice, and corresponding increase in density (ca. 2%).

We wish to emphasize the close similarity of the crystal structure of all the members of the polyacene series, from Nph to Pc. This is of great importance for the comparison of optical and electronic properties of polyacene crystals.[332,351]

Polymorphism of Polyacene Crystals. Polyacenes, due to relatively high molecular symmetry and regular form of the Van der Waals surfaces (cf. Fig. 1.3) do not exhibit a tendency to formation of polymorphic structures.

However, atom-atom potential calculations, as well as direct experimental observations, provide evidence that under specific conditions metastable phases can be formed in an Ac crystal (see Ref. 332, p. 16, and references therein).

Several authors have reported the existence of phase transitions in Tc crystals at low temperatures (see Ref. 351, p. 38).

Most detailed studies of the phase transition in Tc, based on the analysis of reported data and their own calculations and experimental observations, were performed by the authors of Ref. 366. They have shown that the temperature T_t of phase transition in Tc may depend on the morphology and prehistory of the crystal, size of the grains in case of polycrystalline sample, rate of cooling and other factors. On the whole, the phase transition in Tc usually occurs below ca. 200 K. However, the T_t value may depend on external pressure and internal stress fields of the crystal.

X-ray analysis and calculations show[366] that the low-temperature phase of Tc also belongs to the triclinic structure. The phase transition is connected with rotation of the Tc molecule in position (1/2, 1/2, 0) around its N axis perpendicular to the molecular plane. As a result more dense packing of the molecules in the lattice is obtained, and the symmetry is lowered from space group $P\bar{1}$ to P1.

Concerning Pc, there is no reliable report on observed phase transitions. However, the authors of Ref. 99 report that their electron diffraction studies of thin layers of Pc indicate that at low temperatures higher density of molecular packing in the lattice takes place.

1A.3.2. Pyrene and Perylene. B-type Lattice

Pyrene and perylene (Fig. 1.7a,b) belong to the conjugated, aromatic hydrocarbons also having D_{2h} molecular symmetry.

The packing of molecules in pyrene lattice is unusual: the role of the "crystallographic molecule" is played by molecular pairs having a dimer structure (see Fig. 1.7c). The adjacent parallel molecules of dimers are relatively closely spaced, the separation distance being 3.53 Å (Fig. 1.7d). Such configuration causes considerable overlap of π orbitals, and, consequently, considerable interaction. This promotes exciton self-trapping and excimer formation (see Chapter 2). In other aspects the crystal structure is isomorphic to that of anthracene (see Table 1.3). It is monoclinic, belongs to space group $P2_{1/a}$, except that there are four molecules in a unit cell ($z = 4$).[53,147]

Jones, Ramdas, and Thomas[147] have hypothesized the existence of two modifications of pyrene crystals: the well-known phase at room temperature — pyrene I, stable above 120 K — and pyrene II which is formed in a phase transition below 120 K. Pyrene II retains the space group and structure of pyrene I, with slightly

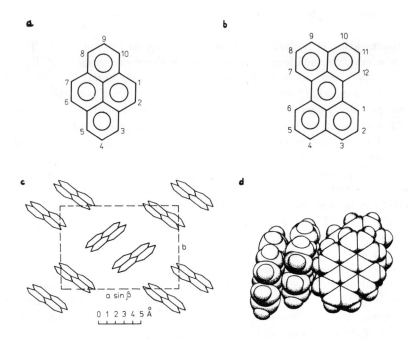

FIGURE 1.7. Pyrene (a) and perylene (b); (c) schematic crystallographic structure of pyrene (projection of the molecules in the **ab** plane); (d) picture of the complementary molecular arrangement of pyrene dimers in the lattice.[147]

TABLE 1.3. *Crystallographic Data on Pyrene (Py) and Perylene (Pl) Crystals[a]*

Parameters	Py $C_{16}H_{10}$[b]	P1 $C_{20}H_{12}$ α-modification[c]	P1 $C_{20}H_{12}$ β-modification[d]
Crystal structure	Monoclinic	Monoclinic	Monoclinic
Space group	$P2_1/a$	$P2_1/a$	$P2_1/a$
a, Å	13.65	11.35	11.27
b, Å	9.26	10.87	5.88
c, Å	8.47	10.31	9.65
β, °	100.28	100.8	92.1
V, Å³	1052.9	1249	394.3
z	4	4	2
d_{calc}, g/cm³	1.288	—	—
d_{oxp}, g/cm³	1.27	—	—
M, daltons	202.2	252.3	252.3

[a]*Notations*: V, unit cell volume (in Å³); z, number of molecules in unit cell, d, density of the crystal (g/cm³); M, molecular weight (in daltons). For the orientation of molecules in the lattice see Table 1.7 in Ref. 332.
[b]Taken from Ref. 53.
[c]Taken from Ref. 91.
[d]Taken from Ref. 380.

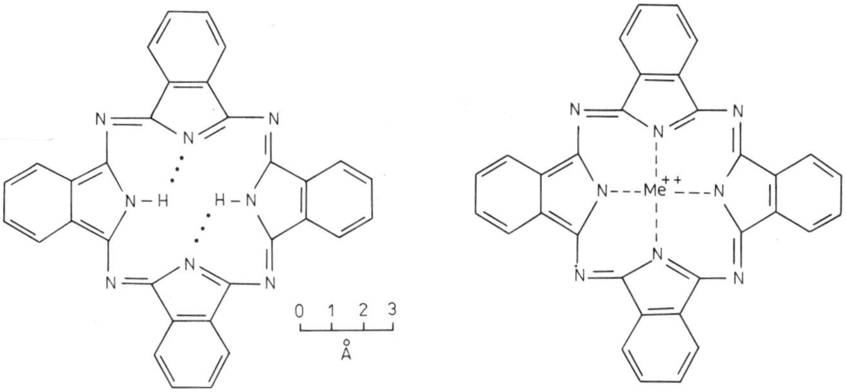

FIGURE 1.8. (a) Metal-free phthalocyanine (H$_2$Phc), (b) metal phthalocyanine (MePhc).

changed lattice parameters. Pyrene II exhibits closer packing of molecules (e.g., the separation of molecular dimer pairs (see Fig. 1.7d) decreases from 3.53 Å to 3.44 Å). Accordingly, the lattice energy value is increased by about 4 kJ/mol.

Perylene exists in two crystallographic modifications: α-perylene[91] having dimer pyrenelike structure with four molecules in a unit cell (B-type of crystal structure) (see Table 1.3), and β-perylene,[380] having an Ac- or A-type of crystal structure with two molecules in a unit cell (Table 1.3).

The formation of α or β structure depends on the conditions of crystal growth (see Refs. 380 and 332, p. 25).

Perylene is thus polymorphic, capable of existence either in the A or B types of aromatic crystal structure. From this point of view it may serve as an excellent model compound.

1A.3.3. Phthalocyanines

Phthalocyanines are derivatives of biologically important porphins. They possess the characteristic structure of porphins and are sometimes called tetrabenzotetraaminoporphins.

The molecule of phthalocyanine forms a practically planar macrocycle with a conjugated π-electron system containing 42 electrons (Fig. 1.8). The presence of the large cyclic polyconjugated π-electron system determines the remarkable optical, electric, photophysical, and related properties of phthalocyanines and, hence, have become, for decades, fascinating objects for active and intense research and applications (see review in monographs[288,357,351] and references therein). At present, besides the metal-free phthalocyanine (H$_2$Phc) (Fig. 1.8a) more than seventy of its metal derivates are known (Fig. 1.8b).

The metalphthalocyanines (MePhc), with metal atom in the center of the ligand, possess D$_{4h}$ molecular symmetry. For the greatest part of MePhc the macrocyclic ring is practically planar (with the deviation from the planar form not more than ca.

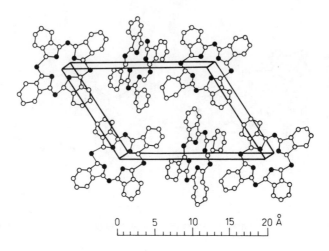

0   5   10   15   20 Å

FIGURE 1.9. Crystallographic structure of phthalocyanine of β-modification. (After Ref. 48.)

0.3 Å).[357] The metal phthalocyanines possess remarkable thermal and chemical stability, they are stable up to 400–500 °C.[357] On the whole, phthalocyanines demonstrate a variety of structural features, including polymorphism and a wide spectrum of optical, electric, photophysical, and other interesting properties.[288,357] The first crystallographic studies of H_2Phc and some MePhc have been performed by Robertson and coworkers in the late thirties (see Ref. 357 and references therein). A refined x-ray analysis of a number of MePhc have been carried out later (see Refs. 46, 47, 261, 311, 395, 144, 120, and 412 and Tables 1.4–1.7).

Recently Uyeda, Kobayashi, and coworkers[187,189,191,397] have developed high-resolution electron microscopy and obtained superb electron micrographs of direct images of phthalocyanine (Phc) molecules and constituent atoms both of the perfect Phc crystals (see Fig. 1.1) and of the ones with defects.[397]

Low-energy electron diffraction (LEED) techniques have been successfully used for the study of Phc monolayers and thin films.[48]

Metal-free Phc (Fig. 1.8a) and a number of MePhc (Fig. 1.8b) reveal *polymorphism* and usually forms two main, most stable structures — α and β-modifications (see Fig. 1.10 and Tables 1.4 and 1.5).

Figure 1.9 shows the configuration of H_2Phc molecules in the unit cell of β-modification. Crystallographic data of β- and α-modifications of H_2Phc and a number of MePhc are presented in Tables 1.4 and 1.5, respectively. The β-modification is the most stable one for H_2Phc and a number of Me phthalocyanines [manganese (Mn), iron (Fe), cobalt (Co), copper (Cu), zinc (Zn), etc.] (see Table 1.4).

As can be seen from Fig. 1.9, the crystallographic structure of β-modification is reminiscent of the symmetry of Ac-type crystals (see Fig. 1.6). The structure is *monoclinic* and possesses the same space group $P2_{1/a}$ with two ($z = 2$) nonequivalent molecules in the unit cell. However, there is one principal difference. The phthalocyanine molecules have the highest density of packing along the **b** axis of

TABLE 1.4. *Crystallographic Data on Phthalocyanines (β-modification)[a]*

Parameters	H_2Phc^b	$MnPhc^c$	$FePhc^b$	$CuPhc^d$	$ZnPhc^d$
Space group	$P2_1/a$	$P2_1/a$	$P2_1/a$	$P2_1/a$	$P2_1/a$
z	2	2	2	2	2
a, Å	19.85	19.36	20.2	19.41	19.27
b, Å	4.72	4.755	4.77	4.79	4.85
c, Å	14.8	14.58	15.0	14.63	14.53
β, °	122.25	120.7	121.6	120.56	120.48
V, Å³	—	1153.9	—	1166	—
d_{calc}, g/cm³	—	1.633	—	1.639	1.614
d_{exp}, g/cm³	—	—	—	1.63	1.62

[a]See Fig. 1.10.
[b]Taken from Ref. 48.
[c]Taken from Ref. 261.
[d]Taken from Ref. 46.

the crystal (b = 4.7–4.9 Å) and form a stack (or column) along this axis (see Figs. 1.10 and 1.11).

The α-modification of the above-mentioned Phc crystals is less stable (metastable) and can be obtained under specific conditions of crystal growth and treatment. As an exception to this, one should mention PtPhc for which the most stable form is the α-modification (see Table 1.5).

According to Ref. 47, in case of PtPhc a new metastable modification (called the γ-modification) is observed. This modification has only slightly different lattice parameters (see Table 1.5), however, it belongs to another space group, namely, $B2_{1/a}$. The possible existence of the β-modification of PtPhc is controversial.[47]

As may be seen from Tables 1.4 and 1.5, for both α- and β-modifications of Phc there exist strongly pronounced *isomorphism* of the crystallographic structure. This is confirmed by the very close values of lattice parameters for the H_2Phc and MePhc series. However, one may observe a strict empirical rule — the parameter b increases with increasing atomic weight of the central metal atom. Due to the

TABLE 1.5. *Crystallographic Data on Phthalocyanines (α-modification) (Fig. 1.10b)[a]*

Parameters	H_2Phc^b	$FePhc^b$	$CuPhc^b$	$PtPhc^c$
Space groups	C2/c	C2/c	C2/c	C2/c
z	4	4	4	4
a, Å	26.14	25.90	25.90	26.29
b, Å	3.182	3.765	3.79	3.82
c, Å	23.97	24.97	23.10	23.92
β, 0	91.1	90.0	90.4	94.6

[a]*Notation*: For PtPhc V = 2393 Å³; d_{cal} = 1.963 g/cm³.
[b]Taken from Ref. 48.
[c]Taken from Ref. 47.

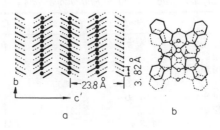

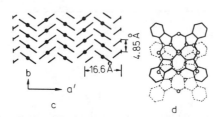

FIGURE 1.10. Schematic arrangement of molecules in the lattice of α- and β-modifications of phthalocyanine (after Ref. 188). (a, b) shows the α-modification of ZnPhc; (c, d) the β-modification of PtPhc; (a) projection on the (100) plane; (c) projection on the (001) plane; (b, d) superposition of molecules connected by the translation along the **b**-axis.

structural isomorphism one may expect similar isomorphism in electrical and photoelectric properties of described phthalocyanines.

The main difference in crystallographic structure of α- and β-modifications emerges in the molecular arrangement and orientations in the stack along the **b** axis of the crystal (see Fig. 1.10).

In case of the α-modification the molecules are positioned in the stack along the *b*-axis at an angle 25.3°, the intermolecular distance *b* having the value $b \approx 3.8$ Å (see Fig. 1.10a). In case of the β-modification the angle equals ca. 45.8° and the packing is of lower density, i.e., $b = 4.7–4.8$ Å (Fig. 1.10c) (see also Tables 1.4 and 1.5). Consequently, the intermolecular interaction along the **b** axis should be higher in the α-modification. Hence, one may expect corresponding anisotropy in optical and electric properties of both modifications. Thus, e.g., the infrared (IR) spectra of α- and β-forms differ considerably allowing the study of the kinetics of the phase transition $\alpha \rightarrow \beta$.[357]

The electron micrograph of the images of ZnPhc molecules perpendicular to the *b*-axis of molecular stacks (Fig. 1.11) confirms the structural arrangement as obtained by x-ray diffraction techniques (cf. Fig. 1.10c).

For the greatest part of metal Phc, formed by the transition metals, having the radius of the ion of the order of ca. 0.7–0.8 Å, the molecule is practically planar.[357] In this case the metal ion fits the "hole" in the center of the ligand and thus becomes positioned in the plane of the Phc molecule.

However, in case of PbPhc the ionic radius is ca. 1.20 Å and does not fit the

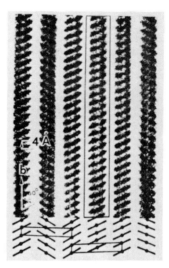

FIGURE 1.11. Electron micrograph of the images of the ZnPhc molecules perpendicular to the **b**-axis of the molecular stacks according to Kobayashi (Ref. 187, p. 42). The micrograph also shows a stacking fault of one misoriented molecular column.

central "hole" of the ligand and is shifted outside the molecular plane.[357]

The PbPhc appears in two modifications — monoclinic and triclinic (Table 1.6). As may be seen from Table 1.6, these differ considerably in lattice parameter values, especially, in the **c**-direction of the lattice. For the monoclinic modification the distance between the lead (Pb) atoms along the c-direction is only 3.73 Å and thus forms a chain of Pb atoms possessing amazingly high, almost semimetallic conductivity, which is ca. 10^6 times higher than the conductivity of the dielectric triclinic form[144] (see Table 1.6).

In the family of metal phthalocyanines there is a remarkable member, namely, vanadyl phthalocyanine (VOPhc) (Fig. 4.26a). In this case the central V atom is bounded covalently with an oxygen atom. VOPhc, like the parent phthalocyanine and many of its metal derivatives, possessing interesting semiconductor and photo-

TABLE 1.6. *Lattice Parameters of Lead (PbPhc) Phthalocyanine[a–c]*

Crystal structure	Space group	z	a, Å	b, Å	c, Å	α, °	β, °	γ, °	Specific electric conductivity, $(\Omega^{-1}cm^{-1})$
Monoclinic	$P2_{1/b}$	4	25.49	25.49	3.73	90	90	90	10^{-2}
Triclinic	$P\bar{1}$	4	13.2	16.3	12.89	94.22	96.20	114.19	10^{-8}

[a]*Notation*: For monoclinic form $d = 1.98$ g/cm³; for triclinic one $d_{calc} = 1.95$ g/cm³; $d_{exp} = 1.93$ g/cm³.
[b]Taken from Ref. 395.
[c]Taken from Ref. 144.

TABLE 1.7. *A Comparison of Lattice Parameters of Vanadyl (VOPhc) and Tin (SnPhc) Phthalocyanines*[a]

Phthalocyanine	Crystal structure	Space group	z	a, Å	b, Å	c, Å	α, °	β, °	γ, °
VOPhc (II)	Triclinic	$P\bar{1}$	2	12.027	12.571	8.690	96.04	94.80	68.20
SnPhc	Triclinic	$P\bar{1}$	2	12.060	12.618	8.675	95.89	95.08	68.17

[a]Taken from Refs. 120 and 412.

conductor properties. It has been shown, e.g., that VOPhc is a photosensitive pigment useful in both photoelectrophoretic and xerographic imaging;[412] recently amphiphilic derivatives of VOPhc have been used for designing LB multilayers (see Section 4.7 and Fig. 4.26).

The crystallographic structure of VOPhc has been studied by Griffiths *et al.*[120,412] (Table 1.7). It is interesting that VOPhc forms an isomorphic crystal structure with the tin phthalocyanine (SnPhc).[120] The stable VOPhc form (II) has triclinic structure and belongs to the $P\bar{1}$ space group with $z = 2$. It is noticeable that in case of VOPhc there is no lattice direction with preferably small intermolecular distance (see Tables 1.4, 1.5, and 1.7); according to Ref. 120 it is prone to systematic disorder in the **ac** plane.

The data in Tables 1.4–1.7 illustrate the great variety of structural forms and pronounced *polymorphism* of the phthalocyanine family.

It should also be mentioned that germanium (Ge) and silicon (Si) phthalocyanines easily form poly-phthalocyanines via an oxygen bridge, e.g., poly-GeOPhc (see Ref. 187 and references therein).

1A.4. SOME DYNAMIC PROPERTIES OF ORGANIC MOLECULES AND CRYSTALS

In organic molecular crystals there are two kinds of vibrations: intramolecular, high-frequency ones, i.e., the normal "internal" vibrational modes of molecules and intermolecular, low-frequency ones, connected with the lattice vibrational modes, namely, with lattice acoustic and optical phonons.

Since these dynamic components play a very important role in interaction phenomena of charge carriers and excitons in organic crystals we shall briefly discuss both molecular and lattice vibrations.

1A.4.1. Intramolecular Vibrations

Let us consider for illustrative and tutorial purposes the intramolecular vibrational modes of our main model compound *anthracene* (see Section 1A.3.1). Since the Ac molecule belongs to the D_{2h} symmetry group it has $3N-6$ normal vibrational modes, where N is the number of atoms in the molecule. Hence, for $C_{14}H_{10}$ we have $3\times24-6 = 66$ vibrational modes. Table 1.8 presents the whole spectrum of the

TABLE 1.8. *Intramolecular Vibrational Modes, Symmetry Types, and Corresponding Vibron Wave Numbers of an Anthracene Single Crystal*[a,b]

Vibrational mode	Symmetry type	Wave number (cm⁻¹)	Vibrational mode	Symmetry type	Wave number (cm⁻¹)	Vibrational mode	Symmetry type	Wave number (cm⁻¹)
ω_1	b^*_{1g}	247/244.5	ω_{23}	b_{3g}	1376	ω_{45}	a^*_u	870
ω_2	b^*_{2g}	290	ω_{24}	a_g	1403	ω_{46}	b^*_{3u}	883
ω_3	b_{3g}	390	ω_{25}	a_g	1482	ω_{47}	b_{1u}	903
ω_4	a_g	395	ω_{26}	a_g	1557	ω_{48}	b^*_{3u}	954
ω_5	b_{3g}	478.5/481	ω_{27}	b_{3g}	1576	ω_{49}	a^*_u	
ω_6	b^*_{1g}	477	ω_{28}	b_{3g}	1634	ω_{50}	b_{2u}	998
ω_7	b^*_{2g}	577	ω_{29}	b_{3g}		ω_{51}	b_{2u}	1068
ω_8	a_g	622	ω_{30}	b_{3g}		ω_{52}	b_{1u}	1145
ω_9	b_{1g}	747	ω_{31}	a_g	3066	ω_{53}	b_{2u}	1163
ω_{10}	a_g	753	ω_{32}	a_g	3088	ω_{54}	b_{1u}	1270
ω_{11}	b^*_{2g}	762.5/764	ω_{33}	a_g	3108	ω_{55}	b_{1u}	1314
ω_{12}	b^*_{2g}	896	ω_{34}	b^*_{3u}	129/111	ω_{56}	b_{2u}	1346
ω_{13}	b^*_{1g}	904	ω_{35}	a^*_u	172	ω_{57}	b_{2u}	1398
ω_{14}	b_{3g}	918/916	ω_{36}	b_{1u}	235	ω_{58}	b_{1u}	1447
ω_{15}	b_{3g}	954/959	ω_{37}	b^*_{3u}	475	ω_{59}	b_{2u}	1462
ω_{16}	b^*_{2g}	980/978.5	ω_{38}	a^*_u		ω_{60}	b_{2u}	1533
ω_{17}	a_g	1008	ω_{39}	b^*_{3u}	603	ω_{61}	b_{1u}	1616
ω_{18}	b_{3g}	1103	ω_{40}	b_{2u}	600	ω_{62}	b_{1u}	3024
ω_{19}	a_g	1163	ω_{41}	b_{1u}	650	ω_{63}	b_{2u}	3050
ω_{20}	b_{3g}	1187	ω_{42}	b^*_{3u}	727	ω_{64}	b_{1u}	3050
ω_{21}	a_g	1261	ω_{43}	a^*_u		ω_{65}	b_{2u}	3093
ω_{22}	b_{3g}	1274	ω_{44}	b_{2u}	808	ω_{66}	b_{1u}	3108

[a]*Note:* The asterisk (*) denotes out-of-plane vibrations. For assignment of symmetry types see the text.
[b]Taken from Ref. 198.

intramolecular vibrational modes, their symmetry types and wave numbers determined in Ac single crystal from the vibrational structure of electronic transitions, Raman scattering, and IR absorption spectra[198] (see also Ref. 154, p. 113).

b_{1u} are vibrational modes, polarized parallel to the short (M), b_{2u} — to the long (L) axis of the molecule, but b_{3u} — parallel to the axis (N) normal to the plane of the molecule. These odd (u) types of vibrational symmetry are connected with the electric dipole moment P_i change and appear in the IR absorption spectra. The even (g) types of vibrational symmetry are connected with the polarizability tensor α_{ik} change and are active in the Raman scattering spectra.

a_u vibrations are forbidden for the point group D_{2h} of the free molecule but they may occur in the crystal spectra (see Table 1.8).

As can be seen from Table 1.8, one-half (33) of the 66 vibrational modes belong to the even (g), the other half (33) to the odd symmetry. These vibrational modes are distributed among the symmetry types in the following way:

$$12a_g+5a^*_u+11b_{3g}+6b^*_{3u}+4b^*_{1g}+11b_{1u}+6b^*_{2g}+11b_{2u}.$$

The asterisk denotes out-of-plane vibrations; to this type of vibrations belong 21 modes. It should be mentioned that due to the space group symmetry $C^5_{2h}-P2_{1/a}$, the factor group (Davydov) splitting occurs through the interaction of two symmetry-related equivalent molecules in the unit cell (Fig. 1.6) (see Table 1.8). As may be seen from Table 1.8, the intramolecular vibrations lie in the high-frequency region and cover the energy range from 120 to 3000 cm^{-1}.

The interaction of charge carriers with the intramolecular vibrational modes and its implications will be discussed in Chapter 3.

1A.4.2. Lattice Vibrations

There are two types of lattice vibrations: *acoustic* and *optical*. The optical lattice vibrations may be observed by the low-frequency IR and Raman spectroscopy.

We shall give below the characteristic optical lattice vibration frequencies for linear polyacenes and α- and β-perylene crystals.

Naphthalene. Reported optical lattice vibration frequencies of naphthalene are classified by Suzuki, Yokoyama, and Ito[373] as rotational modes (librons) at 125, 109, 74, 71, 51, and 46 cm^{-1}, and translational modes at 98, 73, and 39 cm^{-1}. Dows et al.[93] have compiled the reported and their own data in naphthalene lattice frequencies and compared them with calculated values. These authors give following assignments of lattice mode symmetry: Raman active modes A_g at 109; 74 and 51 cm^{-1} and B_g at 125; 71 and 46 cm^{-1}; IR active modes A_u at 102 and 49 cm^{-1} and B_u at 65 cm^{-1}.[332,351]

Anthracene. Reported optical lattice vibration frequencies of anthracene are classified according to Cruickshank[79] as rotational modes (librons) at 120; 68 and 48 cm^{-1} and transitional mode at 59 cm^{-1}. Dows et al.[93] give the following assign-

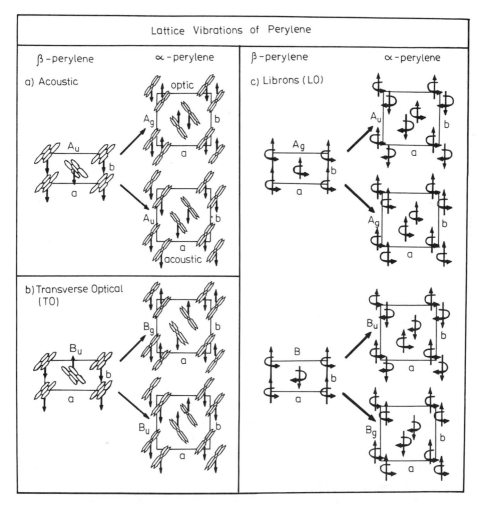

FIGURE 1.12. Schematic illustration of characteristic lattice vibrational modes in α- and β-perylene according to Ref. 194.

ments of lattice mode symmetry: Raman active modes A_g at 121; 70 and 39 cm^{-1} and B_g at 125; 65 and 45 cm^{-1}; IR active modes A_u at 101 and 44 cm^{-1} and B_u at 60 cm^{-1}. Hadni et al.[123] have identified three translational modes at 120, 101, and 61 cm^{-1}. [332,351]

Tetracene. Tomkievicz, Groff, and Avakian[384] have reported the following Raman active lattice vibration frequencies of Tc: 133, 123, 94, 61, and 45 cm^{-1}. Later Jankowiak et al.[146] have detected the following Raman active frequencies of Tc: 132, 120, 94, 61, 49, and 41 cm^{-1}, in good agreement with earlier data of Ref. 384.

TABLE 1.9. *Intermolecular Lattice Vibrational Modes, Their Symmetry Types, and Corresponding Phonon Wave Numbers of α- and β-Perylene Crystals[a,b]*

Translational phonons				Rotational phonons (librons)			
β-perylene		α-perylene		β-perylene		α-perylene	
Symmetry type	Wave number (cm^{-1})	Symmetry type	Wave number (cm^{-1})	Symmetry type	Wave number (cm^{-1})	Symmetry type	Wave number (cm^{-1})
A_u	106*	A_g	200*	A_g	120*	A_u	256*
		A_u	63			A_g	80*
		B_g	170	B_g	101	B_u	180*
B_u	68*	B_u	46			B_g	72
		A_g	161*			A_u	125*
A_u	50	A_u	33	A_g	94*	A_g	56*
B_u(acoustic)	—	B_g	128			B_u	116*
		B_u(acoustic)	—	B_g	84	B_g	50
		A_g	104*			A_u	100*
A_u(acoustic)	—	A_u(acoustic)	—	A_g	53*	A_g	33*
		B_g	95*			B_u	80*
B_u(acoustic)	—	B_u(acoustic)	—	B_g	40	B_g	27

[a]*Note*: The asterisk (*) denotes reported experimental values which have been identified by calculation data. The optical phonons of A_u and B_u symmetry are active in the IR absorption spectra, while those of A_g and B_g symmetry in the Raman scattering spectra.
[b]Taken from Ref. 194.

Perylene. Peculiar features of the crystallographic structures of α- and β-perylenes (see Table 1.3) emerge in a distinct way in the lattice vibrations, i.e., in the spectra of acoustic and optical lattice phonons.[351]

The rich diversity of the lattice vibrational modes of α- and β-perylenes is illustrated, according to Kosic, Schosser, and Dlott[194] in Fig. 1.12. Table 1.9 presents the spectra of optical and acoustic lattice phonons and their symmetry types of α- and β-perylene.

The optical vibrational modes of odd (u) symmetry A_u and B_u are dipole-active and occur in IR absorption spectra, the A_u modes are polarized along the **b**-axis; the B_u modes are polarized in **ac**-plane.

The optical modes of even (g) symmetry A_g and B_g are active in Raman scattering spectra; the A_g modes are observed in **aa**, **bb**, and **cc** spectra, the B_g modes in **ab** and **ac** spectra.[194]

As may be seen from Table 1.9 in case of β-perylene there are 12 branches of lattice phonon vibrations: nine optical modes (six libronic, Raman-active and three translational, IR active) and three acoustic modes.

The spectrum of lattice vibrations of α-perylene, as may be seen from Table 1.9, is similar to that of β-perylene. However, in this case there appear new modes connected with frequency doubling due to in- and out-of-phase vibrations of the dimers (see Fig. 1.12). As the result, in case of α-perylene there emerge 24 branches of lattice phonon vibrations: 21 optical modes and 3 acoustic ones (see Table 1.9).

The vibrations of the dimers in opposite phase leads to increased frequency and may be regarded as excitation of "dimer molecules" of the crystal.

The authors of Ref. 194 indicate that the data presented in Table 1.9 should be regarded as a first approximation, due to possible thermal shifts of reported frequencies, as well as possible errors in assignment of the vibrational modes.

It should be noticed that the frequency range of optical lattice phonons lies well below that of intramolecular vibrations (see Table 1.8). This fact actually reflects the large energy difference of inter- and intramolecular forces characteristic of organic molecular crystals.

The data on lattice phonon spectra presented above will be useful later for analysis of thermalization and relaxation processes of hot charge carriers which are mainly due to inelastic scattering on the optical lattice phonons (see Chapter 7).

1A.5. MOLECULAR INTERACTION FORCES IN OMC. MOLECULAR AND SOLID-STATE APPROACHES

As already mentioned, the molecular interaction in OMC mainly manifests itself via Van der Waals forces which represent a specific kind of nonvalent interaction between electrically neutral particles. The most general type of Van der Waals forces are so-called dispersion forces which give rise to mutual attraction of neutral molecules. This kind of molecular interaction is of special importance for aromatic crystals consisting of nonpolar molecules. A quantum-mechanical treatment of dispersion forces was first given by London.[207] It is based on the concept of electrical interaction between fluctuating multipole moments of molecules.[184] A simple static approach yields zero interaction between neutral nonpolar molecules. However, even in atoms and molecules with zero mean value of multipole moments there exist dynamically fluctuating multipole moments depending on momentary states of electron motion. The momentary electric field created by these moments induces multipole moments in adjacent atoms or molecules. Interaction between the electric moments of the initial molecule and induced moments of the adjacent ones, averaged over the total set of molecules, produces forces of attraction among them which are responsible for formation of molecular solids.

As shown by London,[207] this effect is primarily due to external, more weakly bonded electrons such as valence σ and π electrons of aromatic molecules. These electrons are also responsible for the dispersion of light. That is why London called this phenomenon of attraction between neutral molecules dispersion forces, and this term has been generally accepted.

Using quantum-mechanical second-order perturbation theory, London[207] derived an approximate, but quite adequate expression for the dispersion interaction energy $U_{dis}(r)$ as a function of the distance r between two molecules. The potential energy of the interacting molecular pair is expressed in the form of multipole series, of which only the dipole–dipole terms are taken into account. This yields an expression of interaction energy $U_{dis}(r)$, averaged over various orientations of the dipoles with respect to the vector **r**, connecting the molecules:[184,332]

$$U_{dis}(r) = -A/r^6, \qquad (1A.1)$$

where A is a constant, equal to

$$A = \frac{2}{3} \sum_{i'} {}' \sum_{j'} {}' \frac{|\langle i|\mu_1|i'\rangle|^2 \, |\langle j|\mu_2|j'\rangle|^2}{E_{i'}+E_{j'}-E_i-E_j},$$

where i, j denote ground states of nonperturbated molecules; i', j' denote excited states of molecules; E_i, E_j and $E_{i'}$, $E_{j'}$, are energies of molecules in the ground and excited states, respectively; $\langle i|\mu_1|i'\rangle$ is the matrix element of the dipole moment between states i and i' for molecule 1; $\langle j|\mu_2|j'\rangle$ is the matrix element of the dipole moment between states j and j' for molecule 2.

According to Eq. (1A.1) $U_{dis}(r)$ is negative and describes the Van der Waals attractive interaction between neutral molecules. Since formula (1A.1) is not expedient for practical calculations, London proposed the following approximate formula for estimation of dispersion force for spherically symmetric molecules:[207]

$$U_{dis}(r) = -\frac{3}{2} \, \alpha_a \alpha_b \frac{I_a I_b}{I_a + I_b} \cdot \frac{1}{r^6}, \qquad (1A.2)$$

where α_a and α_b denote the mean polarizability of molecules a and b, respectively, and I_a and I_b denote their ionization potentials.

Formula (1A.2) illustrates the direct influence, according to the diagram on Fig. 1.2, of the main molecular parameters for the molecular interaction processes.

One should mention that the London dispersion forces are a particular case of more general electronic polarization phenomena in OMC. It will be shown in the framework of Hamiltonian description of interaction phenomena in OMC (see Section 1B) that the molecular interaction forces, responsible for the formation of unexcited Van der Waals crystals, are of the same physical nature as many-electron interaction forces of excited quasiparticles, i.e., charge carriers or excitons, with the electronic subsystem of the surrounding molecular lattice. All these very fast electronic interaction forces, with the characteristic femtosecond polarization time scale $\tau_e = 10^{-16}$–10^{-15} s, may be described in terms of a generalized approach in the framework of unified electronic polarization model (see Section 1B).

Thus, it should be emphasized here that the electronic polarization phenomena are the most important feature of molecular crystals responsible for formation of the energy structure of electronic and excitonic polaron states, of the local states of structural origin for charge carriers and excitons (see Chapter 4), as well as for the very existence of the molecular lattice itself in an unexcited state.

London's theory of dispersion forces provides a satisfactory description of molecular interaction at distances of the order of Van der Waals radii (see Fig. 1.3). At closer intermolecular distances considerable overlapping of molecular orbitals takes place (see Fig. 1.4), leading to strong mutual repulsion between the molecules. Quantum mechanical treatment of the repulsion potential U_{rep} yields a complicated dependence of U_{rep} on distance r. For this reason $U_{rep}(r)$ is usually described by means of empirical formulas. Most frequently an exponential approximation of the following type is used:[332]

$$U_{rep} = B \exp[-Cr], \tag{1A.3}$$

where B and C are empirical constants.

At a certain distance r_0 the attractive interaction U_{dis} and repulsive interaction U_{rep} become equal coresponding to the potential energy minimum of the system. The full potential curve of intramolecular interaction can be described by the empirical Buckingham formula:[49,50]

$$\varphi(r) = -\frac{A}{r^6} + B \exp[-Cr]. \tag{1A.4}$$

The potential $\varphi(r)$ in Eq. (1A.4) is often called the Buckingham (6-exp) potential.

Another empirical formula, proposed by Lennard-Jones, is also frequently used for $\varphi(r)$ calculations:[217]

$$\varphi(r) = -\frac{A}{r^6} + \frac{B'}{r^{12}}, \tag{1A.5}$$

B' also being an empirical constant.

Potential $\varphi(r)$ in Eq. (1A.5) is usually known as the Lennard-Jones (6-12) potential.

In the case of polar molecules other types of Van der Waals interaction, apart from dispersion forces, can take place, namely, inductive and orientational interaction of mixed electronic polarization and electrostatic origin. Induction forces appear, if molecule **a** with permanent dipole moment μ_a polarizes a nonpolar molecule **b**, thus inducing a dipole moment in the latter. The mean induction interaction energy (called also Debye polarization effect) is in this case:[332]

$$U_{ind} = -\frac{2\mu_a^2 \alpha_b}{r^6}, \tag{1A.6}$$

where α_b is the polarizability of molecule **b**.

The energy of orientational or dipole-dipole interaction between molecules **a** and **b**, with dipole moments μ_a and μ_b, respectively, equals:[332]

$$U_{d-d} = -\frac{2}{3kT} \cdot \frac{\mu_a^2 \mu_b^2}{r^6}. \tag{1A.7}$$

If we have molecule **a** with dipole moment μ_a interacting with molecule **b**, possessing permanent quadrupole moment Θ_b, then the interaction energy equals:[332]

$$U_{d-Q} = -\frac{1}{kT} \cdot \frac{\mu_a^2 \Theta_b}{r^8}. \tag{1A.8}$$

The mean energies of induction and dipole-dipole interactions are of the same order as that of dispersion interaction. All three types of Van der Waals interaction have similar dependence on the distance r, i.e., $\sim 1/r^6$.

Consequently, for organic crystals consisting of polar, e.g., heterocyclic molecules, the total molecular interaction energy U_{vdW} is an additive sum of all three Van der Waals interaction energies:

$$U_{\text{vdW}} = U_{\text{dis}} + U_{\text{ind}} + U_{\text{d-d}}. \qquad (1A.9)$$

This means that in case of polar molecular crystals the interaction energy may be ca.2–3 times higher than that of nonpolar ones.

The theory of London-type Van der Waals forces, as well as corresponding formulas for molecular interaction potential (1A.4) and (1.A5), are based on the assumption of a field of central forces. Hence, quantitative calculations of intermolecular interaction energy, using the Buckingham or the Lennard-Jones formula, were first carried out on the simplest kind of molecular crystals formed by noble-gas atoms. The high symmetry of these crystals and the spherical symmetry of the atoms suit excellently the main requirements of London's theory.[332] As regards more complex molecular systems, such as crystals of aromatic hydrocarbons consisting of large polyatomic molecules, direct interaction energy calculation using potential functions of types (1A.4) or (1A.5) is not feasible. A more detailed empirical method of calculation has to be employed in this case, known as the atom-atom potential method, proposed and developed by Kitaigorodsky.[182–184]

According to the atom-atom potential approach, the interaction energy of molecules is obtained as the sum of interaction energies between the atoms of neighboring molecules. In such approximation the forces of interaction between atoms can be considered as central ones conforming with the basic concepts of London's theory and thus permitting the use of an interaction potential of type (1A.4) or (1A.5).

The method is based on the assumption that the energy U of molecular interaction may be calculated as the additive sum of interaction energies φ_{ij} of atoms constituting the molecule

$$U = \frac{1}{2} \sum_{i,\,j} \varphi_{ij}, \qquad (1A.10)$$

where i, j are the symbols for all types of *nonvalent* interaction between atoms of neighboring molecules (e.g., interaction between atoms C . . . C, C . . . H, and H . . . H in hydrocarbons).

The method of atom-atom potentials had been successfully used for calculation of the arrangement of molecules, lattice energies, etc., both in perfect and imperfect OMC.

However, here we are interested in another aspect of the problem. It follows from formula (1A.10) that the molecular interaction energy U will be higher if there are more atom-atom contacts within the range of Van der Waals radii (see Fig. 1.3). This means that the molecular interaction will be greater in case of pronounced complementarity of adjacent molecules, i.e., in case of the possibly closest packing. And vice versa, the more dense packing of the molecules will take place with increasing total interaction energies. Thus, molecular arrangement and molecular interaction are mutually interrelated. This phenomenon is symbolized by the double arrow connecting both blocks in the diagram in Fig. 1.2. In this case one can speak about positive "feedback" providing the highest possible regularity and density of packing of molecules in the crystal.

The diagram in Fig. 1.2 illustrates the types of interaction that are involved in the treatment of optical and electronic properties of molecular crystals. Energy structure and electronic properties of molecular crystals are determined both by crystalline and molecular structure. Thus, for instance, appreciable electronic conductivity and photoconductivity can be found mainly in crystals containing molecules with polyconjugated bond systems, such as aromatic hydrocarbons and conjugated heterocyclic compounds. These molecules contain delocalized, weakly bonded conjugated π-electron systems, or heteroatoms with lone pairs of n electrons. These are composed of π- and n-electrons, which are the potential sources of free charge carriers formed through action of light or temperature.

This direct impact of the properties of molecules on those of the crystal is illustrated in the diagram on Fig. 1.2 by the vertical downward arrow.

In summing up this introductory part 1A of the first Chapter let us reiterate that organic molecular crystals possess a number of specific features making them essentially different from covalent or ionic crystals (see Table 1.1). Hence, they can not be described in the framework of traditional solid-state physics.

In the second part of this chapter (1B) there is an attempt to formulate a more generalized theory of organic molecular crystals based on a Hamiltonian description of possible interactions. This generalized treatment further serves as conceptual basis for analysis and interpretation of more specific cases.

To a reader not inclined to theory, we recommend that the second part (1B) of this chapter be omitted on first reading.

Part 1B
Hamiltonian Description of Interaction Phenomena in Organic Molecular Crystals

Before turning attention to the Hamiltonian formulation of the theory of molecular crystals, one should realize the importance of adequate theoretical background and a good balance between theoretical treatment and experimental evidence in the text.

The physics of organic molecular crystals relates the properties of molecules and that of solids (see Fig. 1.2). For the sake of simplicity we will first start from isolated molecules. The aim of part 1B of Chapter 1 is to derive the necessary theoretical background of most of the topics treated in the book, starting from the

Hamiltonian formulation of the molecule and of the molecular crystal. In later chapters, we shall often return to theory in connection with the experimental material presented. The reader should judge how successful this approach is for understanding the physics of such complex systems like organic molecular crystals.

1B.1. HAMILTONIAN OF AN ISOLATED MOLECULE; RIGID (CRUDE) AND DYNAMIC (ADIABATIC) REPRESENTATIONS

To introduce some basic notions, let us first consider an *isolated* molecule in nonrelativistic approximation. Its Hamiltonian H contains the *electronic* part

$$H_{EE}(r) = -\sum_j \frac{\hbar^2}{2m} \Delta_{\mathbf{r}_j} + \frac{1}{2} \sum_{i \neq j} \frac{e^2}{|\mathbf{r}_i - \mathbf{r}_j|} \qquad (1B.1.1)$$

($\mathbf{r}$ represents the set of electron variables $x_j = \mathbf{r}_j \sigma_j$ where σ_j is the spin of the jth electron), the *nuclear* part

$$H_{NN}(R) = -\sum_k \frac{\hbar^2}{2M_k} \Delta_{\mathbf{R}_k} + \frac{1}{2} \sum_{k \neq l} \frac{Z_k Z_l e^2}{|\mathbf{R}_k - \mathbf{R}_l|} \equiv T_N + E_{NN} \qquad (1B.1.2)$$

($\mathbf{R}$ representing the set of nuclear coordinates), and the *electron–nuclear* interaction part

$$H_{EN}(r, \ R) = -\sum_{i, k} \frac{Z_k e^2}{|\mathbf{r}_i - \mathbf{R}_k|} . \qquad (1B.1.3)$$

Ignoring for a while the nuclear kinetic energy [the first term on the right-hand side of Eq. (1B.1.2)], one may introduce the electronic Hamiltonian

$$H_{EN}(r, \ R) + H_{EE}(r) = h + v^{(0)},$$

$$h = -\sum_j \frac{\hbar^2}{2m} \Delta_{\mathbf{r}_j} - \sum_{i,k} \frac{Z_k e^2}{|\mathbf{r}_i - \mathbf{R}_k|},$$

$$v^{(0)} = \frac{1}{2} \sum_{i \neq j} \frac{e^2}{|\mathbf{r}_i - \mathbf{r}_j|} \qquad (1B.1.4)$$

encountered in the adiabatic approximation in which $\mathbf{R}$ stands as a parameter. In the second quantization in a fixed (R-independent) basis of spinorbital states $|P\rangle, |Q\rangle \ldots$, usually called the rigid or crude basis,

$$H_{EN}(r, \ R) + H_{EE}(r) = \sum_{PQ} \langle P|h|Q\rangle a_P^\dagger a_Q + \frac{1}{2} \sum_{PQRS} \langle PQ|v^{(0)}|RS\rangle a_P^\dagger a_Q^\dagger a_S a_R$$

$$= \sum_{I} h_{II} + \frac{1}{2} \sum_{IJ} (v_{IJIJ}^{(0)} - v_{IJJI}^{(0)}) + \sum_{PQ} h_{PQ} \, \mathcal{N}[a_P^\dagger a_Q]$$

$$+ \sum_{PQI} (v_{PIQI}^{(0)} - v_{PIIQ}^{(0)}) \, \mathcal{N}[a_P^\dagger a_Q]$$

$$+ \frac{1}{2} \sum_{PQRS} v_{PQRS}^{(0)} \, \mathcal{N}[a_P^\dagger a_Q^\dagger a_S a_R]. \tag{1B.1.5}$$

Here we have introduced the usual notion of the normal product $\mathcal{N}[..]$; and $I, J, \ldots$ designate occupied states.

Now, by superscript (0), we denote quantities for R fixed at some (e.g., equilibrium) positions R_0; while prime will refer to deviations from these values depending on $(R - R_0)$. Then

$$H_{EN}(r, \ R) + H_{EE}(r) = E_{SCF}^{(0)} + E_{SCF}' + \sum_{P} \epsilon_P^{(0)} \, \mathcal{N}[a_P^\dagger a_P] + \sum_{PQ} h_{PQ}' \, \mathcal{N}[a_P^\dagger a_Q]$$

$$+ \frac{1}{2} \sum_{PQRS} v_{PQRS}^{(0)} \mathcal{N}[a_P^\dagger a_Q^\dagger a_S a_R] \tag{1B.1.6}$$

provided that the fixed (crude) set $|P\rangle, |Q\rangle \ldots$ has been chosen as solutions of the Hartree-Fock equations, i.e., these are eigenstates of

$$\sum_{PQ} |P\rangle \left[h_{PQ}^{(0)} + \sum_{I} (v_{PIQI}^{(0)} - v_{PIIQ}^{(0)}) \right] \langle Q| \tag{1B.1.7}$$

with single-particle eigenenergies $\epsilon_P{}^{(0)}$. Further

$$E_{SCF}^{(0)} = \sum_{I} \epsilon_I^{(0)} - \frac{1}{2} \sum_{IJ} (v_{IJIJ}^{(0)} - v_{IJJI}^{(0)}) \tag{1B.1.8}$$

is the Hartree-Fock energy of the ground state of the molecule calculated at R_0; E'_{SCF} is its change due to the shift $R - R_0$,

$$h_{PQ}' = \langle P| \sum_{ik} \left[-\frac{Z_k e^2}{|\mathbf{r}_i - \mathbf{R}_k|} + \frac{Z_k e^2}{|\mathbf{r}_i - \mathbf{R}_k^{(0)}|} \right] |Q\rangle. \tag{1B.1.9}$$

Expanding h_{PQ} and E_{NN} in the Taylor's series we get

$$E_{NN}(R) = E_{NN}^{(0)} + E_{NN}' = \sum_{i=0}^{+\infty} E_{NN}^{(i)},$$

$$h_{PQ}(r) = h_{PQ}^{(0)} + h_{PQ}' = \sum_{i=0}^{+\infty} h_{PQ}^{(i)},$$

$$E_{SCF}' = \sum_{i=1}^{+\infty} E_{SCF}^{(i)}, \quad E_{SCF}^{(i)} = \sum_I h_{II}^{(i)} \qquad (1B.1.10)$$

where $^{(i)}$ denotes the order in $R - R_0$. Then the total Hamiltonian H may be written as

$$H = H_0 + W \qquad (1B.1.11)$$

where

$$H_0 = H_e + H_{\text{vibr}}, \qquad (1B.1.12a)$$

$$H_e = E_{NN}^{(0)} + E_{SCF}^{(0)} + \sum_P \epsilon_P^{(0)} \mathcal{N}[a_P^\dagger a_P] + \frac{1}{2} \sum_{PQRS} v_{PQRS}^{(0)} \mathcal{N}[a_P^\dagger a_Q^\dagger a_S a_R],$$

$$(1B.1.12b)$$

$$H_{\text{vibr}} = T_N + E_{NN}' + E_{SCF}' \approx \sum_r \hbar \omega_r \left(b_r^\dagger b_r + \frac{1}{2} \right) \qquad (1B.1.12c)$$

and

$$W = \sum_{i=1}^{+\infty} \sum_{PQ} h_{PQ}^{(i)} \mathcal{N}[a_P^\dagger a_Q]. \qquad (1B.1.12d)$$

Here, clearly, H_e is the full electronic Hamiltonian with fixed nuclear coordinates $(R \rightarrow R_0)$. H_{vibr} is the vibrational Hamiltonian. In the second equality of Eq. (1B.1.12c), we have assumed that R_0 has been chosen in such a way that

$$E_{NN}^{(1)} + E_{SCF}^{(1)} = 0. \qquad (1B.1.13)$$

Further, we have disregarded the rotational and translational motion and used a harmonic approximation. Then vibrational creation (annihilation) operators $b_r^\dagger(b_r)$ have appeared.

Eq. (1B.1.12a) describes the motion of electrons and nuclear vibrations. Electrons feel the nuclei at R_0 while the nuclei in H_{vibr} feel an averaged influence of the electrons. This is not the adiabatic but the crude (or rigid) *representation* (see Ref. 137) (the term "crude" originates from the fact that in the framework of this representation it is impossible to describe changes of the molecular orbitals due to deviations of the nuclei from their equilibrium (R_0) positions while working with just *one* spinorbital.)

The coupling W in Eq. (1B.1.12d) then starts from $i = 1$, i.e., from the first-order terms in the shifts $R - R_0$ (nuclear displacements) so that, introducing the normal coordinates $Q_r \sim b_r + b_r^\dagger$,

$$h_{PQ}^{(i)} \sim (b_{r_1} + b_{r_1}^\dagger) \ldots (b_{r_i} + b_{r_i}^\dagger). \tag{1B.1.14}$$

One usually takes into account only the first term (with $i = 1$) on the right side of Eq. (1B.1.12d). We should, however, stress here that this approximation is only good for tutorial purposes. The point is that in reality, electrons usually follow the adiabatic motion of nuclei quite well; the simplest way to see it is to take the translations or rotations of molecules as examples. It is just the interaction term W that is (in the crude representation) responsible for the quick accommodation of electrons to the instantaneous position of nuclei (i.e., to the whole molecular position, orientation, and instantaneous deformation). Thus, the coupling term W cannot be regarded as a small perturbation using the crude basis (see Chapter 5). For reasons connected with following facts, in particular that

1. Working with the fixed (crude or rigid) basis is technically simple.
2. The usual adiabatic approximation intimately connected with the adiabatic representation may fail (see below).
3. Modern methods of introducing representation of working in the adiabatic (i.e., moving with the skeleton of nuclei) electronic states beyond the adiabatic approximation are well developed just for molecules (see Ref. 137) but not yet for solids like molecular crystals;

we will work in the field of molecular crystals (see Section 1B.2) with the crude (rigid) states, i.e., in the terms of crude representation. In this connection, one should realize what happens in case of either excitation or ionization of a given molecule (by adding or removing one or more electrons). Because of our way of choosing R_0 (equilibrium nuclear positions) [see Eq. (1B.1.13)] and because of term W in Eq. (1B.1.11), it would immediately lead to a change of mean nuclear positions. So, excited (or ionized) states are not only electronically excited (ionized) but are also deformed from the point of view of the molecular skeleton; nuclei then vibrate around new positions even possibly with new frequencies. A theory of such vibronic states of molecules in connection with molecular solids may be found elsewhere (see, e.g., Ref. 85).

1B.2. HAMILTONIAN OF A MOLECULAR CRYSTAL; MOLECULAR EXCITONS

As we have indicated above, we will use the crude (rigid) basis, i.e., in the framework of representation based on spinorbitals which do not change with the deviations $R - R_0$ of nuclei from their fixed positions R_0. Thus, we try to avoid the adiabatic approximation which often is regarded as a standard theoretical tool in the molecular as well as solid-state physics. Since the reader who is acquainted with standard arguments supporting the adiabatic approximation (e.g., that the ratio of the electron and nuclear masses $m/M_k \ll 1$) might be surprised, we would like to dwell on this point a bit.

The fact that electrons are lighter than nuclei by several orders of magnitude is certainly reflected in their typical velocities, i.e., the characteristic times necessary to accommodate to a changed position of their partner. In molecules and molecular crystals, this reasoning leads to the standard inequality $\tau_e \approx \hbar/\Delta E_{exc} \approx 10^{-16}$ -10^{-15} s $\ll \Omega^{-1}$ where Ω is a typical phonon frequency and ΔE_{exc} is identified with the lowest electronic molecular excitation energy ($\gtrsim 1$ eV). Then τ_e (typical time needed for reconstruction of the electron orbitals) is certainly much shorter than the characteristic period of the nuclear vibration and the adiabatic approximation seems to be well justified (see Fig. 3.9). On the other hand, one should realize that ΔE_{exc} might be any electron excitation energy, not only the one specified above. Thus, roughly speaking, we have in fact infinitely many typical electronic relaxation times extending in molecular solids from the above value to infinity. One of the reasons for this is that in molecular solids we have finite, narrow energy bands. Since the estimated bandwidths are typically quite narrow ($\lesssim 0.02$ eV) (see Chapter 2), the difference of two such in-band energies (i.e., ΔE_{exc}) may become arbitrarily small, which makes the above arguments (supporting the adiabatic approximation) dubious for organic molecular crystals. Thus, arbitrarily small electron energy changes ΔE_{exc} may require corresponding arbitrarily large relaxation times τ_e, according to the relation $\tau_e \approx \hbar/\Delta E_{exc}$. This is the reason why we prefer to avoid the adiabatic approximation (i.e., why we work in terms of the crude representation) here. The interested reader will find more convincing arguments and references against the seemingly plausible adiabatic approximation in molecular solids in Chapter 2.

In principle, it would be possible to regard the whole molecular crystal as one big molecule, then applying the formalism of the previous section. However in this case the spin-orbital basis $|P\rangle$, $|Q\rangle$ ... would be delocalized in the whole crystal, having little to do with individual molecules. For periodic solids, such an approach would therefore, lead to the usual extended single-particle (Bloch) wave function description (with correlation terms owing to the electron-electron coupling). This is good for traditional semiconductors and metals. In our case, for molecular solids, we need to preserve notions like molecular orbitals and molecular wave functions, i.e., to preserve both the molecular and the solid-state aspects in our treatment (see Section 1A).

Let us introduce a new notation. Let P, Q, ... (in $|P\rangle$, $|Q\rangle$ ...) be replaced by $p\pi$, $q\kappa$, ... where $|p\pi\rangle$, $|q\kappa\rangle$, ... designate the π-th Hartree-Fock orbital of the pth molecule. Then, for the whole (finite but arbitrarily large) crystal, Eq. (1B.1.5) reads

$$H_{EN}(r, R) + H_{EE}(r) = \sum_{p\iota} h_{p\iota, p\iota} + \frac{1}{2} \sum_{p\iota_1, q\iota_2} (\nu^{(0)}_{p\iota_1, q\iota_2, p\iota_1, q\iota_2}$$

$$- \nu^{(0)}_{p\iota_1, q\iota_2, q\iota_2, p\iota_1}) + \sum_{p\pi, q\kappa} h_{p\pi, q\kappa} \mathcal{N}[a^\dagger_{p\pi} a_{q\kappa}]$$

$$+ \sum_{p\pi, q\kappa, l\iota} (\nu^{(0)}_{p\pi, l\iota, q\kappa, l\iota} - \nu^{(0)}_{p\pi, l\iota, l\iota, q\kappa}) \mathcal{N}[a^\dagger_{p\pi} a_{q\kappa}]$$

$$+ \frac{1}{2} \sum_{p\pi, q\kappa, r\rho, s\sigma} \nu^{(0)}_{p\pi, q\kappa, r\rho, s\sigma} \mathcal{N}[a^\dagger_{p\pi} a^\dagger_{q\kappa} a_{s\sigma} a_{r\rho}].$$

(1B.2.1)

Here summation $\sum_\iota$ (or $\sum_{\iota 1}$ etc.) denotes summation over the occupied Hartree-Fock spinorbitals (in the ground state) of the corresponding molecule. [In Eq. (1B.2.1) and below, we ignore the fact that the spinorbit states $|p\pi\rangle$, $|q\kappa\rangle$,.. cannot be taken directly as molecular Hartree-Fock solutions of isolated molecules but should additionally be orthogonalized on neighboring molecules. In molecular crystals with very small overlap between neighboring molecules this may be regarded a negligible correction.] One should realize that $\mathcal{N}$ (sign of the normal ordering) reorders creation and annihilation operators so that the creation operators always precede the annihilation ones.

Here, however, for purposes of normal ordering $a^\dagger_{p\pi}$ (or $a_{p\pi}$) is taken as a creation (or annihilation) operator if π is an unoccupied spinorbital of molecule p in its ground state; if it is occupied, $a^\dagger_{p\pi}$ (or $a_{p\pi}$) should be taken as an annihilation (or creation) operator of a hole. Thus, for example, in the third term on the right-hand side of Eq. (1B.2.1),

$$\sum_{p \neq q, \, \iota_1 \iota_2} h_{p\iota_1, \, q\iota_2} \mathcal{N}[a^\dagger_{p\iota_1} a_{q\iota_2}] = - \sum_{p \neq q, \, \iota_1 \iota_2} h_{p\iota_1, \, q\iota_2} a_{q\iota_2} a^\dagger_{p\iota_1}$$

(1B.2.2a)

represents transfer of a hole from $p\iota_1$ to $q\iota_2$ while

$$\sum_{p \neq q; \, \pi\kappa \text{ unoccup}} h_{p\pi, \, q\kappa} \mathcal{N}[a^\dagger_{p\pi} a_{q\kappa}] = \sum_{p \neq q; \, \pi\kappa \text{ unoccup}} h_{p\pi, \, q\kappa} a^\dagger_{p\pi} a_{q\kappa}$$

(1B.2.2b)

represents transfer of an electron. The first two terms on the right-hand side of Eq. (1B.2.1) yield the self-consistent electronic ground state energy E_{SCF}.

One must be careful because the site-off-diagonal ($p \neq q$ in the second term) contributions to E_{SCF} do not still yield a contribution of dispersion forces to the total energy of the molecular crystal keeping the crystal together (see Section 1A.5). We return to the problem of the dispersion forces later. The fourth term yields corrections to the electron and hole-hopping term (the third term) owing to the Coulomb interaction, while the last term gives the electron–electron interaction beyond the Hartree–Fock approximation.

A procedure completely analogous to that in Section 1B.1 yields the total Hamiltonian

$$H = H_0 + W, \tag{1B.2.3}$$

where the unperturbed (by the electron-phonon coupling) Hamiltonian is

$$H_0 = H_e + H_{\text{vibr}}. \tag{1B.2.4a}$$

In the above the electronic Hamiltonian (including holes, as well as excitons) is

$$H_e = E_{NN}^{(0)} + E_{SCF}^{(0)} + \sum_{p\pi,\,q\kappa} h_{p\pi,\,q\kappa}^{(0)} \mathcal{N}[a_{p\pi}^{\dagger} a_{q\kappa}]$$

$$+ \sum_{p\pi,q\kappa,l\iota} (v_{p\pi,l\iota,q\kappa,l\iota}^{(0)} - v_{p\pi,l\iota,l\iota,q\kappa}^{(0)}) \, \mathcal{N} \, [a_{p\pi}^{\dagger} a_{q\kappa}]$$

$$+ \frac{1}{2} \sum_{p\pi,q\kappa,r\rho,s\sigma} v_{p\pi,q\kappa,r\rho,s\sigma}^{(0)} \mathcal{N}[a_{p\pi}^{\dagger} a_{q\kappa}^{\dagger} a_{s\sigma} a_{r\rho}]. \tag{1B.2.4b}$$

The vibrational Hamiltonian is given by

$$H_{\text{vibr}} = T_N + E'_{NN} + E'_{SCF} \approx \sum_r \hbar \omega_r (b_r^{\dagger} b_r + 1/2). \tag{1B.2.4c}$$

The electronic-vibrational interaction Hamiltonian then reads

$$W = \sum_{i=1}^{+\infty} \sum_{p\pi,\,q\kappa} h_{p\pi,\,q\kappa}^{(i)} \mathcal{N}[a_{p\pi}^{\dagger} a_{q\kappa}]. \tag{1B.2.4d}$$

Since $h^{(i)} \sim (R - R_0)^i \sim (b_1 + b_1^{\dagger})^{\alpha_1} \ldots (b_N + b_N^{\dagger})^{\alpha_N}$, $\alpha_1 + \alpha_2 + \ldots \alpha_N = i$, Eq. (1B.2.4d) contains not only linear ($i = 1$) but also *quadratic* ($i = 2$) and higher-order couplings. In the second equality in Eq. (1B.2.4c), we have assumed as usual that the equilibrium values of nuclear coordinates (R_0) have been chosen in such a way that the linear term in $u = R - R_0$ disappears; further, the expansion in powers of u has been terminated at the second order (the harmonic approximation).

Hamiltonian (1B.2.4a–d) is sufficiently general. Before simplifying it for our purposes, let us introduce several important notions.

First, we introduce the Frenkel, i.e., molecular exciton, creation ($\bar{a}^{\dagger}$) and annihilation ($\bar{a}$) operators. Let, for example, $\pi = 0$ be the highest occupied spinorbital and $\pi = 1$ the lowest unoccupied spinorbital of the pth molecule. Then we introduce the exciton creation and annihilation operators as

$$\bar{a}_p^\dagger = a_{p1}^\dagger a_{p0}, \quad \bar{a}_p = a_{p0}^\dagger a_{p1}. \tag{1B.2.5}$$

(Similarly, we could introduce the exciton creation and annihilation operators for other occupied as well as unoccupied spinorbitals; in this way, we then get the whole sequence of singlet and triplet exciton operators. If unnecessary, we will not mention this possibility explicitly.) So if

$$|\psi_0\rangle = \prod_{q\iota} a^\dagger_{q\iota}|\text{vac}\rangle = \prod_q |\varphi_q^{(0)}\rangle \tag{1B.2.6}$$

is the Hartree-Fock ground state of H_e (i.e., that of the rigid crystal) with each (e.g., the qth) molecule in its own ground state $|\varphi_q^{(0)}\rangle$,

$$|\psi_p\rangle = \bar{a}^\dagger_p|\psi_0\rangle = |\varphi_p^{(e)}\rangle \prod_{q(\neq p)} |\varphi_q^{(0)}\rangle \tag{1B.2.7}$$

is an excited state of the whole molecular crystal in which all the molecules are in their ground states except for the pth one which is in its excited state $|\varphi_p^{(e)}\rangle$. In other words, we say that $\bar{a}^\dagger_p|\psi_0\rangle$ describes the state of the crystal with one exciton localized at molecule p. Clearly, operators (1B.2.5) fulfil the Pauli commutation relations

$$[\bar{a}_p, \bar{a}^\dagger_{p'}] = [\bar{a}_p, \bar{a}_{p'}] = |\bar{a}^\dagger_p, \bar{a}^\dagger_{p'}| = 0, \quad p \neq p'$$

$$\{\bar{a}_{p'}, \bar{a}_p\} = \{\bar{a}^\dagger_{p'}, \bar{a}^\dagger_p\} = 0$$

$$\{\bar{a}_p, \bar{a}^\dagger_p\} = \begin{cases} 1 & \text{for neutral molecular states} \\ 0 & \text{for ionized molecular states} \end{cases} \tag{1B.2.8}$$

(where $\{\ldots\}$ designates an anticommutator), i.e., the Frenkel excitons are neither bosons nor fermions. If their concentration is low (specially, for only one exciton), they can be treated as bosons.

One can easily see that operators (1B.2.5) appear in Eq. (1B.2.4d), too. Thus, excitons interact with vibrations. For a while, let us ignore that (which might be a good approximation for low temperature range), i.e., let us treat the crystal as rigid with Hamiltonian H_e in Eq. (1B.2.4b). The primary question is about the exciton energy.

First, it is clear that $|\psi_p\rangle$ in Eq. (1B.2.7) is not an eigenstate of H_e while making [in Eq. (1B.2.4)] the exciton operators explicit. To see it in detail, H_e can be rewritten as

$$H_e = H_{\text{exc}} + H_{\text{el}} + H_h + H_{\text{exc-el}} + H_{\text{exc-h}} + H_{\text{el-h}} + H_{\text{exc-el-h}} + E_{NN}^{(0)} + E_{SCF}^{(0)} \tag{1B.2.9}$$

with

$$H_{\text{exc}} \approx \sum_p [(h^{(0)}_{p1,p1} - h^{(0)}_{p0,p0}) \bar{a}^\dagger_p \bar{a}_p + h^{(0)}_{p1,p0} \bar{a}^\dagger_p + h^{(0)}_{p0,p1} \bar{a}_p] + \frac{1}{2} \sum_{p \neq q} [v^{(0)}_{p0,q0,p1,q1} \bar{a}_p \bar{a}_q$$

$$+ v^{(0)}_{p1,q1,p0,q0} \bar{a}^\dagger_p \bar{a}^\dagger_q + v^{(0)}_{p0,q1,p1,q0} \bar{a}^\dagger_q \bar{a}_p + v^{(0)}_{p1,q0,p0,q1} \bar{a}^\dagger_p \bar{a}_q]$$

$$+ \frac{1}{2} \sum_{p \neq q} [v^{(0)}_{p0,q0,p0,q0} + v^{(0)}_{p1,q1,p1,q1} - v^{(0)}_{p0,q1,p0,q1} - v_{p1,q0,p1,q0}] \bar{a}^\dagger_p \bar{a}_p \bar{a}^\dagger_q \bar{a}_q$$

$$+ \frac{1}{2} \sum_{p \neq q} \left\{ [v^{(0)}_{p0,q1,p1,q1} - v_{p0,q0,p1,q0}] \bar{a}^\dagger_q \bar{a}_q \bar{a}_p + [v^{(0)}_{p1,q1,p0,q1} \right.$$

$$- v^{(0)}_{p1,q0,p0,q0}] \bar{a}^\dagger_q \bar{a}_q \bar{a}^\dagger_p + [v^{(0)}_{p1,q0,p1,q1} - v^{(0)}_{p0,q0,p0,q1}] \bar{a}^\dagger_p \bar{a}_p \bar{a}_q + [v^{(0)}_{p1,q1,p1,q0}$$

$$\left. - v^{(0)}_{p0,q1,p0,q0}] \bar{a}^\dagger_p \bar{a}_p \bar{a}^\dagger_q \right\} + \sum_p \sum_{l(\neq p)} \sum_\iota [v^{(0)}_{p1,l\iota,p1,l\iota} - v^{(0)}_{p0,l\iota,p0,l\iota}] \bar{a}^\dagger_p \bar{a}_p$$

$$+ \sum_p \sum_{\iota(\neq 0)} [v^{(0)}_{p1,p\iota,p1,p\iota} - v^{(0)}_{p1,p\iota,p\iota,p1} - v^{(0)}_{p0,p\iota,p0,p\iota} + v^{(0)}_{p0,p\iota,p\iota,p0}] \bar{a}^\dagger_p \bar{a}_p$$

$$+ \sum_p \sum_\iota \left\{ \left[\sum_l v^{(0)}_{p1,l\iota,p0,l\iota} - v^{(0)}_{p1,p\iota,p\iota,p0} \right] \bar{a}^\dagger_p \right.$$

$$\left. + \left[\sum_l v^{(0)}_{p0,l\iota,p1,l\iota} - v^{(0)}_{p0,p\iota,p\iota,p1} \right] \bar{a}_p \right\}. \tag{1B.2.10a}$$

Here, double-overlap integrals like $v^{(0)}_{p0,\,q0,\,q1,\,p1}$, $p \neq q$ have been omitted assuming, as typical of OMCs, that spinorbitals on neighboring molecules practically do not overlap. Further,

$$H_{\text{el}} = \sum_p h^{(0)}_{p1,p1} (a^\dagger_{p1} a^\dagger_{p0} a_{p0})(a_{p1} a^\dagger_{p0} a_{p0})$$

$$+ \sum_{p \neq q} h^{(0)}_{p1,q1} (a^\dagger_{p1} a^\dagger_{p0} a_{p0})(a^\dagger_{q0} a_{q0} a_{q1}) + \sum v^{(0)}_{\cdots} \cdots , \tag{1B.2.10b}$$

$$H_{\text{h}} = - \sum_p h^{(0)}_{p0,p0} (a_{p0} a_{p1} a^\dagger_{p1})(a^\dagger_{p0} a_{p1} a^\dagger_{p1})$$

$$- \sum_{p \neq q} h^{(0)}_{p0,q0} (a_{q0} a_{q1} a^\dagger_{q1})(a_{p1} a^\dagger_{p1} a^\dagger_{p0}) + \sum v^{(0)}_{\cdots} \cdots , \tag{1B.2.10c}$$

$$H_{\text{exc-el}} = \sum_{p \neq q} [h_{p0,q0}^{(0)} \bar{a}_q^\dagger a_{p1}^\dagger a_{q1} \bar{a}_p - h_{p1,q0}^{(0)} \bar{a}_q^\dagger (a_{p1}^\dagger a_{p0}^\dagger a_{p0})(a_{q1} a_{q0}^\dagger a_{q0})$$

$$- h_{p0,q1}^{(0)} (a_{p1}^\dagger a_{p0}^\dagger a_{p0})(a_{q0}^\dagger a_{q0} a_{q1})\bar{a}_p] + \sum v_{\ldots}^{(0)} \cdots , \qquad \text{(1B.2.10d)}$$

$$H_{\text{exc-h}} = \sum_{p \neq q} [-h_{p1,q1}^{(0)} \bar{a}_p^\dagger a_{q0} a_{p0}^\dagger \bar{a}_q - h_{p1,q0}^{(0)} \bar{a}_p^\dagger (a_{q0} a_{q1} a_{q1}^\dagger)(a_{p0}^\dagger a_{p0} a_{p0}^\dagger)$$

$$- h_{p0,q1}^{(0)} (a_{q0} a_{q1} a_{q1}^\dagger)(a_{p1} a_{p1}^\dagger a_{p0}^\dagger)\bar{a}_q] + \sum v_{\ldots}^{(0)} \cdots , \qquad \text{(1B.2.10e)}$$

$$H_{\text{el-h}} = \sum_{p \neq q} [h_{p1,q0}^{(0)} (a_{p1}^\dagger a_{p0}^\dagger a_{p0})(a_{q1} a_{q1}^\dagger a_{q0})$$

$$+ h_{p0,q1} (a_{p0}^\dagger a_{p1} a_{p1}^\dagger)(a_{q0}^\dagger a_{q0} a_{q1})] + \sum v_{\ldots}^{(0)} \cdots \qquad \text{(1B.2.10f)}$$

and

$$H_{\text{exc-el-h}} = \sum_{p \neq q} [-h_{p0,q0}^{(0)} \bar{a}_q^\dagger (a_{p0}^\dagger a_{p1} a_{p1}^\dagger)(a_{q1} a_{q0}^\dagger a_{q0})$$

$$- h_{p0,q0}^{(0)} (a_{p1}^\dagger a_{p0}^\dagger a_{p0})(a_{q0} a_{q1} a_{q1}^\dagger)\bar{a}_p + h_{p1,q1} \bar{a}_p^\dagger (a_{p0}^\dagger a_{p1} a_{p1}^\dagger)$$

$$\times (a_{q1} a_{q0}^\dagger a_{q0}) + h_{p1,q1}^{(0)} (a_{p1}^\dagger a_{p0}^\dagger a_{p0})(a_{q0} a_{q1} a_{q1}^\dagger)\bar{a}_q$$

$$- h_{p1,q0}^{(0)} \bar{a}_p^\dagger \bar{a}_q^\dagger (a_{p0}^\dagger a_{p1} a_{p1}^\dagger)(a_{q1} a_{q0}^\dagger a_{q0}) - h_{p0,q1}^{(0)} (a_{p1}^\dagger a_{p0}^\dagger a_{p0})$$

$$\times (a_{q0} a_{q1} a_{q1}^\dagger)\bar{a}_p \bar{a}_q] + \sum v_{\ldots}^{(0)} \cdots .$$

$$\text{(1B.2.10g)}$$

In Eqs. (1B.2.10b–g), we have only indicated terms resulting from the electron-electron Coulombic interaction; we will return to interpretation of individual terms later. Referring to Eq. (1B.2.7) and using Eqs. (1B.2.9–10), we obtain

$$\langle \psi_p | H_e | \psi_q \rangle = \delta_{pq}(E_{NN}^{(0)} + E_{SCF}^{(0)} + E_p^{\text{exc}} + D_p) + (1 - \delta_{pq}) M_{pq} ,$$

$$E_p^{\text{exc}} = h_{p1,p1}^{(0)} - h_{p0,p0}^{(0)} + \sum_{\iota (\neq 0)} [v_{p1,p\iota,p1,p\iota}^{(0)} - v_{p1,p\iota,p\iota,p1}^{(0)} - v_{p0,p\iota,p0,p\iota}^{(0)}$$

$$+ v_{p0,p\iota,p\iota,p0}^{(0)}],$$

$$D_p = \sum_{l (\neq p)} \sum_{\iota} [v_{p1,\ l\iota,\ p1,\ l\iota}^{(0)} - v_{p0,\ l\iota,\ p0,\ l\iota}^{(0)}],$$

$$M_{pq} = \nu^{(0)}_{p1,\ q0,\ p0,\ q1}.$$ (1B.2.11)

Hitherto we have not used the periodicity of the crystal, i.e., all our formulas in this section apply even for arbitrary molecular aggregates. In an infinite periodic crystal, $\langle \psi_p | H_e | \psi_q \rangle$ in Eq. (1B.2.11) can be diagonalized. With only one molecule per unit cell (implying that $D_p = D$ and $E_p^{exc} = E^{exc}$ are p-independent), it yields eigenstates in the form of plane waves

$$|\psi_{\mathbf{k}}\rangle \approx \frac{1}{\sqrt{N}} \sum_p \exp(i\mathbf{k}\mathbf{r}_p)|\psi_p\rangle = \frac{1}{\sqrt{N}} \sum_p \exp(i\mathbf{k}\mathbf{r}_p)\bar{a}_p^\dagger|\psi_0\rangle,$$ (1B.2.12)

with the $\mathbf{k}$ vector lying inside the first Brillouin zone while the eigenenergies are given by

$$E_{\mathbf{k}} \approx E_{NN}^{(0)} + E_{SCF}^{(0)} + \epsilon_{\mathbf{k}},$$ (1B.2.13a)

$$\epsilon_{\mathbf{k}} = E^{exc} + D + \sum_q M_{pq} \exp[-i\mathbf{k}(r_p - r_q)].$$ (1B.2.13b)

As $E_{NN}^{(0)} + E_{SCF}^{(0)} = \langle \psi_0 | H_e | \psi_0 \rangle$ approximates the crystal ground state energy, $\epsilon_{\mathbf{k}}$ represents the exciton energy. It consists of several terms; E^{exc} is the excitation energy of the isolated molecule (i.e., the molecule in the gas phase) and D is the change of interaction of one given molecule with its surroundings on its excitation. The third term in Eq. (1B.2.13b) is the $\mathbf{k}$-dependent contribution to the exciton energy (yielding the exciton energy dispersion under conditions for which Eq. (1B.2.13) applies — see below) owing to the resonance interaction M_{pq} describing the possibility of the excitation to be transferred in space ($q \to p$) without transferring electrons (see Fig. 2.1). In connection with the dipole-dipole interaction, we find yet another contribution to D (or D_p) as given below. Typical experimental values of D then lie somewhere in the interval $(-0.2, -0.4)$ eV (see Section 2.1).

The validity of Eq. (1B.2.13), i.e., the applicability of the coherent plane-wave description of the exciton states, is certainly limited by the exciton-exciton interaction (see the terms $\sim \bar{a}_p^\dagger \bar{a}_p \bar{a}_q^\dagger \bar{a}_q$ in H_{exc}) which, however, becomes unimportant for low exciton concentrations. A further limitation is imposed by the exciton-electron and exciton-hole scattering (H_{exc-el}, H_{exc-h}, $H_{exc-el-h}$) as well as by the exciton-vibration coupling. These form parts of terms included in W in Eq. (1B.2.4d) but so far ignored here. The importance of such processes increases with increasing electron, hole, and phonon populations, i.e., with increasing temperature. Thus, one may expect that the above coherent regime (i.e., the plane-wave description of excitons) may be applied only at very low temperatures. With increasing temperature, the incoherent regime gradually starts to appear; for its description, we need to use the exciton density matrix (see Chapter 5).

Two comments are worth making. First, in Eq. (1B.2.13), we have assumed one molecule per unit cell. In case of M molecules (generally), one obtains (for any $\mathbf{k}$)

M solutions for the plane-wave states and corresponding energies, i.e., we have M exciton branches. Splitting between corresponding states at $\mathbf{k} \approx 0$ is known as *Davydov splitting*.[85] We have used the fact that, owing to translational symmetry, the wave vector of the created singlet exciton must equal that of the impinging photon, up to a possible vector of the reciprocal lattice $\mathbf{K}$. As the magnitude of the wave vector of the visible light is less by several orders of magnitude than the dimensions of the first Brillouin zone, $\mathbf{K} = 0$ and the exciton wave vector $\mathbf{k}$ is practically zero, too. (For details see any textbook on solid-state physics.) Here, it should be realized that the long-range character of the dipole-dipole coupling, dominating in M_{pq} in Eq. (1B.2.11) causes nonanalyticity of the exciton energies at $\mathbf{k} \approx 0$ in Eq. (1B.2.13b). The effect is, on the other hand, relatively small as the total bandwidth of the singlet exciton in, e.g., Ac is just about 400 cm^{-1}.[85,87]

The second comment concerns terms in H_{exc} that do not conserve the number of excitons. (Similar terms whose importance increases with increasing temperature, result from W in Eq. (1B.2.4d); for a while, we will ignore these.) The most important part of these terms (see H_{exc}) is contained in

$$\Delta H_{\text{exc}} = \frac{1}{2} \sum_{p \neq q} [\, v^{(0)}_{p0,q0,p1,q1} \bar{a}_p \bar{a}_q + v^{(0)}_{p1,q1,p0,q0} \bar{a}^\dagger_p \bar{a}^\dagger_q$$

$$+ v^{(0)}_{p0,q1,p1,q0} \bar{a}^\dagger_q \bar{a}_p + v^{(0)}_{p1,q0,p0,q1} \bar{a}^\dagger_p \bar{a}_q]. \qquad (1B.2.14)$$

For a while, we will assume singlet excitons (i.e., the spinorbital 1 is not spin-orthogonal to 0 even when relativistic effects are omitted) whereby all the matrix elements in Eq. (1B.2.14) are nonzero. The crucial point is their dependence on $\mathbf{r}_p - \mathbf{r}_q$ (difference of the positions of centers of gravity of molecules p and q). Expanding the Coulombic interaction in the standard multipole expansion

$$\frac{e^2}{|\mathbf{r} - \mathbf{r}'|} = \sum_{n=0}^{+\infty} \frac{1}{n!} \, [(\mathbf{r} - \mathbf{r}_p - \mathbf{r}' + \mathbf{r}_q)\nabla_{\mathbf{r}_p}]^n \, \frac{e^2}{|\mathbf{r}_p - \mathbf{r}_q|} \qquad (1B.2.15)$$

and using orthogonality of $p\alpha$, $\alpha = 0, 1$ (as well as $q\beta$, $\beta = 0,1$) spinorbitals, we find that the lowest contribution to Eq. (1B.2.14) comes from $n = 2$ yielding

$$\Delta H_{\text{exc}} \approx \frac{1}{2} \sum_{p \neq q} \frac{1}{|\mathbf{r}_p - \mathbf{r}_q|^5}$$

$$\cdot \sum_{i,j=1}^{3} [(d_{ip})_{01} \bar{a}_p + (d_{ip})_{10} \bar{a}^\dagger_p][(d_{jq})_{01} \bar{a}_q + (d_{jq})_{10} \bar{a}^\dagger_q]$$

$$\cdot [3(\mathbf{r}_p - \mathbf{r}_q)_i (\mathbf{r}_p - \mathbf{r}_q)_j - \delta_{ij}(\mathbf{r}_p - \mathbf{r}_q)^2]. \qquad (1B.2.16)$$

Here i, $j = 1,2,3$ designate the Cartesian coordinates (i.e., d_{ip} is the ith coordinate of $\mathbf{d}_p$) and

$$\mathbf{d}_p = e \sum_{\alpha\beta} \langle p\alpha|\mathbf{r}-\mathbf{r}_p|p\beta\rangle a^\dagger_{p\alpha}a_{p\beta}+d^{nucl}_p \tag{1B.2.17}$$

is the dipole moment of the pth molecule. Clearly, Eq. (1B.2.16) is the dipole-dipole interaction of molecules forming the crystal provided that the molecules are nonpolar in both the ground and excited states. In the opposite case, the missing [in Eq. (1B.2.16)] part of the dipole-dipole interaction enters via $E_{SCF}^{(0)}$. In the lowest order, ΔH_{exc} (yielding dispersion of the exciton energies) then makes no contribution to the ground state energy as

$$\langle\Psi_0|\Delta H_{exc}|\Psi_0\rangle = 0 \tag{1B.2.18}$$

with $|\Psi_0\rangle$ being given in Eq. (1B.2.6). In the second order, it gives the standard Van der Waals (dispersion) contribution to the ground state energy which keeps the crystal together (see Section 1A.5). The simplest way to see this is to use the usual nondegenerated perturbation theory. For simplicity, we ignore the coupling to the vibrations (i.e., the term W in Eq. (1B.2.4d), as well as all the terms in our Hamiltonian H in Eq. (1B.2.9) except for H_{exc}. Then our Hamiltonian can be split into two terms

$$H_{exc} = \mathcal{H}_0 + \mathcal{H}_1,$$

where

$$\begin{aligned}
\mathcal{H}_1 = \Delta H_{exc} &+ \sum_p \left[h^{(0)}_{p1,p0}+\sum_\iota\left(\sum_l v^{(0)}_{p1,l\iota,p0,l\iota}-v_{p1,p\iota,p\iota,p0}\right)\right]\bar{a}^\dagger_p \\
&+ \sum_p \left[h^{(0)}_{p0,p1}+\sum_\iota\left(\sum_l v^{(0)}_{p0,l\iota,p1,l\iota}-v^{(0)}_{p0,p\iota,p\iota,p1}\right)\right]\bar{a}_p \\
&+ \frac{1}{2}\sum_{p\neq q}[(v^{(0)}_{p0,q1,p1,q1}-v^{(0)}_{p0,q0,p1,q0})\bar{a}^\dagger_q\bar{a}_q\bar{a}_p+(v^{(0)}_{p1,q1,p0,q1} \\
&- v^{(0)}_{p1,q0,p0,q0})\bar{a}^\dagger_q\bar{a}_q\bar{a}^\dagger_p+(v^{(0)}_{p1,q0,p1,q1}-v^{(0)}_{p0,q0,p0,q1})\bar{a}^\dagger_p\bar{a}_p\bar{a}_q \\
&+(v^{(0)}_{p1,q1,p1,q0}-v^{(0)}_{p0,q1,p0,q0})\bar{a}^\dagger_p\bar{a}_p\bar{a}^\dagger_q].
\end{aligned} \tag{1B.2.19}$$

Clearly, with $|\Psi_0\rangle$ from Eq. (1B.2.6),

$$\mathcal{H}_0|\Psi_0\rangle = E_0|\Psi_0\rangle,$$

$$E_0 = E_{NN}^{(0)}+E_{SCF}^{(0)}, \tag{1B.2.20}$$

i.e., $|\Psi_0\rangle$ is the unperturbed ground state. (It can be verified that it is nondegenerate.) The second-order perturbation (in $\mathcal{H}_1$) theory yields for the perturbed ground state

$$|\Psi_{\text{ground}}\rangle = |\Psi_0\rangle \left[1 - \frac{1}{2} \sum_{m,n(\neq 0)} \frac{\langle \Psi_m| \mathcal{H}_1|\Psi_n\rangle\langle \Psi_n| \mathcal{H}_1|\Psi_0\rangle}{(E_0-E_n)(E_0-E_m)} \right.$$

$$+ \sum_{m(\neq 0)} |\Psi_m\rangle \left[\frac{\langle \Psi_m| \mathcal{H}_1|\Psi_0\rangle}{E_0-E_m} \right.$$

$$\left. + \sum_{n(\neq 0)} \frac{\langle \Psi_m| \mathcal{H}_1|\Psi_n\rangle\langle \Psi_n| \mathcal{H}_1|\Psi_0\rangle}{(E_0-E_n)(E_0-E_m)} \right] + \cdots \qquad (1B.2.21a)$$

and the corresponding energy

$$E_{\text{ground}} = E_0 + \langle \Psi_0| \mathcal{H}_1|\Psi_0\rangle + \sum_{n(\neq 0)} \frac{|\langle \Psi_0| \mathcal{H}_1|\Psi_n\rangle|^2}{E_0-E_n} + \cdots .$$

$$(1B.2.21b)$$

Here $|\Psi_m\rangle$ and E_m are, respectively, the unperturbed eigenstates of $\mathcal{H}_0$ and its eigenenergies describing situations with an excited single exciton [see Eq. (1B.2.7)], as well as pairs, triplets, and so on of excitons. The true ground state $|\Psi_{\text{ground}}\rangle$ in Eq. (1B.2.21a) is thus a superposition of states with such virtually excited excitons which are correlated on neighboring molecules according to the structure of $\mathcal{H}_1$. In Eq. (1B.2.21a), we have already used the fact that

$$\langle \Psi_0| \mathcal{H}_1|\Psi_0\rangle = 0. \qquad (1B.2.22)$$

Then, except for higher-order corrections only the second term in Eq. (1B.2.21b) remains as a correction to the zeroth order energy $E_0 = E_{NN}^{(0)} + E_{SCF}^{(0)}$. Let us now analyze its structure. First, we analyze a contribution to E_{ground} in Eq. (1B.2.21b) from the number of excitons nonconserving terms beyond ΔH_{exc} in Eq. (1B.2.19). Clearly, the contribution comes from $|\Psi_n\rangle$ in form of the single-exciton states (1B.2.7) with

$$\langle \Psi_p| \mathcal{H}_1 - \Delta H_{\text{exc}}|\Psi_0\rangle = h_{p1,p0}^{(0)} + \sum_{\iota} \left(\sum_l v_{p1,l\iota,p0,l\iota}^{(0)} - v_{p1,p\iota,p\iota,p0}^{(0)} \right)$$

$$= \sum_{l(\neq p)} \langle p1| - \sum_{k\in l} \frac{Z_k e^2}{|\mathbf{r}_k - \mathbf{r}|} + \sum_{\iota} \langle l\iota| \frac{e^2}{|\mathbf{r}' - \mathbf{r}|} |l\iota\rangle |p0\rangle.$$

$$(1B.2.23)$$

In the second equality here, we have used the orthogonality of our spinorbitals on molecule p; $\Sigma_{k\in l}$ means summation over nuclei of the lth molecule, the integration in the single-particle matrix element $\langle l\iota|\ldots|l\iota\rangle$ is over $\mathbf{r}'$ while that in $\langle p1|\ldots|p0\rangle$ is over $\mathbf{r}$. Let us now perform expansion of arguments in Eq.

(1B.2.23) in powers of $\mathbf{r}-\mathbf{r}_p$ and $\mathbf{r}-\mathbf{r}_l$. Because of neutrality of individual molecules, the zeroth-order terms cancel. As far as the molecules are not polar, their mean total (electronic plus nuclear) dipole moment is zero whereby the first nonvanishing term in Eq. (1B.2.23) (i.e., the contribution of the number-of-excitons nonconserving terms) comes from the dynamic interaction of the dipole moment of the pth molecule (off-diagonal elements $\langle p1|\ldots|p0\rangle$) with the mean electronic quadrupole of the lth molecule, decreasing as $|\mathbf{r}_p-\mathbf{r}_l|^{-4}$. The corresponding contribution to E_{ground} is therefore, in accordance with the usual opinion, negligible. Thus, the dominating contribution to E_{ground} in the second-order theory comes from ΔH_{exc}. The matrix elements $\langle \Psi_n|\Delta H_{\text{exc}}|\Psi_0\rangle$ are nonzero as far as $|\Psi_n\rangle$ are two-exciton states, i.e.,

$$E_{\text{ground}} \approx E_0 - \frac{1}{2} \sum_{p \neq q} \frac{1}{|\mathbf{r}_p-\mathbf{r}_q|^6} \frac{1}{2E_{pq}}$$

$$\cdot \left[3 \frac{[(\mathbf{d}_p)_{10}(\mathbf{r}_p-\mathbf{r}_q)][(\mathbf{d}_q)_{10}(\mathbf{r}_p-\mathbf{r}_q)]}{|\mathbf{r}_p-\mathbf{r}_q|^2} - (\mathbf{d}_p)_{10}(\mathbf{d}_q)_{10} \right]^2$$

$$\text{(1B.2.24a)}$$

where

$$2E_{pq} = E_p^{\text{exc}} + E_q^{\text{exc}} + D_p + D_q \approx 2E^{\text{exc}} \qquad \text{(1B.2.24b)}$$

is the energy needed in the zeroth order in $\mathcal{H}_1$ to excite two excitons at the pth and qth molecules. Clearly, the second term in (1B.2.24a) originating from ΔH_{exc} is the usual dispersion (Van der Waals) interaction energy keeping the molecular crystal together. This again illustrates the importance of the term ΔH_{exc} which cannot be omitted. As a by-product, we have illustrated a minor role of the number-of-excitons nonconserving terms in H_{exc} beyond ΔH_{exc} (in at least the ground state energy), in accordance with the usual belief.

A comment is worth mentioning here regarding formula (1B.2.11). Term D (or D_p if we have more than one molecule per elementary cell) has a meaning of *shift of energy* (of a localized exciton) owing to change of interaction of the excited molecule with its surroundings provided that ΔH_{exc} (i.e., mostly the dipole-dipole coupling) is negligible. If this is not the case (in particular, when the polarizability α^* of the excited molecule exceeds that of α, of the nonexcited one), one might try to formally apply the perturbation series to the state of standing localized exciton in the same way as above. This yields an additional contribution (i.e., renormalization)

$$D_p \rightarrow D_p + \Delta D_p,$$

$$\Delta D_p \approx - \sum_{q(\neq p)} \frac{1}{|\mathbf{r}_p - \mathbf{r}_q|^6} \frac{1}{2E^{\text{exc*}}}$$

$$\cdot \left[3 \frac{[(\mathbf{d}_p)_{21}(\mathbf{r}_p - \mathbf{r}_q)][(\mathbf{d}_q)_{10}(\mathbf{r}_p - \mathbf{r}_q)]}{|\mathbf{r}_p - \mathbf{r}_q|^2} - (\mathbf{d}_p)_{21}(\mathbf{d}_q)_{10} \right]^2.$$

(1B.2.25)

Here, state 2 is the next (second) excited spinorbital determining the polarizability of the excited molecule p; $2E^{\text{exc*}}$ is the corresponding excitation energy of a pair of molecules with one of them already being in the first excited state. Here, however, $p \rightarrow q$ virtual excitation transfer processes have been ignored.

1B.3. ELECTRONS AND HOLES IN MOLECULAR CRYSTALS

Both the Hamiltonians H_{el} [(Eq. 1B.2.10b)] and H_h [Eq. (1B.2.10c)] for the electrons and holes, respectively as well as H_{exc} in Eq. (1B.2.10a) and all other terms in Eqs. (1B.2.10d–g) result from the electronic Hamiltonian H_e [Eq. (1B.2.9)]. In this way we have separated electronic processes in the conduction band (called electronic processes further on), electronic processes in the valence band (reformulated in terms of holes), and coupled-electron-hole processes on the same molecule (reformulated in terms of the molecular exciton). The technique of separation underlying Eq. (1B.2.9) is illustrated for the third term of H_e on the right-hand side of Eq. (1B.2.4b). This reads

$$\sum_{p\pi} \sum_{q\kappa} h_{p\pi, q\kappa} \mathcal{N}[a_{p\pi}^{\dagger} a_{q\kappa}] = \sum_{p} [h_{p1,p1}^{(0)} a_{p1}^{\dagger} (a_{p0} a_{p0}^{\dagger} + a_{p0}^{\dagger} a_{p0}) a_{p1}$$

$$- h_{p0,p0}^{(0)} a_{p0} (a_{p1} a_{p1}^{\dagger} + a_{p1}^{\dagger} a_{p1}) a_{p0}^{\dagger} + h_{p1,p0}^{(0)} a_{p1}^{\dagger} a_{p0}$$

$$+ h_{p0,p1}^{(0)} a_{p0}^{\dagger} a_{p1}] + \sum_{p \neq q} \sum_{\pi\kappa} \cdots . \qquad (1B.3.1)$$

The parentheses (...) in Eq. (1B.3.1) are in fact equal to unity; on the other hand, term by term, the square bracket in Eq. (1B.3.1) yields, on using Eq. (1B.2.5), the first term in Eq. (1B.2.10a), the first term in Eq. (1B.2.10b), and the second and third terms in Eq. (1B.2.10b) as indicated. Therefore

$$A_p^{\dagger} = a_{p1}^{\dagger} a_{p0}^{\dagger} a_{p0}, \qquad A_p = a_{p1} a_{p0}^{\dagger} a_{p0} \qquad (1B.3.2)$$

serve as new creation and annihilation operators of electrons in the conduction band (i.e., the electron in the electron-hole nomenclature). The statistical factors $a_{p0}^{\dagger} a_{p0}$ ensure that the electron is created or annihilated in the excited spinorbital assuming that the lower spinorbital is occupied; otherwise, this would be the exciton. Similarly

$$B_p^\dagger = a_{p0} a_{p1} a_{p1}^\dagger, \quad B_p = a_{p0}^\dagger a_{p1} a_{p1}^\dagger \qquad (1B.3.3)$$

are the hole creation and annihilation operators in the valence band. Thus

$$H_{el} = \sum_p h_{p1,\,p1}^{(0)} A_p^\dagger A_p + \sum_{p \neq q} h_{p1,\,q1}^{(0)} A_p^\dagger A_q + \sum \nu_{\ldots}^{(0)} \ldots \qquad (1B.3.4a)$$

and

$$H_h = -\sum_p h_{p0,\,p0}^{(0)} B_p^\dagger B_p - \sum_{p \neq q} h_{p0,\,q0}^{(0)} B_q^\dagger B_p + \sum \nu_{\ldots}^{(0)} \ldots . \qquad (1B.3.4b)$$

For simplicity, let us ignore the terms $\sum \nu_{\ldots} \ldots$ in Eqs. (1B.3.4a and b) describing electron-electron and hole-hole scattering, respectively, assuming that the electron and hole concentrations are sufficiently low. The remaining parts of H_{el} and H_h can be diagonalized in the plane-wave representation. For example, with one molecule per elementary cell

$$H_{el} \approx \sum_{\mathbf{k}} \epsilon^{el}(\mathbf{k}) A_{\mathbf{k}}^\dagger A_{\mathbf{k}}, \qquad (1B.3.5a)$$

and

$$H_h \approx \sum_{\mathbf{k}} \epsilon^h(\mathbf{k}) B_{\mathbf{k}}^\dagger B_{\mathbf{k}} \qquad (1B.3.5b)$$

with

$$\epsilon^{el}(\mathbf{k}) = h_{p1,p1}^{(0)} + \sum_{p(\neq q)} h_{p1,q1}^{(0)} \exp[-i\mathbf{k}(\mathbf{r}_p - \mathbf{r}_q)],$$

$$\epsilon^h(\mathbf{k}) = -h_{p0,p0}^{(0)} - \sum_{p(\neq q)} h_{p0,q0} \exp[i\mathbf{k}(\mathbf{r}_p - \mathbf{r}_q)],$$

$$A_{\mathbf{k}}^\dagger = \frac{1}{\sqrt{N}} \sum_p \exp(i\mathbf{k}\mathbf{r}_p) A_p^\dagger,$$

$$B_{\mathbf{k}}^\dagger = \frac{1}{\sqrt{N}} \sum_p \exp(i\mathbf{k}\mathbf{r}_p) B_p^\dagger. \qquad (1B.3.6)$$

N is the number of $\mathbf{k}$-vectors in the first Brillouin zone, i.e., the number of elementary cells in the basic region of the crystal. With more molecules per unit cell and/or more excited and/or occupied spinorbitals involved, the whole band structure may be obtained. So the question arises about the justification of the band model of molecular crystals (see Chapter 3, Section 3.2.3).

There is often confusion regarding the term *"band model."* This term may be used either for the energy band structure with the electron and hole energies depending on the wave vector **k** or for the bandlike Boltzmann theory of the electron (hole, or exciton) transport. For the latter meaning of the word, it is always necessary that the mean free path $\bar{l}$ be much greater than the lattice constant, a, i.e.

$$\bar{l} \gg a \qquad (1B.3.7a)$$

and, also, that

$$\hbar/\tau_r(\approx \hbar/\tau) \ll k_B T. \qquad (1B.3.7b)$$

Here, τ is the quasiparticle lifetime with respect to any scattering which, except for the carrier scattering on unscreened charged impurities, is of the order of the relaxation time τ_r. Criteria (1B.3.7) are the so-called Landau-Peierls criteria for the validity of the bandlike Boltzmann theory of electron (hole) transport. In principle, on the right-hand side of Eq. (1B.3.7b), any relevant quantity of the dimension of energy (with respect to which the energy broadening $\hbar/\tau$, which is ignored in the Boltzmann theory, should be negligible) may be included. In particular, this is true for the relevant band width ΔE. Thus condition

$$\hbar/\tau_r(\approx \hbar/\tau) \ll \Delta E \qquad (1B.3.7c)$$

should apply. In organic molecular crystals, these criteria are currently not very well (or not at all) fulfilled (see Chapter 3, Section 3.2.3). This makes the Boltzmann-like theory of the electron and hole conduction questionable, and more general approaches are needed. We shall return to this question in Chapter 5. One should say, however, that violation of conditions [Eq. (1B.3.7)] does not necessarily mean the opposite, i.e., that the hopping character of the charge carrier transport is described as an incoherent motion of a localized species (see Chapter 5). Moreover, any sufficiently general theory of electron or hole propagation can be formulated in terms of both extended and localized states. Even when the Boltzmann bandlike theory of the particle transport becomes uncertain (as is usual in molecular crystals at higher temperature, see Chapter 7), the single-particle Bloch-functions, see Eq. (1B.3.6), may serve as input parameters for more general approaches.

We now return to the term "band model" in the first sense, i.e., regarding the single-particle energy band structure and its correspondence with real molecular crystals. At this time, we will not speak about the polaron band structure which may remain meaningful even when that of bare electrons and holes is becoming obscure, with increasing temperature.

Only a few calculations were performed in the above scheme (see Ref. 87); other methods were mostly used.[163,213,328,382] (Also see Section 3.2 for details.) The general conclusion is that the band widths ΔE of electrons and holes in, e.g., Nph and Ac are always ≤ 0.2 eV, sometimes even less than of the order of $k_B T$.

Inclusion of the electronic polarization effects reduces ΔE further by 40–50% (see Fig. 3.2 and Refs. 117 and 284). At room temperatures $l \approx 3$–4 Å, so that condition (1B.3.7a) is violated. This makes the Boltzmann theory inapplicable but it says very little about the physical meaning and significance of the single-particle band structure obtained via standard methods of solid-state physics. In our opinion, unless we perform a concrete physical experiment, the magnitude of ΔE itself can hardly rule out the single-particle band structure except that the position of narrow bands in the energy diagram becomes uncertain. On the other hand, what makes this band structure in molecular crystals meaningless are many-body *polaron phenomena*, especially those connected with the polarization of π-shells of molecules neighboring to a particular carrier — see below. In particular, one can hardly believe that energy shifts by the electron polarization ≥ 1 eV could preserve any meaning to the single-particle band structure ignoring these effects (see Refs. 332 and 351 and discussion in Section 3.2.3). All these facts have now made the band structure theory of molecular crystals obsolete.

A few words are worth mentioning here regarding the polaron (i.e., not the single-particle) band structure. We shall treat the problem in detail later. Here, the reader should be warned that, strictly speaking, the polaron formation is actually a *many-particle* phenomenon. So the band structure becomes a quasiparticle band structure where the dispersion curves $\epsilon(\mathbf{k})$ are broadened. As we shall see in Chapters 3–7, various *polaron* states may play different roles in different situations. Thus, their band structure should be investigated in connection with specific experiments.

1B.4. ELECTRON–EXCITON COUPLING; ELECTRONIC POLARON AROUND THE ELECTRON

The first term on the right-hand side of Eq. (1B.2.10d) gives the exciton-electron scattering. This type of scattering is elastic (kinetic energy conserving) and may correspondingly be important. On the other hand, in real molecular crystals, it is ineffective if the electron or exciton concentration is low. In addition, it is also suppressed, when $h^{(0)}_{p0, q0}$, $p \neq q$ is negligible which corresponds to a narrow single-particle hole band (see Fig. 3.3). The second and third terms in Eq. (1B.2.10d) correspond to inelastic processes of the electron-induced exciton creation and annihilation. Their ineffectiveness is due to the smallness of $h^{(0)}_{p1, q0}$ and $h^{(0)}_{p0, q1}$. Thus, one should turn attention to terms in $H_{\text{exc-el}}$ that are due to Coulomb interaction, and are indicated as $\Sigma \nu [\dots]^{(0)} \dots$ in Eq. (1B.2.10). We will designate these terms as $\Delta H_{\text{exc-el}}$.

In $\Delta H_{\text{exc-el}}$, we take into account those processes that are of the zeroth order in the presumably very small overlap S of molecular spinorbitals located on different molecules. This is an approximation, of course, but it simplifies further reasoning and becomes exact in the limit $S \to 0$ (notice that the electron bandwidth ΔE is $\sim S$). Thus

$$\Delta H_{\text{exc-el}} = \frac{1}{2} \sum_{p \neq q} \left[(\nu^{(0)}_{p1,q1,p1,q1} - \nu^{(0)}_{p0,q1,p0,q1}) \bar{a}^\dagger_p \bar{a}_p A^\dagger_q A_q \right.$$

$$+ (\nu^{(0)}_{p1,q1,p1,q1} - \nu^{(0)}_{p1,q0,p1,q0}) A^\dagger_p A_p \bar{a}^\dagger_q \bar{a}_q$$

$$+ (\nu^{(0)}_{p0,q1,p1,q1} \bar{a}_p + \nu^{(0)}_{p1,q1,p0,q1} \bar{a}^\dagger_p) A^\dagger_q A_q$$

$$\left. + (\nu^{(0)}_{p1,q0,p1,q1} \bar{a}_q + \nu^{(0)}_{p1,q1,p1,q0} \bar{a}^\dagger_q) A^\dagger_p A_p \right] + O(S). \quad (1B.4.1)$$

Here $A_r^\dagger$ (A_r) are creation (annihilation) operators of electron in the conduction band introduced in Eq. (1B.3.2). Because the processes described by Eq. (1B.4.1) do not preserve the number of excitons, the presence of at least one (in our approximation, immobile) electron leads necessarily to creation of a cloud of virtual excitons around it. Such an object is called an electronic polaron, introduced by Toyozawa.[386] A qualitative picture of an electronic polaron (see Figs. 3.4 and 3.5) and its phenomenological treatment will be given in Chapter 3, Section 3.3. For a more rigorous description of the electronic polaron model we need to introduce the corresponding excitonic Hamiltonian. Thus, the simplest Hamiltonian of the electron polaron model is the following (in the $S \rightarrow 0$ zero overlap limit and for one electron for the sake of simplicity):

$$H_{\text{el.pol}} = [H_{\text{el}} + H_{\text{exc}} + H_{\text{exc-el}}]_{S=0} + O(S)$$

$$= \sum_p h^{(0)}_{p1,p1} A^\dagger_p A_p + \sum_p \Delta E_p \bar{a}^\dagger_p \bar{a}_p$$

$$+ \Delta H_{\text{exc}} + \sum_{p \neq q} [\nu^{(0)}_{p1,q0,p1,q1} \bar{a}_q + \nu^{(0)}_{p1,q1,p1,q0} \bar{a}^\dagger_q] A^\dagger_p A_p + O(S),$$

$$\Delta E_p = E^{\text{exc}}_p + D_p. \quad (1B.4.2)$$

Here we have assumed that all molecules (except the central one bearing the additional electron), are neutral and all are nonpolar. Finally, as in Eq. (1B.2.23), we have transformed the exciton non-conserving terms to the dipole-quadrupole interaction term which drops out owing to the last simplifying assumption used here that mean values of all multipoles (of individual molecules) higher than dipoles have a negligible effect. Thus, if the electron is created at the molecule 0 (central molecule at origin), one has to solve the following Schrödinger equation

$$\left[\sum_{p(\neq 0)} \Delta E_p \bar{a}_p^\dagger \bar{a}_p - e \sum_{p(\neq 0)} [(\mathbf{d}_p)_{01} \bar{a}_p + (\mathbf{d}_p)_{10} \bar{a}_p^\dagger] \frac{\mathbf{r}_p}{|\mathbf{r}_p|^3} \right.$$

$$+ \frac{1}{2} \sum_{p \neq q(p,q \neq 0)} \frac{1}{|\mathbf{r}_p - \mathbf{r}_q|^5} \sum_{i,j=1}^{3} [(d_{ip})_{01} \bar{a}_p + (d_{ip})_{10} \bar{a}_p^\dagger][(d_{jq})_{01} \bar{a}_q$$

$$\left. + (d_{jq})_{10} \bar{a}_q^\dagger] \cdot [3(\mathbf{r}_p - \mathbf{r}_q)_i (\mathbf{r}_p - \mathbf{r}_q)_j - \delta_{ij} (\mathbf{r}_p - \mathbf{r}_q)^2] \right] |\Psi\rangle = E|\Psi\rangle, \quad \epsilon < 0.$$

$$(1B.4.3)$$

Here we have again expanded the Coulomb matrix elements into the multipole series omitting multipoles higher than dipole. Having solved Eq. (1B.4.3) [with energy E taken with respect to $h^{(0)}{}_{p1, p1}$ in Eq. (1B.4.2)], we obtain the total state of the electronic polaron around the central localized electron (negatively charged molecular ion) with energy E as

$$|\Psi^{\text{el}}{}_{\text{el.pol}}\rangle = A_0^\dagger |\Psi\rangle. \qquad (1B.4.4)$$

We deal later with Eq. (1B.4.3). Here, we would like to say that the three terms on the left-hand side of Eq. (1B.4.3) reduce to Hamiltonians of the energy needed to polarize neutral molecules, interaction energy of the additional charge at the central molecule with induced dipoles around it, and, finally, the energy of the induced-dipole vs. induced-dipole interaction. Taking the second and the third terms on the left-hand side of Eq. (1B.4.3) as a perturbation, one can easily see the structure of the electronic polaron: Around the central charged molecule, there is a cloud of molecular excitons whose phases are correlated. Taking as a variational trial state the product of the ground states of individual molecules [see $|\Psi_0\rangle$ in Eq. (1B.2.6)], one gets zero mean energy. This is certainly a poor trial state; thus, E will become negative for the true ground state (or with any better trial state) which is the source of the electron polarization energy in molecular crystals.

More detailed dynamic approaches to the electronic polarization and microelectrostatic methods of electronic polarization energy calculations are given in Sections 3.4 and 3.5.

1B.5. HOLE–EXCITON COUPLING: ELECTRONIC POLARON AROUND THE HOLE

As a matter of fact, one can simply repeat all the reasoning of the last Section; with $e < 0$ being the electronic charge, the only change with respect to Eq. (1B.4.3) is the change of sign of the second term on the left hand side. Having then found $|\Psi\rangle$, the state of the electronic polaron around hole reads

$$|\Psi^h{}_{\text{el.pol}}\rangle = B_0^\dagger |\Psi\rangle. \qquad (1B.5.1)$$

More complicated differences would appear if we did not neglect higher-order multipoles.

1B.6. COMBINED EXCITON–ELECTRON-HOLE INTERACTION

Leading terms (those which do not result from the electron-electron Coulomb interaction) of this combined interaction are given in Eq. (1B.2.10g). These terms describe the following simple processes: el+h → exc (the first and third terms), exc → el+h (the second and fourth terms), el+h → exc+exc (the fifth term) and exc+exc → el+h (the sixth term). Therefore $H_{\text{exc-el-h}}$ is important for dynamic exciton creation and annihilation processes which will not be considered here.

1B.7. COUPLING OF EXCITONS, ELECTRONS, AND HOLES TO PHONONS

The coupling Hamiltonian of electrons (i.e., excitons, electrons, and holes in our nomenclature) to phonons [see Eq. (1B.2.4d] can be rewritten as

$$
\begin{aligned}
W = \sum_{i=1}^{+\infty} & \left[\sum_{p} \sum_{\pi\kappa} h^{(i)}_{p\pi,\,p\kappa} \, \mathcal{N}[a^{\dagger}_{p\pi} a_{p\kappa}] + \sum_{p \neq q} \sum_{\pi\kappa} h^{(i)}_{p\pi,q\kappa} \, \mathcal{N}[a^{\dagger}_{p\pi} a_{q\kappa}] \right] \\
= \sum_{i=1}^{+\infty} & \left[\sum_{p} \left[h^{(i)}_{p1,p1} a^{\dagger}_{p1}(a_{p0} a^{\dagger}_{p0} + a^{\dagger}_{p0} a_{p0}) a_{p1} - h^{(i)}_{p0,p0} a_{p0}(a_{p1} a^{\dagger}_{p1} + a^{\dagger}_{p1} a_{p1}) a^{\dagger}_{p0} \right.\right. \\
& + h^{(i)}_{p1,p0} a^{\dagger}_{p1} a_{p0} + h^{(i)}_{p0,p1} a^{\dagger}_{p0} a_{p1} \Big] + \sum_{p \neq q} \Big[h^{(i)}_{p1,q1} a^{\dagger}_{p1}(a_{p0} a^{\dagger}_{p0} + a^{\dagger}_{p0} a_{p0})(a_{q0} a^{\dagger}_{q0} \\
& + a^{\dagger}_{q0} a_{q0}) a_{q1} - h^{(i)}_{p0,q0} a_{p0}(a_{p1} a^{\dagger}_{p1} + a^{\dagger}_{p1} a_{p1})(a_{q1} a^{\dagger}_{q1} + a^{\dagger}_{q1} a_{q1}) a^{\dagger}_{q0} \\
& + h^{(i)}_{p1,q0} a^{\dagger}_{p1}(a_{p0} a^{\dagger}_{p0} + a^{\dagger}_{p0} a_{p0})(a_{q1} a^{\dagger}_{q1} + a^{\dagger}_{q1} a_{q1}) a_{q0} \\
& \left.\left. + h^{(i)}_{p0,q1} a^{\dagger}_{p0}(a_{p1} a^{\dagger}_{p1} + a^{\dagger}_{p1} a_{p1})(a_{q0} a^{\dagger}_{q0} + a^{\dagger}_{q0} a_{q0}) a_{q1} \right] \right].
\end{aligned}
\tag{1B.7.1}
$$

Using definitions of new creation and annihilation operators of excitons, electrons (in the conduction band) and holes (in the valence band), i.e., Eqs. (1B.2.5), (1B.3.2), and (1B.3.3), we can reorganize Eq. (1B.7.1) as

$$
W = W_{\text{exc-ph}} + W_{\text{el-ph}} + W_{\text{h-ph}} + W_{\text{exc-el-ph}} + W_{\text{exc-h-ph}} + W_{\text{el-h-ph}} + W_{\text{exc-el-h-ph}}
\tag{1B.7.2}
$$

with

$$W_{\text{exc-ph}} = \sum_{i=1}^{+\infty} \sum_{p} [[h^{(i)}_{p1,p1} - h^{(i)}_{p0,p0}]\bar{a}^{\dagger}_p\bar{a}_p + h^{(i)}_{p1,p0}\bar{a}^{\dagger}_p + h^{(i)}_{p0,p1}\bar{a}_p],$$

(1B.7.3a)

$$W_{\text{el-ph}} = \sum_{i=1}^{+\infty} \sum_{p,q} h^{(i)}_{p1,q1}A^{\dagger}_p A_q,$$ (1B.7.3b)

$$W_{\text{h-ph}} = -\sum_{i=1}^{+\infty} \sum_{p,q} h^{(i)}_{p0,q0}B^{\dagger}_q B_p,$$ (1B.7.3c)

$$W_{\text{exc-el-ph}} = \sum_{i=1}^{+\infty} \sum_{p \neq q} [h^{(i)}_{p0,q0}\bar{a}^{\dagger}_p A^{\dagger}_q A_p \bar{a}_q - h^{(i)}_{p1,q0}A^{\dagger}_p \bar{a}^{\dagger}_q A_q - h^{(i)}_{p0,q1}A^{\dagger}_p \bar{a}_p A_q],$$

(1B.7.3d)

$$W_{\text{exc-h-ph}} = \sum_{i=1}^{+\infty} \sum_{p \neq q} [-h^{(i)}_{p1,q1}\bar{a}^{\dagger}_p B^{\dagger}_q B_p \bar{a}_q - h^{(i)}_{p1,q0}\bar{a}^{\dagger}_p B^{\dagger}_q B_p - h^{(i)}_{p0,q1}B^{\dagger}_q B_p \bar{a}_q],$$

(1B.3.7e)

$$W_{\text{el-h-ph}} = \sum_{i=1}^{+\infty} \sum_{p \neq q} [h^{(i)}_{p1,q0}A^{\dagger}_p B^{\dagger}_q + h^{(i)}_{p0,q1}B_p A_q],$$ (1B.7.3f)

$$W_{\text{exc-el-h-ph}} = \sum_{i=1}^{+\infty} \sum_{p \neq q} [h^{(i)}_{p1,q1}(A^{\dagger}_p B^{\dagger}_q \bar{a}_q + \bar{a}^{\dagger}_p B_p A_q) + h^{(i)}_{q0,p0}(B^{\dagger}_p A^{\dagger}_q \bar{a}_q + \bar{a}^{\dagger}_p A_p B_q)$$

$$+ h^{(i)}_{p1,q0}\bar{a}^{\dagger}_p \bar{a}^{\dagger}_q A_q B_p + h^{(i)}_{p0,q1}B^{\dagger}_q A^{\dagger}_p \bar{a}_p \bar{a}_q].$$ (1B.7.3g)

Most of these terms are expected. Possibly surprising is the absence of the site-off-diagonal exciton-phonon interaction term which is, on the other hand, common in standard approaches (see, e.g., Ref. 253). A few words about this problem are, therefore, worth mentioning.

One should first of all realize that here we work in so-called *crude* representation (i.e., with molecular electronic states in form of the Hartree–Fock states with equilibrium, i.e., rigid, positions of nuclei). This means that applying the usual many-body field theory (in deriving, e.g., Hamiltonian [Eq. (1B.2.1)] or other ensuing forms), we have expanded the electron-field operator

$$\hat{\psi}(\mathbf{x}) = \sum_{p\pi} \langle \mathbf{x}|p\,\pi\rangle a_{p\pi},$$

$$\langle \mathbf{x}|p\,\pi\rangle \equiv \langle \mathbf{x}|p\,\pi(R_0)\rangle, \quad a_{p\pi} \equiv a_{p\pi}(R_0)$$ (1B.7.4a)

(**x** denotes the electron position in space with spin). Here neither $\langle \mathbf{x}|p\,\pi\rangle$ nor $a_{p\pi}$ depends on nuclear deviations $(R - R_0)$ from their equilibrium values (R_0). To work in the adiabatic representation, the same operator would have to be written as

$$\hat{\psi}(\mathbf{x}) = \sum_{p\pi} \langle \mathbf{x}|p\,\pi(R)\rangle a_{p\pi}(R). \tag{1B.7.4b}$$

Here $R \equiv \{\mathbf{R}_k\}$, $k = 1,2,\ldots$ represents a set of "moving" (on the lattice and intramolecular vibrations or librations) nuclear coordinates. While the left-hand sides of Eqs. (1B.7.4a) and (1B.7.4b) coincide, the individual corresponding terms on the right-hand sides coincide when all $\mathbf{R}_k$ are equal to their equilibrium values. Now the problem is — and it is often forgotten — that $a_{p\pi}$ in Eq. (1B.7.4b) *do* depend on R and can be taken as R-independent only in terms of the adiabatic approximation. As very little is known in dynamic theories about nonadiabatic corrections, (i.e., beyond the standard stationary quantum chemistry), it is advisable to work (as we mostly do here) in the framework of *crude* representation, i.e., with electronic states corresponding to rigid nuclear positions. In Chapter 2, some argument will be given that in solids, the adiabatic approximation is at best a very unreliable tool which strongly supports our choice of the crude representation.

On the other hand, in reality, the electrons react almost instantaneously to changing nuclear coordinates, i.e., they move (vibrate, librate) with individual molecules. Such shifted molecular states may be expanded in series in terms of the crude states with rigid nuclear positions. Thus, in our case, one should in principle work with higher molecular states even when describing low-temperature equilibrium situations. Technically, this causes the necessity of introducing additional indices corresponding to these states. On the other hand, in many standard problems like transport in space, traces over such indices appear, reflecting the fact that we are then interested in position of electrons or excitations in space but not in their distribution over these excited states. Since nothing physically new appears, we mostly ignore these higher excited states for the sake of simplicity here. In any case, the relation between the crude and adiabatic representations deserves some attention here. Comparing Eqs. (1B.7.4a–b), we have

$$a_{p\pi} = \sum_{p'\pi'} \langle p\,\pi|p'\,\pi'(R)\rangle a_{p'\pi'}(R)$$

$$a_{p\pi}(R) = \sum_{p'\pi'} \langle p\,\pi(R)|p'\,\pi'\rangle a_{p'\pi'}. \tag{1B.7.5}$$

Thus, the operator of the number of excitons entering $W_{\text{exc-ph}}$ in Eq. (1B.7.3a) may be expressed as

$$\bar{a}_p^\dagger \bar{a}_p = a_{p1}^\dagger a_{p0} a_{p0}^\dagger a_{p1}$$

$$= \sum_{q_1,\, q_2,\, q_3,\, q_4} \sum_{\kappa_1,\, \kappa_2,\, \kappa_3,\, \kappa_4} \langle q_1\, \kappa_1(R)|p1\rangle\langle p0|q_2\, \kappa_2(R)\rangle\langle q_3\, \kappa_3(R)|p0\rangle$$

$$\times\langle p1|q_4\, \kappa_4(R)\rangle a_{q_1\, \kappa_1}^\dagger(R) a_{q_2\, \kappa_2}(R) a_{q_3\, \kappa_3}^\dagger(R) a_{q_4\, \kappa_4}(R). \qquad (1B.7.6)$$

It also contains, besides contributions of operators of the number of excitons at one site in the adiabatic representation

$$\bar{a}_q^\dagger(R)\bar{a}_q(R) = a_{q1}^\dagger(R) a_{q0}(R) a_{q0}^\dagger(R) a_{q1}(R) \qquad (1B.7.7)$$

($q_1 = q_2 = q_3 = q_4 = q$ and $\kappa_1 = \kappa_4 = 1$, $\kappa_2 = \kappa_3 = 0$), other terms including those describing the exciton transfer in the adiabatic representation

$$\bar{a}_q^\dagger(R)\bar{a}_{q'}(R) = a_{q1}^\dagger(R) a_{q0}(R) a_{q'0}^\dagger(R) a_{q'1}(R), \quad q \neq q' \qquad (1B.7.8)$$

($q_1 = q_2 \equiv q \neq q_3 = q_4 = q'$ and $\kappa_1 = \kappa_4 = 1$, $\kappa_2 = \kappa_3 = 0$) or electron-electron interaction terms having nothing to do with excitons in the adiabatic representation. The same applies when transforming back from the adiabatic to the crude representation.

Thus, summarizing, we may conclude that:

1. There is no real contradiction between the adiabatic representation (handicapped physically in connection with the adiabatic approximation) and the crude representation (handicapped technically).

2. The definition of individual terms in Eq. (1B.2.9) and Eq. (1B.7.2) depends on the representation used; only their sums are representation-independent.

3. Creation and annihilation operators of excitons, electrons, and holes are R-dependent in the adiabatic representation; this dependence (causing technical problems) disappears in the adiabatic approximation whose validity is, as will be shown in Chapter 2, at best uncertain.

One should keep these arguments in mind in connection with the exciton or carrier transport phenomena discussed in Chapter 5.

1B.8. LATTICE POLARON AROUND ELECTRONS, HOLES, AND EXCITONS

Since we will discuss this kind of polaron in some detail in Chapter 3, we should like to introduce it here briefly. For most of the following discussion, it is not important whether we speak about the lattice polarons around electron, hole, or exciton. Thus, instead of operators $A_p^\dagger(A_p)$, $B_p^\dagger(B_p)$, and $\bar{a}_p^\dagger(\bar{a}_p)$, we shall use $a_p^\dagger(a_p)$ specifying in the text, if necessary, what kind of species we have in mind. One should recall that for low electron, hole, or exciton concentration, it does not matter which kind of species we have in mind. Let us assume the simplest electron (or hole or exciton)-phonon Hamiltonian in the form

$$H = \sum_{m \neq n} J_{mn} a_m^\dagger a_n + \sum_k \hbar \omega_k \left(b_k^\dagger b_k + \frac{1}{2} \right)$$

$$+ \frac{1}{\sqrt{N}} \sum_n \sum_k \hbar \omega_k g_k \exp(i\mathbf{kr}_n) a_n^\dagger a_n (b_k + b_{-k}^\dagger)$$

$$\equiv H_{el} + H_{ph} + H_{el\text{-}ph}, \quad g_k = g_{-k}^*. \tag{1B.8.1}$$

This is simply the sum of Eqs. (1B.3.4a), (1B.2.4c), and (1B.7.3b) provided that:

1. We have one molecule per unit cell.

2. $h^{(0)}{}_{p1,\,p1}$ may be taken as zero by a proper choice of zero energy.

3. We have only a single branch of vibrations for which the harmonic approximation can be used.

4. Only the linear term $\sim b + b^\dagger$ is considered in the electron-phonon coupling (1B.7.3b) (i.e., $i = 1$).

5. The site-off-diagonal electron-phonon coupling is negligible in Eq. (1B.7.3b); for excitons in the crude (rigid) basis, this is not an assumption since no such coupling exists.

These assumptions, though not fully corresponding to reality, are sufficient for a qualitative discussion. One should realize that the term with $i = 1$ (point 4) means keeping (in the electron-phonon coupling) terms linear in the nuclear displacements $u_{\alpha_1 \alpha_2}(l)$. In general in perfect crystals with periodic boundary conditions,

$$u_{\alpha_1 \alpha_2}(l) = \sum_{r=1}^{3s} \sum_k \sqrt{\frac{\hbar}{2MN\omega_{rk}}} A^{(r)}_{\alpha_1 \alpha_2}(\mathbf{k}) \exp(i\mathbf{kr}_l)[b_{rk} + b_{r,-k}^\dagger],$$

$$\alpha_1 = 1, 2, 3 \quad \alpha_2 = 1, 2, \ldots s. \tag{1B.8.2}$$

Here $\omega_{rk} = \omega_{r-k}$ is the phonon frequency in branch r, s is the number of atoms (not molecules) in the unit cell, N is the number of elementary cells in the basic region of the crystal, l designates these unit cells, $\mathbf{r}_l$ is the position vector of the lth unit cell, M is an arbitrary (e.g., average molecular or atomic or total for one unit cell) mass, and $A^{(r)}{}_{\alpha_1 \alpha_2}(\mathbf{k})$ are vibrational amplitudes obtained from classical equations of motion in the harmonic approximation. The normalization is expressed by

$$\sum_{\alpha_1 = 1}^{3} \sum_{\alpha_2 = 1}^{s} \sum_{l} M_{\alpha_2} [A^{(r)}_{\alpha_1 \alpha_2}(\mathbf{k}) \exp(i\mathbf{k}\mathbf{r}_l)][A^{(r')}_{\alpha_1 \alpha_2}(\mathbf{k}') \exp(i\mathbf{k}'\mathbf{r}_l)]^*$$

$$= NM \delta_{rr'} \delta_{\mathbf{k}\mathbf{k}'}. \tag{1B.8.3}$$

In Eq. (1B.8.3), M_{α_2} is the mass of the α_2-th atom in any unit cell. According to the point 3 above, the summation Σ_r, as well as index$^{(r)}$ (or $_r$) drops out in Eq. (1B.8.2).

Let us first assume that the band width of electrons is zero, i.e., $J_{mn} = 0, m \neq n$. Then Eq. (1B.8.1) can be exactly diagonalized with eigenfunctions

$$|m\{\mu_\mathbf{k}\}\rangle = a^{\dagger}_m \prod_\mathbf{k} \left[\frac{1}{\sqrt{\mu_\mathbf{k}!}} \left(b^{\dagger}_\mathbf{k} + \frac{1}{\sqrt{N}} g^m_\mathbf{k} \right)^{\mu_\mathbf{k}} \right] \cdot \exp\left(\frac{1}{\sqrt{N}} \sum_\mathbf{k} g^m_\mathbf{k}(b_\mathbf{k} \right.$$

$$\left. - b^{\dagger}_{-\mathbf{k}}) \right) |vac\rangle,$$

$$g^m_\mathbf{k} = g_\mathbf{k} \exp(i\mathbf{k}\mathbf{r}_m), \quad \mu_\mathbf{k} = 0,1,2 \dots \tag{1B.8.4a}$$

and eigenenergies

$$E_{m\{\mu_\mathbf{k}\}} = -\frac{1}{N} \sum_\mathbf{k} \hbar\omega_\mathbf{k}|g_\mathbf{k}|^2 + \sum_\mathbf{k} \hbar\omega_\mathbf{k}\left(\mu_\mathbf{k} + \frac{1}{2} \right). \tag{1B.8.4b}$$

As the momentum $p_{\alpha_1 \alpha_2}(l)$ (α_1-th component of the momentum of the α_2-th atom in the lth cell) conjugate to $u_{\alpha_1 \alpha_2}(l)$ reads

$$p_{\alpha_1 \alpha_2}(l) = \sum_r \sum_\mathbf{k} (-i) \sqrt{\frac{\hbar\omega_{r\mathbf{k}}}{2MN}} M_{\alpha_2} A^{(r)}_{\alpha_1 \alpha_2}(\mathbf{k}) \exp(i\mathbf{k}\mathbf{r}_l)[b_{r\mathbf{k}} - b^{\dagger}_{r,-\mathbf{k}}],$$

$$\tag{1B.8.5}$$

one can see that the exponential in Eq. (1B.8.4a) forms the Taylor expansion producing shift of the equilibrium nuclear coordinates caused by the electron (hole or exciton) located at the mth molecule. Another equivalent way to see that is to introduce new creation operators for phonons as follows:

$$B^{\dagger}_\mathbf{k} = b^{\dagger}_\mathbf{k} + \frac{1}{\sqrt{N}} g^m_\mathbf{k}. \tag{1B.8.6}$$

It creates in Eq. (1B.8.4a) phonons in the lattice already deformed by the presence of the electron located at molecule m. Thus, in Eq. (1B.8.2)

$$u_{\alpha_1 \alpha_2}(l) = u'_{\alpha_1 \alpha_2}(l) + \Delta u_{\alpha_1 \alpha_2}(l),$$

$$u'_{\alpha_1 \alpha_2}(l) = u_{\alpha_1 \alpha_2}(l)|_{b \to B, b^{\dagger} \to B^{\dagger}} \tag{1B.8.7}$$

where (in terms of our model)

$$\Delta u_{\alpha_1 \alpha_2} = -\sum_{\mathbf{k}} \frac{2}{N} \sqrt{\frac{\hbar}{2M\omega_{\mathbf{k}}}} A_{\alpha_1 \alpha_2}(\mathbf{k}) g_{-\mathbf{k}} \exp(i\mathbf{k}(\mathbf{r}_l - \mathbf{r}_m)) \quad (1B.8.8)$$

are static shifts induced by the electron at molecule m. Here we call an electron (or, correspondingly, a hole or an exciton) surrounded by a cloud of the *lattice* (as well as *intramolecular*) deformation a *lattice polaron*. In our case, assuming $J_{mn} = 0$, $m \neq n$, Eq. (1B.8.4a) represents a state of a localized standing polaron. Its energy, in addition to the usual harmonic lattice terms, includes the polaron shift, i.e., energy gain:

$$\Delta E_{\text{pol}} = -\frac{1}{N} \sum_{\mathbf{k}} \hbar \omega_{\mathbf{k}} |g_{\mathbf{k}}|^2 < 0, \quad (1B.8.9)$$

which is due to this deformation. One should realize that in our model, the polaron deformation described by transformation (1B.8.6) causes no changes in the nuclear momenta (1B.8.5) and no changes in the phonon frequencies.

Assume now for simplicity that we deal with only one intramolecular branch of vibrations which is completely dispersionless. Thus

$$\omega_{\mathbf{k}} = \omega_0, \quad g_{\mathbf{k}} = g_0 \quad (1B.8.10)$$

in Eq. (1B.8.1). Hence, introducing local (molecular) creation and annihilation phonon operators

$$b_n^\dagger = \frac{1}{\sqrt{N}} \sum_{\mathbf{k}} \exp(-i\mathbf{k}\mathbf{r}_n) \cdot b_{\mathbf{k}}^\dagger,$$

$$b_n = \frac{1}{\sqrt{N}} \sum_{\mathbf{k}} \exp(i\mathbf{k}\mathbf{r}_n) \cdot b_{\mathbf{k}}, \quad (1B.8.11)$$

one obtains

$$H_{\text{el-ph}} = \hbar \omega_0 g_0 \sum_n a_n^\dagger a_n (b_n + b_n^\dagger). \quad (1B.8.12)$$

Therefore the deformation should be just at the place at which the electron is located. Indeed, taking Eq. (1B.8.10), $A_{\alpha_1 \alpha_2}(\mathbf{k})$ becomes $\mathbf{k}$-independent so that we have from Eq. (1B.8.8)

$$\Delta u_{\alpha_1 \alpha_2}(l) = -\delta_{lm} \sqrt{\frac{2\hbar}{M\omega_0}} g_0 A_{\alpha_1 \alpha_2}, \quad (1B.8.13)$$

i.e., a local deformation. This leads to the notion of a small polaron where small (in the molecular crystal context) means localization on just one molecule. Because in reality, however, every phonon branch (including that of intramolecular vibrations) has a finite though possibly weak dispersion, we understand that the real deformation extends over a few neighboring molecules (see Fig. 3.10 and discussion in Section 3.6.1). Notice that without dispersion which is able to remove additional lattice energy, there would be no molecular skeleton relaxation possible around a standing electron. So we are led to the notion of the nearly small molecular polaron (see Section 3.6) which is needed for interpretation of various experimental data (see Chapters 4, 6, and 7).

The question is, what happens with the small or the nearly small molecular polaron once we allow the electron to move (i.e., when $J_{mn} \neq 0$, $m \neq n$). Without going into detail, we can foresee the main results as follows.

Clearly, as far as

$$| J_{mn} | \ll \hbar \bar{\omega}, \quad m \neq n \tag{1B.8.14a}$$

where $\bar{\omega}$ is a typical phonon frequency for a given branch, the electron motion is sufficiently slow to allow the lattice (molecular skeleton) to accommodate to the presence of the electron (hole or exciton) at any given molecule. Thus, in contrast to the opposite case

$$| J_{mn} | \gg \hbar \bar{\omega} \tag{1B.8.14b}$$

where the phonons are able to scatter but not to get in phase to produce a deformation around the fast electron, the polaron cloud becomes well developed, almost as in the case of the standing electron. Thus, in this "slow electron limit," the propagating species is the electron plus the lattice polaron cloud (we are not speaking about the electron polarization mentioned above and treated in more detail in Chapter 3). The role of J_{mn}, which is proportional to the probability amplitude per second for the bare electron to be transferred to the neighboring molecule, is then effectively replaced by

$$\tilde{J}_{mn} = \langle m\{\mu_{\mathbf{k}}\}| \sum_{r \neq s} J_{rs} a_r^\dagger a_s |n\{\mu_{\mathbf{k}}\}\rangle. \tag{1B.8.15}$$

Here — means thermal averaging with respect to the phonon occupation numbers $\mu_{\mathbf{k}}$ at a given temperature T. In Eq. (1B.8.15), we have ignored the possibility of transfer with the change of phonon occupation numbers. Such phonon-assisted processes do exist; for their influence on transport characteristics see Chapter 5. A straightforward calculation yields

$$\tilde{J}_{mn} = J_{mn} \exp\left(-\frac{1}{N} \sum_{\mathbf{k}} |g_{\mathbf{k}}|^2 (1 - \cos \mathbf{k}(\mathbf{r}_m - \mathbf{r}_n))[1 + 2n_B(\hbar \omega_{\mathbf{k}})] \right) \tag{1B.8.16}$$

where

$$n_B(z) = [\exp(\beta z) - 1]^{-1}, \quad \beta = 1/(k_B T) \qquad (1B.8.17)$$

is the Bose-Einstein distribution representing the mean occupation number of a harmonic phonon state at equilibrium with temperature T. Clearly, with $T \to 0$, $\tilde{J}_{mn} \to$ const which remains in magnitude less than J_{mn}. When $T \to +\infty$ on the other hand, $\tilde{J}_{mn} \to 0$ exponentially as for $k_B T \gg \mathrm{Max}(\hbar \omega_k)$, $n_B(\hbar \omega_k) \approx k_B T / \hbar \omega_k$. Thus, if an effective mass $m^*(T)$ of the polaron can be introduced, we have

$$m^*(T) = m^*(0)\exp(T/T_0), \quad T \gg \mathrm{Max}(\hbar \omega_k)/k_B. \qquad (1B.8.18)$$

This $m^*(T)$ dependence, obtained from theory, is very important for the interpretation of simulation data of charge carrier generation and transport processes which yield empirically the above exponential formula of $m^* = m^*(T)$ (see Chapters 6 and 7).

Thus, summing up, we may conclude that the Hamiltonian description of interaction phenomena in molecular crystals allows us to describe in a generalized treatment, according to Eq. (1B.2.9) and its following expansions, practically all kinds of electronic and excitonic interactions including the formation of electronic and excitonic polarons. This generalized approach also includes the formation of the Van der Waals interaction forces which keep the molecules together in the crystal. In addition, one can describe by Eq. (1B.7.2) and its expansions the interaction of electrons, holes, and excitons with the inter- and intramolecular phonons of the lattice and formation of lattice polarons, including the nearly small molecular polaron.

In the following chapters, we shall see applications of these basic ideas.

CHAPTER 2

Exciton Interaction with Local Lattice Environment

May it not be universally true that the concepts produced by the human mind, when formulated in a slightly vague form, are roughly valid for reality, but that when extreme precision is aimed at, they become ideal forms whose content tends to vanish away.

LOUIS DE BROGLIE
IN NOBEL PRIZE ADDRESS

2.1. EXCITONS IN ANTHRACENE-TYPE CRYSTALS

Neutral excited states in a molecular crystal can be described in terms of *molecular exciton* theory developed by Davydov.[82,84–87] Davydov extended the model of a small-radius exciton, proposed earlier by Frenkel[110] to Ac-type molecular crystals and developed a consistent theory of light absorption, exciton formation, and coherent motion in such crystals.

The Frenkel exciton model proved to be the most appropriate for describing the behavior of neutral excited states in OMC characterized by weak intermolecular forces.

The molecular exciton theory was further extended and developed by Agranovich,[7,8] Craig and Walmsley,[75] and many other authors (see Refs. 45, 204, 288, 332, and 351 and Chapter 5).

Let us first discuss, for tutorial purposes, a simplified treatment of exciton state energetics.

If an excited molecule is embedded in a polarizable isotropic molecular solid, its self-energy decreases by the term D due to electronic polarization of surrounding molecules (see Fig. 2.1a). D is the gas to solid state energy shift term arising from Van der Waals dispersion forces between the excited molecule, having polarizability α^*, and the surrounding unexcited molecules with a mean polarizability α; usually $\alpha^* > \alpha$ (see formula 3.5.4).

The term D may be represented in the following simplified form:[332]

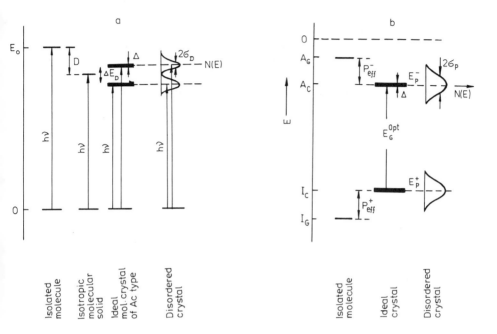

FIGURE 2.1. Schematic energy diagram of exciton (a) and ionized (b) states in an Ac-type crystal (modified after Ref. 15). D is the gas to solid-state shift term, i.e., the electronic polarization energy term of an exciton; ΔE_D is Davydov splitting. I_G, A_G, and I_C, A_C are ionization potentials and electron affinity of a molecule and crystal, respectively; P_{eff}^+, P_{eff}^- are energy of electronic polarization of charge carrier; E_P^+, E_P^- are energy levels of electronic polarons; E_G^{Opt} is the optical energy gap; σ_P and σ_D are the corresponding dispersion parameters of Gaussian distribution of electronic states.

$$D = -A^* \sum_{i \neq k} 1/r_{ik}^6, \qquad (2.1.1)$$

where A^* is an empirical constant for the given excited state which may be estimated from crystal spectra, and r_{ik} is the distance between the excited molecule i and the kth neutral molecule. Thus, the term D is obtained as a sum of the excited molecule's interaction energy with all $N-1$ surrounding neutral molecules of the crystal. For dipole-allowed singlet transition the term D is proportional to the oscillator strength f and the mean polarizability of the medium α:[15]

$$D = -f\alpha \sum_{i \neq k} 1/r_{ik}^6. \qquad (2.1.2)$$

Thus, the excitonic state in OMC is actually a many-electron interaction state of electronic polarization origin and similar in nature to electronic polarization of the lattice by a charge carrier (see Fig. 2.1b as well as theoretical treatment of the problem in Section 1B).

It should be stressed that in both cases the total polarization energy is strongly dependent on the topology of the surrounding molecules in the crystal. However, the r-dependence is different [see formulas (3.5.5) and (2.1.2)] and the exciton state is neutral as compared to the electronic polaron. Since the exciton is actually a polaron-type quasiparticle, it should be called *excitonic polaron* to emphasize its polaronic nature.

In addition to dispersion (electronic polarization) type of interaction, exciton self-energy is also determined by the resonance interaction (exciton transfer) term I_{mn}. The term I_{mn} may be described, in dipole-dipole approximation as follows (see Ref. 351, p. 225):

$$I_{mn}^{st} = d^s d^t / r_{mn}^3 \cdot F_{mn}^{[k, k']}, \tag{2.1.3}$$

where d^s and d^t are the dipole lengths of the corresponding s and t transitions; F_{mn} characterizes the relative orientation of the molecules m and n. Formula (2.1.3) can be derived from Eq. (1.B.2.11).

According to the molecular exciton theory,[84,85] in a crystal containing several molecules in a unit cell several branches of excited states of the crystal correspond to each excited electronic state of a molecule.

In the case of Ac-type crystals containing two molecules in a unit cell, two bands of excited states correspond to each molecular term. These form the so-called Davydov doublet.[45,84,85] Davydov splitting is due to the symmetry properties of the crystal, i.e., to the nonequivalence of the positions of two otherwise identical molecules in the unit cell (see Fig. 1.6).

Taking into account these exciton resonance effects the total change, e.g., of the singlet exciton energy δE_{ex}^s in an Ac crystal, due to dispersion D and resonance I interaction, may be written as:[45,85]

$$\delta E_{ex}^s = E^s(\mathbf{k}) - E_o^s = D^s + I_{11}^s(\mathbf{k}) \pm I_{12}^s(\mathbf{k}), \tag{2.1.4}$$

where $E^s(\mathbf{k})$ is the total energy of the singlet exciton, E_o^s is the singlet state energy of an isolated molecule, and $I_{11}^s(\mathbf{k})$ and $I_{12}^s(\mathbf{k})$ are the resonance interaction terms. The value of the Davydov splitting is $2I_{12}^s(\mathbf{k}) = \Delta E_D$ (see Fig. 2.1a).

It may be seen that formula (2.1.4) is a specific case of a more general expression for exciton self-energy [see Eq. (1B.2.13b)].

It is interesting to compare the energies of the two interaction terms in the Ac crystal. For Ac $E_o^s = 27685$ cm^{-1} = 3.43 eV. The **a** and **b** polarized exciton transitions in an Ac crystal occur at 3.14 and 3.12 eV, respectively.

Thus, $\Delta E_D = 0.02$ eV and $I_{12} \approx 0.01$ eV. Since $I_{12} \approx I_{11}$ it follows that $D + I_{11} \approx D = 0.3$ eV.

Consequently, the contribution of the resonance term I_{11} is negligible — less than 1% of the total δE_{ex}^s value. This means that the self-energy of the excitonic polaron

TABLE 2.1. *Experimental Values of the Term D of Solid-State Shift of the First Singlet Absorption Band in Polyacene Crystals and the Corresponding Davydov Splitting ΔE_D According to Ref. 351*

Crystal	D (eV)	ΔE_D (eV)	Crystal	D (eV)	ΔE_D (eV)
Naphthalene	0.43	0.001 (10 cm^{-1})	Tetracene	0.20	0.078–0.087 (620–700 cm^{-1})
Anthracene	0.29	0.024 (190 cm^{-1})	Pentacene	0.19	0.127 (1030 cm^{-1})

in Ac is practically determined by the dispersion forces, i.e., by the electronic polarization of the lattice:

$$\delta E_{ex}^s \approx D^s. \tag{2.1.5}$$

The reported D and ΔE_D values for other polyacene crystals are shown in Table 2.1. Since the ΔE_D value increases and the term D slightly decreases for higher polyacenes, the contribution of resonance term cannot be neglected in this case.

From an energetic point of view it is important to estimate the possible value of the exciton bandwidth. Exciton band structure calculations for the first singlet excited state in an Ac crystal were performed by Davydov and Sheka.[87] The calculated exciton energy $\varepsilon(\mathbf{k})$ dependences in the wave vector $\mathbf{k}$ for both exciton bands ($\mu = 1,2$) in three crystalline directions ($\mathbf{c},\mathbf{a},\mathbf{b}$) are presented in Fig. 2.2.

As may be seen from Fig. 2.2, the singlet exciton bands in Ac are extremely narrow, the bandwidth Δ being ca.100–300 cm^{-1} (i.e., 0.01–0.04 eV) in the **ab**-plane and only ca.10–20 cm^{-1} in the $\mathbf{c}'$ direction, thus showing similarly expressed anisotropy as in the case of calculated charge carrier bands (see Fig. 3.2).

Excitonic polaron energy levels may therefore be regarded as quasidiscrete compared to the D term shift. In addition, there is overwhelming experimental evidence that singlet exciton motion in OMC is coherent, bandlike only at very low temperatures ($T < 60$ K) (see Chapter 5). At higher temperatures non-coherent, diffusive motion of excitons prevails, similar as in case of charge carriers. The contemporary transport theories describe both the coherent wave-like and noncoherent diffusive motion of excitons (and charge carriers) in OMC in a framework of unified single models. They will be discussed in detail in Chapter 5 which is devoted to transport theories.

Let us now briefly discuss the properties of excitons in terms of some basic concepts presented in Section 1B of Chapter 1.

First, let us remind that all that has been said about the *electronic* polaron in Section 1B.4 remains feasible also for an *excitonic* polaron. The differences are only qualitative, or may be due to exciton nonconserving terms, i.e., due to finite lifetime of excitons.

Qualitative differences are characteristic, in particular, for triplet excitons. For such excitons, the terms contained in ΔH_{exc} [see Eq. (1B.2.14)] yield zero in the

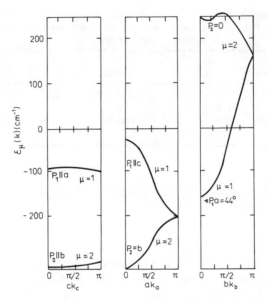

FIGURE 2.2. Calculated energy $\varepsilon_\mu(k)$ dependence on the wave vector **k** for the two first singlet exciton bands ($\mu = 1.2$) in an anthracene crystal.[87]

nonrelativistic limit because, e.g., $|p1\rangle$ and $|p0\rangle$ spinorbitals then become spin-orthogonal. This means that then, e.g., $v_{p0,\, q1,\, p1,\, q0}^{(0)} \to 0$. So for the triplet excitons, the possibility to migrate coherently is almost zero and practically the only way for them to move in space is via very weak exchange and virtual processes, i.e., incoherently. If for singlet excitons the zero-temperature effective mass $m^*(0)$ remains finite and the bandwidth Δ, though small, nevertheless appreciable (see Fig. 2.2), $m^*(0)$ becomes effectively infinite in case of triplet excitons and the corresponding bandwidth approaches zero.

This situation is reflected in typical values of diffusion coefficients D of triplet and singlet excitons in polyacene crystals (see Tables 2.2 and 2.3). The corresponding D values of triplet excitons are smaller by an order of magnitude (or more) than those of singlet excitons.

The data in Tables 2.2 and 2.3 also demonstrate the highly pronounced anisotropy of diffusive motion of excitons in polyacene crystals. The motion is practically two-dimensional and takes place preferably in the **ab**-plane of the crystal. In this aspect, there is qualitative difference as compared to motion of charge carriers. For them one observes only slight anisotropy of mobilities (see Chapter 7). The physical cause of this difference is discussed in Section 3.6.2.

Let us here briefly mention the possible role of the number-of-excitons nonconserving terms in H_{exc} [see Eq. (1B.2.14)]. As such terms contained in ΔH_{exc} have been sufficiently discussed in Section 1B of Chapter 1, let us now treat only the linear (in $\bar{a}_p$ and $\bar{a}_p^\dagger$) terms in Eq. (1B.2.10a). Clearly, as in Eq. (1B.2.23),

TABLE 2.2. *Diffusion Coefficients D of Triplet Excitons in Several Aromatic Crystals at T = 300 K* [a]

Crystal	D_{aa} (cm^2s^{-1})	D_{bb} (cm^2s^{-1})	$D_{c'c'}$ (cm^2s^{-1})	Reference
Naphthalene	$(3.3\pm0.4)\times10^{-5}$	$(2.7\pm0.4)\times10^{-5}$	—	104
Anthracene	$(1.5\pm0.2)\times10^{-4}$	$(1.8\pm0.2)\times10^{-4}$	$(1.2\pm0.5)\times10^{-5}$	103
Tetracene	—	$(4\pm1)\times10^{-3}$	—	9
Pyrene	$(3\pm1)\times10^{-5}$	$(1.25\pm0.3)\times10^{-4}$	$(3\pm1)\times10^{-5}$	14

[a]*Note:* The mean life-time of triplet excitons in high-purity anthracene may be of the order of 20 ms (Ref. 288, p. 138).

$$h^{(0)}_{p1,\,p0} + \sum_{\iota}\left[\sum_{l} v^{(0)}_{p1,\,1\iota,\,p0,\,1\iota} - v^{(0)}_{p1,\,p\iota,\,p\iota,\,p0}\right]$$

$$\approx \sum_{ijk\,=\,1}^{3}(d_{pi})_{10}\sum_{l(\,\neq\,p)}Q^l_{jk}/|\mathbf{r}_{pl}|^4$$

$$\cdot\left[3\,\frac{(\mathbf{r}_{pl})_k\,\delta_{ij} + (\mathbf{r}_{pl})_j\,\delta_{ik} + (\mathbf{r}_{pl})_i\,\delta_{jk}}{|\mathbf{r}_{pl}|} - 15\,\frac{(\mathbf{r}_{pl})_i(\mathbf{r}_{pl})_j(\mathbf{r}_{pl})_k}{|\mathbf{r}_{pl}|^3}\right],$$

$$\mathbf{r}_{pl} = \mathbf{r}_p - \mathbf{r}_l. \tag{2.1.6}$$

TABLE 2.3. *Diffusion Coefficients D, Diffusion Length L_s, and Mean Lifetime of Singlet Excitons τ^0_s in Several Aromatic Crystals at 300 K*

Crystal	D (cm^2s^{-1})	Range of reported L_s (nm)	τ^0_s (ns)	Method of measurement	Reference
Naphthalene	$D_{aa} = (2+1)\times10^{-4}$ $D_{c'c'} = 5\times10^{-5}$	—	—	—	131
Anthracene	$D_{aa} = 1.8\times10^{-3}$ $D_{bb} = 5.1\times10^{-3}$ $D_{c'c'} = 6.6\times10^{-4}$	60 100 36	10	Surface quenching of luminescence	290
	$D = 4\times10^{-3}$	90	10	Bimolecular quenching of luminescence	290
	$D_{aa} = 5.1\times10^{-3}$ $D_{bb} = 3.4\times10^{-3}$	100 130	10	Photoconductivity	290
	$D_{bb} = 1.9\times10^{-3}$	—	6	—	385
Tetracene	$D = 1.7\times10^{-3}$	12	0.4	Surface quenching of luminescence	227
	$D = 1.1\times10^{-2}$ $D_{aa} = 4\times10^{-2}$	29 —	0.4 —	Photoconductivity	227
α-Perylene	$D = 3.2\times10^{-8}$	0.2	12	Bimolecular	54
Pyrene	$D = 3.6\times10^{-9}$	0.3	112	Quenching of luminescence	

Here Q^l_{jk} is the quadrupole tensor (of the total charge including that of nuclei) of the lth molecule. This term [i.e., (2.1.6)] is, owing to the high power of $|\mathbf{r}_{pl}|$ in the denominator, very small, i.e., that part of the exciton-nonconserving terms which is due to the crystal surroundings may be practically neglected.

Other aspects of finite-lifetime effects, particularly those of singlet excitons (see Table 2.3), discussed in Section 5.10. In Chapter 5 the reader will also find a detailed theoretical treatment of exciton transport problems, including those of possible change of exciton motion with temperature, coupling strength, etc.

2.2. EXCITONIC POLARON: ADIABATIC VERSUS CRUDE REPRESENTATION. EXCITON SELF-TRAPPING

2.2.1. Traditional Treatment in Terms of Adiabatic Approximation

It is often stressed (with good reason) that the crude representation using rigid molecular spinorbitals (not moving with vibrations or librations of molecules) is technically handicapped since, in reality, the excitations really move with the molecules in which they reside. In fact, for the crude representation, this objection means only that one has to take into account a sufficient number of rigid excited spinorbitals to describe the moving (unexcited or excited) molecular states. On the other hand, application of the adiabatic representation (for correspondence between the adiabatic and the crude representation see Section 1B.7 above) yields technical complications in solving the Schrödinger equation, which disappear just when we neglect the R-dependence of the creation and annihilation operators $a^{\dagger}_{p\pi}(R)$ and $a_{p\pi}(R)$. This actually means introduction of the *adiabatic approximation*. As far as it is allowed, omission of the exciton-nonconserving as well as the exciton-exciton interaction terms yields the exciton Hamiltonian in the standard form

$$H_{\text{exc}}(R) \approx E_o(R) + \sum_p [E^{\text{exc}}_p + D_p(R)]\bar{a}^{\dagger}_p\bar{a}_p + 1/2 \sum_{p \neq q} M_{pq}(R)\bar{a}^{\dagger}_p\bar{a}_p$$

(2.2.1)

[compare to Eq. (1B.2.13b)]. Notice that not only $D_p(R)$ (i.e., the local exciton energy with respect to the excitation energy of an isolated molecule) but also the resonance (hopping) interaction integral $M_{pq}(R)$ depend on the lattice point displacements owing to the motion and dynamical deformation of the molecules. The greater the R-dependence of $D_p(R)$, the greater the tendency to exciton localization. On the other hand, the greater is the $|M_{pq}(R)|$ dependence, the greater the tendency for exciton delocalization. Expanding in powers of $R - R_o$, we obtain for the exciton-phonon Hamiltonian in the adiabatic approximation

$$H \approx E_o(R_o) + \sum_p [E_p^{exc} + D_p(R_0)]\bar{a}_p^{\dagger}\bar{a}_p + (1/2) \sum_{p \neq q} M_{pq}(R_0)\bar{a}_p^{\dagger}\bar{a}_q + \sum_k [P_k^2/2M$$

$$+ (1/2)M\omega_k^2 Q_k^2] + \sum_p \sum_k \partial D_p/\partial Q_k|_{Q_k = 0} Q_k \bar{a}_p^{\dagger}\bar{a}_p$$

$$+ (1/2) \sum_{p \neq q} \sum_k \partial M_{pq}/\partial Q_k|_{Q_k = 0} Q_k \bar{a}_p^{\dagger}\bar{a}_q. \tag{2.2.2}$$

Here Q_k and P_k are the normal (harmonic lattice) coordinates and momenta. Two limiting cases now may be considered:

1. The coupling to vibrations [the fifth and sixth terms on the right hand side of Eq. (2.2.2)] is negligible. Then the solution is a free exciton wave as in Eq. (1B.2.12), accompanied by harmonic oscillations of the lattice uninfluenced by the exciton. This is the case of a completely delocalized exciton.

2. The exciton delocalization terms [the third term in Eq. (2.2.1) and the sixth term in Eq. (2.2.2)] are zero. Then Eq. (2.2.2) may be rewritten as

$$H \approx E_0(R_0) + \sum_p [E_p^{exc} + D_p(R_0)]\bar{a}_a^{\dagger}\bar{a}_p$$

$$+ \sum_k \left[P_k^2/2M + (1/2)M\omega_k^2 Q_k + \left(1/M\omega_k^2 \sum_p \partial D_p/\partial Q_k|_{Q_k = 0}\bar{a}_p^{\dagger}\bar{a}_p \right)^2 \right]$$

$$- \sum_k \sum_{p1,p2} (1/2M\omega_k^2)\partial D_{p1}/\partial Q_k|_{Q_k = 0} \partial D_{p2}/\partial Q_k|_{Q_k = 0} \cdot \bar{a}_{p1}^{\dagger}\bar{a}_{p1}\bar{a}_{p2}^{\dagger}\bar{a}_{p2}. \tag{2.2.3}$$

The last term here is the exciton-polaron vs. exciton-polaron coupling induced by vibrations (for $p_1 \neq p_2$) and the exciton-polaron energy shift (when $p_1 = p_2$). Clearly, the single-exciton states can be written as either localized states

$$|\Psi_p\rangle = \bar{a}_p^{\dagger}|\Psi_{exc.vac}\rangle \prod_k \varphi_{\{\mu_k\}}(Q_k - \chi_k^p),$$

$$\mu_k = 0,1,2,\dots, \quad \chi_k^p = -(1/M\omega_k^2)\partial D_p/\partial Q_k|_{Q_k = 0} \tag{2.2.4}$$

(created by a localized exciton state and shifted lattice oscillator functions) with energies

$$E_p \approx E_0(R_0) + [E_p^{exc} + D_p(R_0)] + \sum_k \hbar\omega_k(\mu_k + 1/2) - \sum_k (1/2)M\omega_k^2(\chi_k^p)^2 \tag{2.2.5}$$

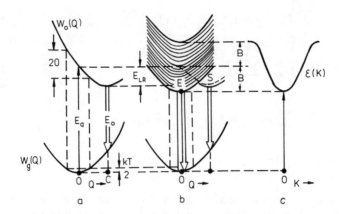

FIGURE 2.3. Configuration coordinate models for localized excitation (a), exciton (b), and excitonic band in rigid lattice (according to Ref. 389). $W_g(Q)$ and $W_e(Q)$ — potential energy functions as dependent on configuration coordinate Q for ground and excited states, respectively; E_a and E_e — energies of absorbed and emitted photons; E_{LR} — energy of lattice relaxation connected with exciton self-trapping; B — half-width of the exciton band; $\varepsilon(\mathbf{k})$ is dependence of the exciton energy on wave vector $\mathbf{k}$ in the exciton band.

or as arbitrary linear (possibly also extended) combinations of such states [Eq. (2.2.4)] which have the same energy [Eq. (2.2.5)].

Passing between these two limits of delocalized and localized excitons means investigating the phenomenon of self-trapping.

The theory of exciton self-trapping in organic crystals has been developed and elaborated in the framework of adiabatic approximation by Toyozawa[389,390] and Rashba.[142,143,294,295] The basic ideas of this approach are illustrated according to Toyozawa, in Fig. 2.3.

As may be seen from Fig. 2.3, for a local molecular excitation in a crystal there are two competitive pathways of evolution: creation of a free (F) exciton in the exciton band of rigid lattice or formation of a self-trapped (S) exciton as a result of local lattice deformation. The first pathway leads to delocalization, the second one to localization of the exciton, according to limiting cases discussed above.

The energy gain at the formation of a localized S-exciton, equals the energy of lattice relaxation E_{LR} (see Fig. 2.3). On the other hand, the energy gain in the formation of a free F-exciton equals the half-width B of the exciton band (i.e., B equals the kinetic energy needed for the localization of an already delocalized exciton).[389]

The exciton-phonon interaction (coupling) constant g can be determined phenomenologically as the following ratio:[389]

$$g = E_{LR}/B, \tag{2.2.6}$$

where E_{LR} is the lattice relaxation energy due to the exciton self-trapping. The ratio (2.2.6) characterizes the relative energy gain for both stable exciton states. If

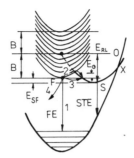

FIGURE 2.4. A double minimum adiabatic potential energy curve applicable to strong exciton-phonon coupled system (according to Matsui[227,232]). F and S denote the bottom of free F-exciton band and the lowest-energy state of self-trapped S-exciton, respectively. O denotes the energy of the lowest singlet-excited level of free molecule; E_B is the height of self-trapping barrier; other notations as in Fig. 2.3.

$E_{LR} < B$ and, correspondingly, $g < 1$ (as in Fig. 2.3), it is energetically more favorable for the exciton to exist in the form of an F-exciton, and, vice versa, if $E_{LR} > B$ and $g > 1$ formation of an S-exciton is more probable. Since parameter B is proportional to the resonance interaction while parameter E_{LR} depends on the term D, condition (2.2.6) may be regarded as a physically more instructive criterion of delocalization or localization tendencies than conditions presented by Eqs. (2.2.2) and (2.2.3).

Another important parameter for the dynamics of exciton processes is that of exciton self-trapping depth:[229]

$$E_{SF} = B - E_{LR}. \tag{2.2.7}$$

It follows from Eq. (2.2.7) that self-trapping is possible only if $E_{LR} > B$ and $E_{SF} < 0$.

In the case of a strong exciton-phonon coupled system, a double-minimum adiabatic potential is formed[227,232] in which the free F-exciton state is a metastable and the self-trapped S-exciton state is the stable one (see Fig. 2.4). The F-exciton and S-exciton states are separated by a so-called self-trapping barrier E_G. When an exciton is created in the exciton band, it usually relaxes down to the bottom F of the free exciton band. Further, the relaxed F-exciton can be annihilated via the following channels (see Fig. 2.4).

First, it can return to the ground state in radiative (1) or some nonradiative (2) annihilation processes. Second, it may form a self-trapped S-exciton either thermally "climbing" the barrier (3) or (4) tunneling through it.[227]

Possible tunneling mechanisms of exciton self-trapping have been discussed in terms of adiabatic approximation by Nasu and Toyozawa[267] and Rashba.[294,295] Fig. 2.5 illustrates, according to Toyozawa,[389] the delocalization and localization processes and formation of a double minimum adiabatic energy curve of F- and S-exciton states in the phonon field.

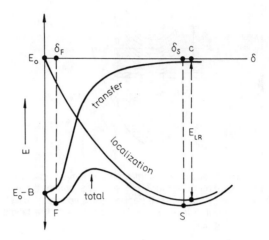

FIGURE 2.5. Energy of an exciton moving (with wave vector $k \neq 0$) in the phonon field, as a function of the accompanying lattice distortion δ chosen as variational parameter.[389]

Figure 2.6 shows the presumed effective mass m^* dependence on the phonon-dressed exciton as a function of coupling constant g. At some critical value of $g = g_c$ exciton self-trapping prevails and formation of a phonon-dressed S-exciton with considerably larger effective mass m^* value. The critical coupling constant $g_c > 1$ may be regarded, figuratively, as a "watershed" between domains of free and self-trapped exciton states. As may be seen from Table 2.4, expressed self-trapping with $g_c > 1$ is characteristic namely for the B-type of aromatic crystals (see Section 1A.3), such as pyrene and α-perylene, having parallel dimer pairs of

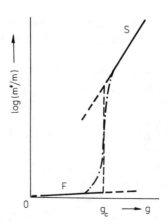

FIGURE 2.6. Effective mass of a phonon-dressed exciton as a function of coupling constant g.[389]

TABLE 2.4. *Parameters of the Exciton Self-trapping Model (see Fig. 2.3) Evaluated from Experimental Data: Exciton-Phonon Coupling Constant g, Self-Trapping Depth E_{SF}, Width of the Exciton Band ΔE_{ex}, and Lattice Relaxation Energy E_{LR} for Some Aromatic Crystals According to Refs. 227 and 229.*

Crystal	g	E_{SF} meV	E_{SF} cm^{-1}	$\Delta E_{ex} = 2B$ meV	$\Delta E_{ex} = 2B$ cm^{-1}	E_{LR} meV	E_{LR} cm^{-1}
Anthracene	0.85	+7	+60	90	750	39	315
Tetracene	0.95	−7	−60	> 80	> 670	> 49	> 395
β-perylene	1.09	−31	−250	> 120	> 960	> 90	> 730
α-perylene	1.61	−80	−650	260	2130	210	1715
Pyrene	1,5–3,2	−82	−660	—	—	—	—
Dichlor- anthracene	1,5–2,7	−101	−813	—	—	—	—

molecules in the unit cell. The exciton self-trapping, due to strong coupling with dimer vibrations, leads to formation of an *excimer* state.

The exciton self-trapping phenomena, particularly in pyrene and α-perylene crystals, have been extensively studied by Matsui *et al.*[113,227–233,239,240,241] who applied picosecond and ordinary spectroscopic techniques. A complete list of related references is given in the recent review paper by Matsui.[227]

Figure 2.7 shows, according to Ref. 229, a schematic diagram of potential energy curves of free (F) and self-trapped (S) exciton (in this case excimer) states. If an exciton resides on one of the molecules of a dimer pair the intramolecular distance

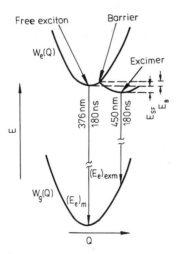

FIGURE 2.7. Schematic diagram of potential energy curves for the ground and excited exciton and eximer states in a pyrene crystal.[229] $(E_e)_m$ — luminescence of free exciton; $(E_e)_{exm}$ — eximer luminescence.

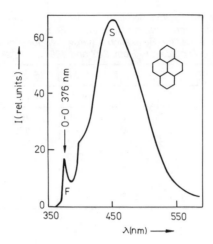

FIGURE 2.8. The luminescence spectra of a pyrene crystal at 300 K.[229] *F* — free exciton band; *S* — eximer (self-trapped exciton) band.

of the pair decreases due to resonance interaction. This process is equivalent to local lattice deformation.

As a result, a new potential curve minimum is formed which leads to self-trapping of the exciton, i.e., to formation of an excimer state (see Fig. 2.7).

Experimentally exciton self-trapping exhibits itself in the excimer luminescence band (see Fig. 2.8). The wide Gaussian luminescence band with a maximum at $\lambda = 450$ nm is caused by emission from the excimer (S) state while the narrow band at $\lambda = 376$ nm should be assigned to the free (F) exciton luminescence. The corresponding transitions are shown in Fig. 2.7.

It has been shown that free (F) excitons, yielding the corresponding F-band of luminescence, are in a thermal equilibrium with the S-excitons.[229] First, the intensity of the F-band decreases exponentially with decreasing temperature; second, the characteristic emission time of both F- and S-exciton states are equal, i.e., 180 ns (see Fig. 2.7). Thus, the intensity vs. temperature dependence of free (F) exciton luminescence $I_F \sim \exp(-E_{SF}/kT)$ allows us to evaluate the self-trapping depth E_{SF}.[227,229]

Detailed experimental studies of self-trapping phenomena, carried out over more than a decade by Matsui and coworkers, demonstrate that the process is actually more complicated than discussed above. The exciton self-trapping conditions and performance may be dependent on temperature range, possible phase transitions, pressure, and other factors affecting the arrangement of molecules in the lattice.[113,227] Thus, it has been found[113] that in pyrene above 120 K, the structural phase transition point (see Section 1A.3 in Chapter 1) most of the photogenerated hot excitons relax directly to the self-trapped state (shown by arrow in Fig. 2.4); only a small fraction of these excitons relax to the bottom of the free-exciton band. This fraction almost completely disappears at $T = 60$ K.

The possible mechanisms of hot exciton self localization have been considered theoretically in Ref. 143.

Recent experimental studies of exciton self-trapping in pyrene crystals[113,227] give following values of the main self-trapping parameters (see Fig. 2.4): at the high T region $E_{SF} = 660$ cm^{-1} and $E_B = 260$ cm^{-1}; at the low T region $E_B = 360$ cm^{-1}.

The main parameters of the exciton self-trapping model (Fig. 2.4), as evaluated from experimental data, are presented for some aromatic crystals in Table 2.4.

As can be seen from Table 2.4, for polyacenes (Ac,Tc) the coupling constant $g < 1$, and the self-trapping depth E_{SF} is very small (for Tc) or even positive (for Ac); therefore for these polyacenes the exciton self-trapping probability should be very small indeed. Both representatives of the B-type of lattice (see Section 1A.3), namely, α-perylene and pyrene, yield $g > 1$ and considerable self-trapping depth. According to these criteria they are sure candidates for strong exciton self-trapping as confirmed by intense excimer luminescence.

Additional evidence of singlet exciton self-trapping effect in α-perylene and pyrene crystals is given in Table 2.3. As may be seen from Table 2.3 the diffusion coefficients D of singlet excitons in these crystals are extremely small — of the order of 10^{-8}–10^{-9} cm^2/s. Also the corresponding diffusion length ($L_s = 0.2$–0.3 nm) is practically equal to the intermolecular spacing and the self-trapping act of an exciton presumably occurs already at the very first encounter with the dimer pair in the lattice. That this self-trapping effect is specific only for singlet excitons is demonstrated by the fact that for triplet excitons the D value of pyrene is similar to that of a naphthalene crystal (see Table 2.2).

A strong tendency for exciton self-trapping is also seen in such a polar compound as dichloranthracene. An intermediate state between these two groups is occupied by β-perylene with $g = 1.09$ and, consequently, a slight tendency to exciton self-trapping may be observed here.

It has been shown[239] that in Ac single crystals, an inversion of the positions of minima of F- and S-states takes place under the influence of high hydrostatic pressure ($p = 32$ kbar).

As a result, the positive value of the E_{SF} (see Table 2.4) changes to a negative one, forming a minimum of self-trapped S-exciton state below the F-state minimum, as seen in Fig. 2.4. The influence of hydrostatic pressure on exciton dynamics in α-perylene crystals is described in Ref. 233.

Table 2.5 presents some criteria and perspectives of localization versus delocalization of an exciton due to interaction with the phonon field of the crystal, as reflected in absorption and emission spectra.

In the case of emission spectra of a relaxed exciton state, such a criterion is set by the coupling constant g, as discussed above and shown in Table 2.4.

In the case of absorption spectrum of a nonrelaxed exciton state one might use as a criterion the parameter σ of the Urbach-Martienssen's rule describing the low-energy tail of the spectrum of the crystal. This rule and its implications will be discussed in Section 2.4.

TABLE 2.5. *Perspective of Localization and Delocalization of an Exciton Due to Interaction with the Phonon Field, as Reflected in Optical Spectra of OMC, Modified after Toyozawa*[389]

Nonrelaxed exciton	Strong scattering (exciton localized)	Weak scattering (exciton delocalized)
Shape of absorption band	Gaussian	Lorentzian
Half-width of the band	H_G	H_L
Urbach rule $\sigma_0 \approx 1.5/g$	$\sigma_0 < 1.5$	$\sigma_0 > 1.5$
	Increase of the B value $\longrightarrow$	

Relaxed exciton	Strong coupling	Weak coupling
$g = E_{LR}/B$	$g > 1$	$g < 1$
Emission band	Stokes-shifted Gaussian	Resonant
Shape of emission band	Wide Gaussian	Narrow band

The statements and interpretations, discussed above, were carried out along the lines of traditional adiabatic approximations.

However, the validity and applicability of the adiabatic approximation have been recently seriously questioned.[401–403]

We shall present below some critical comments and an alternative approach based on the crude representation.

2.2.2. Exciton Self-trapping in the Crude Representation

It follows from the principle of degenerate perturbation theory, and this is of special importance to periodic crystals, that extended linear combinations of Eq. (2.2.4) [rather than the localized states (2.2.4) themselves] are proper zeroth order unperturbed states with respect to the exciton delocalization terms [the third and sixth terms in Eq. (2.2.2)]. Thus, the exciton self-trapping phenomenon is generally *not* the transition from extended to localized states (as it is often misinterpreted) of the excitonic polaron. It is rather a transition to extended states with appreciably lowered mobility. Or, in other words, the exciton self-trapping phenomenon is a transition to equilibrium with appreciably lowered values of the off-diagonal matrix elements of the single-exciton density matrix in the site (localized molecular) representation. Moreover, if we optimize (using the variational principle) the energy of the excitonic polaron in any localized state, it follows that extended linear combinations of such optimized localized states yield still lower energy.[63,404] Thus, in terms of this approach, there is likely to be no problem of an abrupt transition from an extended to localized state.

As discussed above, a considerable amount of theoretical work has been done in the field of the exciton self-trapping phenomena, in the framework of adiabatic approximation (see Refs. 142, 143, 294, 295, 389, and 390). Before further discussing this phenomenon, we would like to stress that the adiabatic theory, on which most of the earlier theories rely, taking the self-trapping as a transition across an adiabatic barrier, is probably inapplicable in solids, in particular in the case of self-trapping dynamics. The first warning in this direction was pronounced by Wagner in 1985.[401,402] A detailed discussion of the problem, in particular that of divergences connected with an extreme violation of the adiabatic principle is given in Ref. 403. In particular, it was concluded[403] that

1. The adiabatic wave function is highly inadequate in the barrier region where the barrier must be overcome by the nuclear system on transition from the extended to the self-trapped state.

2. Since the adiabatic wave function (and the adiabatic approximation itself) is indispensable for the very definition of the process of overcoming the adiabatic barrier, practically all the contemporary theory of the process based on these notions is rather unjustified.

Below we will give some theoretical arguments supporting the idea in the crude representation. On the other hand, it must be concluded (in contrast to the usual opinion but in view of what has been said above) that the proper dynamic theory of the time-dependence of self-trapping, free of unjustified notions, is still lacking.

As argued above, one should be, at least at the present stage of knowledge, critical of any theory of the exciton-phonon dynamics that is formulated in the adiabatic representation and is connected with the adiabatic approximation. This controversy makes us uncertain as far as the dynamics of the exciton self-trapping process is concerned. On the other hand, mostly for technical reasons, many authors simply ignore the site-off-diagonal exciton-phonon coupling [the last term on the right-hand side of Eq. (2.2.2)] in such well-defined problems (mostly stationary) as, e.g., the variational determination of the wave function (see Refs. 107 and 389). One should realize that the crude representation Hamiltonian of the exciton-phonon system may be expressed as in Eq. (1B.8.1) (with just the substitution $a_p \rightarrow \bar{a}_p$, etc.) except for exciton-nonconserving and many-exciton terms. This equation has, however, the same form as Eq. (2.2.2) provided that the site (molecule)-off-diagonal exciton-phonon coupling is ignored. Thus, in the crude representation, we can simply follow these works.

Possibility of an abrupt transition (with a continuous change of the coupling strength) from an effectively free to a "self-trapped" state was, for the first time, suggested by Toyozawa[387] in 1961. Since there is no exact solution of the problem and because the continuous or abrupt (as a function of the coupling strength) character of the transition strongly depends on the type of the variational function used, it is difficult to make a rigorous statement. At present, it is well established that the local accommodation of the lattice around the standing (or slowly moving) exciton certainly yields a gain in energy.[135,263,398] On the other hand, it is equally well established that the delocalized (i.e., extended) combinations of such states (with the localized exciton and the lattice accommodated to it) yield still lower energy.[63,404] Thus, the proper variational wave function relevant to the problem must fulfil the following requirements:

1. It should be extended in periodic systems like periodic crystals.

2. When expanded into components with the localized exciton, it should describe the lattice distortion around the localized exciton relative to its position in any of these components [e.g., as in Eq. (1B.8.4a)].

Not all wave functions found in the current literature obey these requirements. Probably the most detailed recent treatment that so does is found in Ref. 404. According to its authors, in case of a model of an exciton in a molecular chain interacting with phonons via short-range forces (the case for which the abrupt transition was originally predicted in Ref. 387):

1. Not an abrupt but a continuous behavior (with increasing coupling strength) of the effective bandwidth was found in a qualitative correspondence with our reasoning leading (at $T = 0$ and for excitons) to Eq. (1B.8.16).

2. The total energy of an extended state was found to be less than the energy of the corresponding localized (solitonlike) state (at least for not very broad unperturbed bandwidths — in the opposite case, more complicated variational wave functions are needed).

3. The rate (probability per unit time) of the self-trapping process was argued to be much less intense in the strong-coupling regime than that in earlier approaches.

4. At least in the one-dimensional model investigated, any abrupt (discontinuous) dependence of physical quantities on the coupling strength may be excluded.

For a recent contribution in this field, see Ref. 192.

As far as the dynamics (i.e., the time-dependence) is concerned, very little is known beyond the adiabatic approximation treatments. A recent nonadiabatic approach by Nasu[266] may be qualitatively correct. Though it works with notions inherent to the adiabatic theory, it goes beyond that. The tunneling rate between the "free" and "self-trapped" state Γ_{FS} was found to be determined by the nonadiabatic overlap between the "free" state and that at the top of the potential barrier, in contrast to the standard adiabatic theory. For pyrene at $T = 0$ Γ_{FS} is calculated to be about 200 ps and the sliding time about 1 ps, in rough agreement with experimental results.

2.3. ABSORPTION AND SCATTERING OF LIGHT BY THE EXCITON-PHONON SYSTEM

In modeling these physical processes, one usually works with a so-called standard molecular crystal model

$$H = B/z \sum_{m} \sum_{\delta} \bar{a}_m^\dagger \bar{a}_{m+\delta} + \sum_m \hbar \omega b_m^\dagger b_m + \sqrt{S} \sum_m \hbar \omega \bar{a}_m^\dagger \bar{a}_m (b_m + b_m^\dagger).$$

$$(2.3.1)$$

Here $\sum\limits_{\delta}$ is summation over the nearest neighbors only. Simultaneously with the exciton absorption curve, one might also be interested in other quantities of possible experimental interest, like the exciton density of quasiparticle states or the exciton energy band structure [broadened by the exciton–phonon coupling — see the third term on the right-hand side of Eq. (2.3.1)]. That is why sometimes these processes are also considered.

Because the problem connected with the Hamiltonian [Eq. (2.3.1)] is not exactly soluble, one should first mention the most reliable approximate methods. The first (and still the basic one) is the dynamic coherent potential approximation (DCPA) suggested by Sumi.[368-370] This procedure (with an independent alternative approach suggested in Ref. 67) proved to be numerically relatively simple, while being a reliable and physically well justified method. With the single-particle Green function, it is applicable to the optical absorption, density of states, and exciton band structure calculations. For emission, it may be used indirectly.[369] Extension of the approach to the case of the second-order optical response (including emission spectra with Raman scattering, hot as well as ordinary luminescence) has been given by Miyazaki and Hanamura.[236-238] These authors, using an approximate form of the Generalized Master Equations,[238] included a dynamic damping of the intermolecular Einstein phonons [see Eq. (2.3.1)]. In Ref. 238 an extensive discussion of various quantities in various regimes is given. More recent calculations, usually omitting the latter damping, extended the approach to treat time-resolved phenomena.[2-4] The method yields excellent qualitative results, mostly in agreement with experiment in molecular crystals. Unfortunately, the DCPA method is still an approximate one, so that we are not sure whether possible qualitative or quantitative amendment should concentrate on including effects lying beyond the DCPA or beyond the very model of the Eq. (2.3.1).

A very good prospect in this direction, yields another, so-called recursion method which is purely computational but very accurate so that in fact, it relies just on approximations connected with the Hamiltonian (2.3.1). It has been developed by Sherman and published in a series of papers. The final version of the method may be found in Refs. 325 and 326. Agreement with experiment is fairly good; difference between results for the ordinary luminescence with excitation above the fundamental absorption edge obtained by this method and by the DCPA method may be ascribed to either the approximate character of the DCPA (as suggested in Ref. 326) or, at least in principle, to the Einstein phonon damping ignored in Ref. 326. To exclude the latter possibility (which is likely), an extension of the recursion method[325,326] to models including the damping of the Einstein phonons is highly desirable.

2.4. THE URBACH–MARTIENSSEN RULE

This rule was found experimentally first by Urbach[396] in silver halides and later by Martienssen[224] in alkali halides that the low-energy tail of the exciton absorption spectra depends exponentially on the photon energy $h\nu$, with a constant inversely proportional to temperature:

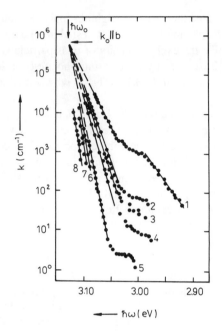

FIGURE 2.9. The dependence of the absorption coefficient k on the photon energy $\hbar\omega$ at low-energy exciton absorption tall in a semilog plot with temperature as a parameter.[351] The spectral curves 1-8 correspond the following temperatures: 300, 178, 156, 135, 77, 35, 20, and 4.2 K.

$$k(h\nu, T) = k_0 \exp[-\sigma(h\nu_0 - h\nu)/k_B T], \qquad (2.4.1)$$

where k is the absorption coefficient, k_0 and $h\nu_0$ are constants (see Fig. 2.9); the ratio $\sigma/k_B T$ determines the steepness of the low-energy spectral tail. The low-energy exponential spectral dependence (2.4.1) is known in literature as the Urbach or Urbach-Martienssen rule. The Urbach rule has been later observed in various organic and inorganic solids, including organic molecular crystals (see the review[201]).

According to the formula (2.4.1) spectral curves of $k(h\nu)$ of the low-energy absorption tail form in a semilog plot a "fan" of straight lines which converge at a single point with coordinates k_0 and $h\nu_0$ (see Fig. 2.9). The temperature range at which the exciton absorption follows the dependence (2.4.1) usually lies above the characteristic Debye temperature of the crystal. At lower temperatures there may be deviations from the rule [Eq. (2.4.1)] caused by the peculiarities of low temperature absorption.

Experimental data show that the steepness coefficient σ in Eq. (2.4.1) is dependent on temperature according to the following empirical formula:[351]

$$\sigma(T) = \sigma_0 \frac{2k_B T}{\hbar\omega_{ph}} th \frac{\hbar\omega_{ph}}{2k_B T}, \qquad (2.4.2)$$

where σ_0 is a constant that characterizes the interaction of excitons with local lattice environment. It has been shown that the parameter σ_0 is related to the exciton-phonon coupling constant g:[389]

$$\sigma_0 = s/g,$$

where s is the so so-called "steepness index."

Parameter $\hbar\omega_{ph}$ in Eq. (2.4.2) is the energy of phonons that take part in formation of the low-energy absorption tail.

Thus, the Urbach-Martienssen rule yields important parameters that may serve as criteria of the strength of exciton-phonon interaction and of the exciton tendency for localization versus delocalization (see Table 2.5).

The simplicity and universality of this empirical rule has evoked a number of theoretical and experimental studies (see, e.g., Refs. 83, 201, 351, and 389).

The earlier theories based on the idea of the band edge fluctuation due to thermal phonons[363] and those invoking the idea of the exciton ionization by electric field of phonons[92] seem to be ruled out at present (see Ref. 372). Contemporary theories are actually based on the ideas of Toyozawa on exciton lattice interaction, going back to the early sixties.[388]

Davydov[83] developed a more general approach according to which the shape of the low-energy absorption tail is determined by quantum transitions from lattice vibronic sublevels to the first electronically excited state. Later it was well recognized that the effect is due to exciton-phonon coupling and that the constant σ in Eq. (2.4.1) is inversely proportional to the square of the exciton-phonon coupling constant [i.e., to S with $\sqrt{S}$ from Eq. (2.3.1)]. It corresponds well to the experimental observation that, to extend the validity of Eq. (2.4.1) to very low temperatures, a substitution

$$k_B T \rightarrow k_B T^* = (1/2)\hbar\omega \coth(\hbar\omega_{ph}/2k_B T) \qquad (2.4.3)$$

with a typical phonon frequency ω_{ph} should be made [compare Eq. (2.4.2)]. (Then T^* is proportional to the square of the amplitude of the mode with frequency ω_{ph}.) Thus, the effect should have something to do with photon absorption to, possibly (even self-localized) excitonic polaron states. As we have seen above, the contemporary theory of the exciton self-localization within terms of the adiabatic approximation is questionable. Theory beyond the adiabatic approximation, on the other hand, is only at the stage of development. In this connection, it is therefore remarkable that (without specifying in detail the final exciton states after photon absorption) both the DCPA[2-4,67,236-238,368-370] and the recursion method[325,326] at finite temperature fit the Urbach-Martienssen rule with great accuracy. In this sense, we can say that the origin of this rule Eq. (2.4.1) is theoretically well understood. In our opinion, the question about the character of the final exciton states is redundant and not very well posed, since in principle, any complete system of states may be equally well used to describe the phenomenon.

In organic molecular crystals, many experimental studies have also been done. One can find details in Refs. 112, 199, 202, 205, 206, 226, 242, 264, and 351. A survey of reported experimental parameters σ_0, $h\nu_0$, k_0, and $\hbar\omega_{ph}$, estimated according to the Urbach-Martienssen rule for some aromatic crystals, is shown in Table 2.6.

As can be seen from Table 2.5, the parameter σ_0 may serve as an additional criterion of exciton localization. In this case the critical σ_0 value, separating the exciton delocalization and localization domains, is approximately equal $(\sigma_0)_c \approx 1.5$.

The data in Table 2.6 show that for polyacenes and β-perylene the σ_0 value lies (with few exceptions) a little below this critical value $\sigma_0 \lesssim 1.5$ in practically all lattice directions. This means that, according to this criterion, these crystals of A-type (see Section 1.A.3.1) show a slight tendency toward localization (see Table 2.4). On the other hand, both representatives of the B-type of α-perylene and pyrene crystal (see Section 1A.3.2), as well as a polar derivative of anthracene, i.e., 9,10-dichloranthracene, show a strong tendency toward exciton localization (see also Table 2.4).

The value of the parameter $\hbar\omega_{ph}$ in Eqs. (2.4.2), (2.4.3) corresponds to the typical magnitudes of optical lattice phonon energies (see Section 1A.4) or their multiples. This is in accordance with the exciton-phonon interaction theories. In general, the agreement of the Urbach rule formulas (2.4.1)–(2.4.3) with experiment is quite good.

Figure 2.10 shows the temperature dependence of the parameter $\sigma = \sigma(T)$ for polyacene crystals.

One may notice that the $\sigma = \sigma(T)$ dependences of tetracene and particularly pentacene are qualitatively different from those of Nph and Ac (see Fig. 2.10). These effects may be connected with different symmetry groups of the crystals. It was indicated earlier[388] that the crystals of the C_{2h}^5 ($P2_{1/a}$) symmetry should obey the Urbach rule, while in case of the C_i ($\bar{P}1$) symmetry one may expect violations of the rule.

These and other problems of the applicability and physical meaning of the Urbach-Martienssen rule deserve further studies.

In concluding this chapter on exciton interaction with local lattice environment, we should emphasize that the above-presented interaction (g and σ_0), and transfer (D) parameters, as well as the self-trapping depth E_{SF} values for aromatic crystals, unequivocally indicate that real singlet exciton *self-trapping* takes place only in case of dimer-type crystals like pyrene and α-perylene (and possibly also in case of some polar derivatives of polyacenes). Concerning nonpolar polyacene crystals of Ac type, these data indicate only the presence of relatively strong exciton localization at room temperatures in a sense that the excitons are moving as corpuscular quasiparticles in an *incoherent*, diffusive way of transfer from one lattice site to another (see Chapter 5).

The theoretical interpretation of these empirical data, especially concerning the self-trapping phenomena, is still controversial. At present, the validity of the traditional treatment in terms of an adiabatic approximation has been seriously questioned as shown above. That means that further theoretical studies of the problem

TABLE 2.6. Experimental Parameters σ_0, $h\nu_0$, and k_0 of the Low-energy Tail of Exciton Absorption in Several Aromatic Crystals, Estimated According to the Urbach-Martienssen Rule[a] [351]

Crystal	Symmetry	$(\sigma_0)_{xx}$	$h\nu_0$ (eV)	k_0 (cm⁻¹)	$\hbar\omega_{ph}$ meV	$\hbar\omega_{ph}$ cm⁻¹	Temperature range (K)	Reference
Naphthalene	C_{2h}^5 ($P2_{1/a}$)	$(\sigma_0)_{bb}$ = 1.32	3.93	10^4	15.6	125.7	20–190	351
		$(\sigma_0)_{c'c'}$ = 1.28	3.92	10^5	10.6	85.4	50–250	351
Anthracene	C_{2h}^5 ($P2_{1/a}$)	$(\sigma_0)_{bb}$ = 1.76	3.13	4×10^5	33	266.0	79–346	226
		$(\sigma_0)_{c'c'}$ = 1.73	3.17	3×10^5	38	306.3	79–346	226
		$(\sigma_0)_{aa}$ = 1.48	3.15	10^5	17.3	139.4	4–300	205
		$(\sigma_0)_{bb}$ = 1.50	3.13	5×10^5	14.8	119.3	4–300	205
		$(\sigma_0)_{c'c'}$ = 1.30	3.14	5×10^4	25	201.5	4–300	205
Tetracene	C_i ($P\bar{1}$)	$(\sigma_0)_{aa}$ = 1.3	2.45	3×10^4	55.8	449.7	77–300	202
		$(\sigma_0)_{aa}$ = 1.37	—	—	—	—	13–300	292
		$(\sigma_0)_{bb}$ = 1.58	—	—	—	—	13–300	292
Pentacene	C_i ($P\bar{1}$)	$(\sigma_0)_{bb}$ = 1.4	1.855	—	—	—	4–300	206
9,10-dichlor-anthracene	C_{2h}^5 ($P2_{1/a}$)	$(\sigma_0)_{aa}$ = 0.56	2.92	10^4	18.7	150.7	4–300	112
		$(\sigma_0)_{bb}$ = 1.0	2.94	3×10^4	26.9	216.8	4–300	112

[a]Note: Estimated value of parameter σ_0 in α-perylene equals 0.93,[389] in pyrene 1.0 > σ_0 > 0.47,[229] in β-perylene σ_0 = 1.38.[389]

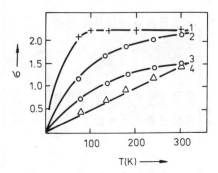

FIGURE 2.10. The temperature dependence of the parameter $\sigma = \sigma(T)$ for naphthalene (1), anthracene (2), tetracene (3), and pentacene (4) crystals.[351]

are required. However, in this connection we should remember the prophetic words by Louis de Broglie (see the epigraph of this chapter) that "when extreme precision is aimed at, they" (i.e., concepts produced by the human mind) "become ideal forms whose content tends to vanish away."

CHAPTER 3

Charge Carrier Interaction with Local Lattice Environment

The observed system is required to be isolated
in order to be defined,
yet interacting in order to be observed.

HENRY STAPP

Isolated material particles are abstractions,
their properties being definable and observable
only through their interactions with other systems.

NIELS BOHR

This chapter deals with charge carrier interaction phenomena in OMC. First, we discuss the charge carrier interaction with the electronic subsystem of the surrounding lattice, i.e., electronic polarization processes leading to the formation of new quasiparticles—the electronic polarons.

The discussion will be made on two levels, similar to Chapters 1 and 2.

We begin with a brief introduction to traditional band model approach of a single-electron approximation and with a critical review of the limitations of this approach. This will be followed by a phenomenological, qualitative description of electronic polarization phenomena. On the second level of presentation, dynamic approaches to electronic polarization will be discussed, based on theoretical treatment of the electronic polaron concept as presented in Chapter 1B, Section 1B.4.

As a connecting link between the phenomenological and quantum physical treatments a detailed description of microelectrostatic methods of calculation of electronic polarization energies will be given.

In the second part of this chapter we will consider charge carrier interaction with the nuclear subsystem of the surrounding lattice, i.e., interaction with intra- and intermolecular vibrations of local lattice environment. First, we give a qualitative picture of molecular and lattice polaron formation which will be followed by a Hamiltonian description of the formation of these quantum quasiparticles of many-body interaction, based on extension and refinement of a more general lattice polaron concept as introduced in Section 1B.8 of Chapter 1.

3.1. CONDITIONS OF CHARGE CARRIER LOCALIZATION AND DELOCALIZATION

In general, the problem of localization of charge carriers is similar to that of excitons (see Section 2.2). However, there are peculiarities related with the physical nature of charge carriers in OMC; in addition to the presence of an electric charge q, the carrier emerges as a polaron-type quasiparticle. Namely, a polarization-type interaction of a charge carrier with local lattice environment is the main cause of carrier localization.

As in the case of excitons, the dynamics of charge carrier motion in solids depends on two competing trends, namely, on *delocalization* of the carrier in form of a Bloch wave, and on *localization* of the wave packet as a result of interaction with the local surroundings (with electronic or nuclear subsystems of the lattice).

In general, the delocalization of a charge carrier in a solid is determined by its resonance-type interaction energy with neighboring atoms or molecules in the lattice, characterized by the transfer integral J_{nm} of the carrier between sites n and m. The value of J_{nm} actually determines the effective width of permitted energy bands for charge carriers in a solid. If the J_{nm} value is of the order of an electron-volt, as it is in the case of a metal or a covalent crystal, then the band width is of the same order. In such broad bands the charge carriers are completely delocalized and move almost coherently in form of a Bloch wave with a definite wave vector **k**, their effective mass m_{eff} being close to that of a free electron ($m_{eff} \approx m_e$). It may be noted that delocalization of a charge carrier in a crystal is accompanied by a certain energy gain. This effect manifests itself, for instance, in band formation in the case of covalent crystals. The energy band gap (forbidden energy band) E_G between the top of the valence band E_v and the bottom of the conduction band E_c in which the charge carriers move, is smaller than the energy interval ΔE between the highest occupied and the lowest unoccupied levels of atoms or molecules. Splitting of the latter forms, respectively, the valence band E_v and the conduction band E_c. Accordingly, for the localization of a charge in wavelike motion a certain energy is required, the so-called localization energy δE_{loc} which is a magnitude with positive sign ($\delta E_{loc} > 0$). The fact that localization of the charge wave packet demands energy consumption follows directly from the uncertainty principle.[288,351]

On the other hand, in the dynamic process of delocalization the charge carrier interacts with the electronic or nuclear subsystems of the lattice, causing local polarization. Any form of polarization of the lattice on the part of a charge carrier (electronic, vibronic, or lattice) is accompanied by gain in energy in form of polarization energy δE_{pol}. In addition, the carrier may form local bonds with a definite molecule, or atom, or group of atoms or molecules of the lattice in the process of localization. In the case of an OMC, for instance, a charge localized on a molecular site may form a molecular ion or a nearly small-radius molecular polaron (see Section 3.6). Such a process is, naturally, accompanied by a gain in charge bonding energy δE_b. Hence, the terms δE_{pol} and δE_b have a negative sign (δE_{pol}; $\delta E_b < 0$). The fate of the charge carrier in the crystal is determined by the energy balance of the above terms.

Under conditions when

$$\delta E_{loc} + \delta E_{pol} + \delta E_b = \delta E > 0 \qquad (3.1.1)$$

the charge carrier is *delocalized*. Such a situation occurs if the terms δE_{pol} and δE_b are small. This may be the case if parameter J_{nm} is large and the valence and conduction bands are broad. With increasing polarization the bands become narrower and, accordingly, the charge mobility decreases. One may still retain here the band model of a delocalized charge carrier if one ascribes to it a suitable increased effective mass m_{eff}. If the contribution of polarization is due to the interaction with the lattice, then the state of the delocalized carrier may be described by introducing a quasiparticle — a lattice polaron in terms of a large-radius[111] or a nearly small-radius polaron models.[97] Also a small-radius lattice polaron may be delocalized at low temperatures, when it is possible to use concepts of a polaron band. On delocalization of a charge carrier in polaron theory representation see the review by Appel[12] and monographs[247,288,351] (see also Section 3.6.1).

On the other hand, if we have

$$\delta E_{loc} + \delta E_{pol} + \delta E_b = \delta E < 0 \qquad (3.1.2)$$

the *localized* state of the charge carrier will be energetically more favorable. The representation of the charge carrier in form of a Bloch wave with a definite wave vector **k** has to be replaced by a corpuscular concept of a carrier moving stochastically by hopping without conservation of the wave vector **k**. In other words, the localized carrier may be considered, in this case, as a Brownian particle, the motion of which can be described by the diffusion equation. It is, however, necessary to emphasize that the inequalities (3.1.1) and (3.1.2) cannot be applied as strict quantitative criteria. The same applies to the above picture of the localized-delocalized transition of charge carrier. One should, rather, consider them as qualitative illustrations of the conditions for domination of one or the other form of motion of the charge carrier. Thus, in the formulas (3.1.1) and (3.1.2) there is no explicit presence of such an important factor as *temperature*. In this connection one should stress that the increase in temperature is one of the main factors causing localization of the particle, due to its strong interaction with the phonon field, as we have been able to convince ourselves in Chapter 2 in the case of excitons, and as it will be shown in Chapter 5 for charge carriers.

The other important factor is *time*. At short time intervals the motion of quasiparticle may be coherent, while at longer time intervals it may become incoherent leading to its localization. It should be pointed out that in the description of the state of the quasiparticle in terms of a density matrix both types of approximation (localized or delocalized) may be represented as equivalent, within the framework of a single generalized model (see Chapter 5).

Since the density of states is larger in a narrow band than in a broad one, the tendency of the wave packet towards localization is more pronounced in narrow-band solids.[288] In other words, the term δE_{loc} in narrow-band crystals, such as anthracene, is comparatively small, owing to weak intermolecular interaction, i.e., owing to a small value of the parameter J_{nm} (see Section 3.2.3). Therefore a detailed analysis of polarization processes, determining the localization of charge carriers is a problem of paramount importance in the physics of OMC, which will be the leading topic of this chapter.

The next section (3.2) of the present chapter deals with attempts to apply the band model of single-electron approximation for the description of the energy structure and transfer mechanisms of charge carriers in Ac-type organic crystals. Such a brief historical background is rather instructive. It illustrates the inertia of traditional approach which had hindered the development of organic solid-state physics for more than a decade.

In a single-electron approach one does not take into account the effects of *polarization* phenomena. In other words, we assume that in the inequality (3.1.1) $\delta E_{loc} > |\delta E_{pol} + \delta E_b|$, and, accordingly, the carrier may be delocalized in form of a Bloch wave. Such an assumption is quite adequate, from the point of view of electronic polarization, in the case of broad-band, e.g., covalent crystals, for which the condition (3.1.1) is obviously valid and localization of the carrier is energetically unfavorable.

The band model of single-electron approximation has been found valid for the description of charge carrier transport in a number of inorganic solids, such as diamond-type covalent crystals (Ge, Si) and others. It forms the basis of traditional solid-state physics (see, e.g., Refs. 85, 129, and 185).

However, the single-electron band approach has been found to be inadequate for the description of other kinds of solids, in which one cannot neglect the polarization effects and the trend toward localization of charge carriers (cf., e.g., Ref. 247). Among such solids are ionic crystals, amorphous solids, polymers, as well as Van der Waals molecular crystals, including organic crystals.[288,332,351] As will be shown in Section 3.2.3, in the case of OMC the single-electron approach must be replaced by more complex models of multielectron and multiparticle interaction between the charge carrier and the local lattice environment. In other words, one must consider both the electronic, as well as the molecular (vibronic) and lattice polarization in a framework of extended polaron theory approach.[332,351]

For systematic development of the problem we start with the simplest approximation, i.e., with the single-electron band theory. This is followed by a brief history of calculations of the band structure of Ac-type crystals, and an analysis of the limits of applicability of the single-electron approximation for the description of the energy structure of ionized states in OMC.

3.2. BAND MODEL OF SINGLE-ELECTRON APPROXIMATION

3.2.1. General Scheme of the Model. Bloch Function

The simple band model is based on the adiabatic approximation, according to which all nuclei of the system are considered as fixed and immobile. The multielectron problem is reduced to a single-electron one assuming that the wave function ψ of the electron in the rigid crystal lattice can be represented in form of a Slater determinant of single-electron wave function $\varphi_i(j)$[351] as follows:

$$\psi(1, 2, \ldots, N) = \frac{1}{\sqrt{N!}} \begin{vmatrix} \varphi_1(1), & \varphi_1(2), & \ldots, & \varphi_1(N) \\ \varphi_2(1), & \varphi_2(2), & \ldots, & \varphi_2(N) \\ & \ldots & & \\ \varphi_N(1), & \varphi_N(2), & \ldots, & \varphi_N(N) \end{vmatrix}, \qquad (3.2.1)$$

where $j \equiv (\mathbf{r}_j, \sigma_j)$ denotes the spatial $\mathbf{r}_j$ and the spin σ_j variables of the jth electron, while the index i of the wave function gives the number of the solution of the single-electron Schrödinger equation

$$\left[\frac{\hbar^2}{2m} \Delta + U(1) \right] \varphi_i(1) = E_i \, \varphi_i(1) \qquad (3.2.2)$$

with a definite (specific for the given type of approximation) periodic potential U of the crystal. Due to the periodicity of the crystal and of the boundary conditions, the index i becomes a multiplicity of indices. Apart from the spin index σ and the band index n it includes the wave vector $\mathbf{k}$ of the first Brillouin zone expressing the symmetry properties of the Bloch wave function (3.2.3) with respect to translational displacements $\mathbf{R}_m = \sum_{i=1}^{3} m_i \mathbf{a}_i$ which transform the crystal into itself (where $\mathbf{a}_i$ are the primitive translation vectors and m_i are integers), i.e.

$$\varphi_i(1) = \psi_{kn\sigma}(\mathbf{r}_1, \boldsymbol{\sigma}_1) \equiv e^{i\mathbf{k}\mathbf{r}_1} u_{kn\sigma}(\mathbf{r}_1, \boldsymbol{\sigma}_1). \qquad (3.2.3)$$

Since the wave function $u_{kn\sigma}$ is periodic with respect to translation $\mathbf{R}_m$, one can easily show that

$$\psi_{kn\sigma}(\mathbf{r}_1 + \mathbf{R}_m, \sigma_1) = e^{i\mathbf{k}\mathbf{R}_m} \psi_{kn\sigma}(\mathbf{r}_1, \sigma_1). \qquad (3.2.4)$$

Figure 3.1 shows schematically the electronic Bloch-type wave function in the crystal.[129]

The corresponding energy eigenfunctions $E_{n\sigma}(\mathbf{k})$ do not depend on σ (if the crystal is not ferromagnetic); hence this index may be omitted. Further, if $\mathbf{k}$ passes through all the values in the first Brillouin zone, allowed by the periodicity conditions, then the dispersion energy dependence on the wave vector $\mathbf{k}$ produces on the energy axis, bands of permitted energies (numbered by the band index n), the so-called permitted bands. These bands either overlap, or are separated by a forbidden energy gap E_G. Thus, the functions $E_n(\mathbf{k})$ describe the *band structure* of the crystal.

The band model discussed above has considerable shortcomings arising from the approximations used. First, the adiabatic approximation does not take into account the vibrations of the nuclei of the system. Yet we have a rich spectrum of nuclear vibrations in OMC, both high-frequency (intramolecular) ones, as well as low-frequency ones (optical and acoustical intermolecular lattice vibrations) (see Chapter 1, Section 1A.4). Second, in the framework of a single-electron approximation the complex problem of multielectron Coulombic interaction is reduced to

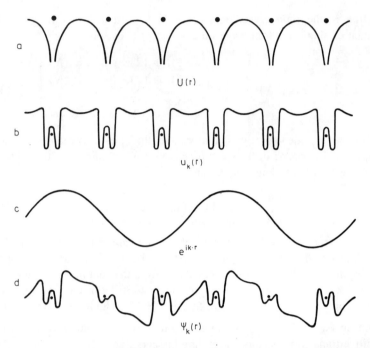

FIGURE 3.1. Schematic representation of electronic wave functions in a crystal, after Ref. 129. (a) Periodic potential $U(r)$ along a chain of atoms in a crystal; (b) corresponding Bloch wave function $u_k(r)$ possessing the lattice periodicity; (c) wave function of a free electron; (d) eigenfunction $\psi_k(r)$ of an electron in a crystal, as represented in the form of a product of a Bloch function (b) and a planar free electron wave (c) (see formula 3.2.3). The function is a complex one. The figure presents only its real part.

that of a single-electron one, in which this interaction is accounted for by means of corrections to the effective periodic potential of the crystal $U(r)$ modulating the wave function of the free electron (see Fig. 3.1). In this case the applicability of the model depends on the shape and width of the electronic bands.

Traditionally, many investigators consider that the presence of translational symmetry of a crystal is a sufficient condition for the applicability of the band theory. However, this condition is necessary, but not sufficient. A condition which is of no less importance is the presence of strong interaction between the particles forming a solid, i.e., the parameter J_{nm} must be sufficiently large. The periodicity of the lattice in a solid permits the use of Bloch wave functions (3.2.3) for the description of the motion of a delocalized electron. But the nature of the interaction between particles in the crystal determines the formation of a band structure $E_n(\mathbf{k})$, the effective bandwidths δE, and other parameters, thus determining the natural limits of the applicability of the band model.[332,351]

The weak interaction forces of Van der Waals type in molecular crystals (see Section 1A.5), i.e., the small value of the parameter J_{nm}, causes a low value of localization energy E_{loc}. As a result, condition (3.1.2) is satisfied, leading to localization of the carrier due to prevailing polarization effects of the local lattice surroundings. The dominating type of polarization, in this case, as will be shown below is the polarization of the *electronic* subsystem of the surrounding, easily polarizable, molecules in Ac-type crystals. This precludes completely the applicability of the simple band theory of single-electron approximation (see Section 3.2.3). The latter has to be replaced by the *polaron theory of multielectron* interaction (see, e.g., Refs. 228, 332, and 351). However, as we shall see in Section 3.2.2, the calculations of the band structure of Ac-type OMC in terms of simple band theory yield results, which, apart from the above-mentioned considerations, are in contradiction with the fundamental notions of the model itself. These are not internally consistent.

3.2.2. Results of Calculations of the Band Structure of Ac-Type Crystals

The first calculations of the band structure of Ac-type crystals were performed by LeBlanc in 1962.[213] He used tight-binding approximation with the aim of a more or less adequate representation of the weak intermolecular interaction forces which cause marked localization of the charge carriers on individual molecules. The single-electron wave functions $\psi_k(r)$ of the crystal were constructed on the basis of linear combinations of single-electron molecular wave functions χ_n:

$$\psi_k(\mathbf{r}) = N^{-1/2} \sum_{n=1}^{N} \exp(i\mathbf{k}\mathbf{r}_n)\chi_n(\mathbf{r}-\mathbf{r}_n). \qquad (3.2.5)$$

Here, the vector $\mathbf{r}_n$ locates the geometrical center of molecule n, and summation is carried out over all N molecules of the crystal. The molecular wave functions, in turn, are constructed according to molecular orbital (MO) theory, namely, as linear combinations of $2p_z$ — atomic orbitals of the Slater type, using Hueckel coefficients of the lowest antibonding and of the highest bonding π-orbital for the electron and the hole band, respectively. The final results of LeBlanc's[213] calculations were very controversial and are difficult to interpret in terms of conventional band model concepts.

The calculated effective electron and hole bandwidths δE in anthracene proved to be extremely small, ~ 0.015 eV, i.e., 0.6 kT for electrons and 0.5 kT for holes at room temperature.

Thaxton *et al.*[382] used a similar approach for calculating the band structure of Nph, Tc, and Pc crystals. They obtained the following effective bandwidths δE for electrons and holes: (0.2–0.3) kT for Nph and (0.5–0.6) kT for Pc at room temperature.

The question now was, whether the band theory concepts are in principle applicable for such narrow bands. LeBlanc himself was critical of the viability of band theory in case of Ac-type crystals. In his opinion the narrowness of the calculated bandwidths speaks in favor of the model of a localized charge carrier rather than a delocalized one in the band.

There have, however, been suggestions that the narrowness of the calculated bands might be due to rather crude approximation in the work of LeBlanc[213] and Thaxton, Jarningen, and Silver.[382]

At a later stage Katz et al.[163] refined LeBlanc's method[213] using, instead of Slater orbitals, more accurate atomic orbitals obtained within the framework of MO self-consistent field theory (MO-SCF). They also improved the accuracy of the calculations. Silbey et al.[328] refined the calculation procedure still further taking into account effects of intermolecular electron exchange and vibronic coupling in the weak coupling scheme.

The refined calculation methods of Katz et al.,[163] as well as those of Silbey et al.[328] produced slightly higher values of the effective bandwidth, lying between 0.05 and 0.2 eV. However, in this case other difficulties arose which cast doubts on the applicability of the approximation itself. An estimate of the mean free path of the electron $\bar{l}$ yielded a value of 3–4 Å for Ac, i.e., turned out to be even smaller than the lattice parameters a_o of the unit cell. These results have a clear physical meaning. The charge carriers are scattered at every site of the lattice, i.e., they are really strongly localized. Such a localization automatically excludes the applicability of traditional band scheme concepts, since electron motion cannot any more be adequately described by means of the wave vector **k** which loses in this case a sound physical meaning.[332,351]

3.2.3. Applicability of a Single-Electron Band Model Approximation

The inconsistency of band structure calculations of Ac-type crystals of Refs. 163 and 328 induced the authors of Ref. 117 to perform a detailed analysis of the limits of applicability of the band model of one-electron approximation. The results of this analysis have been thoroughly discussed in Refs. 332 and 351.

In addition to electron exchange effects (the importance of accounting for them has already been demonstrated by Silbey et al.[328]), Glaeser and Berry[117] consider also the effects of electronic polarization of molecules next to the excess charge to the transfer processes.

According to Glaeser and Berry[117] the wave function of the crystal ψ_1 in localized representation may be constructed as an antisymmetrized product of molecular wave functions. In this approach, one of the molecules is considered as a negative or a positive ion, and the rest as neutral molecules perturbed (polarized) by the molecular ion. Symbolically the wave function ψ_1, corresponding to an electron or hole positioned on the molecule i in the lattice, can be represented in the following form:

$$\psi_l = A\psi_i(2a\pm 1) \prod_{j(\neq i)} \psi_j^{(i)}(2a), \qquad (3.2.6)$$

where a denotes occupied orbitals of neutral molecules, $\psi_i(2a \pm 1)$ is the wave function of the corresponding molecular ion, but $\psi_j^{(i)}$ is the wave function of the jth neutral molecule in the field of the given ion.

In the case of Bloch representation the tight-binding wave functions may be constructed as linear combinations of localized crystal wave functions ψ_1:

$$\psi_\pm(\mathbf{k}) = \sum_l (\pm 1)^l \exp(i\mathbf{k r}_l)\psi_l. \qquad (3.2.7)$$

As mentioned earlier, the Glaeser-Berry approach shows certain features of a multi-electron approximation; the wave functions of neutral molecules take into account the effect of electronic polarization on charge carrier transfer and are, in essence, multielectron wave functions.

The charge carrier eigen-energy dependence on the wave vector $\mathbf{k}$ in the conduction band is given by

$$E_k^\pm(\mathbf{k}) = \sum_j \exp(i\mathbf{k r}_j)J_{ij}, \qquad (3.2.8)$$

where J_{ij} is the charge carrier transfer integral which may be represented as follows:

$$J_{ij} = (E_R + \Delta E_R + E_{es})S. \qquad (3.2.9)$$

Here E_R is the resonance term (exchange interaction) between neighboring molecules i and j; ΔE_R is a correction to the resonance term that results from the use of polarized electron orbitals of molecules i and j; E_{ST} is the term of long-range electrostatic interaction between the excess charge and the dipoles induced on neighboring molecules and S is the overlap factor of the polarized electron orbitals of neighboring molecules.

Calculated band structures for an excess electron and a hole in anthracene are presented in Figs. 3.2 and 3.3. These bands are seen to possess reasonable band-width δE, as compared to kT, i.e., of the order of 0.1–0.2 eV (except the c' direction for an electron). The band structure of Ac reveals strongly pronounced anisotropy, depending on crystallographic direction, as well as on the sign of the charge carrier. These results are in good agreement with earlier work (see Refs. 163 and 328). However, in the light of our discussion on localization other results of calculations in Ref. 117 are more important. As shown in Figs. 3.2 and 3.3, electronic polarization diminishes the effective bandwidth δE by ca. 40–50%. In the physical sense this means increased localization of the charge carrier. Petelenz[284] also reports in his studies of model Ac dimers that polarization actually lowers the transfer integral value by about 50%.

Čápek[57] investigated the effect of electronic polarization on the dependence of energy dispersion of a localized electron $E(\mathbf{k})$ using a simple isotropic molecular model. The author considers electronic polarization self-consistent with electron

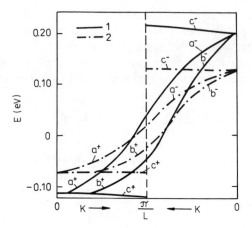

FIGURE 3.2. Calculated structure for an excess electron in an anthracene crystal according to Ref. 117. The figure shows the energy E dependences of an electron on the wave vector **k** in the a^{-1}, b^{-1}, and c^{-1} crystallographic directions of the lattice: - without taking into account the electronic polarization; -.- with taking into account electronic polarization (calculated for oscillator strength $f = 2.0$).

motion and shows that in molecular solids possessing narrow conduction bands ($\delta E \ll \Delta E_{ex}$, where ΔE_{ex} is the excitation energy of an electronic exciton) polarization narrows the effective bandwidth. Tentative estimates of the author show that the correction factor ΔW on the given narrowing equals $\sim 1/2 \, \delta E$ in Ac-type crystals, i.e., it has a value of ca. 0.03–0.1 eV.

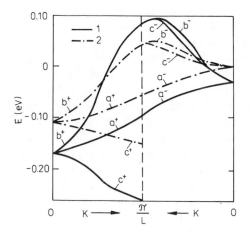

FIGURE 3.3. Calculated band structure for an excess hole in an anthracene crystal according to Ref. 117. The figure shows the energy E dependence of a hole in the a^{-1}, b^{-1}, and c^{-1} directions of the lattice: - without taking into account of the electronic polarization; - · · · · - taking into account the electronic polarization (calculated for oscillator strength $f = 2.0$).

Thus we see that in narrow-band molecular solids polarization effects favor further localization of charge carriers. On the other hand, in broad-band solids, in which we have $\delta E \geqslant \Delta E_{ex}$, the polarization shell does not begin to form around the charge carrier during its motion, and it retains fully its state of delocalization in accordance with the inequality (3.1.1).[57] To check the applicability of the band model of single-electron approximation the following criteria are often used.[117,332,351]

In the first place, the calculated mean free path of the charge carrier $\bar{l}$ must be considerably larger than the lattice parameter a_o, i.e.

$$\bar{l} \gg a_o. \tag{3.2.10}$$

Second, the calculated relaxation time for the free path of the carrier τ_0 must considerably exceed the critical lifetime τ which is limited by the uncertainty principle:

$$\tau_0 \gg \tau \approx \hbar/\delta E. \tag{3.2.11}$$

To check the adequacy of the band model, the $\bar{l}$ and τ_0 values, calculated by Glaeser and Berry, were used. First, the calculated value of the mean free path of the carrier, $\bar{l}$ is actually smaller than the lattice parameter a_o. Hence, condition (3.2.10) is obviously not valid. Criterion (3.2.11) is practically also invalid, since the relaxation time τ_0 for the mean free path exceeds by a factor of only 2–3 the critical value of τ according to the uncertainty principle (for details see Ref. 351). On the basis of the above analysis, using their own data as well as those of earlier calculations, see Ref. 163 and 328, Glaeser and Berry[117] came to the conclusion of the inapplicability of the band model to molecular crystals of the Ac type. These results testify that the charge carrier interacts so strongly with lattice vibrations that the wave vector **k** is not conserved and the coherent wavelike motion of the carrier is replaced by noncoherent hopping motion of a Brownian-type particle. In other words, the motion of the carrier is, in this case, randomized by scattering from each site of the lattice, and the stochastic hopping model is more suitable for the description of the physical picture than the band model. Thus in Ac-type crystals electronic polarization, in conjunction with the interactions of the charge carrier with inter- and intramolecular vibrations (molecular and lattice polarization), determines the localization of the carrier, in accordance with inequality (3.1.2).

As indicated in Section 1B.3, there is often confusion regarding the term "band model." This term may be used either to denote the energy *band structure* $E = E(\mathbf{k})$ (see Figs. 3.2 and 3.3) or the *bandlike* Boltzmann theory of charge carrier *transport*. The inequality (3.2.10) is actually the most important criterion for the validity of the bandlike transport model [see Eq. (1B.3.7a)].

In this aspect, attempts to estimate the $\bar{l}$ value from band structure calculations in Refs. 117 and 328 seem not correct. However, it was proved later (see Chapter 7, Figs. 7.15 and 7.16) that in several aromatic crystals at higher temperatures ($T \geqslant 150$ K) the mean free path $\bar{l}$ of the charge carrier actually becomes practically equal to the lattice constant a_o, namely,

$$\bar{l} \approx a_o. \tag{3.2.12}$$

Therefore it would be more correct to say that in the temperature range where the condition (3.2.12) is fulfilled, the bandlike transport model is not applicable. On the other hand, fulfillment of the condition (3.2.12) does not necessarily mean that a simple hopping model is adequate.

Historically, the hopping model was handicapped with respect to the band theory approach by a number of circumstances. In the case of charge carrier hopping, at least a dozen different hopping models have been proposed, based on different physical principles and approximations. Unfortunately, the majority of these models give only a qualitative description without the possibility of computational treatment.[332]

As an alternative to the band model Glaeser and Berry[117] consider an improved variant of the hopping model. As a rule, under conditions of strong scattering of charge carriers, when we have $\bar{l} \approx a_o$, preference is given to the traditional hopping model of a small polaron, according to which charge carrier transfer from one site of the lattice to another one proceeds by a thermally activated process[12,247,285,288] (see also Section 3.6.1). This model, developed by Holstein[136] for simple molecular systems, requires a mobility that increases exponentially with temperature, i.e., $\mu \sim \exp[-E_a/kT]$ (see also Ref. 136a). However, in Ac-type crystals carrier mobility exibits a different kind of dependence, namely, $\mu \sim T^{-n}$, where $n > 1$ (or, in some cases constant within a certain temperature range) (see Ref. 288, p. 337). Hence, an activated hopping model is, on principle, inapplicable for Ac-type OMC (for details see Chapter 7).

Instead of an activation model Glaeser and Berry[117] offer a dynamic model of charge carrier tunneling through intermolecular barriers. Here the "static" potential barriers, introduced by Eley, Inokuchi, and Willis,[102] (see also Ref. 122), are replaced by "dynamic" intermolecular barriers. According to this alternative hopping model the probability of the carriers' penetration through the barrier is modulated by intermolecular vibrations. It has been shown[117] that the probability of tunneling or resonance charge transfer between neighboring molecules is a sensitive function of the vibrational state of the lattice. Thus, for instance, in excited vibrational state, the distance between neighboring molecules may considerably decrease, as compared to that in ground state. Since the transfer integral J_{ij} depends exponentially on intermolecular distance, its effective value will increase with increasing vibrational energy of the lattice. Such a kind of tunneling may be formally described by a probability function. In the latter not only the distribution of coordinates of neighboring molecules around the localized charge carrier must be given, but also the quantum numbers of intermolecular vibrations. To be able to consider such a range of probabilities a detailed information on the phonon spectrum is necessary, as well as estimated data on the extent of changes in the corresponding transfer integrals.

Owing to lack of the necessary quantitative data the authors of Ref. 117 used the approximation of discrete probability distribution, based on a transfer integral, as calculated for the case of equilibrium configuration in the crystal. In this approach the problem is reduced to that of a simple hopping theory of resonance transfer, in which one considers stochastic charge carrier hopping from one localized state to another.

In addition to the proof of the inapplicability of the simple band model in the case of OMC there are some more aspects in the analysis of the work of Glaeser and Berry that we will use further.

First, the applicability of a hopping model of the type of tunneling transfer of a localized charge carrier in Ac-type crystals has been used as an argument in favor of the adequacy of the modified Sano-Mozumder's model (see Refs. 344, 355, and 356). The latter is based on the concept of charge carrier as a classical quasiparticle for computer simulation of carrier thermalization and transport processes in OMC crystals (see Chapters 6 and 7). As will be shown in Chapters 6 and 7, one of the most important results of this work consists in the fact that it is necessary to consider a charge carrier as a quasiparticle of the molecular nearly small-radius polaron type. The latter presumably moves, in the above description, by stepping via tunneling, thus presenting qualitatively a close analogue to the Glaeser's and Berry "hopping" model.

Second, the results of calculations of transfer integrals J_{ij} in Ac[117] allow us to estimate the mean localization (residence) time τ_h of a charge carrier on a separate lattice site and to compare it to the average time τ_e necessary for electronic polarization of the surrounding molecules.

The transfer integral J_{ij} determines the mean time interval τ_h between separate hopping acts, i.e., the time interval of localization in accordance with the quantum mechanical uncertainty principle:

$$\tau_h \approx \hbar/J_{ij}. \tag{3.2.13}$$

A noteworthy circumstance consists in the extremely small value of J_{ij} (see Ref. 117) which generally does not exceed 0.03 eV and directly reflects the most characteristic properties of the molecular crystal, namely, the weak intermolecular interaction forces (see Chapter 1A). As a result, the time interval between separate hoppings which, according to Eq. (3.2.13), is inversely proportional to J_{ij}, is considerable compared with the typical time scale of fast electronic processes, namely, $\tau_h \gtrsim 10^{-14}$ s (see Fig. 3.9).

The value of τ_h can also be estimated experimentally from measured microscopic mobility μ_0 values of charge carriers. Such an estimate may be obtained from the formula[299]

$$\tau_h = \frac{e}{kT} \frac{r_{ij}^2}{6\mu_0}. \tag{3.2.14}$$

The calculated and experimental estimates of τ_h coincide satisfactorily in the case of hopping onto the nearest neighboring molecules, i.e., from the site (0,0,0) onto (1/2,1/2,0) and (0,1,0) and is of the order of $\tau_h \approx 10^{-14}$ s. However, in the directions **a** and **c**, where $r_{ij} > 6$ Å the calculated τ_h values are obviously overestimated ($\tau_h > 10^{-12}$ s) and do not correspond to the experimental ones which do not reveal such an anisotropy (see for details Ref. 351, p. 124).

The above discussion suggests that the experimental τ_h values appear to be more reliable. One may therefore consider that in Ac crystals $\tau_h = 10^{-14}–10^{-13}$ s (see Fig. 3.9).

The relaxation time necessary for the polarization of the electronic orbitals τ_e of the molecules of the crystal is given by[186a,332]

$$\tau_e \approx \hbar / \Delta E_{\text{ex}}, \qquad (3.2.15)$$

where ΔE_{ex} is the excitation energy of the electronic exciton.

In the first approximation in molecular crystals ΔE_{ex} corresponds to the excitation energy ΔE_{ex} of a singlet S_1-exciton. Since ΔE_{S_1} has a value of 2–4 eV in typical molecular crystals, we get $\tau_e = 10^{-16}–10^{-15}$ s.[332,351]

Thus, the mean hopping time of τ_h of a charge carrier in an anthracene crystal is larger by two orders of magnitude than the time τ_e necessary for polarization of the electronic subsystem of the molecules surrounding the localized charge carrier in the crystal, i.e.,

$$\tau_h \gg \tau_e. \qquad (3.2.16)$$

The inequality (3.2.16) is a necessary and sufficient condition for electronic polarization of the crystal by the charge carrier. It is also a criterion for the necessity of considering electronic polarization in the investigation of charge carrier transfer, and in the determination of the energy structure of conduction states in a solid.

As may be seen from the expression (3.2.14), only at $\mu_0 \geqslant 100$ cm^{2}Vs and the value $\tau_h \leqslant \tau_e$ may electronic polarization be neglected. The charge carrier is sufficiently delocalized under these conditions.

3.3. ELECTRONIC POLARIZATION IN OMC. PHENOMENOLOGICAL REPRESENTATION

In Section 1B.4 the electronic polarization phenomena in organic crystals and the formation of an electronic polaron around a charge carrier were treated in terms of a quantum field theory.

For the general reader in this section we give a phenomenological representation of electronic polarization in OMC based on visual models and on a visualized mode of thinking. This approach, we hope, will also serve as a tutorial link between dynamic theories (Section 3.4) and microelectrostatic methods of calculation of electronic polarization energies (Section 3.5).

A visual picture of the process of electronic polarization of a molecular crystal, i.e., the formation of induced dipoles $\mathbf{d}_i$ on neutral molecules of the crystal in the field of a localized positive charge carrier, is schematically shown in Fig. 3.4b. This process is determined by the condition $\tau_h \gg \tau_e$ (Fig. 3.4a) which allows to consider the localized charge carrier as "standing" during the polarization process and actu-

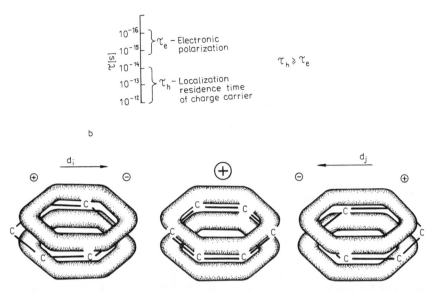

FIGURE 3.4. Schematic representation of electronic polarization of a molecular crystal by a localized charge carrier:[332,351] on the lower part (b) of the figure the process of electronic polarization — formation of induced dipoles d_j on neutral molecules of the crystal in the field of a localized positive charge — is visualized. The electronic polarization on the figure is conditionally represented as displacement of the π-orbitals of neutral molecules in the field of the charge: as a result induced dipole moments d_j are formed. On the upper part of the figure typical time scales of electronic polarization τ_e and charge carrier localization τ_h times in Ac-type crystals are shown (see Fig. 3.9).

ally reduces the problem to a microelectrostatic one. Figure 3.5 shows a schematic representation of the process of formation of an electronic polarization "cloud" around the localized charge carrier.

Thus, during its motion the charge carrier moves together with its polarization cloud either coherently in form of a polaron wave, or by incoherent hopping. Such a state of an electron was first described by Toyozawa[386] and was named *electronic polaron* (see Section 1B.4). It ought to be stressed that in the case of electronic polarization only electronic orbitals of surrounding molecules are "deformed." Accordingly, electronic polarization is sometimes called deformational polarization. "Displacement," i.e., deformational polarization of the nuclear subsystems inside the molecules (i.e., molecular polarization) or of the nuclear subsystem inside the lattice as a whole (i.e., lattice polarization) require considerably longer time (see Fig. 3.9). These polarization phenomena are discussed in Section 3.6.

Electronic polarization is the most important electronic process determining the self-energy of charge carriers in organic solids. Since the electronic polarization energy P is of the order of 1–1.5 eV, it is the basic factor determining the position of ionized-state levels in the energy diagram of OMC (see Chapter 4). For this reason this and the following two sections will be devoted to electronic polarization and to its energy calculations and measurements.

FIGURE 3.5. Schematic representation of the process of formation of an electronic polarization "cloud" around a localized negative charge carrier[332,351] in a hypothetic molecular crystal consisting of benzene-type molecules. Polarization of π-orbitals of neutral molecules in the charge field is conditionally shown as displacement of the "aromaticity ring" of the given molecules; r_i is the distance between the center of the molecule with a localized charge carrier and the molecule i with the induced dipole moment d_i.

As already mentioned the electronic polarization of a molecular crystal by a charge carrier may be treated in terms of dynamic and microelectrostatic approaches.[332,351] The static approximation is based on the inequality (3.2.16) which is obviously valid in Ac-type crystals (see Figs. 3.4 and 3.9). Hence, in evaluations of electronic polarization energy P, it is reasonable to regard the problem of calculating the interaction of the charged molecule with the electrons of the surrounding neutral molecules as a static one.

A localized charge carrier mainly affects the π-orbitals of conjugating molecules which have high polarizability (cf. Fig. 3.4); in other words, a dipole moment d_i is induced on the neighboring neutral molecules. This dipole moment is proportional to $1/r_i^2$, where r_i is the distance between the center of the molecule with the localized charge and that of the ith neutral molecule (see Fig. 3.5). The energy of attraction between the charge and the dipole, in its turn, is also proportional to $1/r_i^2$. Hence, the energy of interaction between the charge and the induced dipoles, i.e., the polarization energy P is proportional to $1/r_i^4$. In addition, in molecular crystals interaction of type charge (q)-induced dipole (d_i), i.e., term W_{q-d} is proportional to the mean polarizability α of neutral molecules and constitutes the basic part of the total electronic polarization energy P (see Section 3.5.1).

The basic concepts of microelectrostatic approximation originated in the classical paper by Mott and Littleton.[249] The excess carriers were treated as static point charges localized on definite ions, and the corresponding electronic polarization of the valence electrons of neighboring ions was calculated by classical electrostatics methods, using ionic polarizability values, dielectric constants, and lattice constants. Subsequently static approximation methods were successfully applied in the calculations of electronic polarization energy in ionic crystals[300] and in noble-gas crystals.[94] The first serious attempts to calculate electronic polarization energy P in molecular crystals in terms of microelectrostatic approximation were made by Lyons and coworkers in the early sixties[221,223] (see also Ref. 122). They were later refined by a number of authors (see Refs. 288 and 332). Achievements in more recent years in this field will be discussed in detail in Section 3.5.

Before we discuss in greater detail refined microelectrostatic calculation methods we must first consider the theoretical viability and limits of validity of the microelectrostatic approximation itself.

The dynamic approximation is more strict and better founded theoretically. It is, however, a rather complicated method, since it reduces the problem to that of quantum electrodynamics of multielectron systems (see Section 1B.4).

If the motion of the excess charge is not too fast in the solid, it manages to interact with the electronic subsystem of the lattice, and electrons of the surrounding atoms or molecules screen the excess charge off. This kind of reaction of the electronic subsystem may be described by a frequency (or wave vector) dependence of the dielectric constant. In metals an infinitely small energy is required for excitation of the electronic subsystem, and, correspondingly, screening of the excess charge is practically ideal. In the case of insulators to which the greatest part of molecular crystals belong, the situation is more complex. In pure insulators the electronic polarization is determined, in quantum mechanics, by virtual transitions, i.e., dielectric screening is a result of virtual transition of atoms or molecules of the crystal from their ground states to excited quantum states and vice versa. In the most general form the dynamic theory of electronic polarization was proposed by Toyozawa in 1954.[386] His theory is based on the assumption that electronic polarization of an insulator is determined by virtual transition of the system of N valence electrons to excited states which may be approximately represented as the first singlet exciton band.

Such dynamic approximation thus permits delocalization of the charge carrier, "clothed" by the polarization cloud (see Fig. 3.5) in form of a Bloch wave. Under such delocalization conditions, the charge carrier carries the polarization field with it, forming a quasiparticle — an *electronic polaron*.[386] In other words, owing to Coulomb interaction between the charge carrier and the valence electrons of the crystal quantized excited states of atoms or molecules are created, i.e., electronic (Frenkel) *excitons*. We therefore say that the excess carrier interacts in its motion with the quantum field of electronic excitons, forming an electronic polaron (see Section 1B.4) in Chapter 1.

A number of earlier dynamic theories on electronic polarization in insulators have been discussed in detail in the review by Wang, Mahutte, and Matsura.[405]

The dynamic electronic polaron theories[386,405] present, both from the point of view of terminology as well as in mathematical formalism, an analogue to the well-developed lattice polaron theories (cf., e.g., Refs. 12, 247, and 285). The basic difference consists in the circumstance that, in the case of lattice polaron theory, atoms or molecules are displaced from equilibrium position (see Fig. 3.11), i.e., the charge interacts with the phonon field of the crystal. On the other hand, in the case of electronic polarization only the electronic orbitals of the molecules forming the crystal are "deformed" (see Figs. 3.4 and 3.5), i.e., the charge interacts with the field of electronic excitons. The characteristic relaxation times also differ accordingly, equaling $\tau_e \approx 10^{-16}-10^{-15}$ s in the case of electronic polarization, and $\tau_l > 10^{-14}$ s in the case of lattice polarization in Ac-type crystals (see Fig. 3.9).

The question of the feasibility of static approximation in calculations of electronic polarization has been dealt with in detail by Fowler.[109] The author applied Toyozawa's electronic polaron theory[386] and developed on its basis the theory of the r-dependent dielectric function.[127] Calculating the electronic polarization of a number of high-ohmic solids, such as alkali- and silver halides and noble-gas molecular crystals, Fowler[109] showed that the dynamic theories discussed above may actually be reduced to the Mott-Littleton static approximation.[249] Consequently, the classical electrostatic approach is fully justified for description of electronic polarization in materials with low mobility of charge carriers. In any case, the "classical" electronic polarization energy values may serve as the upper limit of the true values of P, since in the case of fast motion of the excess charge carrier the electronic polarization energy decreases. In the limiting case, when the excess carrier moves so fast that the electronic orbitals of the neighboring molecules are not polarized appreciably, the electronic polarization energy may approach zero value.[109] The inequality (3.2.16) might therefore be considered as a suitable criterion of the applicability of the static approximation. If τ_h is substantially larger then τ_e, one may regard electronic polarization as practically "inertialess" and apply static approximation in the calculation of P.

In Ac-type molecular crystals these conditions are apparently always valid for thermalized and epithermal charge carriers. Only in the case of "hot" charge carriers does the value of τ_h approach the same order as that of τ_e, and the static approximation has to be replaced by a more complex quantum dynamic approach. In another limiting case, when $\tau_e \geq \tau_h$, the multielectron approximation reduces itself to a single-electron one.

As may be seen, theoretically the most difficult case is that of the intermediate "hot" (but not too "hot") charge carriers which may form a broad "band" of polaron state densities of the order of 1–2 eV (see Fig. 4.11).

It is noteworthy that, in the static limit, the results of the dynamic theories of Toyozawa and of Haken-Schottky coincide with the value of the static Mott-Littleton zero-order approximation. Static approximations of higher order usually yield more precise results than dynamic theories. This is easily understandable, since, as already pointed out, it is difficult, within the framework of dynamic theories, to account for certain more refined effects, such as dipole-dipole and charge-quadrupole interactions.

This becomes particularly evident if one applies Toyozawa's theory to electronic polarization in molecular crystals.[28] A treatment based on the tight-binding approximation of an excess carrier localized on a particular molecule in the crystal, automatically reduces the problem to the quasistatic approach.

At a later stage the problem of the applicability of the microelectrostatic approximation in molecular solids was studied by Čápek.[55] He showed that the quantum mechanical treatment of electronic polarization in molecular crystals (with full account for dipole-dipole interaction) leads (if the small quantum mechanical correction term W_{quant} is considered[56,340]) to expressions that are identical with the corresponding expressions of microelectrostatic approximation. He thus presented additional support for the equivalence of the dynamic and the microelectrostatic approximations for molecular crystals.

In the next section we will discuss the dynamical theory of electronic polarization in some detail, as well as the problem of the correspondence between the dynamic and microelectrostatic approaches.

3.4. DYNAMIC APPROACHES TO ELECTRONIC POLARIZATION IN OMC

As a starting point for the present discussion we shall use the results of a more general treatment of electronic polarization around negative (Sect. 1B.4) and positive (Sect. 1B.5) charge carriers in organic molecular crystals.

3.4.1. Correspondence Between Dynamic and Microelectrostatic Approaches

As previously mentioned, Toyozawa[386] introduced the notion of the *electronic polaron* as a cloud of polarization of π electron orbitals of neighboring molecules (described in the crude basis as virtual excitations of Frenkel molecular excitons) around a "standing" charge. This quasiparticle (the central charge plus the cloud of virtual excitons) can, in principle, move but here, for the sake of simplicity, we will take the central charge as immobile. Our goal is to link these modern methods of the quantum field theory (as applied to molecular crystals) with standard microelectrostatic approaches which have been very successfully used for electronic polarization calculations in OMC.

These microelectrostatic approaches go back to Mott and Littleton[249] who used the method for polar solids. In Refs. 55 and 56 this classical method was shown to be equivalent with a variational estimate of the ground state energy of the electronic polaron around an electron or a hole as treated in Sections 1B.4–5. We extend this theory for solution of the problems discussed in the present section.

First, let us refer to Sections 1B4–5 to show that for the dynamical approach to the electronic polaron formation, one must solve the Schrödinger equation

$$H|\Psi\rangle = E|\Psi\rangle \qquad (3.4.1)$$

with

$$H = H_1 + H_2 + H_3 = \sum_{p(\neq 0)} \sum_\lambda E_\lambda \bar{a}^\dagger_{p\lambda} \bar{a}_{p\lambda} + q \sum_{p(\neq 0)} \sum_\lambda \varphi_{p\lambda}(\bar{a}_{p\lambda} + \bar{a}^\dagger_{p\lambda})$$

$$+ \frac{1}{2} \sum_{p \neq q} \sum_{\lambda\lambda'} w_{p\lambda,\,q\lambda'}(\bar{a}_{p\lambda} + \bar{a}^\dagger_{p\lambda})(a_{q\lambda'} + \bar{a}^\dagger_{q\lambda'})$$

$$(3.4.2)$$

[see Eq. (1B.4.3)]. Here, for simplicity, we have [as compared to Eq. (1B.4.3)] assumed that molecules are equivalent, i.e., the local excitation energy of a single exciton (ΔE_p) is independent of p. However we have included the possibility of excitons in higher excited states (denoted by λ). Thus, instead of ΔE_p in Eq. (1B.4.3), we have E_λ in Eq. (3.4.2). Further, we have assumed that our spinorbitals are real (i.e., we have neglected the spin-orbit coupling) and set

$$\varphi_{p\lambda} = -\mathbf{d}_\lambda \mathbf{r}_p / |\mathbf{r}_p|^3,$$

$$w_{p\lambda,\,q\lambda'} = \left[\mathbf{d}_\lambda \mathbf{d}_{\lambda'} - 3 \frac{[\mathbf{d}_\lambda(\mathbf{r}_p - \mathbf{r}_q)][\mathbf{d}_{\lambda'}(\mathbf{r}_p - \mathbf{r}_q)]}{|\mathbf{r}_p - \mathbf{r}_q|^2} \right] \frac{1}{|\mathbf{r}_p - \mathbf{r}_q|^3}, \quad (3.4.3)$$

where $\mathbf{d}_\lambda = \langle p0|\mathbf{d}|p_\lambda\rangle = \langle p_\lambda|\mathbf{d}|p0\rangle$ is the transition dipole moment $p0 \leftrightarrow P_\lambda$ (off-diagonal matrix element of the dipole moment) for any p. We also assume, as usual, that the central charge q (which might be equal to $\pm e$) is located on the molecule at the origin $(0,0,0)$. So, for any molecule p (except for $p = 0$),

$$\mathbf{d}_p \approx \sum_\lambda \mathbf{d}_\lambda(\bar{a}_{p\lambda} + \bar{a}^\dagger_{p\lambda}) \qquad (3.4.4)$$

is its *electric dipole moment* [provided that, for simplicity, (i) we neglect its matrix elements between two different excited spinorbitals and (ii) assume that we have nonpolar molecules].

For solving Eq. (3.4.1) by a variational method, let us choose our trial function as

$$|\Psi\rangle = \prod_{m(\neq 0)} \left(u_m|\varphi_m^{(0)}\rangle + \sum_\lambda v_{m\lambda}|\varphi_m^{(\lambda)}\rangle \right) |\varphi_0^q\rangle \qquad (3.4.5)$$

with $|\varphi_m^{(\lambda)}\rangle$, $\lambda = 1,2,\ldots$ being excited spinorbitals of molecule m, and $|\varphi_0^q\rangle$ being the ground state of the molecular ion at the origin. Here, u_m and $v_{m\lambda}$ are normalized variational parameters with the condition

$$|u_m|^2 + \sum_\lambda |v_{m\lambda}|^2 = 1. \qquad (3.4.6)$$

This gives (with real $u_m \geqslant 0$, $v_{m\lambda}$) the mean value of the energy of the system

$$E = \langle \Psi | H | \Psi \rangle = \sum_{n(\neq 0), \lambda} E_\lambda v_{n\lambda}^2 - q \sum_{n(\neq 0)} \langle \mathbf{d}_n \rangle \mathbf{r}_n / |\mathbf{r}_n|^3 + \frac{1}{2} \sum_{m \neq n} \left[\langle \mathbf{d}_m \rangle \langle \mathbf{d}_n \rangle \right.$$

$$\left. - 3 \frac{[\langle \mathbf{d}_m \rangle (\mathbf{r}_m - \mathbf{r}_n)][\langle \mathbf{d}_n \rangle (\mathbf{r}_m - \mathbf{r}_n)]}{|\mathbf{r}_m - \mathbf{r}_n|^2} \right] \Big/ |\mathbf{r}_m - \mathbf{r}_n|^3 \qquad (3.4.7)$$

where

$$\langle \mathbf{d}_m \rangle = 2 \sum_{\lambda(\neq 0)} \mathbf{d}_\lambda v_{m\lambda} \left[1 - \sum_{\lambda'} v_{m\lambda'}^2 \right]^{1/2}. \qquad (3.4.8)$$

Further, the variational condition $\delta E = 0$ yields

$$E_\lambda v_{m\lambda} = \left[q\mathbf{r}_m / |\mathbf{r}_m|^3 - \sum_{n(\neq m)} \left[\langle \mathbf{d}_n \rangle - 3 \frac{(\mathbf{r}_m - \mathbf{r}_n)[\langle \mathbf{d}_m \rangle (\mathbf{r}_m - \mathbf{r}_n)]}{|\mathbf{r}_m - \mathbf{r}_n|^2} \right] \Big/ |\mathbf{r}_m \right.$$

$$\left. - \mathbf{r}_n|^3 \right] 1/2 \frac{\partial \langle \mathbf{d}_m \rangle}{\partial v_{m\lambda}}. \qquad (3.4.9)$$

These are a complicated set of nonlinear equations for $v_{m\lambda}$. From Eqs. (3.4.8–9), it follows, however, that $\langle \mathbf{d}_m \rangle \sim q$, $q \to 0$. So, to the linear order in q (which is well fulfilled for $q = \pm e$ and means negligible influence of higher nonlinear polarizabilities)

$$1/2 \frac{\partial \langle \mathbf{d}_m \rangle}{\partial v_{m\lambda}} \approx \mathbf{d}_\lambda. \qquad (3.4.10)$$

As a result we obtain

$$\langle \mathbf{d}_m \rangle_i = \sum_{j=1}^{3} \alpha_{ij} (\vec{E}_m)_j. \qquad (3.4.11)$$

Here

$$\vec{E}_m = q\mathbf{r}_m / |\mathbf{r}_m|^3 - \sum_{n(\neq m)} \left[\langle \mathbf{d}_n \rangle - 3 \frac{(\mathbf{r}_m - \mathbf{r}_n)[\langle \mathbf{d}_n \rangle (\mathbf{r}_m - \mathbf{r}_n)]}{|\mathbf{r}_m - \mathbf{r}_n|^2} \right] \Big/ |\mathbf{r}_m - \mathbf{r}_n|^3$$

$$(3.4.12)$$

is the total electric field at molecule m at $\mathbf{r}_m$ as a combination of the field of the molecular ion at the origin $(0,0,0)$ and that due to induced dipoles of other molecules. The quantity

$$\alpha_{ij} = 2 \sum_{\lambda} (\mathbf{d}_\lambda)_i (\mathbf{d}_\lambda)_j / E_\lambda \qquad (3.4.13)$$

is the linear molecular polarizability tensor.

It is interesting to notice that a classical interpretation may be given to the second and the third terms on the right-hand side of Eq. (3.4.7). It would be satisfying to show that it is possible to ascribe a classical meaning also to the first term of Eq. (3.4.7):

$$\sum_{n(\neq 0),\,\lambda} E_\lambda v_{n\lambda}^2 \approx 1/2 \sum_{n(\neq 0)} \sum_{i,\,j=1}^{3} \langle \mathbf{d}_n \rangle_i (\alpha^{-1})_{ij} \langle \mathbf{d}_n \rangle_j. \qquad (3.4.14)$$

Unfortunately, in general, one cannot verify Eq. (3.4.14) as an exact equality. However, it may be regarded as a highly plausible assumption, and thus it gives the Eq. (3.4.7) the meaning of E_{class}, equal to the sum of classical microelectrostatical energy and the standard energy gained in polarization of the surrounding molecules [see the right-hand side of Eq. (3.4.4)]:

$$E_{\text{class}} = \sum_{n(\neq 0)} 1/2 \sum_{i,j=1}^{3} (\mathbf{d}_n)_i (\alpha^{-1})_{ij} (\mathbf{d}_n)_j - q \sum_{n(\neq 0)} \mathbf{d}_n \mathbf{r}_n / |\mathbf{r}_n|^3$$
$$+ 1/2 \sum_{m \neq n} \left[\mathbf{d}_m \mathbf{d}_n - 3 \frac{[\mathbf{d}_m (\mathbf{r}_m - \mathbf{r}_n)][\mathbf{d}_n (\mathbf{r}_m - \mathbf{r}_n)]}{|\mathbf{r}_m - \mathbf{r}_n|^2} \right] \Big/ |\mathbf{r}_m - \mathbf{r}_n|^3.$$

$$(3.4.15)$$

The second term on the right-hand side of Eq. (3.4.15) gives the interaction energy of charge q with the induced dipoles (term W_{q-d}); the third term gives the dipole-dipole interaction energy (W_{d-d}); while the first term describes the self-energy of induced dipoles; α^{-1} being the reciprocal molecular polarizability tensor [reciprocal to α in Eq. (3.4.13)].

Now, if we minimize the Eq. (3.4.15) under the assumption that we identify the classical induced dipole moment $\mathbf{d}_m$ with the mean value of the quantum dipole moment $\langle \mathbf{d}_m \rangle$:

$$\mathbf{d}_m \equiv \langle \mathbf{d}_m \rangle, \qquad (3.4.16)$$

the minimization yields Eqs. (3.4.11) and (3.4.12).

The results obtained justify the application of the classical microelectrostatic methods of electronic polarization energy calculations in the limit of a "standing" localized charge carrier and confirm the correspondence between the dynamic and microelectrostatic approaches.

Simultaneously, these results also yield a possibility of estimating quantum corrections as corrections to the simple variational estimate with our trial function (3.4.5). One should realize that the variational estimate, the classical value of energy determined in the above way, yields the energy difference

$$P \equiv P_q = E(q) - E(q = 0) \equiv E|_q - E|_{q=0} \qquad (3.4.17)$$

called the polarization energy P of charge q. According to our definition, (3.4.17) it is always *negative* ($P < 0$), having the meaning of the difference of the ground state energy of the molecular ion in the lattice and that of the same molecular ion in the gas phase. One should also notice that changing the sign of q and, simultaneously, that of all d_m does not change $E(q)$. Thus, the energy of electronic polarization around electron and hole gives (within our treatment, neglecting all higher order multipoles) the same value of P. This symmetry with regard to the change of the sign of a charge carrier is a characteristic feature of the charge-induced dipole interaction term W_{q-d} (see Section 3.5.1).

3.4.2. Quantum Corrections to the Polarization Energies

Čápek et al.[56,340] have studied the possible impact of some quantum effects on the value of the static "classical" polarization energy P_{class} and the necessity to introduce a quantum correction term ΔP_{quant} to P_{class}.

There may be several dynamic quantum effects giving some contribution to the quantum correction term ΔP_{quant}. First, there may be effects due to perturbation (polarization) of the Van der Waals coupling of neutral molecules in the vicinity of the molecular ion. However, it has been shown[56,340] that the main contribution to ΔP_{quant} is due to the dynamical polarization of the molecular ion itself, i.e., due to dynamical fluctuations of the dipole moment of the molecule with the localized charge carrier. The physical origin of this correction is following.

In symmetric configurations, the state of the molecular ion and surrounding neutral molecules may be considered as a linear combination (superposition) of several local configurations. In any of these configurations the dipole moment of the molecular ion is either zero or is directed to one of the neighboring molecules with dipole moments of these neutral molecules being accommodated to this nonsymmetrical situation. Taking the above linear combination, subject to the point symmetry of the crystal around the molecular ion, the mean value of the dipole moment of the ion gives zero but accommodation of the neighboring dipole moments in any of these configurations leads to nonzero values of correlation functions

$$\langle(\mathbf{d}_m)_i(\mathbf{d}_0)_j\rangle \equiv \langle(\mathbf{d}_m)_i(\mathbf{d}_{ion})_j\rangle, \quad m \neq 0 . \tag{3.4.18}$$

In this respect, the dynamical quantum correction ΔP_{quant} to the classical interaction term P_{class} may be considered as a result of dynamical distortion of the Van der Waals interaction in the vicinity of the localized charge q.

The quantum correction ΔP_{quant} may be represented by the following general expression;

$$\Delta P_{quant} = P_{quant} - P_{class} = [E_{quant}(q) - E_{quant}(q = 0)]$$
$$- [E_{class}(q) - E_{class}(q = 0)] . \tag{3.4.19}$$

Čápek et al.[340] have attempted to calculate the value of ΔP_{quant} in linear polyacenes applying the approach suggested in Ref. 56. The authors used a simplified model considering only four neutral molecules surrounding the central one with the localized charge. These estimates showed that in the polyacene series from Nph to Pc the ΔP_{quant} value increases approximately by an order of magnitude. It was concluded that

1. ΔP_{quant} in Nph can be regarded as negligible ($\lesssim 0.05$ eV).

2. ΔP_{quant} in Ac is of the order of experimental or calculational errors (≈ 0.1 eV).

3. ΔP_{quant} in Tc and Pc increases considerably, having the value ≈ 0.3 eV and ≈ 0.4 eV, respectively.

It should be emphasized that the ΔP_{quant} is positive ($\Delta P_{quant} > 0$) and thus lowers the absolute value of P_{class} ($P_{class} < 0$). Thus, if these quantum corrections are taken into account, the calculated static P_{class} values for tetracene and pentacene will be considerably decreased. This problem is discussed later in connection with introduction of the empirical submolecular model of polarization energy calculations (see Section 3.5.4).

3.5. MICROELECTROSTATIC METHODS OF ELECTRONIC POLARIZATION ENERGY CALCULATIONS IN OMC

Microelectrostatic methods of evaluating electronic polarization energies are semiempirical, since they employ experimental data for calculations such as molecular polarizability, molecular dipole or quadrupole moments, crystallographic lattice parameters, and optical dielectric constants of the crystal.[332] Microelectrostatic approximation offers the advantage of presenting a whole choice of methods of numerical or analytical calculation providing various degrees of accuracy. Depending on the desired accuracy, it may suffice to obtain a rough numerical estimate of P using various direct zero-order approximations. Otherwise one can resort to more refined methods of self-consistent polarization field or Fourier transformation. In addition, there practically always exists a possibility of assessing the error involved using one or the other of approximations. The first pioneering polarization energy calculation was performed for polyacene crystals by Lyons and coworkers in the late sixties.[23,122,221,223] A survey of these earlier rough calculation methods and a historical development of their refinement is given in the monograph.[332] Here we shall give only the basic notions and some formulae necessary for further references.

3.5.1. Direct Non-self-consistent Methods

In microelectrostatic approximation the electronic polarization energy P of the crystal may be represented as a sum of all kinds of *electrostatic* interaction between an excess charge carrier localized on a particular molecule (which in a static approach can be regarded as a molecular ion), and the surrounding neutral molecules of lattice. In non-self-consistent approximation this sum may be presented in the following way:[122]

$$P = W_{q-d} + W_{d-d} + W_{q-d_0} + W_{q-Q_0} + W_{q-Q} + W_M + W_S, \qquad (3.5.1)$$

where W_{q-d} is the interaction energy between a localized excess charge carrier (ion) with electrical dipoles induced on surrounding molecules (see Fig. 3.4), W_{d-d} is the energy of dipole-dipole interaction between induced dipoles, W_{q-d_0} is the interaction energy between the charge carrier (ion) and permanent dipoles of surrounding molecules, W_{q-Q_0} is the interaction energy between the charge carrier (ion) and permanent quadrupoles of surrounding molecules, W_{q-Q} is the same with quadrupoles induced on surrounding molecules, W_M is the energy contribution from higher-order multipoles, and W_s is the interaction energy due to hyperpolarization effects, i.e., deviation of molecular polarizability from linear in strong electric fields.

A detailed analysis of all the terms of Eq. (3.5.1) contributing toward electronic polarization P of a molecular crystal has been carried out by Lyons et al.[23,122] and by Hug and Berry.[139] (See detailed survey in Ref. 332). It may be shown that the main contribution in the P value of nonpolar molecular crystals of the Ac-type is due to the first term of the sum (3.5.1), viz. the term of charge-induced dipole interaction W_{q-d}.

The term W_{q-d} in Ac-type crystals may be determined as follows.[122] Let the main axes of the kth molecule L, M, and N (L being the longer, M the shorter, and N being the axis perpendicular to the plane of the molecule) be inclined to the x-, y-, and z-axes of the crystal (x corresponding to the $\mathbf{a}$, y to the $\mathbf{b}$, and z to the $\mathbf{c}'$ axes of the crystal) with the corresponding cosines of the inclination angles being respectively (l_1, l_2, l_3), (m_1, m_2, m_3), and (n_1, n_2, n_3). Then the polarization energy of the kth molecule $W_{q-d}(k)$ by a charge carrier localized on molecule at the origin $(0,0,0)$ is given by

$$W_{q-d}^{(k)} = e^2/2r_k^6 [b_1(l_1 r_x + l_2 r_y + l_3 r_z)^2 + b_2(m_1 r_x + m_2 r_y + m_3 r_z)^2$$
$$+ b_3(n_1 r_x + n_2 r_y + n_3 r_z)^2], \qquad (3.5.2)$$

where b_i ($i = 1,2,3$) are the main components of the tensor of molecular polarizability along the main axes L, M, and N of the molecules, respectively, r is the distance from the molecule with a localized carrier $(0,0,0)$ to the kth molecule.

The total energy of the term W_{q-d} is obtained as the sum of $W_{q-d}^{(k)}$ over all $(N-1)$ surrounding neutral molecules of the crystal:

$$W_{q-d} = \sum_{k=1}^{N-1} W_{q-d}^{(k)} = - \sum_{k=1}^{N-1} (e^2/2r_k^6) \sum_{i,j} b_i(k_{ij} r_{jk})^2, \quad i, j = 1, 2, 3$$

$$(3.5.3)$$

where k_{ij} are the respective cosines of the inclination angles.

For a rough estimate of the term W_{q-d} the tensor of polarizability b_i is sometimes replaced by the mean isotropic polarizability α of the molecule:

$$\alpha = 1/3 \sum_{i=1}^{3} b_i. \tag{3.5.4}$$

In this case the expression (3.5.3) becomes considerably simpler:

$$W_{q-d} = - \sum_{k=1}^{N-1} e^2 \alpha / 2 r_k^4. \tag{3.5.5}$$

The interaction term between induced dipoles W_{d-d} may be evaluated according the following formula:[122]

$$W_{d-d} = \sum_{k,l} (e^2 \alpha^2 / r_k^3 r_l^3)[\mathbf{r}_l \mathbf{r}_k - 3 \mathbf{r}_{kl}^{-2} (\mathbf{r}_k \mathbf{r}_{kl})(\mathbf{r}_l \mathbf{r}_{kl})], \tag{3.5.6}$$

where $\mathbf{r}_k, \mathbf{r}_l$ is the mean distance between the ion and the neutral molecules k and l, respectively, and $\mathbf{r}_{kl}$ is the distance between molecules k and l.

Dipole-dipole interaction partly screens off the field of the localized charge, hence the sign of the term $W_{d-d}(W_{d-d} > 0)$ is opposite to that of the term $W_{q-d}(W_{q-d} < 0)$, i.e., the term W_{d-d} reduces the total energy of polarization P.[332]

The contribution of the term W_{d-d} may be considerable and can be as high as 20–30% in the case of aromatic hydrocarbon crystals.[23] Hence, one cannot neglect the contribution of this term even in rough estimates of the electronic polarization energy. Early estimates of the possible contribution of other terms of Eq. (3.5.1), performed by Lyons et al.[23,122] and by Hug and Berry[139] indicated that the contribution of terms (IV-VII) to the total polarization energy of nonpolar crystals of Ac-type (for which the third term $W_{q-d_0} = 0$) does not exceed 0.1 eV and may be regarded as negligible. Thus, in non-self-consistent calculations of P usually only the first two terms are taken into consideration:[332]

$$P \approx W_{q-d} + W_{d-d}. \tag{3.5.7}$$

This approach is often called zero-order approximation as it may serve as the zeroth order step in iterative self-consistent solutions or other more advanced calculations. In these early stages of non-self-consistent P calculations some important refinements were proposed. Thus, Batley, Johnston, and Lyons[23] and later Hug and Berry[139] showed that the number N of molecules involved in the formation of an electronic polaron must be taken sufficiently large ($N \approx 7000$) in order to obtain an uncertainty in P less than 0.1 eV.[139] Hence, an electronic polaron should be treated in terms of a *large-radius* polaron model (see Fig. 3.5).

Moreover, Batley, Johnston, and Lyons[23] estimated the effect of charge distribution in the molecular ion in comparison with the point charge-induced dipole interaction in the series of polyacene crystals from Nph to Pc. The effect was pronounced and increased with increasing size of the molecular ion. The point-charge model overestimated and the distributed charge model underestimated the value of P, especially in case of Tc and Pc, in disagreement with experimental data.

The inconsistency of the zero-order approximation is evident.[332] Such an approach assumes that the charge-induced dipoles remain unchanged in the field of other dipoles, and these initial zero-order dipole moments $d_i^{(0)}$ are used for calculating the term of dipole-dipole interaction W_{d-d}. This is, in fact, a very rough approximation. The initial charge-induced zero-order dipole moments $d_i^{(0)}$ change in the field of other neighboring dipoles, and the problem of finding the real value of d_i has to be approached by means of self-consistent polarization fields of interacting dipoles.[332]

3.5.2. Method of Self-consistent Polarization Field (SCPF)

The idea of self-consistency in calculations of electronic polarization energy was first proposed in the classical paper by Mott and Littleton[249] (see also Ref. 248).

This approach was later used by Rittner, Hunter, and De Pre[300] for polarization energy calculations in ionic crystals and by Druger and Knox[94] in rare-gas (argon, krypton, and xenon) solids.

Jurgis and Silinsh[149,342] extended the method of self-consistent polarization energy calculations in anisotropic Ac-type crystals (see also Ref. 332, p. 72).

The essence of the SCPF method consists in considering every induced dipole as being in the field of a localized charge carrier (ion) and in the self-consistent field of other induced dipoles.

In an isotropic approximation the polarization energy P of a localized charge carrier can be calculated according to Druger and Knox[94] by means of the following expression:

$$P = -\sum_i \left[\mathcal{E}_0(\mathbf{r}_i)\mathbf{d}_i - \mathbf{d}_i^2/2\alpha - \sum_{j>i} \mathbf{d}_i \lambda_{ij}\mathbf{d}_j \right] = -1/2\sum_i \mathcal{E}_0(\mathbf{r}_i)\mathbf{d}_i ,$$

$$(3.5.8)$$

where $\mathcal{E}_0(\mathbf{r}_i)$ is the electric field intensity created by the charge carrier (ion) at the center of molecule i; $\mathbf{d}_i$, $\mathbf{d}_j$ are the values of induced dipole moments i and j, respectively; α is the mean isotropic polarizability of the molecule [see, e.g., Eq. (3.5.4)]; λ_{ij} is the operator of dipole-dipole interaction:

$$(\lambda_{ij})^{\mu v} \equiv \frac{\mathbf{r}_{ij}^2 \delta_{\mu v} - 3(\mathbf{r}_{ij})^v (\mathbf{r}_{ij})^\mu}{\mathbf{r}_{ij}^5} ,$$

$$(3.5.9)$$

and $\mathbf{r}_{ij}$ is the distance between dipoles.

The first term of Eq. (3.5.8) describes the interaction energy between the localized charge carrier (ion) and the induced dipole d_i; the second one corresponds to the self-energy of the induced dipole d_i; and the third term to the interaction between the induced dipoles d_i and d_j.

The SCPF calculations of P value of anisotropic crystals of Nph and Ac in Refs. 149 and 342 were performed in terms of Eqs. (3.4.15) and (3.4.17), derived in Section 3.4.1, in which the mean isotropic polarizability α is replaced by a molecular polarizability tensor α_{ij} [see Eq. (3.4.13)].

For practical SCPF calculations the following formula was used in Refs. 149 and 324 (see also Ref. 332, p. 72) which can be easily obtained from expressions (3.4.15) and (3.4.17):

$$P = -1/2\sum_n \mathcal{E}_{0n}\mathbf{d}_n. \tag{3.5.10}$$

Here $\mathcal{E}_{0n}$ denotes the electric field strength on the molecule n created by the localized charge carrier at the origin $(0,0,0)$; $\mathbf{d}_n$ is the induced dipole moment on the molecule n calculated numerically by an iterative procedure in the self-consistent field of other induced dipoles.

The SCPF calculations of P in Refs. 149 and 342 were performed in the framework of the point-charge model using the experimental values of the main components of molecular polarizability tensors for Nph and Ac crystals (see Ref. 322, p. 81). The total energy P of electronic polarization was calculated in two stages.

First, using the SCPF method, the polarization energy P' was calculated for a sphere of radius R containing more than one hundred molecules. After that the contribution ΔP of the molecules outside the self-consistency region were estimated in a macroscopic approximation as a polarizable continuum possessing an isotropic mean optical dielectric permeability ϵ using the following formula:

TABLE 3.1. *Comparison of Electronic Polarization Energies P in Naphthalene and Anthracene Crystals, Calculated by the Methods of SCPF and Fourier Transformation Using Different Sets of Polarizability Tensor Components[a]*

Crystal	Polarizability tensor used	SCPF method[149,332] P (eV)	FT method[36] P (eV)
Naphthalene	Experimental, molecular[b]	−1.27	−1.26
	Calculated for crystal[c]	—	−1.27
Anthracene	Experimental, molecular[d]	−1.53	−1.55
	Experimental, molecular[e]	−1.50	−1.51
	Calculated for crystal[c]	—	−1.42

[a]*Note.* For experimental methods of molecular polarizability determination see Ref. 332, p. 81. The effective polarizability of molecules in a crystal has been calculated in terms of point-ring approximation (see Ref. 36).
[b]Taken from Ref. 214.
[c]Taken from Ref. 36.
[d]Taken from Ref. 215.
[e]Taken from Ref. 216.

$$\Delta P = (-e^2/2R) \cdot (1 - 1/\epsilon). \tag{3.5.11}$$

The total energy P of electronic polarization is then

$$P = P' + \Delta P. \tag{3.5.12}$$

The calculated P values for Nph and Ac crystals are shown in Table 3.1. These data are in satisfactory agreement with experimental P values and in good agreement with values calculated by other advanced methods of self-consistency.

In the SCPF calculations of P for Nph and Ac crystals[149,342] the convergence in the iteration procedure was more than satisfactory. The estimated convergence error in this case did not exceed 0.02 eV.

However, convergency of the method becomes worse with increasing size of the self-consistency region in higher polyacenes.[149,342] Moreover, for higher polyacenes (Tc, Pc), the magnitude of the quantum corrections becomes greater (see Section 3.4.2). Finally, increasing the size of the molecules leads to a slower convergency of the multipole expansion which may make the limitation to charge- and dipole- (i.e., to the zeroth and first-order) terms ambiguous. Thus, for higher polyacenes, other modified self-consistence methods, e.g., those that treat the excess charge as distributed over extended molecules, should be used.

3.5.3. Method of Fourier Transformation

Bounds and Munn[36] have developed a new method which gives an explicit self-consistent algebraic expression for the calculation of polarization energy P of a localized point charge in terms of Fourier-transformed lattice multipole sums.

This method has been applied by Munn and coworkers in different contexts and modifications.[36-38,100,101] The idea of the approach is to transform the real space problem to that of the reciprocal space which allows then to express the resulting polarization energy explicitly. The second advantage of the method is connected with efficient application of the Ewald summation method to Fourier-transformed lattice sums; this method was found to be effective.[36]

We will give here only a brief description of the method referring the interested reader to the original papers.[36-38,100,101]

As a starting point, refer to Eqs. (3.4.11) and (3.4.12) in which, according to microelectrostatic approach, the mean quantum dipole moment $\langle \mathbf{d}_m \rangle$ is identified with the induced classical dipole moment $\mathbf{d}_m$; namely, $\mathbf{d}_m \equiv \langle \mathbf{d}_m \rangle$ [see Eq. (3.4.16)].

To retain the notations of Ref. 36 let us change the index m in Eq. (3.4.11) and (3.4.12) to lk where l denotes the unit cell and k, a molecule in this cell. Further, let us assume that the excess point charge is localized on the molecule at the origin $(0,0,0)$ in the unit cell $l = 0$.

N is the number of unit cells, z, the number of molecules in a unit cell. For the given notations the vector $\mathbf{r}_m$ in Eq. (3.4.12) is to be changed to $\mathbf{r}_{lk}$.

As shown in Ref. 36, applying the following Fourier transformation:

$$(1 - \delta_{lk,\,0k'})\mathbf{d}_{lk} = \frac{1}{N} \sum_y e^{iy(\mathbf{r}_{lk} - \mathbf{r}_{0k'})}\mathbf{d}_k(\mathbf{y}), \qquad (3.5.13)$$

where

$$\mathbf{d}_k(\mathbf{y}) = \sum_l e^{-iy(\mathbf{r}_{lk} - \mathbf{r}_{0k'})} \cdot (1 - \delta_{lk,0k'})\mathbf{d}_{lk}, \qquad (3.5.14)$$

Eqs. (3.4.11 and 3.4.12) can be solved in an explicit algebraic manner. As a result we obtain the following formula for the calculation of polarization energy P:

$$P = -\frac{1}{2}\frac{q^2}{N} \sum_y \mathbf{u}(-\mathbf{y})^T[\mathbf{1}| - \alpha\mathbf{T}(\mathbf{y})]^{-1}\alpha\mathbf{u}(\mathbf{y}). \qquad (3.5.15)$$

To simplify Eq. (3.5.15) let us use a matrix representation where $\mathbf{1}|$ is the unit matrix $3z \times 3z$ where z is the number of molecules in a unit cell. The following matrixes are of the same rank.

$$(\alpha)_{jk,\,j''k''} = \delta_{kk''}\,\alpha_{jj''}^k; \qquad (3.5.16)$$

$$[\mathbf{T}(\mathbf{y})]_{jk,j''k''} = \sum_l (1 - \delta_{lk,l''k''})\left[\delta_{jj''} - 3\,\frac{(\mathbf{r}_{lk} - \mathbf{r}_{l''k''})_j(\mathbf{r}_{lk} - \mathbf{r}_{l''k''})_{j''}}{|\mathbf{r}_{lk} - \mathbf{r}_{l''k''}|^2}\right]$$

$$\times \frac{1}{|\mathbf{r}_{lk} - \mathbf{r}_{l''k''}|^3}\,e^{-iy(\mathbf{r}_{lk} - \mathbf{r}_{l''k''})}, \qquad (3.5.17)$$

where $j, j'' = 1,2,3$ and $\mathbf{U}(\mathbf{y})$ is a vector with $3z$ components:

$$[\mathbf{u}(\mathbf{y})]_{jk} = \sum_l (1 - \delta_{lk,0k'})\,\frac{|\mathbf{r}_{lk} - \mathbf{r}_{0k'}|}{|\mathbf{r}_{lk} - \mathbf{r}_{0k'}|^3}. \qquad (3.5.18)$$

Index T in Eq. (3.5.15) denotes transposition. As mentioned earlier, the lattice sums (3.5.17) and (3.5.18) are calculated by the Ewald summation method.[36] Although summation with respect to $\mathbf{y}$ is not trivial, nevertheless it is possible to check the intermediate results in a similar way as in the case of the SCPF method. Thus integration inside a small sphere at the origin of the reciprocal space, where $\mathbf{y} = 0$, may be corrected by the term ΔP of the type given by Eq. (3.5.11) in the real space, thus taking into account more distant induced dipoles. In other words, the value of P may be determined as in the case of the value of the SCPF method, i.e., as a sum of two terms [see. Eq. (3.5.12)]:

$$P = P' + \Delta P, \qquad (3.5.19)$$

where P' is calculated according to Eq. (3.5.15) but the macroscopic correction term ΔP is estimated according to the formula:[36]

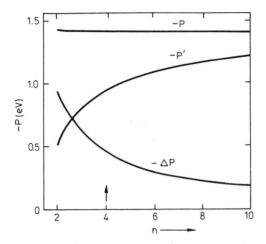

FIGURE 3.6. Calculated dependences of the total polarization energy P and its components P' and ΔP [see formula (3.5.19)] on the value of $n = \sqrt[3]{N}$ in an anthracene crystal, where N is number of unit cells.[36] In calculations the components of the effective molecular polarizability of crystal have been used (the arrow shows the P, P', and ΔP values calculated by the SCPF method in Refs. 149 and 332; see Table 3.1).

$$\Delta P = -\frac{1}{2}\left(\frac{q^2\rho}{\pi\epsilon_0}\right)\left\{1 - \frac{F(\Phi, k)}{[\epsilon_2(\epsilon_3 - \epsilon_1)]^{1/2}}\right\}, \qquad (3.5.20)$$

where ρ is the radius of the sphere Ω in the reciprocal space. $F(\Phi, k)$ is the elliptic integral of the first type with the following parameters:

$$\Phi = \tan^{-1}\left(\frac{\epsilon_3 - \epsilon_1}{\epsilon_1}\right)^{1/2}; \quad k = \left[\frac{\epsilon_3(\epsilon_2 - \epsilon_1)}{\epsilon_2(\epsilon_3 - \epsilon_1)}\right]^{1/2},$$

where ϵ_1, ϵ_2, and ϵ_3 are the main components of the tensor of the dielectric permeability.

In case of an isotropic approximation one obtains a simple formula:[36]

$$\Delta P = -\frac{1}{2}\frac{q^2\rho}{\pi\epsilon_0}\left(1 - \frac{1}{\epsilon_0}\right), \qquad (3.5.21)$$

where ϵ_0 is the relative scalar dielectric permeability. As indicated in Ref. 36 the formula (3.5.21) is equivalent to the formula (3.5.11) of the SCPF method for evaluation of ΔP, since the region outside the sphere with radius R in the real space is equivalent to the region Ω inside the sphere with the radius ρ in the reciprocal space. However, it should be mentioned that formulas (3.5.20) and (3.5.21) have been directly derived in the framework of microscopic theory.[36] Figure 3.6 illustrates the dependence of the total polarization energy P of an Ac crystal and its components P' and ΔP [see formula (3.5.19)] on the value of $n = N^{1/3}$ in Ac, where N is the number of unit cells. As may be seen, the contribution of the P'

steadily increases and that of ΔP decreases with growing number of n in calculations. At the same time the total value of polarization energy practically remains constant and independent of n. This demonstrates a good convergence on both short-range sum P' and long-range contribution ΔP.

Table 3.1 gives a comparison of numerical values of P obtained by the SCPF method[149,332] and the method of Fourier transformations[36] for Nph and Ac crystals, using the same sets of experimental molecular polarizability tensors. As can be seen from Table 3.1 (see also Fig. 3.6) the agreement between both methods was found to be very good since the difference in P values does not exceed 0.01–0.02 eV. This demonstrates that both approaches are comparable. However, for technical reasons, preference is to be given to the Fourier transformation method. The reason is that it allows avoiding the isotropic approximation in the distant (in the real space) non-self-consistency region in a more efficient manner and provides better convergence for large self-consistency regions.

Table 3.1 illustrates also the Fourier transformation calculation results using the calculated components of the effective polarizability tensors of Nph and Ac crystals[255,260] instead of experimental molecular polarizability tensors.[214–216] As may be seen from Table 3.1, the agreement between these calculated results may be regarded as good for Nph and satisfactory for Ac crystals.

3.5.4. Fourier Transformation Method and Submolecular Approach

Bounds and Munn[38] have extended the Fourier transformation method (FTM) of calculating polarization energies P to treat molecules as sets of *submolecules.*

In aromatic hydrocarbon crystals the submolecules may be identified in a natural way with the *aromatic rings* of the extended molecule. This approach takes into account the molecular size, shape, and orientation, and thus is convenient for obtaining calculated results for highly anisotropic molecules in an internally consistent way. It has considerable physical advantages. First, for more extended molecules, one needs multipole expansions for rather small submolecules instead of those for the entire molecule, which makes the expansion not only quickly convergent but also physically meaningful. Second, for some molecules, like biphenyl, the point-molecule polarizability tensor is highly anisotropic so that no reliable polarization energy can be obtained from point-molecule calculations. Similarly, in p-terphenyl, the point-molecule polarizability tensor entering point-molecule polarization energy calculations is not even physically valid, having a negative principal component implying antiparallel local and macroscopic fields. Finally, the method of submolecular approach is flexible in allowing variations in the excess charge distributions over submolecules forming the molecular ion. (However, this distribution should be determined from first principle calculations.)

In the submolecular approach one may use the formalism of Fourier transformation [Eqs. (3.5.13–3.5.20)], only the notation should be extended. For a unit cell, containing z molecules, denoted by index k, s submolecules are introduced, the

TABLE 3.2. *Calculated Polarization Energies (P) for Several Aromatic Hydrocarbon Crystals*[a]

| | | P (eV) | |
| | | Point rings | |
Crystal	Point molecules	Single charge	Distributed charge
Benzene	−1.16	−1.16	−1.16
Naphthalene	−1.22	−1.28	−1.20
Anthracene	−1.25	−1.37	−1.19
Biphenyl	—	−1.31	−1.10
p-Terphenyl	—	−1.39	−0.95

[a]Taken from Ref. 38.

number of which equals the number of aromatic rings in the molecule and which are denoted by index j; instead of molecular polarizability the corresponding polarizabilities of submolecules are used. In the submolecular treatment Bounds and Munn have used two basic models:[38]

1. *Single-charge model*: All the excess charge is placed on one submolecule. (The center of the inner ring in anthracene and p-terphenyl, the center of an end ring in naphthalene and biphenyl, and the center of the only ring in benzene.)

2. *Distributed charge model*: The excess charge is divided equally over all the submolecules forming the molecular ion.

An internally consistent treatment should use as input data such effective molecular polarizability which reproduces the experimental crystal susceptibility. For Nph and Ac, the susceptibility was obtained from the low-frequency permittivity tensor while for other crystals refractive data were used.

Calculated polarization energies obtained by Bounds and Munn[38] are summarized in Table 3.2.

Detailed experimental comparison of these calculated data (see Ref. 307 and Section 3.5.6) shows that both point molecule and single-charge point ring models give overestimated P values for Ac and other higher polyacenes. The best agreement with experiment is provided by the point ring distributed charge model.

Finally, it should be emphasized that the submolecular approach, though physically sound and phenomenologically plausible, is difficult to justify from the quantum mechanical point of view. The problem lies in the fact that the ring-type submolecules of one molecule are in nature tightly interrelated. Although it seems intuitively true that the empirically introduced charge distribution at the submolecules in some way replaces, even quantitatively, the quantum corrections for extended aromatic molecules (see Section 3.4.2) there is not, however, sufficient theoretical reason to justify such substitution.

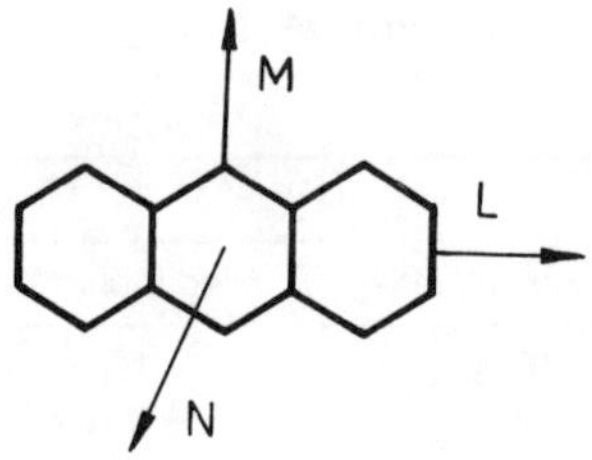

FIGURE 3.7. Orientation of the molecular axes in polyacenes (shown for an anthracene molecule).

3.5.5. Charge Carrier Interaction with Permanent Quadrupole Moments of Surrounding Molecules

Nonpolar organic molecules of polyacenes, possessing D_{2h} symmetry (or D_{6h} symmetry in case of benzene), have no permanent dipole moment. However, due to the planar structure of polyacene molecules, they have a permanent quadrupole moment tensor θ_{AB} the components of which increase with increasing number of aromatic rings in the molecule:[71]

$$\theta_{AB} = \frac{1}{2} e \left\{ \sum_n Z_n [3\mathbf{r}_A(n)\mathbf{r}_B(n) - \mathbf{r}^2(n)\delta_{AB}] - \sum_i \langle 3(\mathbf{r}_i)_A \cdot (\mathbf{r}_i)_B - \mathbf{r}_i^2 \delta_{AB} \rangle \right\},$$

(3.5.22)

where A and B are components of the Cartesian coordinates; e is the charge of a proton; $Z_n e$ is the charge of the nth nucleus in position $\mathbf{r}(n)$; the last term of the equation gives the average distribution of electrons.

Since polyacene molecules possess a high degree of symmetry the main axes of the tensor of the quadrupole moment θ_{AB} may be oriented along the axes L, M, N of the molecule, where L is the long axis, M the short axis, and N is the axis perpendicular to the plane of the molecule (see Fig. 3.7).

In this case the components of the quadrupole moment tensor are related in the system of molecular axes (L, M, N) as follows:

$$\theta_{NN} = -(\theta_{LL} + \theta_{MM}).$$

(3.5.23)

For polyacenes the value of θ_{NN} can be evaluated by the empirical formula:[100]

$$\theta_{NN} = -0.92n_v 10^{-40} \quad C \cdot m^2 = -0.05742n_v e \cdot \text{Å}^2,$$

(3.5.24)

where n_v is the number of valence electrons in the molecule. The first reliable experimental data of electric quadrupole moments θ_{AB} of a benzene molecule in the gaseous phase were only obtained in 1981 (see Ref. 25). Simultaneously Munn and coauthors[71] calculated the quadrupole moment tensors using *ab initio* wave functions, scaled to the benzene values, and similarly for Nph, Ac, and biphenyl molecules. The θ_{AB} values for Tc and Pc were later evaluated[100] using the extrapo-

TABLE 3.3. *Values of the Permanent Electric Quadrupole Moments* Θ_{AB} *of Polyacene Molecules in the System of Molecular Axes L, M, N (see Fig. 3.7) According to Ref. 100* [a]

Molecule	LL	MM	NN	Molecule	LL	MM	NN
Benzene	0.905	0.905	−1.810	Anthracene	1.654	2.163	−3.817
Naphthalene	1.278	1.493	−2.771	Tetracene	1.987	2.836	−4.823
				Pentacene	2.325	3.532	−5.857

[a]*Note:* The values of Θ_{AB} for benzene are based both on experimental[25] and calculated[71] data. For Nph and Ac calculated[100] and for Tc and Pc extrapolated[100] Θ_{AB} values are presented. The electrostatic unit of the quadrupole moment is 10^{-26} $e \cdot cm^2$ (1 buckingham) which equals 3.33564×10^{-40} $C \cdot m^2$.

lation approach of the ratio $-\theta_{LL}/\theta_{NN}$, using the dependence on the number n of aromatic rings in the molecule of the given polyacene.

The values of θ_{AB} for polyacene molecules are shown, according to Ref. 100, in Table 3.3.

As can be seen from Table 3.3, the value of θ_{AB} increases considerably with growing number n of the aromatic rings, i.e., with growing anisotropy of the molecule. Therefore one may expect that for the higher members of the polyacene series (Tc, Pc) the magnitude of the interaction term W_{q-Q_0} should also be considerable.

The electrostatic interaction of a localized charge with permanent quadrupole moment θ of the kth molecule $W_{q-Q_0}(k)$ decreases with distance as $1/r^3$:[122]

$$W_{q-Q_0} = -q\,\theta/r_k^{\,3}. \tag{3.5.25}$$

Thus we see that the sign of the term W_{q-Q_0} depends both on the sign of the quadrupole moment Θ (see Table 3.3) and on that of the charge carrier. That means that the sign of W_{q-Q_0} for a localized electron will be opposite that for a hole. As a consequence quadrupole interaction produces a difference between the apparent effective polarization energies of electrons P_{eff}^- and holes P_{eff}^+, i.e., $P_{\text{eff}}^+ \neq P_{\text{eff}}^-$. This circumstance is of considerable physical importance and creates definite asymmetry in the energy structure of ionized states in OMC (see Chapter 4, Figs. 4.8–4.10).

Rough estimates of the term W_{q-Q_0} are available in Ref. 23. Refined calculation of W_{q-Q_0} were performed later by Munn *et al.*[37,100] using the known values of the elements of the θ_{AB} tensors (Table 3.3). The term W_{q-Q_0} in Refs. 37 and 100 was calculated according to a general formula:

$$W_{q-Q_0} = \frac{q}{3\epsilon_0 V} \sum_k \theta(k)/\mathbf{L}(k,k'), \tag{3.5.26}$$

where $\mathbf{L}(k, k')$ is the Lorentz-factor tensor elements, and V is the volume of an elementary cell; k' and k denotes the molecule with the localized charge carrier and that with the quadrupole moment θ, respectively.

It has been assumed that the tensor θ_{AB} of an isolated molecule remains the same in the crystal.

TABLE 3.4. *Calculated Values of Charge-Induced Dipole P_{id} and Charge-Permanent Quadrupole $W_{Q_0}^+$ and $W_{Q_0}^-$ Interaction Energies, as Well as Corresponding Effective Electronic Polarization Energies of Positive $P_{eff}^+ = P_{id} + W_{Q_0}^+$ and Negative $P_{eff}^- = P_{id} + W_{Q_0}^-$ Charge Carriers in Polyacene Crystals according to Munn and Coworkers[37,100] (All Values in eV)[a]*

Crystal	Coordinates of localized charge carrier	P_{id}	$W_{Q_0}^+$	P_{eff}^+	$W_{Q_0}^-$	P_{eff}^-
Benzene	(0,0,0)	−1.16	−0.04	−1.19	+0.04	−1.12
Naphthalene	(0,0,0)	−1.21	−0.16	−1.38	+0.16	−1.05
Anthracene	(0,0,0)	−1.19	−0.18	−1.38	+0.18	−1.01
Tetracene[a]	(0,0,0)	−1.14	−0.22	−1.36	+0.22	−0.93
	(1/2,1/2,0)	−1.14	−0.23	−1.38	+0.23	−0.91
Pentacene[a]	(0,0,0)	−1.08	−0.22	−1.29	+0.22	−0.86
	(1/2,1/2,0)	−1.07	−0.24	−1.31	+0.24	−0.83

[a]Tc and Pc crystals, belonging to the triclinic conformation, yield slightly different values of W_{Q_0} for charge carriers localized on the nonequivalent molecules.

The results of the calculation of the term W_{q-Q_0} for a number of polyacenes are given in Table 3.4 according to Refs. 37 and 100.

To simplify the notation in Table 3.4 we shall henceforth use the following indexing: P_{id} is the self-consistent electronic polarization energy of charge carrier interaction with *induced dipoles*; $W_{Q_0} = W_{q-Q_0}$ is the energy of charge carrier interaction with permanent quadrupole moments; and P_{eff} is the apparent effective electronic polarization energy; P_{eff} for holes $P_{eff}^+ = P_{id}^+ + W_{Q_0}^+$ and for electrons $P_{eff}^- = P_{id}^- - W_{Q_0}^-$. One should remember that, according to definition, P_{eff} and P_{id} are negative ($P_{eff} < 0$; $P_{id} < 0$) while W_{Q_0} may be either negative or positive (see Table 3.4). Further, for the sake of convenience, the modulus of the terms P_{id} and P_{eff}, i.e., $|P_{id}|$ and $|P_{eff}|$, are used.

Table 3.4 shows the P_{id} values obtained in Ref. 100 by the FTM using the submolecular approach (see Sections 3.5.3 and 3.5.4). Since, in case of charge-induced dipole interaction, the charge q in the relevant formulas appears as a square q^2 [see Eqs. (3.5.15) and (3.5.20)], the value of P_{id} does not depend on the sign of the charge carrier, thus

$$P_{id}^+ = P_{id}^- = P_{id}. \qquad (3.5.27)$$

As the matter of fact equality (3.5.27) holds for point charge approximation [see Eqs. (5.3.3) and (5.3.5)], submolecular approach of distributed charge (see Section 3.5.4), or other form of charge distribution of a molecular ion (see Ref. 23). Quantum chemical calculations, and nuclear resonance measurements show (see Ref. 133) that, for polyacenes, the charge distribution over an anion and cation are practically similar.

On the other hand, due to the asymmetry of the charge-quadrupole interaction (see Table 3.4), the absolute value of the effective electronic polarization energy for holes is greater than that for electrons, thus

$$|P_{\text{eff}}^{+}| > |P_{\text{eff}}^{-}| . \tag{3.5.28}$$

The difference ΔP_{eff} is given by

$$\Delta P_{\text{eff}} = |P_{\text{eff}}^{+}| - |P_{\text{eff}}^{-}| = 2|W_{Q_0}| . \tag{3.5.29}$$

As may be seen from Table 3.4, the value of W_{Q_0} increases almost by an order of magnitude in the series benzene-pentacene and reaches the value of 0.22–0.24 eV, i.e., practically about 25% of the calculated charge-induced dipole interaction energy P_{id} for Pc.

Such an increase in W_{Q_0} is caused, first, by the growing anisotropy of molecules in the polyacene series benzene-Pc, reflecting in increased values of quadrupole tensors (see Table 3.3); second, by the growing anisotropy of the crystal lattice the influence of which may be seen through the Lorentz-factor tensors.[37]

As a result $\Delta P_{\text{eff}} = 2|W_0|$, in case of pentacene, reaches the value 0.43–0.48 eV. It should be emphasized that the charge-quadrupole interaction term $2|W_0|$ does not change the value of the energy gap E_G, but shifts it upward nearer to the vacuum level by ΔP_{eff} (see Chapter 4, Figs. 4.8–4.10).

3.5.6. Evaluation of Total Effective Polarization Energy Using Calculated and Experimental Energy Parameters

So far we have discussed only electronic polarization effects. However, at the present time, there exists direct experimental evidence that the total effective polarization energy of a charge carrier includes, in addition to the electronic polarization energy terms P_{id} and W_{Q_0}, a nuclear (vibronic) polarization term E_b.[307]

A charge carrier created in the bulk of the molecular solids after photogeneration, complementary to the electronic polarization of surrounding molecules, also polarizes the intramolecular vibrational modes of the molecule on which it is located, as well as dipole active modes of the nearest neighboring molecules,[305] thus, forming an extended ionic state in the crystal. As proposed in Refs. 334 and 344, the new quasi-particle of manybody interaction can be described phenomenologically in the framework of a nearly small molecular polaron approach (see Section 3.6).

It has been shown that the relaxation energy E_b, gained in the process of formation of a molecular polaron, can be determined experimentally from the difference between the optical E_G^{Opt} and adiabatic E_G^{Ad} energy gaps of the crystal as follows:[334,344]

$$E_b = (1/2)(E_G^{\text{Opt}} - E_G^{\text{Ad}}). \tag{3.5.30}$$

Reliable values of E_G^{Opt} and E_G^{Ad} have recently been obtained for Ac, Nph, and Pc crystals: E_G^{Opt} from electromodulation spectra and E_G^{Ad} determined from the activation energy spectra $E_a^{ph}(h\nu)$ of intrinsic photogeneration. These methods are described fully in Chapter 4, Sections 4.1 and 4.2; the E_b, E_G^{Opt}, and E_G^{Ad} values as well as corresponding references are given in Tables 3.8, 4.2, and 4.4.

Consequently, to obtain the total effective polarization energy W_{eff}, gained by the charge carrier in the crystal, the E_b value of molecular (vibronic) polarization should be added to the terms of effective electronic polarization terms P_{eff}^+ and P_{eff}^-.

We obtain the resulting final equations for the *total effective polarization energies*:[307]

(i) for the positive charge carrier:

$$W_{\text{eff}}^+ = P_{\text{eff}}^+ + E_b = P_{\text{id}} + W_{Q_0}^+ + E_b \qquad (3.5.31)$$

(ii) for the negative charge carrier:

$$W_{\text{eff}}^- = P_{\text{eff}}^- + E_b = P_{\text{id}} + W_{Q_0}^- + E_b. \qquad (3.5.32)$$

It has been assumed in Eqs. (3.5.31) and (3.5.32) that $E_b^+ = E_b^- = E_b$, where E_b is the mean value obtained according to Eq. (3.5.30) (see Table 3.8).

Independent estimates of W_{eff}^+ and W_{eff}^- are based on experimentally determined energy terms:[307]

(i) for a positive charge carrier:

$$W_{\text{eff}}^+ = I_G - I_C \qquad (3.5.33)$$

where I_G and I_C are the ionization energies of molecule and crystal, respectively.

(ii) for a negative charge carrier:

$$W_{\text{eff}}^- = I_C - A_G - E_G^{\text{Ad}} \qquad (3.5.34)$$

where A_G is the electron affinity of a molecule and E_G^{Ad} is the adiabatic energy gap (see Chapter 4).

Sato, Inokuchi, and Silinsh[307] have performed reevaluation of polarization energies for polyacenes, using both calculated and experimentally determined energy terms. The authors applied Eqs. (3.5.31) and (3.5.32) for a semiempirical evaluation of the total polarization energies W_{eff}^+ and W_{eff}^- using the values P_{id} and W_{Q_0} of refined calculation by Munn *et al.*[37,100] (see Table 3.4) and the values of E_b estimated experimentally according to Eq. (3.5.30)[307] (see also Table 3.8). These values were compared with independent estimates, using reported experimental energy parameters I_G, I_C, A_G, and E_G^{Ad} according to Eqs. (3.5.33) and (3.5.34).

It should be mentioned that in earlier works (see Refs. 122, 288, 332, and 351) $I_G - I_C$ was identified with P_{eff}^+ including only the charge-induced dipole term, i.e., $P_{\text{eff}}^+ - P_{\text{id}}$. As the result the P_{id} value was overestimated. On the other hand, the P_{eff}^- value was usually assumed to be equal to P_{eff}^+, i.e., $P = P_{\text{eff}}^+ = P_{\text{eff}}^-$ under the assumption that the charge-quadrupole interaction term W_{Q_0} can be neglected. This approach has been demonstrated to be invalid in the previous section.

To use Eq. (3.5.33) one needs reliably measured ionization potentials I_G and I_C.

Table 3.5 presents the most reliable data reported on I_G and I_C values for Ac, Tc, and Pc (for references see the review[307]).

TABLE 3.5. *Experimental Values of the Ionization Energies of a Molecule I_G and the Crystal I_C, and Molecular Electron Affinities A_G as Reported for Polyacenes (All Values in eV)[a]*

Crystal	I_G	$\langle I_G \rangle$	I_C	$\langle I_C \rangle$	A_G	$\langle A_G \rangle$
Anthracene	7.47	7.42	5.95	5.77	0.55	0.58
	7.42		5.85		0.57	
	7.41		5.75		0.60*	
	7.40		5.67		0.61*	
	7.38		5.65			
Tetracene	7.04	6.98	5.40	5.26	0.88	1.03
	7.01		5.30		0.95*	
	6.97		5.25		1.03	
	6.88		5.10		1.10*	
					1.18	
Pentacene	6.74	6.64	5.10	5.01	1.19	1.37
	6.64		5.07		1.31	
	6.61		5.00		1.45*	
	6.55		4.85		1.51*	

[a]*Note*: Asterisk indicates calculated A_G values. For references see the review.[307]

The ionization energy I_G can be determined by a number of independent methods, the preferable one being ultraviolet photoelectron spectroscopy (UPS) and photoionization (PI), as well as some special spectroscopy techniques such as Rydberg spectra (see Ref. 332, p. 128). It must be borne in mind that UPS methods yield, as a rule, the vertical ionization potential I_G^V, whereas PI and spectroscopic methods yield the adiabatic ionization potential I_G^{Ad}. However, in polyacenes the difference $\Delta I = I_G^V - I_G^{Ad}$ is supposed to be small, of the order of $\Delta I < 0.05$ eV, i.e., it is smaller than the distribution range of reported values of I_G obtained by the UPS method (cf. Ref. 332). Therefore, one may suggest that the reported value of I_G presented in Table 3.5 have almost the same reliability; for this reason we prefer to use the average value $\langle I_G \rangle$ of these reported I_G values for the evaluation of W_{eff}^+.

The situation is similar, or even more complicated in the case of reported I_C values.[309] The ionization energy of the crystal, I_C, can be determined by two main methods — from the threshold of spectral dependence of photoemission quantum yield (SDQY) and/or from the energy distribution curves (EDC) of photoemitted electrons, according to Einstein's law (see Ref. 332, p. 124). Both methods yield slightly different results. Besides, the measured I_C values may be influenced by the crystalline state of the sample,[307] surface effects, and so on. In order to diminish the role of possible methodological deviations, we have preferred to use the average value $\langle I_C \rangle$ of reported I_C data (see Table 3.5).

Since there are only few reliable experimental data of A_G values of polyacenes, we have used the average $\langle A_G \rangle$ values from both experimental and reliable calculated A_G data reported (see Table 3.5).

The effective polarization energies W_{eff}^+ and W_{eff}^- of positive and negative charge carriers, respectively, in Ac, Tc, and Pc crystals, obtained according to Eqs. (3.5.33)

TABLE 3.6. *Experimental and Calculated Values of Total Effective Polarization Energies for Positive W_{eff}^+ and Negative W_{eff}^- Charge Carriers in Polyacene Crystals (All Values in eV)*[a]

	Experimental			Calculated		
Crystal	W_{eff}^+ according to Eq. (3.5.33)	W_{eff}^- according to Eq. (3.5.34)	$\Delta W_{eff} = W_{eff}^+ - W_{eff}^-$	W_{eff}^+ according to Eq. (3.5.31)	W_{eff}^- according to Eq. (3.5.32)	$\Delta W_{eff} = W_{eff}^+ - W_{eff}^-$
Anthracene	1.65	1.09	0.56	1.52	1.16	0.36
Tetracene	1.72	1.10	0.62	1.52	1.07	0.45
Pentacene	1.63	1.17	0.46	1.48	1.02	0.46
Mean value	1.67	1.12	0.55	1.51	1.08	0.42

[a]Taken from Ref. 307. Note: In formulas (3.5.31), (3.5.32), and Table 3.6 the modulus of the terms P_{eff}, P_{id}, and W_{eff} are used.

and (3.5.34), are presented in Table 3.6. In the evaluation procedure average experimental values of the energy parameters, namely, $\langle I_G \rangle$, $\langle I_C \rangle$, and $\langle A_G \rangle$, were used (see Table 3.5), as well as reported experimental adiabatic gap energies E_G^{Ad} (see Table 4.2). In Table 3.6, theoretically evaluated W_{eff}^+ and W_{eff}^- values obtained from Eqs. (3.5.31) and (3.5.32) are also given. In this case values of parameters P_{id}, $W_{Q_0}^+$, and $W_{Q_0}^-$, as calculated by Munn et al.,[37,100] were used (see Table 3.4), the only experimentally determined parameter being E_b (see Table 3.8).

As can be seen from Table 3.6, both experimental and theoretical data unambiguously demonstrate the *asymmetry* of polarization terms: W_{eff}^+ is considerably, (by $\sim$ 0.4-0.5 eV) larger than W_{eff}^-. The calculated data show the main physical source of this asymmetry in the case of polyacene crystals. This is obviously caused by different signs of the total charge-quadrupole interaction terms W_{Q_0} for opposite charge carriers (see Table 3.4). This means that the so far assumed equality of the polarization terms $P_{eff}^+ = P_{eff}^-$ may be valid only for such OMC as benzene, the molecules of which have a negligible permanent quadrupole moment (cf. Table 3.3).

Results presented in Table 3.6 demonstrate sufficiently good agreement between experimental and theoretically calculated data. Experimental estimates of the parameter W_{eff}^+ yield values $\approx$ 10% higher than the theoretically evaluated ones which might be caused by a systematic error in the determination of I_C values by the threshold approximation, both evaluated values appear to be almost constant for the three compounds. On the other hand, experimental estimates of W_{eff}^- appear to increase slightly, whereas calculated estimates appear to decrease slightly, with increasing number of condensed rings in a molecule. However, the mean values of W_{eff}^- are fairly close in both cases, i.e., $\langle W_{eff}^- \rangle_{exp} = 1.12$ and $\langle W_{eff}^- \rangle_{cal} = 1.08$ eV (see Table 3.6).

Owing to the observed asymmetry of polarization energy, it is necessary to change the conceptual basis for drawing energy level diagrams of ionized states in OMC, and to introduce modified energy diagrams in which asymmetry is taken into account.[307,334] Such modified energy level diagrams for polyacene crystals are shown in Chapter 4, Figs. 4.8–4.10. In order to draw these energy level diagrams,

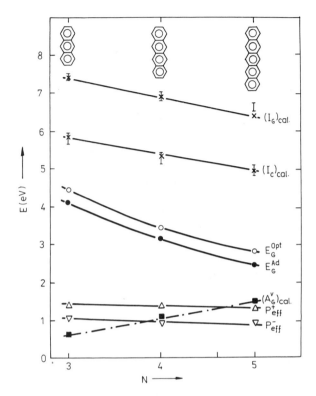

FIGURE 3.8. Dependences of main energy parameters for polyacene crystals[307] (x), (x), values of ionization energies of a molecule I_G and the crystal I_C, respectively, calculated by semiempirical approach. The error bars show the distribution of experimental values of these parameters according to Table 3.5; A_G^V, calculated values of vertical electron affinities of molecule (see Ref. 321). (For other notations see the text.)

experimental gap energies E_G^{Ad} and E_G^{Opt} (see Tables 4.2 and 4.4), calculated electronic polarization terms P_{id}, $W_{Q_0}^+$, $W_{Q_0}^-$ (see Table 3.4), and calculated vertical molecular electron affinity A_G^V (Ref. 321) have been used.

As can be seen from Figs. 4.8–4.10, the asymmetry, due to the quadrupole term W_{Q_0}, does not influence the value of the energy gap but, since $P_{eff}^+ > P_{eff}^-$, shifts the energy gap upwards.

There are some other notable consequences of the reevaluated data. As shown in Figs. 4.8–4.10, additive summation of the energy parameters of A_G^V, P_{eff}^-, E_G^{Opt}, and P_{eff}^+ yields the corresponding values of I_G, and in a similar manner I_C, which can be regarded as semiempirical calculated parameters, since E_G^{Opt} is the only empirical parameter used in the summation.

In Fig. 3.8 calculated terms $\langle I_G \rangle_{cal}$ and $\langle I_C \rangle_{cal}$, obtained in such a semiempirical approach, are compared with the experimental I_G and I_C values. As we see, if one takes account of experimental errors, all calculated values of I_G and I_C for Ac, Nph, and Pc lie within the error bars of the experimental values according to Table 3.5. These results may be regarded as additional proof of the consistency and plausibility of the submolecular approach in the electronic polarization energy calculations by Munn and coworkers[37,100] (see Table 3.4).

Figure 3.8 shows another important characteristic of the effective electronic polarization term, P_{eff}. While the terms I_G, I_C, E_G^{Opt}, and E_G^{Ad} decrease and the term A_G^V increases with increasing number N of condensed rings in polyacene molecules, the values of P_{eff}^+ and P_{eff}^- remain practically constant (see also Tables 3.4 and 3.6). The constant nature of P_{eff}^+ according to Eq. (3.5.33) has been observed also for a number of other polycyclic aromatic hydrocarbons.[309] By examining the data in Table 3.4, it may be seen that the constancy of the effective polarization energy P_{eff} is most probably due to the fact that the decrease in the component P_{id} of the charge-induced dipole interaction with increasing size of the molecule is compensated by the increasing charge-quadrupole interaction term W_{Q_0}.

The constantcy of the effective polarization energy terms P_{eff} and W_{eff} permits estimation of the adiabatic energy gap E_G^{Ad} for other hydrocarbon crystals, when the values of related energy parameters I_C and A_G are known (see Ref. 307 and Chapter 4).

3.5.7. Possible Improvements and Corrections

In our opinion, the introduction and computation of charge carrier *permanent* quadrupole interaction term W_Q by Munn and coworkers[37,100] may be regarded as one of the most important improvements in the electronic polarization energy calculations (see Section 3.5.5). This theoretical refinement affected radical changes in our concepts of ionized state energy structure in OMC (see Chapter 4, Figs. 4.8–4.10). Now the question is whether a possible induced quadrupole moment of neutral molecules could appreciably influence the calculation results. Hug[138] estimated the quadrupole polarization for the neutral molecule next to the charged one and found that the resulting correction in Nph might be of the order 5×10^{-3} eV. This contribution may therefore be ignored.

A further source of error may originate from dipole hyperpolarizability [see Eq. (3.5.1)]. Significant only at high electric fields, these may yield some contributions just from nearest neighbors of the molecular ion. The first (quadratic) hyperpolarizability β is zero in centrally symmetrical molecules, so that the second (cubic) hyperpolarizability comes into operation. Gutmann and Lyons[122] have shown that, for usual nearest neighbor molecules, the contribution does not exceed 0.003 eV which is certainly negligible. This result was supported by Hug.[138]

The quantum corrections have already been discussed in Section 3.4.2. Unfortunately, at present we do not see how to estimate these for the submolecular model (see Section 3.5.4). It should only be repeated that the above quantum theory reproduces a linear set of equations for dipole moments [see Eqs. (3.4.11 and 3.4.12)] but does not exactly reproduce (in the same classical terms) the energy needed to polarize individual molecules [compare the first terms on the right-hand sides of Eqs. (3.4.7) and (3.4.15)]. Though this most probably, is not, a source of considerable correction, one should allow for this possibility, at least in principle.

Relevant corrections could, however, appear in connection with the possibility of the charge to move. Notice that in Eq. (3.4.2), all the terms proportional to overlaps (S) of spinorbitals localized on neighboring sites have already been omitted. The question is what happens when $S \neq 0$.

Glaeser and Berry[117] attempted to include the electronic polarization to energy band structure calculations, finding that inclusion of polarization reduces the band-width δE by about 40–50% (see Figs. 3.2 and 3.3). The resulting band structure is then the electronic polaron band structure (though strongly shifted to higher energies as far as one ignores the electron polarization energy P). Qualitatively the same results and arguments supporting these were found by the authors of Refs. 57 and 284 (see Section 3.2.3). Simple physical reasoning shows, that for a moving charge, very little happens to the electron polarization cloud if it has enough time to accommodate itself to the new position of excess charge carrier (see Section 3.3). This situation takes place when

$$\hbar / \delta E \gg \hbar / E^{\mathrm{exc}} \approx \tau_e , \qquad (3.5.35)$$

where δE is the polaron band width, E^{exc} is the electron excitation energy, and τ is the shortest time needed for reconstruction of the electronic states ($\approx 10^{-15}$–10^{-16} s) (see Fig. 3.9). In the opposite case the polarization energy becomes greatly reduced.[57] Such simple arguments, however, apply when almost relaxed electron states are concerned. In general, electronic as well as molecular and lattice polarons are many-body concepts (quasiparticles). Corresponding Green's functions yield, in principle, nonzero density of quasiparticle states in the energy regions corresponding to both relaxed and unrelaxed states. Thus, in the language of single-particle concepts, the relaxed and unrelaxed polaron states coexist.

3.6. CHARGE CARRIER INTERACTION WITH INTRA- AND INTERMOLECULAR VIBRATIONS. MOLECULAR AND LATTICE POLARONS

Following previous practice in this book, we shall present our discussions in this section at two levels. First, we give a phenomenological introduction of the concepts of molecular and lattice polarons. This will be followed by a Hamiltonian description of the formation and properties of the molecular and lattice polarons, based on further extension and specification of some more general polaron theory, as presented in Chapter 1, Section 1B.8.

3.6.1. Molecular (Vibronic) and Lattice Polarization Phenomena — a Phenomenological Approach

As we have seen (see Sections 3.2.3 and 3.3), *electronic* polarization may be regarded as a very fast, practically inertialess process, in comparison to the char-

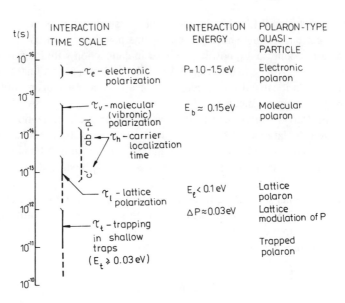

FIGURE 3.9. Various types of interaction between an excess charge carrier and the electronic and nuclear subsystems in anthracene-type crystal (modified according to Ref. 332). Typical interaction time scales, interaction energies, and corresponding polaron-type quasiparticules are shown.

acteristic localization (residence) time τ_h of a charge carrier on separate lattice sites in an Ac-type crystal, according to the inequality $(3.2.16)\tau_h \gg \tau_e$ (see Fig. 3.4a).

The interaction between charge carrier and the *nuclear* subsystem of the surrounding lattice, namely, with intra- and intermolecular vibrations, leads to a nuclear polarization which is a considerably slower process in comparison with exclusively electronic polarization processes.

For a qualitative analysis of such interaction phenomena it is first necessary to estimate the range of characteristic relaxation time scales of the above processes and to compare these with localization times τ_h of the charge carrier.

For such an estimate one may apply, as in the case of electronic polarization, the uncertainty relation [see Eq. (3.2.15)].

Such estimates are schematically shown in Fig. 3.9. This diagram illustrates the hierarchy of interaction time scales and the corresponding interaction energies. Time scales of polaron-type quasiparticle formation as a result of charge carrier interaction with electronic and nuclear subsystems in an Ac crystal are also shown.

However, one should notice that the relaxation time scales in Fig. 3.9, obtained according to the uncertainty relation, actually are as the upper limits of the inter-action rates of given processes. Thus, e.g., in case of electronic polarization, the characteristic relaxation time $\tau_e = 10^{-16} - 10^{-15}$ s corresponds to excitonic excita-tion, having a typical energy $\Delta E_{ex} = 2-4$ eV [see Eq. (3.2.15)]. In electronic processes involving a smaller portion of energy ΔE change the corresponding τ_e

value may be, as indicated in Section 1B of Chapter 1, considerably greater.

Thus, the data presented in Fig. 3.9, and in Table 3.7, should be regarded as tentative, appropriate for qualitative, phenomenological analyses.

3.6.1A. Interaction with Intramolecular Vibrations. Molecular Polaron

For estimating the relaxation time for charge carrier interaction with intramolecular vibrations (vibrons) we may use the vibronic energy range $E_v = \hbar\omega_v$ from 480 to 3000 cm^{-1}, which is characteristic of polyacenes (see Ref. 332 and Section 1A.4.1). We thus obtain the range of the characteristic vibronic relaxation time $\tau_v = \hbar/E_v = 1.7\times10^{-15}-1.1\times10^{-14}$ s (see Table 3.7). A comparison of this time with the interval of change in localization time (hopping time) of the charge carrier τ_h in various crystallographic directions of an Ac crystal (see Fig. 3.9) shows that vibronic relaxation of a molecule with a localized charge carrier is a faster process than hopping of the carrier onto a neighboring molecule, i.e.,

$$\tau_h \gtrsim \tau_v . \tag{3.6.1}$$

In the course of vibronic relaxation of a molecule, changes in bond length and vibrational frequencies take place as a result of redistribution of the localized charge (see, e.g., Refs. 88 and 89). Accordingly, the molecule passes during the charge carrier localization time τ_h from an equilibrium configuration of nuclei in the neutral state into an equilibrium configuration of the ionized state. This process has been schematically illustrated for a single vibrational mode in Fig. 3.10b.

To describe phenomenologically the vibrationally relaxed molecular state, Silinsh and coauthors[334,344,348] have introduced and developed the model of the nearly small molecular polaron (MP). According to this model, a nearly small *molecular polaron* is formed as the result of interaction of a charge carrier with intramolecular vibrations of the molecule on which it is localized during the residence (localization) time τ_h, and also due to interaction with polar IR-active vibrational modes of the nearest-neighbor molecules.[334,348]

The possibility of the existence of this kind of interaction mechanism between a localized charge (molecular ion) and dipole-active vibrational modes of neighboring molecules was suggested by Salaneck et al.[305] to explain the appearance of a temperature-dependent component in the energy distribution spectra of photo-emitted electrons in organic solids (see also Ref. 310).

The formation of a molecular polaron of nearly small radius is schematically illustrated in Fig. 3.10a.

The plausibility of formation of a MP as a result of vibronic relaxation of a charge carrier has been unambiguously confirmed by the ionized state energy structure studies in polyacene crystals (see Ref. 307 and Chapter 4).

The energy gain through vibronic relaxation in the process of molecular polarization (as in that of electronic polarization) lowers the intrinsic energy of the localized charge carrier and determines, as will be shown in Chapter 4, the adiabatic

TABLE 3.7. *Types of Interaction Between an Excess Charge Carrier and the Local Lattice Environment in Anthracene-Type crystals, Hierarchy of Their Interaction Time Scales, Ranges of Typical Interaction Energies and Polaron-Type Quasiparticles formed[a]*

Type of interaction	Interacting subsystem of the crystal	Typical excitation energy of the sub-system	Characteristic interaction time (s)	Typical inter-action energy (eV)	Polaron-type quasi-particles formed
Electronic polarization	Electronic subsystem of the crystal	$h\nu = 2\text{--}4$ eV	$(1.6\text{--}3.3)\times10^{-16}$	$1\text{--}1.5$	Electronic polaron
Molecular (vibronic) polarization	Nuclear sub-system of the molecules. Intra-molecular vibrations (vibrons)	$\hbar\omega = 480\text{--}3000$ cm^{-1} (0.06–0.38 eV)	1.7×10^{-15}– 1.1×10^{-14}	$0.1\text{--}0.2$	Nearly small-radius molecular polaron
Lattice polarization	Lattice sub-system Intermolecu-lar vibra-tions (optic-al and acoustic phonons)	Optical pho-nons 50–120 cm^{-1} (0.006–0.015 eV) Acoustic phonons < 50 cm^{-1} (< 0.006 eV)	4.4×10^{-14}– 1.1×10^{-13} $> 10^{-13}$	< 0.1	Lattice polaron of small radius

[a]Taken from Ref. 351.

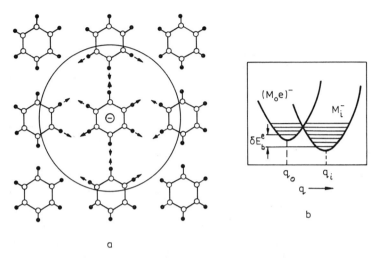

FIGURE 3.10. (a) Schematic representation of molecular **vibronic** polarization (formation of a molecular polaron of nearly small radius) in a molecular crystal[334,351] (for clearness the change of coordinates is conditionally shown for only one vibrational mode, namely, for stretching vibrations of the C-H bond in the molecular ion and nearest neighboring molecules); (b) schematic representation of the binding energy of an electron δE_b^e, i.e., of the vibronic energy relaxation in the formation of the ion M^- (the potential energy curves are shown for only one vibrational mode); q_o, q_i are mean bond length of the equilibrium state of a neutral molecule (M_0) and the ion (M^-), respectively.

energy gap E_G^{Ad} of the crystal (see Figs. 4.8–4.10).

The effective formation energy of a molecular polaron $(E_b)_{eff}$ in polyacene crystals can be determined as the difference between the optical E_G^{Opt} and the adiabatic E_G^{Ad} energy gap of the crystal:[307,334]

$$(E_b)_{eff} = \frac{E_G^{Opt} - E_G^{Ad}}{2}, \tag{3.6.2}$$

where the magnitude $(E_b)_{eff}$ is shown to be the total formation energy of the MP: $(E_b)_{eff} \equiv (E_b)_{MP}$ (see Fig. 3.10).

The reported experimental values of E_G^{Opt} and E_G^{Ad} and the corresponding $(E_b)_{eff}$ values obtained according Eq. (3.6.2) for polyacene crystals are presented in Table 3.8.

It has been shown that the $(E_b)_{eff}$ value is greater (by about 0.05 eV in the case of Ac) than the evaluated magnitude of formation of a molecular ion $(E_b)_i$, namely, $(E_b)_{eff} = (E_b)_{MP} > (E_b)_i$.[351] This confirms the additional contribution of charge carrier interaction with the polar vibrational modes of nearest-neighbor molecules in the term $(E_b)_{eff}$ (see Fig. 3.10). One can assume that about 2/3 of the total effective energy of a MP is supplied by the formation of the ionic state $(E_b)_i \approx 0.10$ eV, and 1/3 through interaction with surrounding molecules (0.05 eV).[351] [The $(E_b)_i$ value ≈ 0.1 eV has been evaluated empirically for Nph and Ac ions found from the

TABLE 3.8. *Experimental Values of Optical E_G^{Opt} and Adiabatic E_G^{Ad} Energy Gaps and Corresponding Formation Energies $(E_b)_{eff}$ of a Molecular Polaron in Polyacene Crystals, according to Ref. 307* [a]

Crystal	E_G^{Opt} (eV)	E_G^{Ad} (eV)	$(E_b)_{eff}$ (eV)
Anthracene	4.40	4.10	0.15
Tetracene	3.43	3.13	0.15
Pentacene	2.83	2.47	0.18

[a]*Note:* For references of reported E_G^{Opt} and E_G^{Ad} values see Tables 4.2 and 4.4; see also Figs. 4.8–4.10.

changes of characteristic vibrational IR and Raman frequencies for ion $(\omega_i)_k$ and neutral molecules $(\omega_0)_k$, respectively; see Ref. 322, p. 95. Earlier reported quantum chemical calculation estimates of $(E_b)_i$ for the Ac ion are 0.07 eV,[327a] 0.09 eV ,[136b] and 0.11 eV.[89] These calculation data can thus be seen to agree satisfactorily with empirically estimated $(E_b)_i$ value.[322] Recently Klimkans and Larsson[185a] have performed refined quantum chemical calculations of so-called reorganization energy λ, equivalent to $(E_b)_i$, in case of formation of Ac and Nph negative molecular ions. These authors used *ab initio* methods with extended basis set including electronic correlation in terms of MP2 (Møller-Plesset) approach. They have obtained the following values of $\lambda = (E_b)_i$: 0.07 eV for Ac and 0.10 eV for Nph ions.

The nearly small MP model will be further extended and extensively applied throughout this book for a phenomenological description of the energy structure of ionized states (Chapter 4), charge carrier photogeneration (Chapter 6), and transport (Chapter 7) processes in OMC.

Thus, a thermalized, vibrationally relaxed charge carrier is a complex quasiparticle "dressed" both in a cloud of electronic (see Fig. 3.5) and vibronic polarization (see Fig. 3.10). The charge carrier in OMC may thus simultaneously form a *dual polaron* — a large-radius *electronic* and nearly small *molecular* one. In other words, a MP is actually a vibrationally relaxed electronic polaron.

Treutlein *et al.*[390a] recently studied the *relaxation dynamics* of nuclear (vibronic) polarization around a pair of separated charge carriers, localized on photoionized special pair of porphyrin molecules in photosynthetic bacterial reaction center (RC) complex of *Rhodopseudomonas viridis*. The authors used the molecular dynamics (MD) simulation techniques employing the stochastic boundary method to limit the simulation region to the central part of the RC. A total of 5726 atoms inside the sphere of 29 Å around the porphyrin ions were included.

The most relevant results of these MD studies are:[390a]

1. The total nuclear (vibronic) polarization energy of the molecules of RC around the charge pair (CP) state was estimated to be ~ 0.37 eV.

2. The estimated relaxation time of the polarization τ_v was about 200 fs $= 2\times10^{-13}$ s.

3. The relaxation time τ_v and polarization energy were found to be temperature

independent from 300 K down to 10 K.

The authors[390a] conclude that the CP induced molecular rearrangements occur without thermal activation. The potential minima of polarized atoms shift almost adiabatically and do not need to overcome energy barriers to achieve the new equilibrium position around the ions.

One may see that the total polarization energy of RC molecules around the CP state is of the same order of magnitude as $2(E_b)_{\text{eff}}$ for polyacene crystals, where $(E_b)_{\text{eff}}$ is the effective formation energy of a single MP, (for Pc $2E_b = 0.36$ eV, see Table 3.8) The estimated τ_v value corresponds a typical intermolecular polarization time (see Table 3.7). These MD simulation results give a new insight in the dynamics of molecular polarization processes.

3.6.1B. Interaction with Intermolecular Vibrations. Lattice Polaron

To estimate the relaxation time scale of charge carrier interaction with intermolecular vibrations, we can use the energy range of optical (translational and libronic) lattice phonons, characteristic of polyacene crystals, i.e., $E_l = \hbar\omega_l$ between 50 and 120 (see Ref. 332 and Section 1A.4.2). As a result we get the range of the characteristic lattice relaxation time for optical phonons $\tau_l = 4.4 \times 10^{-14} - 1.1 \times 10^{-13}$ s (see Table 3.7 and Fig. 3.9). For acoustic phonons $E_l < 50$ cm^{-1} and, correspondingly, $\tau_l > 10^{-13}$ s.

A comparison of the given range of τ_l with the range of the charge carrier localization time τ_h in an Ac crystal (see Fig. 3.9) shows that in the **ab** plane of the crystal $\tau_h < \tau_l$ (only in the **c'** direction we observe $\tau_h \geq \tau_l$).

Accordingly, the process of lattice polaron formation in case of the motion of a charge carrier in the **ab** plane of an Ac crystal is slow, and most likely, the polaron does not manage to form.

It ought to be stressed that the mechanisms of lattice polaron formation in ionic crystals and in Ac-type crystals differ entirely. In ionic crystals the ions get displaced in the process of lattice polaron formation under the effect of direct Coulombic interaction with the charge carrier. In molecular crystals, on the other hand, the displacement of molecules from equilibrium position, as schematically shown in Fig. 3.11a, takes place owing to interaction of the charge (ion) with dipoles induced on neighboring molecules. Charge-dipole interaction in OMC is weaker and acts over smaller distances than Coulomb interaction between a charge and ions in ionic crystals. In addition, one ought to keep in mind that the molecular mass of polyacenes (178.24 for Ac and 278.36 for Pc) considerably exceeds the mass of ions in alkali halides (22.99 for Na$^+$ and 39.1 for K$^+$).

On the basis of these general physical considerations one may conclude that the formation of a small-radius lattice polaron in the **ab** plane of an Ac-type crystal ought to be unlikely. This agrees with the estimates obtained in the calculations by Vilfan[399] who came to the conclusion that the probability of small-radius lattice polaron formation in Ac-type molecular crystals is negligibly small.

There is also other evidence against the possibility of small-radius lattice polaron

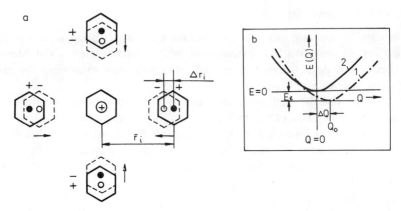

FIGURE 3.11. (a) Schematic representation of lattice polarization (formation of a small-radius lattice polaron in a molecular crystal);[332] r_i is the mean distance between the localized positive charge and the center of the surrounding neutral molecules; Δr_i is the displacement of neutral molecules as a result of the interaction between the charge and the induced dipoles on the neutral molecules. (b) The total energy $E(Q)$ of the small-radius lattice polaron as a function of the configuration coordinate Q (1): E_l is binding energy of the polaron in zero approximation; Q_0 is a coordinate of the equilibrium state of the polaron; (2) energy of elastic deformation of the lattice.[288]

formation in Ac-type crystals, e.g., the temperature dependence of carrier mobility $\mu = \mu(T)$.

In the case of a small-radius lattice polaron (unlike a nearly small radius MP, see Chapter 6) one may expect in the high-temperature range at $T > T_c$ thermally activated mobility of the type $\mu \sim \exp[-E_a/kT]$, where $E_a = 1/2(E_l)_{LP}$ is the activation energy of small-radius polaron mobility (cf. Ref. 288, p. 353, and Fig. 3.11b). Hence, the mobility μ of a small-radius lattice polaron ought to increase with rise in temperature according to exponential law. This obviously is in disagreement with experiment, since we observe in Ac-type crystals just in the high temperature range (up to 300 K) the characteristic dependence $\mu \sim T^{-n}$, where $n > 1$ (see Chapter 7, Section 7.1.3). The $\mathbf{c'}$ direction in the lattice might form an exception. As may be seen from Fig. 3.9, here we have $\tau_h > \tau_l$. However, even in this case the $\mu(T)$ dependences do not testify explicitly in favor of lattice polaron formation (see Section 7.2.3C).

In addition, according to Refs. 118 and 399, the electron-phonon interaction manifests itself chiefly through dynamic modulation of the polarization energy $\Delta P(t)$ (cf. Fig. 3.9). This modulation is of the order of 0.03 eV and is one of the factors that determines the effective conduction level width in OMC at room temperature.

In real molecular crystals the charge carriers may be trapped in shallow traps of structural origin[332] (see also Section 4.6.1). The mean capture time τ_t of carrier in such traps of depth $E_t \geq 0.03$ eV is larger than 10^{-12} s (see Fig. 3.9). Therefore we have $\tau_t \gg \tau_l$ in this case, and local polarization of the lattice is possible, i.e., local deformation due to the approach of the molecules to the trapped carrier (molecular polaron), as a result of interaction of the charge with the dipoles induced on neighboring molecules (see Fig. 3.11a). At such "local" trapping of the charge carrier (molecular polaron) a potential well is formed (see Fig. 3.11b) of depth $\delta E_t = E_l$ of the order of < 0.1 eV (see Refs. 320 and 375). The existing structural trap gets accordingly deeper by this value, i.e., $E_t + \delta E_t$.[332]

3.6.2. Hamiltonian Description of the Formation of Molecular and Lattice Polarons Around Charge Carriers

Pope and Swenberg (Ref. 288, p. 340) as well as Duke and Schein[95] have used a Hamiltonian description as a classification scheme for various limiting types of charge carrier interaction and transport in organic crystals. Čápek and Silinsh (see Ref. 351, p. 177 and Ref. 338) have extended this approach to describe and analyze polaron formation and transport conditions in OMC.

Here we will present an extended Hamiltonian description based on a general treatment of the lattice polaron formation given in Chapter 1, Section 1B.8.

In the introductory portion of Section 1B.8, the lattice polaron concept was presented in a more general way. We discussed its formation around both excitons and charge carriers (electrons and holes). In the present discussion we shall take into account some specific features, characteristics, and peculiarities of charge carriers in OMC.

The most important of these is the presence [even in the crude (rigid) representation, used here] of the site-off-diagonal coupling to phonons which should be taken into account [compare Eqs. (1B.7.3b) or (1B.7.3c) with Eq. (1B.7.3a)].

The second important point is that in the introductory discussion (see Section 1B.8) we did not specifically separate the interaction with intramolecular (high-frequency) and intermolecular (low-frequency) phonons (see Section 1A.4) The quasiparticle formed as a result of these interactions was called simply a "lattice polaron" (see Section 1B.8). In the present discussion we shall distinctly separate these interactions to describe a molecular (vibronic) and a "genuine" lattice polaron (see Figs. 3.10 and 3.11).

For the sake of clarity let us use a simplified notation. Creation and annihilation operators a_n^+ and a_n (related to individual molecules or sites n) will refer to the charge carriers, $b_{\lambda n}^+$ and $b_{\lambda n}$ to the intramolecular (vibronic) quanta while c_{nn}^+ and c_{nn} will refer to the intermolecular lattice phonons. The reader should realize that in the last two cases the transition from the wave vector indices to those identifying individual sites or molecules can be always performed with the help of transformation formulas like (1B.8.11).

On the basis of Eqs. (1B.2.10b) [or (1B.2.10c) in case of holes], (1B.2.4c), and (1B.7.3b) [or (1B.7.3c)], we can write the simplest (for the application to real molecular crystals) Hamiltonian for the charge carriers interacting with intra- and intermolecular phonons in the usual notation as follows:

$$H = H_0 + H_T + H_v + H_v' + H_L + H_T' + H_T'' \tag{3.6.3}$$

where

$$H_0 = \sum_n Ea_n^+a_n + \sum_{\lambda,\,\mathbf{k}} h\nu_\lambda(\mathbf{k})(b_{\lambda\mathbf{k}}^+b_{\lambda\mathbf{k}}+1/2) + \sum_{\mu,\mathbf{k}} h\nu_\mu(\mathbf{k})(c_{\mu\mathbf{k}}^+c_{\mu\mathbf{k}}+1/2).$$

$$\tag{3.6.4}$$

The first term of the equation describes the total energy of the system of charge carriers, the second one that of vibronic, and the third one that of lattice phonons. E is the energy required for the creation of a localized charge carrier in the system.

The second term H_T describes the charge carrier transfer from the site (molecule) m to the site (molecule) n due to the hopping (transfer or resonance) integral J_{nm}

$$H_T = \sum_n \sum_m J_{nm}a_n^+a_m. \tag{3.6.5}$$

The Hamiltonian H_v introduces the site-diagonal part of the charge carrier interaction with *intramolecular* vibrations

$$H_v = \sum_\lambda \sum_n g_{n\lambda}h\nu_\lambda a_n^+a_n(b_{\lambda n}+b_{\lambda n}^+) + \sum_\lambda \sum_n \delta g_{n\lambda}h\nu_\lambda a_n^+a_n(b_{\lambda n}+b_{\lambda n}^+)^2$$

$$\tag{3.6.6a}$$

with

$$\nu_\lambda = 1/N\sum_{\mathbf{k}} \nu_\lambda(\mathbf{k}). \tag{3.6.6b}$$

Here the terms $(b_{\lambda n}+b_{\lambda n}^+)$ and $(b_{\lambda n}+b_{\lambda n}^+)^2$ describe the linear and quadratic (in powers of vibron coordinate of the mode λ) interaction of molecule n, respectively; $g_{n\lambda}$ and $\delta g_{n\lambda}$ are the corresponding interaction constants.

By term H_v' we introduce the site-diagonal interaction of the charge carrier with the *intramolecular* vibrations on neighboring molecules:

$$H_v' = \sum_{n\neq m} \sum_\lambda g_{n\lambda}^m h\nu_\lambda a_n^+a_n(b_{\lambda m}+b_{\lambda m}^+) + \sum_{n\neq m} \sum_\lambda \delta g_{n\lambda}^m h\nu_\lambda a_n^+a_n(b_{\lambda m}+b_{\lambda m}^+)^2.$$

$$\tag{3.6.7}$$

Thus, the terms H_v and H_v' respectively describe the charge carrier interaction with vibronic modes $h\nu_\lambda$ of the same molecule n on which the carrier is localized and with the vibronic modes $h\nu_\lambda$ of the neighboring molecules m. These interactions actually underlie the formation of a quasiparticle — a nearly-small MP (see Fig. 3.10a).

To understand more deeply the dynamics of interaction of the type (3.6.7), and to estimate its role in transfer processes of the molecular polaron, let us briefly discuss (in a way similar to that used in Section 1B.8) the problem of possible delocalization of intramolecular vibrations in the crystal, i.e., the possible existence of a weak dispersion of the phonon frequencies of the intramolecular vibrations.

The existence of a weak dispersion, i.e., $\mathbf{k}$-dependence of the phonon frequency $\nu(\mathbf{k})$, is typical for the lattice vibrations. On the basis of a simple example, we demonstrate that for the appearance of interactions described in Eq. (3.6.7), at least a weak dispersion of the intramolecular vibrations (i.e., $\mathbf{k}$-dependence) is needed.

A generalized Hamiltonian of the carrier-phonon interaction as applied to the intramolecular vibrations can be written as [Eq. (1B.7.3b)]:

$$\bar{H}_v = \sum_\lambda \sum_n \sum_k g_{n\lambda}(\mathbf{k}) h \nu_\lambda(\mathbf{k})[1/\sqrt{N}]e^{i\mathbf{kn}}a_n^+ a_n(b_{\lambda\mathbf{k}}+b_{\lambda,-\mathbf{k}}^+), \quad (3.6.8)$$

where $\mathbf{k}$ is the wave vector of intramolecular vibrations.

Here $\mathbf{n} \equiv \mathbf{r}_n$ designates the equilibrium position in the space of molecule n; for simplicity, we treat just the linear term in the nuclear displacements. For the relation of $b_{\lambda\mathbf{k}}(b_{\lambda\mathbf{k}}^+)$ to $b_{\lambda n}(b_{\lambda n}^+)$, see Eq. (1B.8.11). For simplicity, we also treat just the case of a crystal with a single molecule per unit cell, i.e., we assume that $g_{n\lambda}(\mathbf{k})$ is n-independent.

If we now assume that, in Eq. (3.6.8), $g_\lambda(\mathbf{k})$ and $\nu_\lambda(\mathbf{k})$ are $\mathbf{k}$-independent, i.e., $g_\lambda = const$ and $\nu_\lambda = const$ and, correspondingly, we have zero dispersion, $\bar{H}_v$ becomes equal to the first term on the right-hand side of (3.6.6a). Thus, with zero dispersion of vibronic intramolecular vibration frequencies ν_λ no carrier interaction with intramolecular vibrations of the neighboring molecules of the type (3.6.7) should appear. As a result the term H_v' is zero. One can thus assume that the first (linear) term in H_v' appears actually as a correction to the first (linear) term in H_v because of the $\mathbf{k}$-dependence of $g_\lambda(\mathbf{k})$ and $\nu_\lambda(\mathbf{k})$, i.e., owing to weak dispersion of the coupling constant and frequency. Similarly, the second (quadratic) term in H_v' can be interpreted as a corresponding correction to the second (quadratic) term in H_v. So the existence of at least a weak dispersion of the intramolecular vibration modes is a necessary condition for formation of the nearly small MP (Fig. 3.10a).

Hamiltonian H_L in turn describes the diagonal (in molecular indices) interaction of the charge carrier with the lattice phonons. As mentioned before, the dispersion (i.e., $\mathbf{k}$-dependence) of the phonon frequencies is here more clearly expressed. Thus, the term H_L may be expressed, in analogy with $\bar{H}_v$, as

$$H_L = \sum_\mu \sum_n \sum_k f_{n\mu}(\mathbf{k}) h \nu_\mu(\mathbf{k})[1/\sqrt{N}]e^{i\mathbf{kn}} \cdot a_n^+ a_n(c_{\mu k}+c_{\mu-k}^+)$$

$$+ \sum_{\mu_1 \mu_2} \sum_n \sum_{k_1 k_2} \delta f_{n\mu_1\mu_2}(\mathbf{k}_1\mathbf{k}_2)\sqrt{[h\nu_{\mu_1}(\mathbf{k}_1)h\nu_{\mu_2}(\mathbf{k}_2)]}$$

$$\cdot [1/\sqrt{N}]e^{i(\mathbf{k}_1+\mathbf{k}_2)\mathbf{n}}a_n^+ a_n(c_{\mu_1 k_1}+c_{\mu_1-k_1}^+)(c_{\mu_2 k_2}+c_{\mu_2-k_2}^+). \quad (3.6.9)$$

Here the index μ refers to the corresponding lattice vibrational modes, $c_{\mu k}$ and $c_{\mu k}^+$ are annihilation and creation operators for these modes, and $f_{n\mu}$ and $\delta f_{n\mu_1 \mu_2}$ are the linear and quadratic interaction constants, respectively.

The off-diagonal (in the molecular indices) interaction of the charge carriers with both the intramolecular and the intermolecular vibrations can be viewed as essential terms increasing the carrier transfer (hopping) probability between neighboring molecules — see, e.g., Eq. (1B.7.3b) with $p \neq q$. In an adiabatic picture, this interaction may be regarded as due to the influence of vibrations on the transfer (hopping) integral J_{nm}. The corresponding Hamiltonian H_T' describing the carrier off-diagonal interaction with intramolecular vibronic modes may be presented up to linear terms as

$$H_T' = \sum_{n \neq m} J_{nm} a_n^+ a_m [1/\sqrt{N}] \sum_{k,\lambda} g_{nm\lambda}(\mathbf{k}) \times \exp[i(1/2)\mathbf{k}(\mathbf{n}+\mathbf{m})](b_{\lambda k} + b_{\lambda - k}^+).$$

(3.6.10)

As may be seen, H_T' contributes to the probability of carrier transfer from molecule m to molecule n, due to the coupling to the intramolecular vibrations or, in a more conceptual sense, due to the delocalization of the molecular polaron (MP).

Similarly, the off-diagonal interaction of the carriers with the lattice phonons may be written as

$$H_T'' = \sum_{n \neq m} J_{nm} a_n^+ a_m [1/\sqrt{N}] \sum_{k,m} f_{nm\mu}(\mathbf{k}) \times \exp[i(1/2)\mathbf{k}(\mathbf{n}+\mathbf{m})](c_{\mu k} + c_{\mu - k}^+).$$

(3.6.10a)

As may be seen from Eq. (3.6.10a), term H_T'' yields the lattice phonon-stimulated increase of transfer probability of charge carriers. A similar phonon stimulated transfer mechanism has been proposed by Duke and Schein[95] for explanation of the enlarged electron hopping probability in the c'-direction in naphthalene owing to rotational (libration) motion of molecules in the lattice (see Fig. 3.12).

In terms of this model it is possible to explain qualitatively, why in Ac-type crystals the mobility of an electron in the c'-direction has practically the same value as in the $\mathbf{a}$ and $\mathbf{b}$ directions (e.g., in Ac at 300 K $\mu_{aa}^e = 1.6$; $\mu_{bb}^e = 1.0$ and $\mu_{c'c'}^e = 0.4$ cm^2/Vs; see Ref. 288, p. 338), although the calculated values of the transfer integrals J_{nm} for the c'-direction is considerably smaller than in directions $(0,1,0)$ and $(1/2,1/2,0)$ in the $\mathbf{ab}$ plane (see Ref. 117).

It ought to be stressed here that the above classification scheme considerably simplifies the real picture. First, the interaction (coupling) constants $g_{n\lambda}(\mathbf{k})$, $f_{n\mu}(\mathbf{k})$, $g_{nm\lambda}(\mathbf{k})$, $f_{nm\lambda}(\mathbf{k})$, as well as δg and δf, may differ appreciably for different vibrational modes.

For example, in Nph the localized carrier interacts in different ways with various *intramolecular* modes (see Ref. 332, p. 97): the modes of B_{1u}, B_{2u}, B_{3u}, and B_{3g} symmetry yield a relatively strong change of their characteristic frequencies as the result of the second quadratic term on the right-hand side of Eq. (3.6.6a). On the

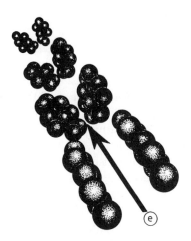

FIGURE 3.12. A diagram of one of the possible mechanisms of charge transport in a naphthalene crystal (after Ref. 95). A localized electron (indicated by an arrow) hops from one molecule to the next when the rotation of the first molecule significantly increases the hopping probability.

other hand, vibrational modes of the A_g type of symmetry are relatively insensitive to the presence of an excess charge on the molecule. In the case of interaction of the localized charge carrier with the *intramolecular* vibrations of the neighboring molecules, the strongest interaction is to be expected with the polar, IR-active modes of odd symmetry, i.e., B_{1u}, B_{2u}, and B_{3u} in case of polyacenes. So particular coupling constants should be ascribed to individual vibrational modes.

The above scheme allows systematization of individual interaction types in a crystal and enables one to estimate their relative contribution, specifying the dominant ones.

For example, in the above scheme, one can specify the dynamic diagonal ($n = m$) Hamiltonians H_v, H_v', and H_L, and the off-diagonal ($m \neq n$ in molecular indices) Hamiltonians H_T' and H_T''.

Correspondingly, the first terms in Eqs. (3.6.6a), (3.6.7), and (3.6.9) describe the linear, while the second terms the *quadratic* interaction in vibrational coordinates (nuclear displacements).

Further, specifying the interaction Hamiltonians with high-frequency intramolecular modes (Hamiltonians H_v, H_v', and H_T') and low-frequency intermolecular (lattice) modes (phonons) (Hamiltonians H_L and H_T'') enables one to analyze conditions of formation of the molecular and *lattice* polarons, and mechanisms of their transfer.

In turn, Hamiltonians H_v, H_v', and H_L determine the tendency of *localization* of charge carriers, due to the diagonal (in the molecular indices) disorder, as well as energetical aspects of this localization.

On the other hand, Hamiltonian H_T determines the carrier's transfer probability between neighboring molecules, i.e., its tendency to *delocalization*. In addition, Hamiltonians H_T' and H_T'', contribute to this delocalization describing the influence of the intramolecular and lattice vibrations, respectively, on the transition probability of carriers.

Thus, both the terms describing the *localization* (H_v, H'_v, H_L) and terms leading to the *delocalization* (H_T, H'_T, H''_T) of the charge carrier are included in the total Hamiltonian [see Eq. (3.6.3)]. The final result of these competing interactions depends in many respects on the relative values of the corresponding coupling constants $g_{n\lambda}$, $g^m_{n\lambda}$, $\delta g_{n\lambda}$, $\delta g^m_{n\lambda}$, $f_{n\mu}(\mathbf{k})$, $g_{nm\lambda}(\mathbf{k})$, and $f_{nm\lambda}(\mathbf{k})$.

In the case when the transfer (hopping) integral J_{mn} dominates over other characteristic quantities of the problem with dimension of energy, the first two terms of the total Hamiltonian (3.6.3) become decisive. Thus, in this case the Bloch single-particle states of a simple band model emerge as dominant ones. This situation is typical of metals and covalent semiconductors where J_{mn} is of the order of 1 eV, and as a result single particle conduction and valence bands are formed. Thus, the corresponding energy eigenvalues are then determined mostly by the dispersion law $E = E(\mathbf{k})$ of the delocalized charge carrier. In this case the dynamic disorder (H_L) terms affect the carrier's transfer via its scattering between these Bloch states, actually causing no localization.

In OMCs, however, the nearest-neighbor transfer (hopping) integrals J_{mn} are considerably smaller: $J_{mn} \leq 0.02$ eV (see Ref. 332, p. 36). In this case the value of J_{mn} is presumably of the same order of magnitude, or even less than the energy of carrier's interaction with intramolecular vibrations, i.e., $|J_{mn}| \leq |g_{n\lambda}| h\nu_\lambda$. As a result, in the total Hamiltonian (3.6.3) the terms H_v and H'_v become dominant over other interaction terms.

Consequently, in this case the charge carrier states should be described in terms of the polaron theory, i.e., in terms of a nearly small MP model (see Fig. 3.10a). In contrast to the lattice polaron (see Fig. 3.11), the molecular centers of gravity remain fixed: only the nuclei of the molecular ion and those of the nearest neighboring molecules are shifted.

The motion of such a charge carrier, "dressed" into a cloud of local deformation of the nuclear subsystem (see Fig. 3.10a), can be phenomenologically described by introducing a temperature-dependent effective mass, i.e., $m_{\mathrm{eff}}(T) > m_e$ [see Eq. (1B.8.18) and simulation results in Chapters 6 and 7].

Although, in this case, a single-electron approximation is in principle inapplicable, the contemporary many-body theory allows one to preserve some single-particle notions introducing the concept of *quasiparticles*. The molecular polaron, formed by one electron, "dressed" into the polarization cloud of the shifted nuclei, is one of the examples of the quasiparticles. This "heavy" quasiparticle, i.e., the nearly small MP, can be physically visualized as a partly delocalized ion in a neutral molecular lattice (see Fig. 3.10a).

Since the localization (residence) time of the charge carrier τ_h in an Ac-type crystal is long enough for establishing the polarization of the nuclear subsystem, namely, $\tau_h \geq \tau_v$ (see Fig. 3.9), we have in fact to deal with a double-polaron; the charge carrier is "dressed" in a cloud of both electronic and nuclear (vibronic) polarization (see Figs. 3.5 and 3.10). Thus, as already mentioned, such a dual polaron may actually be regarded as a vibrationally relaxed electronic polaron.

It will be shown in Chapters 6 and 7, analyzing the results of computer simulation of charge carrier transport processes in polyacene crystals, that the MP can move by hopping (stepping via tunneling) without activation energy, yielding a temperature-dependent mobility $\mu \sim T^{-n}$, with $n > 1$, typical for Ac-type crystals.

In this connection it should be emphasized that the off-diagonal interaction of the charge carrier with the intramolecular vibrations of the nearest-neighbor molecules, according to Eq. (3.6.10), contributes essentially to the hopping probability of the molecular polaron.

Concerning the lattice vibrations, it is very likely that $|J_{nm}| > |f_{nm}(\mathbf{k})| h \nu_\mu(\mathbf{k})$ and $|\delta f_{n\mu_1 \mu_2}(\mathbf{k}_1, \mathbf{k}_2)| \times [h\nu_{\mu_1}(\mathbf{k}_1) \cdot h\nu_{\mu_1}(\mathbf{k}_2)]^{1/2}$ as well as $|f_{nm\mu}(\mathbf{k})| < 1$ for practically all the lattice modes. Consequently, the lattice polaron formation is improbable.

As mentioned before, a typical feature of lattice vibrations is, in addition to low phonon energy, the presence of some dispersion, i.e., the $\mathbf{k}$-dependence of $\nu_\mu(\mathbf{k})$. This effect is to be kept in mind when discussing vibrational interactions in the crystal. The existence of dispersion implies, first, that lattice excitations cannot be localized on any single molecule. As an example, let us consider a crystal with one molecule in the elementary cell. Let $|0\rangle$ be the ground state of the lattice and we excite, at the initial time $t = 0$, one phonon at molecule l in the branch μ (the creation operator then reads $c_{l\mu} = [1/\sqrt{N}] \Sigma_\mathbf{k} e^{-i\mathbf{kl}} c_{\mathbf{k}\mu}^+$. Then the state of the lattice at $t > 0$ can be described by the expression

$$[1/\sqrt{N}] \sum_\mathbf{k} e^{-i\mathbf{kl}} c_{\mathbf{k}\mu}^+ e^{-i\nu_\mu(\mathbf{k})t} |0\rangle. \qquad (3.6.11)$$

Thus, the probability amplitude of finding the excitation at molecule l' is given by projection of Eq. (3.6.11) to

$$c_{1'\mu}^+ |0\rangle = [1/\sqrt{N}] \sum_\mathbf{k} e^{-i\mathbf{kl}'} c_{\mathbf{k}\mu}^+ |0\rangle. \qquad (3.6.12)$$

Consequently, the corresponding probability equals

$$P_{l'} = \left| [1/\sqrt{N}] \sum_\mathbf{k} e^{-i\mathbf{k}(\mathbf{l}' - \mathbf{l}) - i\nu_\mu(\mathbf{k})t} \right|^2. \qquad (3.6.13)$$

It can be shown that $P_{l'}$ tends to $\delta_{ll'}$ if there is no dispersion, i.e., if $\nu_n(\mathbf{k})$ is $\mathbf{k}$-independent.

Consequently, when localizing an electron at a molecule forming a molecular ion in any (excited or ground) state, the phonon escape from the molecular ion is one of the channels of the ion relaxation. In addition to that, lattice phonon absorption processes are also possible, so long as the lattice temperature is nonzero.

As a result of these processes, the energy level of the charge carrier, localized on a molecule, is broadened (thus obtaining the quasiparticle character) even when the charge carrier is immobile. This many-body effect can in principle be one of the sources of small dynamic deviations ΔP of the electronic polarization energy P of the charge carrier in a molecular crystal from a discrete, strictly determined value.

This interpretation agrees well with the results of Ref. 118 leading to the conclusion that electron-phonon interaction mainly causes the electronic polarization energy fluctuations of the order 0.03–0.04 eV (see Fig. 3.9). This type of interaction is likely to be one of the main factors determining the effective width of the conduction levels of molecular polarons (see Chapter 4, Figs. 4.8–4.10).

Attempts have been undertaken to make the picture of the origin of the charge carrier energy in the molecular crystals more rigorous, starting from experimental data on kinetics. They lead to the conclusion that not a single but more energy levels of the carrier (or a broadened energy level) may possibly exist.[57] It should be stressed that from the quasiparticle point of view, introduction of such a broadened level (or a narrow band) does not necessarily require other electronic levels. The excited state can in this case be considered as a state of the charge carrier surrounded by the phonon cloud. At least, calculations of the electron quasiparticle density of states in model molecular crystals within the dynamic coherent potential approximation yield such a picture.[338] Transfer between such levels (sub-bands) seems to correspond to the vibronic relaxation of the charge carrier in the process of formation of the molecular polaron (see Fig. 4.11).

Thus, it may be concluded, that one of the necessary conditions for the formation and transport of the MP is the existence of slight dispersion of intramolecular vibrational modes, i.e., the dependence of the coupling constant $g_{\lambda n}(\mathbf{k})$ and vibronic frequency $\nu_{\lambda}(\mathbf{k})$ on the wave vector $\mathbf{k}$ [see Eqs. (3.6.6), (3.6.7), and (3.6.8) for the Hamiltonians H_v, H_v', and $\bar{H}_v$ and the related arguments above].

The second, but not so strict condition is the existence of weak dispersion of the lattice vibrations, i.e., the $\mathbf{k}$ dependence of lattice phonon frequencies $\nu_{\mu}(\mathbf{k})$ [see Eqs. (3.6.11–3.6.13) and related arguments].

The energy structure arguments, supporting the idea of formation of a MP in OMC, are given in Chapter 4.

Finally, it should be mentioned that in *real* OMC, possessing local states of structural origin (see Section 4.6.1 and Ref. 332) in addition to the local dynamic disorder, one should also take into account a *static* disorder.

This static time-independent disorder may be regarded as local fluctuations of the structure of the crystal caused by different types of structural lattice defects.

Local static disorder affects both the site energy E of molecule n and the charge carrier transfer probability J_{nm} between molecules m and n. It can be described by the following Hamiltonian (see Ref. 288, p. 340):

$$H_s = \sum_n \delta E_n a_n^+ a_n + \sum_n \sum_m \delta J_{nm} a_n^+ a_m, \qquad (3.6.14)$$

where δE_n and δJ_{nm} are the local change of site energy E_n and charge carrier transfer integral J_{nm}, respectively, due to the presence of local state fluctuations of the crystal structure.

The first term is usually called the *static* diagonal disorder, the second the static off-diagonal disorder of the crystal.[288]

In summary, we should like to reemphasize that we regard as the most important result of the discussions in this chapter the introduction of the concepts and models of electronic and molecular polarons. These concepts, presented in terms of many-body interaction of a charge carrier with the electronic and nuclear subsystems of the organic crystal, have been confirmed, in our opinion, both phenomenologically and on a theoretical basis and will be further extended and confirmed in the following chapters.

CHAPTER 4

Energy Structure of Polaron States in OMC

Truth is a running stream the essence
of which remains unchanged.

SUMITRANANDAN PANT

Great truth is a truth whose opposite
is also a great truth.

NIELS BOHR

The discussion of polarization phenomena in Chapter 3 demonstrates that in OMC charge carriers are strongly localized and move in the lattice in the form of polarons, "dressed" in electronic and vibronic polarization "clouds" (see Figs. 3.5 and 3.10). As the result, the corresponding conduction levels of charge carriers in OMC are actually conduction levels of electronic or molecular polarons. The position of these levels in the energy diagram of the crystal are thus determined by the corresponding effective energy of electronic polarization P_{eff} and by the formation energy of the MP. $(E_b)_{eff} = (E_b)_{MP}$.

The conduction levels of nonrelaxed electronic polaron states in the crystal E_P^+ and E_P^- are separated by the energy gap E_G^{Opt}, usually called the optical energy gap.[307,317,318,334,344,348] Conduction levels of relaxed states of MPs M_P^+ and M_P^- are separated by the adiabatic energy gap E_G^{Ad}.[307,317,318,334,344,347,349] Thus, ionized states, i.e., the conduction levels of charge carriers in the energy diagram of an OMC, are determined by the self-energy values for polaron-type quasiparticles — electronic and MPs — and these do not behave as free charge carriers, but heavier quasiparticles—with effective mass m_{eff} which may considerably exceed the mass of a free electron or hole (see Chapters 6 and 7). Such a four-level model of ionized states in Ac-type crystals had already been formulated by Silinsh in 1977[331] as a modification of the initial three-level model of Lyons.[122,220] A more detailed phenomenological four-level model was later developed in the monograph[332] (see p. 95).

However, at development stage of the model relaxed states of charge carriers were treated in a simplified way, as conduction levels of molecular ions $E_e(M)^-$

142

and $E_h(M)^+$,[332] while the concept of a MP of nearly small radius was introduced at a later stage.[334,344,348,355,356]

The modified four-level model of ionized states in OMC gained certain recognition by experts working in the field and was taken as a basis of the energy structure of ionized states of organic crystals by Pope and Svenberg (see Ref. 288, p. 317).

In the present chapter we intend to discuss the physical foundations of the phenomenological model of ionized states in OMC in the framework of the concept of a charge carrier as of a polaron-type quasiparticle[334,344] (see also Ref. 351). From this aspect the present chapter forms a natural continuation and extension of the previous chapter on polarization phenomena in OMC. The application of the polaron concept, in particular that of the MP of nearly small radius, was considerably stimulated by the review by Sir Nevill Mott[246] of the monograph by Silinsh[332] in which he emphasized the fruitful prospects of wider application of the general ideology of polaron theories for the description of charge carrier transport mechanisms in organic crystals.

The present chapter includes a discussion of general physical principles and methodological peculiarities of the determination of the adiabatic and optical energy gaps E_G^{Opt} and E_G^{Ad}, respectively, in OMC, and results of such determinations in Ac-type crystals which are most widely studied from this point of view. We will also discuss energy diagrams of ionized states in a number of other types of OMC.

In Section 4.4 we shall consider *bound ionized* states. Such states are frequently named CP states, and CT states. They are a characteristic feature of organic crystals and are widely discussed in the literature.[122,288,332] Such states are usually formed as intermediate states in processes of photogeneration or recombination of charge carriers in OMC. The photogeneration aspects of the formation of CP-states will be discussed in detail in Chapter 6. In the present chapter our basic attention will be devoted to the energetic aspects of nonrelaxed CT and relaxed CP states. It will be shown that both CT and CP states can be described phenomenologically in the framework of polaron concepts.

In the final sections of this Chapter we will give an overview of local states (traps and antitraps) of polarization origin in imperfect OMC, as well as an example of energy structure of heterogeneous multilayer molecular Langmuir-Blodgett (LB) films.

4.1. ADIABATIC ENERGY GAP

The adiabatic energy gap E_G^{Ad} is the energy interval between the conduction levels of the relaxed charge carriers, i.e., between the conduction levels of the positive M_P^+ and negative M_P^- MPs.

At present the most reliable method for determining the value of E_G^{Ad} in OMC is to use the threshold function of intrinsic photoconductivity.[332]

It has been shown in Refs. 27, 337, and 346 that in anthracene-type crystals the spectral dependence of quantum efficiency of intrinsic photoconductivity in the near-threshold region $\eta = \eta(h\nu)$ can be approximated by the following empirical near-threshold function:

$$\eta(h\nu) \sim (h\nu - E_{th})^n; \quad n = \tfrac{5}{2} \tag{4.1}$$

where E_{th} is the spectral threshold of intrinsic photoconductivity. By means of linear extrapolation of the dependence $\eta = \eta(h\nu)$ in $\eta^{2/5}$, $h\nu$ coordinates, according to expression (4.1), one can obtain graphically the value of E_{th} (a more detailed description of the method may be found in Ref. 332).

It should be noted however that to obtain spectral curves $\eta(h\nu)$ of intrinsic photoconductivity, especially in the near-threshold region, is a complex methodological problem. This is due to the circumstance that the real shape of the curve $\eta(h\nu)$ might be concealed by the superposition of some extrinsic photogeneration mechanism. Hence, measurements of $\eta(h\nu)$ must therefore be performed under conditions of completely blocked injection of charge carriers from the electrodes. In some cases the extrinsic mechanisms of photogeneration might be suppressed by lowering of the temperature (as, for instance, in the case of perylene,[11] tetrathiotetracene, and tetraselenotetracene[20,21]), or by other means. The methodological peculiarities of obtaining reliable $\eta(h\nu)$ curves of intrinsic photoconductivity are discussed in Ref. 332. The methodological difficulties are brought forward, in particular, by the fact that the most reliable $\eta(h\nu)$ dependence for such a widely studied model OMC as anthracene was obtained only in 1980 in the meticulously conducted experiments of Kato and Braun.[160]

As an example we shall present the method of determining E_{th} in case of Pc, the mechanisms of photogeneration for which have been studied in great detail[337,349] (see, Chapter 6.2).

For this purpose experimental spectra of quantum efficiency of intrinsic photoconductivity $\eta(h\nu)$ (Ref. 349) (see Fig. 4.1) were used. Measurements were carried out on thin-layer vacuum-deposited samples of oriented crystallites under conditions of blocking configuration of the electrodes (see Fig. 4.1b). The methodological details are described in Ref. 349.

It may be seen from the Fig. 4.1 that under the given experimental conditions the threshold dependence $\eta(h\nu)$ is clearly visible, changing by over two orders of magnitude within the photon energy range between 2.2 and 3.0 eV.

It was shown in Ref. 349 from $\eta(h\nu)$ dependence studies at various electric field $\mathcal{E}$ and temperature T values that the $\eta(h\nu)$ curves can be approximated in the near-threshold region by the empirical formula

$$\eta(h\nu, \, \mathcal{E}, T) = A(\mathcal{E}, T)(h\nu - E_{th})^{5/2} \tag{4.2}$$

where $\mathcal{E}$ and T are parameters of spectral dependence $\eta(h\nu)$.

In Fig. 4.1a (curves 1' and 2') we have an example of linear approximation of experimental dependences $\eta(h\nu)$ in $(\eta^{2/5}, h\nu)$ coordinates according to expression (4.2), which yields an extrapolated value of $E_{th} = 2.25$ eV. As may be seen, the

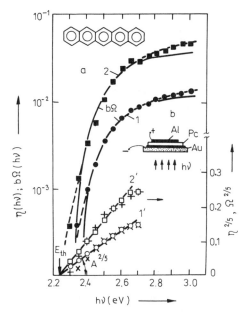

FIGURE 4.1. Spectral dependences of intrinsic photoconductivity of pentacene:[349] (a) ● and ■ are the experimental values of the quantum efficiency spectrum $\eta(h\nu)$ at 205 K and $\mathcal{E} = 10^4$ and 4×10^4 V/cm, respectively; 1, 2 are the calculated $b\Omega(h\nu)$ curves at given T and $\mathcal{E}$ values; 1', 2' are the corresponding $\eta(h\nu)$ curves (○ and □) in ($\eta^{2/5}$,$h\nu$) coordinates, and corresponding $b\Omega(h\nu)$ curves (+ and ×) in ($b\Omega^{2/5}$, $h\nu$) coordinates. (b) Configuration of the thin-layer Au/Pc/Al sample of 1.5 μm thickness.

value of $A^{2/5}$ determines here the slope of the straight line $\eta^{2/5} = \eta^{2/5}(h\nu)$.

On the other hand, Ref. 349 presents quantitative confirmation of the previously proposed suggestion (see Ref. 332, p. 125) that the quantum efficiency of photogeneration in Ac-type OMC is determined by the efficiency of thermal dissociation of Coulombic field-bonded geminal CP-states formed as an intermediate stage of the photogeneration process (see Chapter 6.).

Thus, it has been shown in Ref. 349 (see Fig. 4.1) that the quantum efficiency curves of stationary photoconductivity $\eta(h\nu)$ in the near-threshold spectral region may be quantitatively described by the Onsager formula[277] $\Omega[r_0(h\nu), \mathcal{E}, T]$ [see Eq. (6.7)] using only a single numerical parameter b

$$\eta(h\nu,\ \mathcal{E},T) = b\Omega[r_0(h\nu), \mathcal{E}, T], \qquad (4.3)$$

where $r_0(h\nu)$ is the spectral dependence of the mean charge carrier separation distance in the photogeneration process, i.e., the thermalization length $r_0 = r_{th}$ (see Chapter 6).

Figure 4.1a illustrates the approximation of experimental dependences $\eta(h\nu)$ in Pc by means of calculated curves $b\Omega(h\nu)$ obtained from the Onsager formula [see Chapter 6, formula (6.7)]. It may be seen that the calculated curves describe the shape of experimental dependences $\eta(h\nu)$ in the near-threshold region within a sufficiently wide energy range, as well as their dependence on electric field intensity. The Onsager approximation also describes correctly the temperature dependence of $\eta(h\nu)$ and confirms, on the whole, the adequacy of the Onsager model for describing photogeneration processes in OMC (see Chapter 6).

Parameter b serves in the given case for superimposing the curves $\eta(h\nu)$ and $\Omega(h\nu)$, since it determines, in semilogarithmic coordinates, the vertical shift of the $\Omega(h\nu)$ curve. The physical meaning of parameter b will be discussed in some detail in Chapter 6.

A comparison of Eqs. (4.2) and (4.3) shows that the empirical formula (4.2) is an approximated expression of the Onsager formula in the near-threshold spectral region in which the variables $h\nu$, $\mathcal{E}$, and T are separated into two factors permitting a linear approximation of the $\eta(h\nu,\mathcal{E},T)$ dependences in $(\eta^{2/5}, h\nu)$ coordinates. This is suggested by the data in Fig. 4.1a (curves $1'$, $2'$). The empirical threshold function $A(\mathcal{E},T)(h\nu - E_{th})^{5/2}$ approximates equally well both the $\eta(h\nu)$ dependence, as well as the theoretical Onsager curve $\Omega(h\nu)$ in $(\Omega^{2/5}, h\nu)$ coordinates. An approximation of a large number of experimental $\eta(h\nu,\mathcal{E},T)$ dependences for Pc (Ref. 349) by means of the Onsager formula shows that the coefficient $A(\mathcal{E},T)$ of the threshold function (4.2) determines the slope of the $\eta(h\nu)$ curves in semilogarithmic coordinates, as dependent on $\mathcal{E}$ and T, and may be described by the expression

$$A(\mathcal{E},T) = a\,\mathcal{E}\,\exp(\gamma T), \qquad (4.4)$$

where a and γ are empirical constants. It was shown in Ref. 349 that in the studied Pc samples we have $\gamma = (6\pm3)\times10^{-3}\ \mathrm{K}^{-1}$ (see also Section 6.2.2).

As may be seen, the slope coefficient of the $\eta(h\nu)$ dependences is directly proportional to field intensity $\mathcal{E}$ and grows exponentially with increase in temperature T. Numerical computer simulation of the photogeneration process within the framework of the generalized Fokker-Planck problem confirms these dependences and gives them physical meaning[344] (see Chapter 6).

Table 4.1 presents the threshold values of intrinsic photoconductivity E_{th} in anthracene, tetracene, and pentacene crystals, as obtained by linear extrapolation of experimental $\eta(h\nu)$ dependences after formulas (4.1) and (4.2).

The largest amount of experimental data on E_{th} determination concerns Pc. Owing to exceptionally high photosensitivity of Pc an averaged value of E_{th} has been obtained in Ref. 349 on the basis of $\eta(h\nu)$ dependence measurements on 34 thin-layer samples of various thicknesses (0.8–3.3 μm) over a sufficiently wide range of electric field intensities (2×10^3–2×10^5 V $\cdot$ cm^{-1}) and temperatures (160–300 K). It was found that, in spite of the fact that the slopes of the $\eta(h\nu)$ dependences in semilogarithmic coordinates (see Fig. 4.1a) are strongly $\mathcal{E}$ and T dependent [see expression (4.4)], the value of E_{th} itself is practically insensitive to the changes in parameters $\mathcal{E}$ and T, coinciding with accuracy up to ±0.05 eV at all regimes studied for the large number of Pc samples (see Table 4.1).

TABLE 4.1. *Experimental Threshold Values E_{th} of Intrinsic Photoconductivity in Polyacene Crystals[351]*

Crystal	E_{th} (eV)	Method of determination	Reference
Anthracene	3.88±0.05	Extrapolation of the spectral dependence of quantum efficiency of photoconductivity $\eta(h\nu)$ in $(\eta^{2/5}, h\nu)$ coordinates according to formula (4.2), after Ref. 160	349
	3.91±0.05	Same as above in $(\eta^{1/3}, h\nu)$ coordinates according to formula (4.1) for $n = 3$, after Ref. 116	26
	3.86±0.05	Same as above, after Ref. 24	26
Tetracene	2.90±0.05	Extrapolation of $\eta(h\nu)$ dependence in $(\eta^{2/5}, h\nu)$ coordinates according to formula (4.2). (Averaged value for measurements on four samples.)	349
Pentacene	2.24±0.05	Same as above (averaged over measurements on 34 samples)	349

Although the amount of data of E_{th} values in Tc is smaller, one may consider from the foregoing that the value of $E_{th} = 2.90\pm0.05$ eV (Ref. 349) is sufficiently reliable.

For Ac we present E_{th} values obtained by linear extrapolation of experimental $\eta(h\nu)$ dependences from earlier works,[24,116] as well as from a later paper by Kato and Braun,[160] presenting $\eta(h\nu)$ dependences obtained by more reliable methods.

As shown in Table 4.1, the E_{th} values obtained for Ac show fair coincidence within error range, and their mean value of 3.88 eV ought to be considered as the most reliable.

We must now discuss the question whether one may identify the E_{th} value with the adiabatic gap E_G^{Ad}, i.e., to what extent the equality $E_{th} = E_G^{Ad}$ is justified, as was postulated phenomenologically from general physical considerations,[349] or, at times, from physical intuition (see Ref. 332, p. 120).

On the one hand, in accordance with existing data on photogeneration mechanisms in Ac-type crystals[288,318,332,349,351] one may assume that charge carriers involved in stationary photoconductivity are already completely relaxed, and their conduction levels are separated by the adiabatic energy gap E_G^{Ad}. (As will be shown in Chapter 6, relaxation takes place at early stages of photogeneration, i.e., at the very start of thermalization of the hot charge carrier.)

On the other hand, linear approximation of threshold dependence in $(\eta^{2/5}, h\nu)$ coordinates according (4.2) is a limiting one for determining the minimal value of photon energy $h\nu$ at which there still is, although with very low probability, a possibility to have free charge carrier generation. In the same way as other linearized extrapolations of threshold functions, the given approximation may also cause systematic deviations of a purely methodological nature with respect to the real value of E_G^{Ad}, i.e., we have the inequality $E_{th} \leq E_G^{Ad}$.

Baessler et al.[317,318] have proposed an independent method for determining the

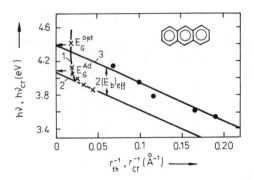

FIGURE 4.2. Curves used for the determination of E_G^{Ad} and E_G^{Opt} values in an anthracene crystal. (1) Photon energy $h\nu$ dependence on reciprocal value of thermalization length $h\nu = f(r_{th}^{-1})$,[318] as constructed from the experimental dependence $r_{th} = f(h\nu)$ after Ref. 160. (2) Limiting asymptotic straight line of the lowest CP states with slope $e^2/4\pi\epsilon\epsilon_0$ which, within the energy range $2(E_b)_{eff}$ is positioned below the straight line 3 of Coulomb bonding of the unrelaxed CP state. Extrapolation of the straight lines 2 and 3 to r_{th}^{-1}, $r_{CT}^{-1} \to 0$ yields the values of the optical and adiabatic energy gaps E_G^{Ad} and E_G^{Opt}, respectively. The straight line 3 representing the dependence of the CT transition energy $E_{CT} = h\nu_{CT}$ on the reciprocal value of CT transition distance (r_{CT}^{-1}) was obtained in Ref. 318 by the electromodulated spectroscopy method (see Section 4.2).

parameter E_G^{Ad} from the experimental dependence of charge separation distance (i.e., from thermalization length value r_{th}) on photon energy $h\nu$ in the photogeneration process (see Chapter 6, Section 6.2.2). In this case the traditional dependence $\langle r_{th}\rangle = f(h\nu)$ is transformed into other coordinates, namely, the dependence $h\nu = f(r_{th}^{-1})$ is taken into consideration. In Ref. 318 it is postulated that at photon energy $h\nu$, corresponding to the most deeply positioned CP state, the excess kinetic energy which can be obtained by the geminate carrier pair, must not exceed the total relaxation energy, i.e., $2(E_b)$. Diminution in $h\nu$ value must lead to asymptotic approach of the dependence $h\nu = f(r_{th}^{-1})$ to a straight line of slope $-e^2/4\pi\epsilon\epsilon_0$, situated by a value of $2(E_b)_{eff}$ below the corresponding straight line of the Coulombic bond of the nonrelaxed geminate pair. An extrapolation of this straight line to $r_{th}^{-1} \to 0$ yields the desired value of E_G^{Ad}. Such an approach was later confirmed by Silinsh and Jurgis.[344] The authors performed computer simulation of charge separation at the photogeneration process in Pc and showed that at photon energy $h\nu$ equal to electron transfer energy on a neighboring molecule, i.e., at $h\nu = h\nu_{CT}$ the excess kinetic energy which may be gained by the geminate CP pair is indeed equal to the total relaxation energy, namely, $E_k = 2E_b$ (Ref. 344) [see Chapter 6, Eq. (6.57) and Fig. 6.21].

This method of determining E_G^{Ad} is illustrated in Fig. 4.2 which shows that the $h\nu = f(r_{th}^{-1})$ dependence curve obtained from the experimental dependence $r_{th} = f(h\nu)$ in an Ac crystal is actually asymptotically a straight line with slope $e^2/4\pi\epsilon\epsilon_0$ of the Coulombic bond of CT state. Figure 4.2 also shows that the straight line 2 lies by an energy interval ΔE below the straight line 3 of the Coulombic bond of the nonrelaxed state. According to the above, the latter may be identified as the

TABLE 4.2. *Experimental Values of the Adiabatic Energy Gap E_G^{Ad} in Polyacene Crystals*[351]

Crystal	E_G^{Ad} (eV)	Method of determination
Anthracene	4.1 ±0.05	Extrapolation of the asymptotic approximation of the $h\nu = f(r_{th}^{-1})$ dependence[318] (see Fig. 4.2)
	4.1 ±0.1	Estimate based on photoinjection data from a metal electrode into an anthracene crystal[18] with correct allowance for the activation energy of decay of intermediate CT-states[17]
Tetracene	3.13±0.05	Extrapolation of the asymptotic approximation of the $h\nu = f(r_{th}^{-1})$ dependence
	3.11 ±0.03	Calculated estimate on the basis of the energy balance at thermalization[17]
Pentacene	2.47±0.04	Extrapolation of the asymptotic approximation of the $h\nu = f(r_{th}^{-1})$ dependence
	2.46±0.04	Calculated value on the basis of computer simulation of the process of charge separation in pentacene[344]

effective relaxation energy of the CP state, i.e., $\Delta E = (2E_b)_{\text{eff}}$. (The method of obtaining the straight line 3 of the Coulombic bond of the unrelaxed CT state will be discussed in Section 4.2.)

Extrapolating the straight line 2 as far as $r_{th}^{-1} \to 0$ one thus obtains a value of $E_G^{Ad} = 4.1$ eV for Ac,[318] the error not exceeding ±0.05 eV (see Table 4.2).

An independent estimate of E_G^{Ad} has been obtained from the spectral dependence of electron and hole injection from a vapor-deposited metallic cerium or magnesium contact into an Ac crystal,[18] taking into account correction for activation energy of the decay of intermediate bonded CP-states formed after injection.[17] This approach also yields $E_G^{Ad} = 4.1 \pm 0.1$ eV (Table 4.2). Similar to Fig. 4.2 extrapolation method has also been used for the determination of E_G^{Ad} value in Pc crystals (see Ref. 351, p. 195). In this case the experimental dependence of $r_{th} = f(h\nu)$ as obtained in Ref. 349 was applied for constructing the $h\nu = f(r_{th}^{-1})$ dependence. The extrapolation yields $E_G^{Ad} = 2.47 \pm 0.05$ eV for Pc (see Table 4.2).

Computer simulation of charge separation processes in Pc at photogeneration[344] yielded $E_G^{Ad} = 2.46 \pm 0.04$ eV. Thus, the experimental and calculated E_G^{Ad} values for Pc crystals coincide within error limit (Table 4.2). It was found, however, in the simulation procedure of charge separation processes in Pc, that the value of E_G^{Ad} reveals slight temperature dependence and actually constitutes a variable parameter for the given problem (see Fig. 6.19). The value $E_G^{Ad} = 2.46$ eV corresponds to $T = 293$ K. The value of E_G^{Ad} decreases slightly with lowering of temperature. Thus, we have a calculated value of $E_G^{Ad} = 2.30$ eV for Pc at 204 K (for more detailed account see Chapter 6).

Extrapolation of the dependence $h\nu = f(r_{th}^{-1})$ for Tc yields $E_G^{Ad} = 3.13$

TABLE 4.3. *Experimental Values of the Adiabatic Energy Gap E_G^{Ad} and Extrapolated Threshold Values of Intrinsic Photoconductivity E_{th}, and the Difference $\Delta E = E_G^{Ad} - E_{th}$ in Polyacene Crystals*

Crystal	E_G^{Ad} (eV)	E_{th} (eV)	ΔE (eV)
Anthracene	4.10	3.88	0.22
Tetracene	3.13	2.90	0.23
Pentacene	2.47	2.24	0.23

$\pm$ 0.05 eV (see Table 4.2). It may be of interest to note that a close value of $E_G^{Ad} = (3.11 \pm 0.03)$ eV was evaluated previously from the energy balance at thermalization.[17]

One may thus assume that the E_G^{Ad} values, as presented in Table 4.2, and coinciding within ± 0.05 eV error, are sufficiently reliable. This conclusion is supported by the constant difference between obtained E_G^{Ad} values and the extrapolated threshold values E_{th} of intrinsic photoconductivity [according to (4.2)], (see Table 4.3). The systematic shift of $\Delta E = 0.22$–0.23 eV is likely to be due to the methodological problems of linear extrapolation in the near-threshold region of the spectral dependence of the quantum efficiency of intrinsic photoconductivity.

This situation is illustrated in Fig. 4.3, showing calculated spectral curves of CP state dissociation efficiency imitating, according to Eq. (4.3) the corresponding spectral dependences $\eta(h\nu)$ at given values of E_G^{Ad}. The calculated curves $\Omega_i(h\nu)$ were obtained by simulating photogeneration in Pc within the framework of a modified Sano–Mozumder model, applying the extended Onsager formula (6.7) according to a method described in Ref. 344 (see Chapter 6). Figure 4.3 shows that linear extrapolation of $\Omega_i(h\nu)$ curves according to Eq. (4.2) yields, in $(\eta^{2/5}, h\nu)$ coordinates, values for E_{th} which are lower by 0.18 eV than the E_G^{Ad} values used in calculations as parameters. The real situation is most precisely imitated by the middle curve 2 which, at $E_G^{Ad} = 2.43$ eV, yields an extrapolated value of $E_{th} = 2.25$ eV. In other words, the calculated values of E_{th} and E_G^{Ad}, and their difference correspond, within error range, to experimental values (see Tables 4.1–4.3). Figure 4.3 reveals the reason for the systematic deviation of E_{th} relative to E_G^{Ad}.

The energy gap E_G^{Ad} is actually positioned in the near-threshold region at $\Omega_i(h\nu) \approx 0.2$ (horizontal arrows), and not at zero value (vertical arrows).

These results are of principal importance. It was considered up to now (see, e.g., Ref. 332) that the extrapolated thresholds E_{th} of the spectral dependences $\eta(h\nu)$ can be identified with the value of the energy gap, and their determination was considered to be the most reliable way of finding E_G^{Ad}. As can be seen from Table 4.3 and Fig. 4.3, the threshold values of intrinsic photoconductivity, as determined by the linear extrapolation method, produce in OMC a value of E_G^{Ad} which is diminished by ca. 0.2 eV. Another peculiarity of the extrapolation method ought to be noted. It follows from the empirical formula (4.2) and from Fig. 4.1 that field intensity and temperature determine the slope of the $\eta(h\nu)$ dependence in semilogarithmical coordinates [as well as in $(\eta^{2/5}, h\nu)$ coordinates]. The parameters $\mathcal{E}$

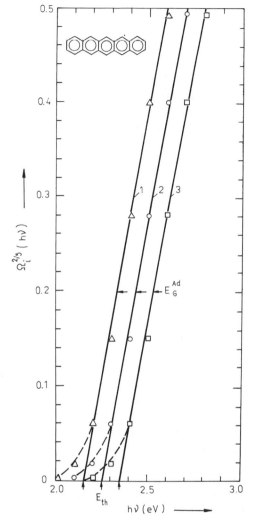

FIGURE 4.3. Calculated spectral curves of the CP-state dissociation efficiency $\Omega(h\nu)$ in $(\Omega^{2/5}, h\nu)$ coordinates, as obtained by computer simulation of photogeneration in pentacene at given values of the adiabatic energy gap E_G^{Ad} (cf. Ref. 344 and Chapter 6); electric field value $\mathcal{E} = 10^4$ V/cm. The vertical arrows indicate the extrapolated threshold values of intrinsic photogeneration E_{th}, the horizontal arrows indicate the corresponding values of E_G^{Ad}.

and T do not, however, practically, affect the extrapolated E_{th} value (see Ref. 349), while E_G^{Ad} reveals, according to computer simulation results, a slight temperature dependence (see Ref. 344 and Fig. 6.19).

The next section deals with methods of determining the optical energy gap E_G^{Opt} in Ac-type crystals, and with its physical meaning.

4.2. OPTICAL ENERGY GAP

A qualitatively new stage in the study of the energy structure of ionized states of anthracene type OMC was achieved in the excellent work of Baessler *et al.* in their investigations of electromodulated absorption in Ac, Tc, and Pc.[317,318] Previously all attempts to determine the energy levels of nonrelaxed charge carriers had been unsuccessful. Thus, Baessler and Vaubel[18] tried to determine the conduction levels of nonrelaxed and relaxed electrons based on the results of electron photoinjection in an Ac crystal from metal electrodes. However, a more detailed study of injection mechanisms from metal electrodes into organic crystals[16,181] showed that in the injection processes a Coulombically bound carrier pair is formed at the initial stage (for instance, an injected hole and its mirror image in the metal). Dissociation of the later into free carriers requires additional thermal activation energy (cf. Ref. 164). As was shown by Baessler and Killesreiter,[17] earlier interpretation of results of spectral dependences of photoinjection in Ac proved to be faulty, and a correct account for the activation energy of the decay of the near-electrode CP state yields a value of $E_G = 4.1 \pm 0.1$ eV, which was interpreted as the adiabatic gap E_G^{Ad} of Ac (see Table 4.2).

An analysis of time scales of vibronic relaxation processes of charge carriers into a molecular polaron state made it clear that the typical time of molecular polarization $\tau_v \leqslant 10^{-14}$ s is too short for the detection of a nonrelaxed state of the carrier in charge carrier injection or transfer processes, since $\tau_v \leqslant \tau_h$ (see Fig. 3.9). Here we must turn to faster processes, i.e., it is necessary to study processes of optical transfer of the carrier, in which relaxation is determined by the Franck-Condon principle. In other words, for the determination of the optical energy gap E_G^{Opt} of nonrelaxed carriers one may, in principle, apply only optical methods of investigation. This is where lies the importance of the work of Baessler *et al.*,[317,318] who succesfully applied the method of electromodulated spectroscopy for determining the optical energy gap in polyacenes.

Considering the great importance of the above mentioned work, we shall briefly discuss the basic principles of electromodulated spectroscopy. The application of this method enabled Abbi and Hanson[1] to detect optical CT-transitions with dipole moment $d = 23D$ in polar crystals of 9, 10-dichloroanthracene, in addition to the conventional Stark effect of intramolecular electronic transitions. These optical CT-transitions correspond to optical electron transfer to the nearest neighboring molecules. Blinov and Kirichenko, using electromodulated spectroscopy methods, detected polar CT-transitions in polycrystalline perylene and phthalocyanine layers.[35] However, these authors did not observe CT-transitions in nonpolar polyacene crystals in which intermolecular interaction is weaker. Baessler *et al.*[317,318] used vacuum-deposited thin-layer *surface* cells in which it was possible to obtain electric field intensities of the order of $10^4 - 10^5$ V · cm^{-1} along the **ab**-plane of oriented polyacene crystallites. They succeded in observing both a Stark-effect in intramolecular transitions, as well as CT-transitions in the **ab**-plane of the crystallites. Blinov *et al.*[35a] recently succeeded in obtaining electroabsorption spectra of CT-transitions in sandwich-type vacuum-deposited Ac and Tc layers. The authors demonstrate that the spectral positions of CT-transitions in Ac and Tc coincide with reported data[317,318] obtained in vacuum-deposited surface cells.

The effect of an electric field $\mathcal{E}$ of the absorption spectrum of a molecule or on a molecular solid manifests itself through field-induced energy change. $\Delta E(\mathcal{E})$ of the electronic transition equals:[317]

$$\Delta E(\mathcal{E}) = E(\mathcal{E}) - E(0) = -(\mathbf{d}_f - \mathbf{d}_i)\mathcal{E} - 1/2\,\mathcal{E}\,\Delta b\,\mathcal{E}, \qquad (4.5)$$

where $E(0)$ is the energy of the electronic transition in the absence of an electric field; $\mathbf{d}_i$ and $\mathbf{d}_f$ are the dipole moments of the initial and final state, respectively; and Δb is the change in molecular polarizability in the excited state. In the case of a transition with transfer of charge q (CT-transition) over a distance r_{CT} we have

$$\mathbf{d}_f = q\mathbf{r}_{CT}. \qquad (4.6)$$

The corresponding change in absorption coefficient $\Delta\alpha$, as dependent on ΔE can be represented in the form of the first two terms of a MacLaurin series:[317]

$$\Delta\alpha = \delta\alpha/\delta E\,\Delta E + 1/2\,\delta^2\alpha/\delta E^2\Delta E^2. \qquad (4.7)$$

Substitution of Eq. (4.5) into expression (4.7) gives us the dependence of $\Delta\alpha$ on the modulating electric field strength $\mathcal{E}$.

Two limiting cases may be considered here:

1. *Electromodulated spectra of intramolecular (neutral) electronic excitations, i.e., of Frenkel excitons in a molecular crystal.* In this case we have $\mathbf{d}_i = 0$ and, presumably $\langle\mathbf{d}_f\,\mathcal{E}\rangle \approx 0$ for nonpolar polyacene-type crystals. The main contribution to the linear ΔE term in Eq. (4.7), in this case, is given by the term involving the polarizability change Δb. Thus we obtain from Eq. (4.5):

$$\Delta\alpha = -1/2\Delta\bar{b}\,\mathcal{E}^2\,\delta\alpha/\delta E, \qquad (4.8)$$

where $\Delta\bar{b}$ is the mean value of the change in the components of the polarizability tensor b_i (see Section 3.5.1). Expression (4.8) describes the so-called quadratic Stark-effect. This effect takes place, as a rule, at electromodulation of absorption spectra of molecules or of molecular crystals, since the polarizability of molecules usually changes on excitation, i.e., $\Delta\bar{b} \neq 0$, as a result of electron density redistribution in the excited state of molecules.

2. *Electromodulated absorption spectra of CT-transitions.* Owing to high dipole moments of polar CT-transitions the dominating part in Eq. (4.7) passes over to the quadratic term ΔE^2. It may be noted, for the sake of comparison, that the optical CT transfer of the elementary charge $q = e$ from the molecule (0,0,0) to the molecule (1,0,0) in Pc over a distance $r_{CT} = 7.9$ Å is connected with a change in dipole moment according to (4.6): $\mathbf{d}_f = 4.8\times10^{-10}\times7.9\times10^{-8}$ e.s.u. $= 37.9D$. At the same time the intramolecular transition $S_0 \rightarrow S_1$ in pentacene, with oscillator strength $f \approx 0.1$, corresponds to a transitions dipole moment of ca. $3D$. Therefore in this case the polar term dominates in Eq. (4.5):

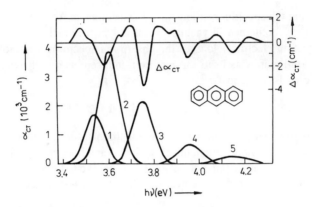

FIGURE 4.4. Spectral dependence of electromodulated CT transitions in anthracene, as obtained after subtraction of the Stark effect signal from the summary curve of electroabsorption $\Delta\alpha$.[318] The lower part of the figure shows the spectral bands of CT transitions, as obtained after twofold integration of the $\Delta\alpha_{CT}(h\nu)$ curve.

$$\Delta E(\mathcal{E}) \approx -\mathbf{d}_f \mathcal{E} \qquad (4.9)$$

($\mathbf{d}_i = 0$ for nonpolar crystals).

Isotropic averaging over randomly oriented charge transfer dipoles (4.6) yields:[317]

$$(\Delta E)^2 = 1/3(q\mathbf{r}_{CT}\,\mathcal{E}).$$

Therefore, we have

$$\alpha(\mathcal{E}) = 1/6(q\mathbf{r}_{CT}\,\mathcal{E})^2\delta^2\alpha/\delta E^2 = 1/6(\mathbf{d}_f\,\mathcal{E})^2\delta^2\alpha/\delta E^2. \qquad (4.10)$$

Summing up, one may conclude that electromodulated spectra of intramolecular (Frenkel) transitions (possessing a small transition dipole moment $\mathbf{d}_f$) must have the nature of the first derivative of the absorption spectra, in accordance with formula (4.8), whereas *polar* CT-transitions must be of the nature of the second derivative of the absorption spectrum curve, in accordance with formula (4.10). It may be seen that in both cases the field dependence of $\Delta\alpha(\mathcal{E})$ is of quadratic nature.

It follows that, if the absorption spectrum $\alpha(E)$ of the crystal is known, then an analysis of the electromodulated spectrum $\Delta\alpha(\mathcal{E})$ or the so-called electroabsorption spectrum makes it possible to determinate the nature of optical excitation (neutral, intramolecular excitation, or CT-transition) and to obtain a quantitative estimate of the parameters Δb and $\mathbf{d}_{CT} = q\mathbf{r}_{CT}$.

Analysis of the electromodulation spectrum of Ac crystal shows, according to Ref. 318, that within the energy range $h\nu < 3.4$ eV the Stark effect of neutral transitions dominates (the first derivative $\delta\alpha/\delta E$ prevails in the spectrum), whereas in the range $h\nu > 3.4$ eV (see Fig. 4.4) one observes polar CT transitions (in the spectrum the second derivative $\delta^2\alpha/\delta E^2$ prevails). From the peaks of CT-transition bands (see Fig. 4.4) it is possible to determine the energies of the corresponding CT-transitions.

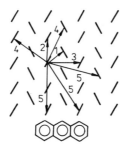

FIGURE 4.5. Schematic picture of the orientation of anthracene molecules in the **ab** plane of the anthracene crystal.[318] Arrows indicate the corresponding CT transitions (see Fig. 4.4).

Figure 4.5 shows schematically the respective CT transitions upon the nearest neighboring molecules in the **ab**-plane of Ac to which the CT bands, presented in Fig. 4.4, are attributed.

As may be seen from Fig. 4.4, the intensity of the CT bands falls drastically with increase of the distance r_{CT} of electron transfer, tending toward zero value at $r_{CT} > 15$ Å. A comparison of total absorption spectral curves $\alpha_{tot}(h\nu)$ and those of CT transition $\alpha_{CT}(h\nu)$ in Ac allows one to estimate the α_{CT}/α_{tot} ratio (Fig. 4.6). Figure 4.6 shows that this ratio lies between 0.08 and 0.03, i.e., the contribution of

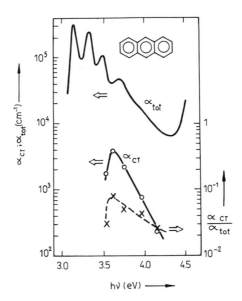

FIGURE 4.6. Comparison of the spectral absorption curve of an anthracene crystal $\alpha_{tot}(h\nu)$ (at excitation by **b**-polarized light),[44] and the spectral dependence $\alpha_{CT}(h\nu)$ of the peaks CT transition bands (see Fig. 4.4[318]); the curve of the ratio α_{CT}/α_{tot} is also presented.

TABLE 4.4. *Extrapolated, According to Expression $E_{CT} = f(r_{CT}^{-1})$ (see Fig. 4.7), and Calculated Values of Optical Energy Gap E_G^{Opt} in Polyacene Crystals*

Crystal	Extrapolated value of E_G^{Opt} (eV)[317,318]	Calculated value of E_G^{Opt} (eV)[42]
Anthracene	4.40±0.05	4.44±0.04
Tetracene	3.43±0.05	3.53±0.05
Pentacene	2.83±0.05	2.78±0.06

CT transitions lies between 8 and 3% of the total absorption and falls sharply with increase in photon energy within the 3.6–4.2 eV range.

After this digression into the field of methodology and peculiarities of experimental determination of CT transitions in polyacene crystals we now approach the most important aspect of the problem.

The CT transition energy E_{CT}, as dependent on distance r_{CT}, may be described by a Coulomb-type law:[318]

$$E_{CT}(r_{CT}) = E_G^{Opt} - e^2/4\pi\epsilon\epsilon_0 r_{CT}. \tag{4.11}$$

Expression (4.11) shows that the $E_{CT}(r_{CT})$ spectrum has a convergence limit at $r_{CT} \to \infty$, which is determined by the width of the optical energy gap E_G^{Opt}. Accordingly, a linear extrapolation of the $E_{CT}(r_{CT})$ dependence in $E_{CT} = f(r_{CT}^{-1})$ coordinates yields at $r_{CT}^{-1} \to 0$ the desired value of E_G^{Opt}.[318] This method of quantitative determination of the value of E_G^{Opt} is illustrated on Fig. 4.2 for the case of an Ac crystal. Extrapolation shown on Fig. 4.2 (curve 3) yields a value of $E_G^{Opt} = 4.40\pm0.05$ eV for Ac.[318] The slope of the straight line 3 corresponds to the mean value of dielectric permeability $\bar{\epsilon} = 3.2$. This value of $\bar{\epsilon}$ corresponds to the isotropic dielectric permeability of an Ac crystal, according to Ref. 159.

The results of E_G^{Opt} determination for Ac, Tc, and Pc (Refs. 317 and 318) by extrapolation approach are presented in Table 4.4.

The authors of Ref. 318 discuss the problem of the adequacy of Coulomb approximation of the energy of CT transitions $E_{CT}(r_{CT})$, i.e., of the feasibility of employing formula (4.11). A comparison of experimental values of E_{CT} with calculated ones, as obtained by Bounds *et al.*[39,40] shows that Coulomb approximation, using an averaged value of isotropic dielectric permeability $\bar{\epsilon}$ adequately describes the $E_{CT}(r_{CT})$ spectrum, small deviations from Coulomb's law being observed only in the case of the deepest CP states of the nearest neighbor molecules. We shall come back to this question in Section 4.4 discussing the calculations of the energy spectrum of $E_{CT}(r)$.

The publication of paper 317 caused wide discussions on the physical meaning of the energy gap, since the value of the gap, as obtained from electromodulated absorption spectra of CT transitions in Tc and Pc, exceeded by 0.5 eV the E_G value, as obtained from the threshold E_{th} value of intrinsic photoconductivity.

Silinsh,[332] considering the problem of the energy structure of ionized states, has already postulated the necessity of accounting for the relaxation energy of charge carriers into ionic states in order to describe the formation of ionic conduction state levels $E_e(M^-)$ and $E_h(M^+)$. Later[349] the relaxation concept was used for the explanation of photogeneration processes in OMC. The idea of nonrelaxed and relaxed states of charge carriers was further developed by Baessler and coworkers[317,318] who introduced the concept of an optical (vertical) energy gap E_G^{Opt}, separating the presumed conduction levels of non-relaxed carriers, and adiabatic (electric) energy gap E_G^{Ad}, separating the conductivity levels of relaxed carriers, as observed in charge carrier photogeneration and transfer experiments. It has now become clear, after a discussion on the processes of electronic and molecular polarization (see Chapter 3), and discussion of characteristic time scales of relaxation in these processes (cf. Fig. 3.9), that the states of nonrelaxed carriers in OMC are, in principle, possible in fast, purely optical processes only, as, for instance, in the previously discussed optical CT-transitions. For slower processes $(\tau \geqslant 10^{-14}$ s) at photo- and field-injection, at multistep photogeneration stages, and in charge carrier transfer processes they manifest themselves solely in relaxed form and must be considered only in the framework of the molecular polaron model (see Refs. 334 and 344 and Section 3.6.1). In this aspect the introduction of the optical gap E_G^{Opt} concept and its experimental determination in Ac, Tc, and Pc by Baessler et al.[317,318] brought considerable clarity to the problem of ionized state energetics in OMC, permitting the distinction of purely optical (vertical) processes of fast polarization of the electronic subsystem from the slower adiabatic processes of adiabatic rearrangement of the nuclear subsystem of the ionized molecule and its surroundings.

To get a clearer idea of the energy parameters of ionized states and of their correlation with spectral curves of quantum efficiency $\eta(h\nu)$ of intrinsic photoconductivity, $\eta(h\nu)$ curves at the near-threshold spectral region of Ac, Tc, and Pc crystals are presented in Fig. 4.7a, as well as the values of these parameters. The E_G^{Opt} values have been obtained by extrapolation of the $h\nu_{CT} = f(r_{CT}^{-1})$ dependences, E_G^{Ad} and E_{th} have been taken from Table 4.3.

Figure 4.7 shows that E_G^{Opt} is positioned almost at the peak of the $\eta(h\nu)$ curves, while E_G^{Ad} is situated at approximately the middle of the near-threshold decline, but E_{th}, according to its definition, is at the very threshold of the $\eta(h\nu)$ curves. The near-threshold region of the $\eta(h\nu)$ dependence occupies a range of ca. 0.5 eV: E_G^{Ad} lies ca. 0.2 eV above the threshold, while E_G^{Opt} accordingly $2(E_b)_{eff} \approx 0.3$ eV above E_G^{Ad}. On the other hand, the observed nonrelaxed CT-transitions lie below E_G^{Opt}, their energies overlapping the tail part of the near-threshold region of the $\eta(h\nu)$ curves (see Fig. 4.7b).

Later Siebrand et al.[42] proposed another version of explaining CT-transitions in Ac, Tc, and Pc as observed in Refs. 317 and 318. In addition to vibrationally nonexcited CT-transitions $(v = 0)$, the authors included in their consideration possible transitions to excited vibrational CT-states. Unlike the data in Refs. 317 and 318, which use possible transition dipole moments $\mathbf{d}_{CT}$ for the identification of corresponding CT-transitions, the authors of Ref. 42 base their conclusions mainly on a spectroscopic analysis of the intensity distribution of the CT-bands. In this case

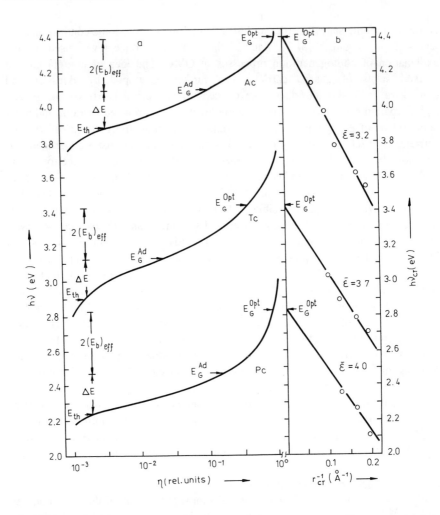

FIGURE 4.7. Spectral curves $\eta(h\nu)$ of the quantum efficiency of intrinsic photoconductivity in the threshold region of the spectrum (a) (for anthracene according to Ref. 160, for tetracene and pentacene according to Refs. 349 and 351) E_G^{Opt} and E_G^{Ad} are the optical and adiabatic energy gaps, respectively, $(E_b)_{eff}$ is the effective formation energy of a molecular polaron; ΔE is the systematic shift of the extrapolated value of the photoconductivity threshold E_{th} with respect to the adiabatic gap E_G^{Ad}. The dependence of CT transition energy on the reciprocal value of the CT transition distance r_{CT}^{-1} and the corresponding extrapolated values of E_G^{Opt} (b) (for anthracene according to Ref. 318, and for tetracene and pentacene according to Ref. 317).

the E_G^{Opt} value is considered as a variable parameter in the problem, and its most probable value is estimated in the process of minimization. The E_G^{Opt} values, thus obtained, are presented in Table 4.4. A comparison of E_G^{Opt} values obtained by extrapolation[317,318] (see Fig. 4.7b), with calculated E_G^{Opt} values,[42] yield satisfactory coincidence within error range.

A graphic representation of calculated E_{CT} values in $E_{CT} = f(r_{CT}^{-1})$ plot, according to the proposed transition scheme,[42] shows, however, that only the most deeply positioned CT-transition, as observed experimentally in Refs. 317 and 318, differs from the Coulombic law (4.11), while other CT transitions, as presented in Refs. 317 and 318 may be satisfactorily approximated by the Coulombic law at appropriate values of the isotropic dielectric constant $\bar{\epsilon}$. One may therefore reasonably assume that the CT-transitions, as proposed in Refs. 317 and 318 (see Figs. 4.4 and 4.7) are more reliable and describe electroabsorption spectra more adequately. We shall therefore use the E_G^{Opt} values obtained by extrapolation[317,318] (see Table 4.4).

At present, the optical energy gap has been reliably determined only for Ac, Tc, and Pc crystals. As to other OMC, such a determination of E_G^{Opt} is practically not possible owing to lack of reliable electromodulated spectra, or owing to difficulties in their interpretation.

4.3. THE ENERGY STRUCTURE OF POLARON STATES IN OMC

The data obtained on the determination of energy gaps E_G^{Ad} and E_G^{Opt} allow one to draw a full energy level diagram of ionized (polaron) states in polyacene crystals. For this purpose one traditionally uses experimental values of the ionization energy I_G of a molecule and I_C of a crystal, electron affinity of the molecule A_G, and the electronic polarization energy P, found, after Lyons,[122,220] from the equation $P = P_h = I_G - I_C$ (see Refs. 228 and 332). The energy gap E_G between the conduction levels of the electron and hole was usually estimated from the extrapolated value of the intrinsic photoconductivity threshold value E_{th}.

At present, it has become necessary to introduce certain corrections into the energy diagrams of ionized states in OMC in view of an analysis of the components of effective polarization energies W_{eff}^+ and W_{eff}^- (see Fig. 3.8), and of the energy gap values E_G^{Opt} and E_G^{Ad} (see Tables 4.2–4.4). This is, in essence, a further refinement of the modified Lyons model, proposed by Silinsh,[332] taking into account the latest experimental and reevaluated computational data[307] (see also Section 3.5).

Considering the absence of reliable experimental data on I_G, I_C, and A_G (see Table 3.5) and, accordingly, the large scattering of effective polarization energy values obtained from these data, we prefer to use the most reliable results of self-consistent calculations of electronic polarization components P_{id} and W_{Q_o}, as obtained by Eisenstein and Munn,[100] in the construction of an energy diagram of ionized states. Since, on the other hand, experimental data for E_G^{Opt}, E_G^{Ad}, and $(E_b)_{eff}$ will be used (see Section 4.1 and 4.2, and Table 3.8) in the construction of energy level diagrams, one may consider the energy diagrams presented below as semiempirically calculated ones.

The total balance of the energy terms in the diagrams is determined by the expressions:

$$P_{\text{eff}}^+ + (E_b^+)_{\text{eff}} + E_G^{\text{Ad}} + (E_b^-)_{\text{eff}} + P_{\text{eff}}^- + A_G^v = (I_G)_{\text{calc}} \qquad (4.12)$$

$$P_{\text{eff}}^+ = P_{\text{id}}^+ + W_{Q_o}^+ \qquad (4.12a)$$

$$P_{\text{eff}}^- = P_{\text{id}}^- - W_{Q_o}^- \qquad (4.12b)$$

$$|W_{Q_o}^+| = |W_{Q_o}^-| \qquad (4.12c)$$

(see Eqs. 3.5.28 and 3.5.29)

$$(E_b^+)_{\text{eff}} = (E_b^-)_{\text{eff}} = \frac{E_G^{\text{Opt}} - E_G^{\text{Ad}}}{2} \qquad (4.12d)$$

(see Tables 3.8, 4.2, and 4.4).

Expression (4.12) is the basic equation of the modified Lyons model, in which the asymmetry of effective electronic polarization P_{eff} is taken into account owing to the different signs of the terms of charge-quadrupole moment interaction W_{Q_o} for a positive and negative charge, as well as the term $(E_b)_{\text{eff}}$ of formation of a molecular polaron.[307,351]

To check these assumptions we shall first draw an energy diagram of ionized (polaron) states of an Ac crystal, which is the favorite model compound in OMC physics. Anthracene has also been chosen for the reason that we have the most reliable calculated data on P_{id} and W_{Q_o} (see Sections 3.5.3–3.5.5), and reliable experimental data on E_G^{Opt} and E_G^{Ad} (see Sections 4.1 and 4.2).

If all the terms of the sum (4.12) are adequate, we ought to obtain a semiempirical value of $(I_G)_{\text{calc}}$ after their additive summation. This value must fit into the range of scattered experimental values of I_G (see Table 3.5).

The diagrams constructed in such a way can also be used for checking the reliability of calculations of the terms P_{id} which, as follows from Table 3.5, cannot be reliably estimated from the difference $I_G - I_C$ owing to the wide scatter of experimental values.

To build up an energy level diagram of ionized states of an Ac crystal (Fig. 4.8b) data on P_{eff}, P_{id}, and W_{Q_o} calculated by Eisenstein and Munn[100] were used (see Table 3.4), and experimental E_G^{Opt}, E_G^{Ad}, and $(E_b)_{\text{eff}}$ values from Tables 3.8, 4.2, and 4.4. Since the experimental data on electron affinity A_G are unreliable, we used calculated data on *vertical* electron affinity A_G^v (Ref. 322) (see also Table 3.5). It should be noted that it is just the vertical affinity A_G^v that contains only the *electronic* component of the affinity of the intrinsic polarization energy of the molecule by a localized charge (see Ref. 332, p. 91).

For comparison, neutral electronic excited states (a) of Ac molecules in the crystal are also shown in Fig. 4.8. Molecular ion state S_0^+ of the highest occupied neutral level of the molecule serves as the level for reference with regard to the ionized state levels of the crystal.

Figure 4.8 shows that the levels of the electronic polaron state of the positive E_P^+ and negative E_P^- charges are uniquely determined by the corresponding energies of

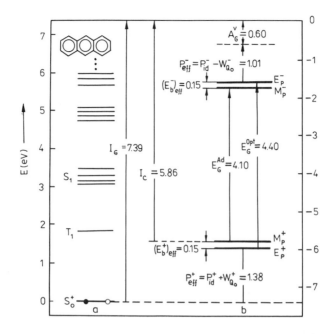

FIGURE 4.8. Energy diagram of neutral (a) and ionized (b) states of an anthracene crystal: E_G^{Opt}, E_G^{Ad} are the optical and adiabatic energy gaps; $(E_b^+)_{eff}$, $(E_b^-)_{eff}$ are the effective formation energies of a positive and negative molecular polaron, respectively; P_{eff}^+, P_{eff}^- are the effective electronic polarization energy values; P_{id}^+, P_{id}^- are the polarization terms for interaction between the charge and the induced dipoles; $W_{Q_0}^+$, $W_{Q_0}^-$ are the polarization terms for the interaction between the charge and the permanent quadrupole moments of molecules; A_G^V is the vertical electron affinity of the molecule; I_G is the ionization energy of the molecule and I_C that of the crystal; E_P^+ and E_P^- are the levels of the self-energy values of the positive and negative electronic polarons; M_P^+ and M_P^- are the conduction levels of the positive and the negative molecular polarons.[307,351] Molecular ion state S_0^+ serves as a level of reference with regard to the ionized state levels of the crystal.

electronic polarization. The latter determine, in their turn, the position of the levels E_P^+ and E_P^- with respect to the level S_0^+ and the vacuum level, thus yielding the value of the optical energy gap E_G^{Opt}. The polaron level E_P^+ lies above the level S_0^+ by a value of P_{eff}^+. The term P_{eff}^+ thus characterizes the electronic component of the affinity of the crystal toward a positive charge (hole).

The polaron level E_P^-, in its turn, lies below vacuum level by a value of A_G^v + P_{eff}^-. The sum $A_G^v + P_{eff}^-$ thus characterizes the electronic component of the affinity of the crystal towards a negative charge (electron), i.e.,

$$A_G^v + P_{eff}^- = A_C^v. \tag{4.13}$$

The MP levels M_P^+ and M_P^- lie, respectively, by a value of $(E_b^+)_{eff}$ above and $(E_b^-)_{eff}$ below the E_P^+ and E_P^- levels and are separated by the adiabatic energy gap E_G^{Ad}.

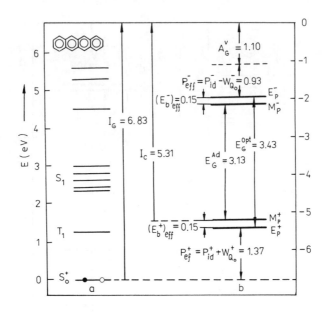

FIGURE 4.9. Energy diagram of neutral (a) and ionized (b) states of a tetracene crystal.[351] Notations as in Fig. 4.8.

It is essential to stress here once more that E_P^+, E_P^- and M_P^+, M_P^- are levels of multielectron (multiparticle) interaction and determine the self-energy of the corresponding quasiparticles — electronic and MPs.

As regards the quantitative aspect of the energy diagram of Ac (see Fig. 4.8), the semiempirically calculated value of $I_G = 7.39$ eV, agrees satisfactorily, as can be seen from Table 3.5 and Fig. 3.8, within the scattered range of reported experimental values $I_G = 7.38-7.47$ eV. This provides a significant support to the adequacy of the calculations of P_{id} and W_{Q_o} by Eisenstein and Munn.[100]

The parameter I_C, as follows from Fig. 4.8, is determined by the value of the terms P_{id}^+, $W_{Q_o}^+$, and $(E_b^+)_{eff}$, i.e., according to the expression

$$I_C = I_G - P_{id}^+ - W_{Q_o}^+ - (E_b^+)_{eff}. \qquad (4.14)$$

In this case, too, the semiempirically calculated value $I_C = 5.86$ eV coincides satisfactorily within the range of experimental values of $I_C = 5.65-5.95$ eV (see Table 3.5 and Fig. 3.8).

Figures 4.9 and 4.10 present energy level diagrams of Tc and Pc crystals. As in the case of Ac (see Fig. 4.8), calculated values of P_{eff}, P_{id}, W_{Q_o},[100] and A_G^v,[332] and experimentally found values of E_G^{Opt} and E_G^{Opt} according to Tables 3.5, 3.8, 4.2, and 4.4 were used.

It follows from Fig. 3.8 that in this case coincidence between semiempirically calculated values $(I_G)_{cal}$ and $(I_C)_{cal}$ with experimental ones are worse. However,

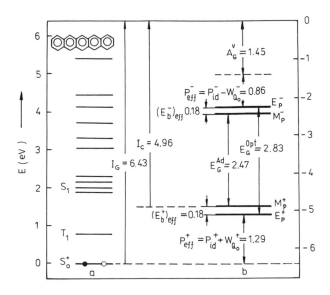

FIGURE 4.10. Energy diagram of neutral (a) and ionized (b) states of a pentacene crystal.[351] Notations as in Fig. 4.8.

such coincidence can still be regarded as satisfactory, since calculated extrapolated values of the tensor components of the molecular polarizability and of the quadrupole moments of the molecules were used in this work.[100]

In postulating the modified four-level Lyons model[332] electronic conduction levels E_h^0 and E_e^0 were introduced, which were already treated as electronic polaron levels at that stage. In the modified model there were also considered levels of localized (relaxed) states which in Ref. 332 were denoted by $E_e(M^-)$ and $E_h(M^+)$ and assumed to form conduction levels of molecular ions (see Ref. 332, p. 95). The levels E_e^0 and E_h^0 were considered to constitute limiting static levels of total electronic polarization of an electron and a hole, but levels $E_e(M^-)$ and $E_h(M^+)$ were treated as limiting levels of vibronic polarization determining the formation of truly ionic conduction states through which, as supposed, stationary conductivity takes place in OMC. Since no data were available at that time on the real value of the optical and adiabatic energy gaps, it was assumed that the difference between the nonrelaxed levels E_e^0 and E_h^0 and relaxed ones in vibronic coordinates $E_e(M^-)$ and $E_h(M^+)$ equals the bonding energy of the charge carrier E_b in the formation of a molecular ion. In other words, it was assumed that $E_e^0 - E_e(M^-) = (E_b^-)_i$, where $(E_b^-)_i$ is the formation energy of a molecular anion, and $E_h^0 - E_h(M^+) = (E_b^+)_i$ where $(E_b^+)_i$ is the formation energy of a molecular cation (see Ref. 332, p. 95).

However, computer simulation of photogeneration processes in Pc later demonstrated[344] that, for the description of the relaxed state of an electron, it is necessary to introduce the concept of a heavy quasiparticle — an adiabatic MP of nearly small radius (see Sections 3.6.1 and 6.3.3). The adequacy of the MP model (see Fig. 3.10) is also supported by purely physical considerations, by an analysis of vibronic (intramolecular) and intermolecular polarization processes, as well as by the fact that we have

$$(E_b)_{\text{eff}} = \frac{E_G^{\text{Opt}} - E_G^{\text{Ad}}}{2} > (E_b^+)_i,$$

(see Section 3.6.1 and Table 3.8).

This is reflected in the changes in notation in the diagrams in Figs. 4.8–4.10, where the eigenvalues of electronic polaron energy levels are denoted by E_P^+ and E_P^-, respectively, while the levels of the eigenvalues of MP energy levels are denoted by M_P^+ and M_P^- to avoid misunderstandings.

Pope and Swenberg discussed in their monograph[288] some dynamic aspects of the Lyons model, as modified by Silinsh[332] (see Ref. 288, p. 317). These authors proposed an energy level diagram representing bands of dynamic states of hot delocalized electrons and holes corresponding to various stages of electronic polarization, as well as bands of "localized" carriers representing, in their turn, stages of vibronic (and, possibly lattice) polarization. A characteristic feature of the diagram consists in the circumstance that, unlike the initial Lyons model,[122,220] but in conformity with the modified model,[332] it is assumed that between the bands of delocalized and localized carriers no energy gap is formed (see Ref. 288, p. 317). Thus, e.g., we have an uninterrupted continuous band of dynamic states for an electron, starting from vacuum level down to the stationary conduction level of a localized electron. In this multielectron interaction model the last rudiments of the concepts of a one-electron band approach have been completely removed.

The energy diagram, as presented in Fig. 4.11 may be regarded as a further refinement of the dynamic model of Ref. 288, since it takes into account the basic principles of polaron approach.

Let us now briefly consider in a visual representation the dynamic aspects of polarization processes that have been discussed in greater detail in Chapter 3. If a free electron is introduced into an Ac-type crystal, first a practically inertialess (relaxation time $\tau_e \leqslant 10^{-15}$ s) polarization of electronic orbitals of the molecule takes place, whitin the precinct of which the electron is localized. Energetically this process is determined by the term A_G^v of the electronic component of the vertical electron affinity of the molecule. At the same rate polarization of electron orbitals of the surrounding molecules takes place (see Fig. 3.5), as well as interaction with the permanent quadrupole moments of the surrounding molecules. This process is energetically represented by the term of effective electronic polarization of the surrounding lattice P_{eff}^-. Thus, the electron crosses the whole dynamic band of states in the process of polarization within the relaxation time range $\tau_{\text{rel}} = \tau_e \leqslant 10^{-15}$ s reaching the "stationary" level of the electronic polaron E_P^- (see Fig. 4.11). Physically, in accordance with Fowler's concept[109] (see Chapter 3), the electron assumes in the process of polarizational relaxation all intermediate dynamic energy values of electronic polarization P_{dyn} in the represented band of dynamic states:

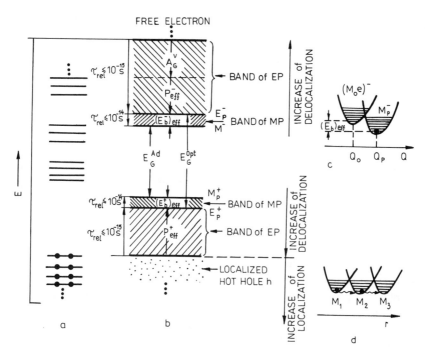

FIGURE 4.11. Schematic energy diagram of dynamic polaron states in anthracene type crystals.[351] (a) Neutral ground (S_0, S_{-1}, S_{-2}, S_{-3}) and excited (S_1, S_2, S_3) states of the molecule in the crystal. (b) Bands of dynamic states of electronic and molecular polarons (notations as in Fig. 4.8). (c) Diagram in generalized configurational coordinates Q of the nuclei showing the relaxation process in the formation of a molecular polaron M_P^- as a result of vibronic polarization. (d) Schematic diagram of an intermolecular tunneling of a molecular polaron M_P^+ in unchanging configurational coordinates of nuclei r.

$$0 \leqslant |P_{dyn}^-| \leqslant |P_{eff}^-| + A_G^v. \tag{4.15}$$

After the dynamic process of *electronic* polarization, which is accompanied by increasing localization of the electron, the next stage of polarization sets in, namely, *vibronic* polarization of the nuclei in the molecule on which the electron is localized, and that of the nuclei of the nearest neighboring molecules (see Fig. 3.10). This dynamic process of nuclear polarization takes place with slower relaxation rate $\tau_{rel} = \tau_v \leqslant 10^{-14}$ s (see Fig. 3.9). During this stage of polarization the electron crosses the narrower energy band of a molecular polaron and finally reaches the stationary conduction level of the molecular polaron M_P^-. The intermediate dynamic states of the molecular polaron $(E_b^-)_{dyn}$ can be described by the expression

$$0 \leqslant (E_b^-)_{dyn} \leqslant (E_b^-)_{eff}. \tag{4.16}$$

It should be emphasized, at this point, that the diagram of dynamic states of the electronic and molecular polaron (Fig. 4.11b) represents only the energetics of the formation of a polaron-type quasiparticle. The dynamics of the process is more complex and remains outside the scope of this simplified conceptual approach. First, the reader should keep in mind that the energy acquired in the process of polarization is converted into the kinetic energy of the new quasiparticle (see Section 6.3). Thus, before the electronic polarization the quasifree electron possesses zero kinetic energy ($E_k = 0$), as a result of polarization a hot quasiparticle is formed, i.e., an electronic polaron with kinetic energy $E_k = A_C^v = A_G^v + P_{eff}^-$. In the case of polyacenes the value of E_k is of the order of 1.6–2 eV (see Figs. 4.8–4.10). As a result of polarization the mean velocity of the polaron quasiparticle also increases. However, the change in velocity is of a more complex nature, since polarization is accompanied by an increase in effective mass m_{eff} of the quasiparticle and its mean velocity v equals ($2E_k/m_{eff}$), where m_{eff} increases in the process of polarizational relaxation.

Second, the process of polarizational relaxation itself, as well as the accompanying process of localization of the quasiparticle is strongly temperature-dependent. With rise in temperature the tendency toward localization of the quasiparticle (see Section 1B.8) increases drastically. Phenomenologically this manifests itself in the temperature dependence of the effective mass m_{eff} of the quasiparticle and is particularly conspicuous in the case of the formation of a MP, the effective mass of which increases exponentially with increase in temperature (see Chapters 6 and 7).

It ought to be stressed that the formation of a hot quasiparticle in the polarization process favors its localization, since, according to some concepts on localization processes (see Chapter 2 and references therein), hot quasiparticles have a higher probability of passing through the potential barrier between delocalized band and localized states than a thermalized particle.

The relaxation process of the formation of a molecular polaron is represented in Fig. 4.11c by means of potential curves in generalized coordinates. In this process the nuclear framework of the molecule with the localized electron $(M_o e)^-$ manages to react to the new electronic configuration and relaxes into a new polaronic equilibrium state with an energy gain $(E_b)_{eff}$ of formation of a MP M_P^-.

The most characteristic peculiarity of an adiabatic MP of nearly small radius consists in the circumstance that, unlike a small-radius lattice polaron, it can move from site to site in the lattice by hopping (or stepping) via tunneling without thermal activation energy (see Chapters 6 and 7 and Refs. 344, 355, and 356). The temperature dependence of mobility $\mu_0(T)$ of a molecular polaron in Ac-type crystals follows a typical dependence $\mu_0(T) \sim T^{-n}$ where $n > 0$.[344,355,356] The physical meaning of this transfer mechanism will be discussed in Chapter 7. Figure 4.11d only presents a schematic picture of tunneling transport of a molecular polaron which illustrates the conservation of nuclear coordinates in the transport process.

Thus, M_P^+ and M_P^- are the real conduction levels in OMC. The electronic polaron levels E_P^+ and E_P^-, as already stated in Section 4.2, manifest themselves only in optical charge transfer; the relaxation time τ_v of the formation of a MP is, as a rule, shorter than the mean hopping time τ_h of the carrier (see Fig. 3.9), and relaxation proceeds faster than the intermolecular hopping.

One ought, however, to keep in mind that the dynamic states of a hot electronic polaron, as well as the higher exciton states are delocalized to a certain extent, and "horizontal" transfer is capable of competing with the "vertical" relaxation process. Such a transfer process of a hot electronic polaron might possibly be represented within the framework of a model similar to that proposed by Kenkre[167] for the description of hot exciton transfer. Čápek[59a] has recently analyzed the interplay of exciton or electron transfer and relaxation during the thermalization process (see Section 5.15).

In addition to optical transitions, hot electronic polarons can also be formed through field- or photoinjection of hot electrons from contacts or by an electron gun. Similarly, one may consider relaxation processes with formation of intermediate electronic and MP states, as well as transfer processes in the case of a hole (see Fig. 4.11b). In this case, however, a number of differences appear, particularly at higher energies.

An electron with an energy $E > 0$ (see Fig. 4.11b) is quasifree. It is characteristic of OMC, however, that the excess energy of the electron (especially in the high temperature range) is quickly lost in inelastic collisions, producing either Frenkel excitons (at $E \geqslant E_{exc}$) or lattice phonons (at $E < E_{exc}$). Effective inelastic collisions of an electron determine its localization, and, as soon as $\tau_h > \tau_e$, the electronic polarization sets in. As mentioned before, the gain in polarization energy increases the kinetic energy of the newly formed polaron quasiparticle, the "hot" polaron exhibits still more pronounced tendency toward localization than a thermalized polaron (see Ref. 143 and Chapter 2). The situation is somewhat different in the case of hot holes (see Ref. 288, p. 317). If the hole is positioned above the energy level S_0 (see Fig. 4.11b), it can be considerably delocalized. The degree of localization increases in the relaxation process, similarly, as it takes place in the case of an electron. If, however, the hole is very hot and is positioned below the S_0 level, it is strongly localized on the respective molecular or atomic orbitals. Localization is more pronounced in this case, the deeper the corresponding level is situated, since the wave-functions of deeply positioned levels of atomic cores practically do not overlap.

A certain asymmetry of the energy diagram of an electron and a hole is also seen. The question naturally arises, why the term A_G^v denoting the electron affinity of the molecule appears in the diagram, while a similar term, which would account for hole affinity of the molecule, is absent. This problem is linked with the intrinsic polarization of the electronic orbitals of the molecule on which the negative or the positive charge is localized. It leads to the problem of nonvalidity of the Koopmans theorem[193] in considering the total energetics in the case of electronic polarization. This problem is discussed in detail in the book (see Ref. 332, p. 91).

In the case of electronic polarization the contribution P_{eff} of the $N-1$ molecules of the crystal surrounding the charge is approximately equal to the contribution of the intrinsic polarization W_P of a molecule with a localized charge and equals 1.2–1.3 eV.[332] From this aspect an electronic polaron in an OMC may be regarded as a polaron of *large* radius.

In the case of a MP the basic contribution to the effective polarization energy $(E_b)_{eff} \approx 0.15$ eV is due (see Table 3.8) to the formation energy of a molecular ion $(E_b)_i \approx 0.1$ eV (see Section 3.6.1). The rest is supplied by the nearest-neighbor molecules of the ion (see Fig. 3.10). In this sense it is perfectly viable to consider the molecular polaron as an *intermediate* form of an adiabatic polaron of nearly small radius. The nearly small MP may be called a *mesopolaron* since it is intermediate between polarons of large and small radius.

A comparison between energy structure diagrams of linear polyacenes — Ac, Tc, and Pc (see Figs. 4.8–4.10) — shows that with increase in the number of condensed benzene rings (or, correspondingly, π-electrons) the energy terms I_G, I_C, E_G^{Opt}, and E_G^{Ad} decrease according to a certain rule, while the term A_G^v increases (see also Fig. 3.8). These data prove the direct effect of the electronic structure of a molecule on the energy structure of ionized states of a crystal and indicate the way how to change this structure by tailored design and synthesis of organic compounds.

On the other hand, as is seen from Fig. 3.8, the effective electronic polarization terms P_{eff}^+ and P_{eff}^- remain practically constant, independent of the number N of condensed rings in polyacene and other polycyclic aromatic molecules (see Section 3.5). The authors of Ref. 307 have used this property of P_{eff}^- to estimate the energy gap E_G^{Ad} values for a number polycyclic aromatic hydrocarbon crystals. For this purpose the following equation can be used[307]

$$E_G^{Ad} = I_C - A_G - \langle W_{eff}^- \rangle,$$

where $W_{eff}^- = P_{eff}^- + (E_b)_{eff}$.

Using the mean value of $\langle W_{eff}^- \rangle$ for polyacenes from Table 3.6 $\langle W_{eff}^- \rangle = 1.12$ eV and the reported I_C and A_G values, one can roughly estimate the energy gap E_G^{Ad} of the crystal. Unfortunately, there are few reliable A_G data (see, e.g., Table 5 in Ref. 307). This approach gives, e.g., for Nph, which has more or less reliable I_C and A_G values, the energy gap value $E_G^{Ad} = 5.13$ eV.

A particularly strong effect on the energy level structure of a molecular crystal is produced by introduction of heteroatoms into the conjugated cycles, or substituents.[332]

We present, as an example, the values of the adiabatic gap E_G^{Ad} for a number of Tc derivatives[351] (Fig. 4.12). The E_G^{Ad} value for crystals of these compounds was determined from the intrinsic photoconductivity threshold E_{th}, applying an appropriate correction for the shift $\Delta = 0.2$ eV (see Section 4.1).

Figure 4.12 shows that in the case of rubrene, although here four additional phenyl rings are added, the E_G^{Ad} value decreases by only 0.13 eV. This may be due to the fact that in the rubrene molecule the phenyl rings are not in plane with the tetracene nucleus.[20,351] The noncoplanarity is still further enhanced by steric hindrance. The effect of the phenyl side-groups on E_G^{Ad} is therefore small. A much stronger effect is produced by inclusion of the chalcogenides sulfur and selenium into the Tc ring (see Fig. 4.12). We see that the value of E_G^{Ad} in tetrathiotetracene and tetraselenotetracene approaches that of typical semiconducting materials.[20] These examples show the unlimited possibilities of organic synthesis of new hetero-

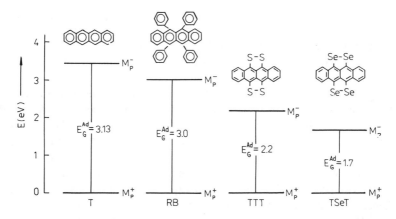

FIGURE 4.12. Adiabatic energy gap values E_G^{Ad} of tetracene (Tc), rubrene (Rb), tetratiotetracene (TTT), and tetraselenotetracene (TSeT); M_P^+, M_P^- are the conduction levels of the positive and negative molecular polarons, respectively.

cyclic compounds with appropriately varied parameters of the energy level structure of the crystal.

In a recent review by Silinsh, Bouvet, and Simon[336] on the determination of energy gap values in OMC, in addition to optical and photoelectric methods of E_G^{Opt} and E_G^{Ad} evaluation, discussed above, determination of the adiabatic energy gap E_G^{Ad} via intrinsic dark conductivity activation energy E_{act} and electrochemical redox potential measurements have been discussed in some detail.

4.4. BOUNDED ELECTRONIC AND MOLECULAR POLARONS. CHARGE TRANSFER (CT) AND CHARGE PAIR (CP) STATES

Polarons of opposite signs are capable of interacting in OMC with attractive Coulomb forces, forming bonded polar, but nonconducting states. In the case of a nonrelaxed electronic polaron pair, formed in optical CT-transitions (see Section 4.2), it is customary to speak of a charge transfer state, or *CT-state*.

As regards a *relaxed* molecular polaron pair, the term charge pair or CP-state has been proposed.[344,349]

CP-states are a characteristic feature of OMC. They manifest themselves as intermediate states in many electronic processes — photogeneration, at field — and photoinjection from the electrodes, in recombination, and in other processes.[288,332] From the terminological aspect bound polarons are sometimes called *bipolarons*.

The contention expressed above regarding the localization of molecular polarons (see Section 4.3) is also valid with respect to a polar pair of CP-states. Both the positive M_P^+, as well as the negative M_P^- polarons of the charge pair are localized on definite molecules of the crystal and do not form analogues of a large-radius Wannier excitons.[186a] The exciton concept cannot be, on principle, applied in the traditional sense to the CP-state, since the localized MPs of the charge pair in the crystal move by noncoherent stochastic hopping and are usually treated within the framework of the Onsager diffusion theory.[288,332]

Unlike relaxed CP-states, nonrelaxed CT-states produced in direct optical CT-transitions can be regarded as CT-excitons.[39,40]

The existence of CP-states as a characteristic feature of the energy structure of ionized states in OMC was already postulated by Lyons in his primary model;[220] (see also Ref. 122). It may be noted that Lyons was the first to postulate the possible existence of CT-states which he conjectured as a Rydberg level series converging to the bottom of a "wide" E_C band of "delocalized" electrons (see Fig. 2.2 in Ref. 332). The energy of the CP-states was first estimated roughly by Choi et al.[74]

The first extended calculations of the energy spectrum of CP-states, as dependent on charge separation distance r in an Ac crystal, were performed by Hug and Berry.[139] However, these authors applied in their calculations non-self-consistent approximations (see Section 3.5.1). As a result they obtained a spectrum of CP-states with overestimated energy values.

The first self-consistent calculations of CP-states spectra in Nph and Ac crystals were carried out[149] and later refined[342] by Jurgis and Silinsh (see also Ref. 332, p. 72). In their calculations of CP-state polarization energy P_{CP}, as dependent on charge separation distance r, they used the SCPF method [see Section 3.5.2] and performed numerical self-consistent iteration employing the following formula (analogous to formula 3.5.10) in calculating P for a single charge:

$$P_{CP}(r) = -(1/2)\sum_n \mathcal{E}_0(r_n)\mathbf{d}_n \qquad (4.17)$$

where $\mathcal{E}_0(r_n)$ is the total electric field strength on the nth molecule, created by the charge pair of the CP-state; $\mathbf{d}_n$ is the dipole moment induced on the nth molecule, as calculated in the iteration procedure by the SCPF method.

The binding energy of the CP-state U_{CP}, i.e., the energy interval between the level of the respective CP-state and the conduction level of the free negative polaron M_P^- (see Fig. 6.1) may be described by the expression

$$U_{CP}(r) = W_C(r) + [P_{CP}(r) - 2P_{id}] = W_C(r) - \Delta P_{CP}(r) \qquad (4.18)$$

where $W_C(r)$ is the term of Coulomb interaction of the charge pair

$$W_C(r) = -e^2/r \qquad (4.19)$$

and P_{id} is the term of self-consistent polarization energy of a single charge; $P_{id}^+ = P_{id}^- = P_{id}$ for linear polyacene crystals.

As may be seen, the U_{CP} value characterizes the bond energy of the CP-state and is equal to the energy of its thermal dissociation, i.e., $U_{CP} = E_{dis}$.[332] Apart from the term U_{CP} the concept of the CP-state energy $E_{CP}(r)$ may be used for characterization of the CP-states:

$$E_{CP}(r) = E_G^{Ad} + W_C(r) + \Delta P_{CP}(r). \tag{4.20}$$

As one can see, E_{CP} determines the energy interval from the conduction level of a free positive polaron M_P^+ to the corresponding level of the CP-state (see Fig. 6.1). Hence

$$|U_{CP}| + E_{CP} = E_G^{Ad}. \tag{4.21}$$

In macroscopic approximation the critical Coulomb radius r_c of the charge separation, at which $E_{CP} \rightarrow E_G^{Ad}$ and the CP-state dissociates into free polarons M_P^+ and M_P^- is determined by the expression

$$r_c = \frac{e^2}{4\pi\bar{\epsilon}\epsilon_0 kT} \tag{4.22}$$

where $\bar{\epsilon}$ is the mean dielectric permeability of the crystal. In polyacene crystals in which $\bar{\epsilon}$ varies from about 3.3 to 4.0, r_c equals 130–160 Å at 300 K. Accordingly, the Coulombic interaction radius in Ac-type crystals is comparatively large, considerably exceeding the mean free path, $\bar{l}$ of a MP which is of the order of the lattice constant, i.e., $\bar{l} \ll r_c$. This determines the exceptionally large role played by CP-states in OMC.

Equations (4.18) and (4.19) of the microelectrostatic approximation may be replaced by expressions in macroscopic (continuum) approximation (4.23) and (4.24) by introducing the effective isotropic dielectric permeability $\bar{\epsilon}$ which allows to account for the polarization effects of screening the Coulombic interaction of the charges.

$$U_{CP}(r) = -e^2/r + \Delta P_{CP}(r) \approx \frac{e^2}{4\pi\bar{\epsilon}\epsilon_0 r} \tag{4.23}$$

and

$$E_{CP}(r) \approx E_G^{Ad} - \frac{e^2}{4\pi\bar{\epsilon}\epsilon_0 r}. \tag{4.24}$$

As will be shown below, this approximation describes the realistic situation sufficiently adequately, and the $U_{CP}(r)$ curve may be considered as a Coulombic one.

Bounds and Siebrand[40] applied the FTM (see Section 3.5.3) for calculating the energy parameters of CP-states in an Ac crystal. This made it possible for the authors to extend considerably the range of r values between the charges and to obtain a rich calculated spectrum of $E_{CP}(r)$ terms for charge separation values r up to ca. 80 Å.

For equal E_G^{Ad} values both SCPF and FTM of calculation yield close values of $E_{CP}(r)$ and $U_{CP}(E)$. The insignificant difference in results might be due to the fact that the authors of Ref. 40 used slightly different sets of components of the molecular polarizability tensor b_i.

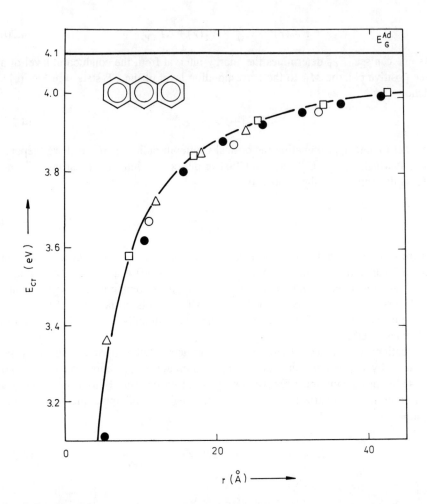

FIGURE 4.13. CP-state energy values E_{CP} for an anthracene crystal, as dependent on charge separation distance r along the **a**($\square$), **b**($\triangle$), **c**($\bigcirc$) axes and in the [110] direction (●) of the crystal according to Ref. 39, as calculated by Fourier transformations in submolecular approximation. Solid line, Coulomb approximation of the calculated values by means of formula (4.24) at $\bar{\epsilon} = 3.23$.

On the whole, comparison of the results of CP-state parameter calculations (similarly to the juxtaposition of the polarization energy of a single charge, see Table 3.1) demonstrates complete equivalence of both methods. However, the FTM has a number of advantages, as discussed in Section 3.5.3.

At a later stage Bounds, Petelenz, and Siebrand[39] introduced further improvements in the method of calculating the $E_{CP}(r)$ spectrum, using Fourier transformation combined with submolecule approach, developed for calculating the electronic polarization energy of a single charge (see Section 3.5.4).

Figure 4.13 presents results of calculations of the $E_{CP}(r)$ spectrum in submolecular approximation for an Ac crystal.[39] We see that in this case the Coulomb approximation by the formula (4.24) with $\bar{\epsilon} = 3.23$ provides adequate description for practically all calculated $E_{CP}(r)$ values in different crystallographic directions; all points lie close to the given Coulombic curve. This leads to an important, in the physical sense, conclusion that the interaction between charges (polarons) is of the Coulombic type in all crystallographic directions of an Ac-type crystal.[39] This statement is of supreme importance for the experimental determination of the bond energy of CP-states. It is based on the fact that the bond energy of the CP-state $U_{CP}(r)$ is equal, according to Eq. (4.23), to the activation energy of thermal dissociation of the given CP-state:

$$U_{CP}(r) \approx -\frac{e^2}{4\pi\bar{\epsilon}\epsilon_0 r} = E_a(r). \tag{4.25}$$

One can therefore use the activation energy of intrinsic photogeneration $E_a = E_a^{ph}$ as a basis for calculating, by means of Eq. (4.25) the separation length r_0 between the charges of a geminate intermediate pair, the dissociation of which causes photogeneration of free charge carriers. This method and the results of its application will be discussed in detail in Chapter 6. As already shown in Section 4.2, the Coulombic dependence of the given kind was used in Refs. 317 and 318 for determining the optical gap E_G^{Opt} on the basis of experimental data on optical CT-transitions.

Since, in accordance with Eq. (3.6.2), we have $E_G^{Opt} = E_G^{Ad} + 2(E_b)_{eff}$, and since the energy of the nonrelaxed CT-state E_{CT} for a given r value is connected with the energy of the relaxed CT-state E_{CP} by the expression (see Fig. 4.2)

$$E_{CT}(r) = E_{CP}(r) + 2(E_b)_{eff} \tag{4.26}$$

we obtain, from (4.24)

$$E_{CT}(r_{CT}) = E_G^{Opt} - \frac{e^2}{4\pi\bar{\epsilon}\epsilon_0 r_{CT}}. \tag{4.27}$$

As a result we have an expression identical with the empirical formula (4.11), used in Ref. 318 for determining E_G^{Opt} by linear extrapolation in $E_{CT}(r) = f(r_{CT}^{-1})$ coordinates.

The authors of Ref. 318 compared their experimental results in the determination of $E_{CT}(r)$ for Ac with the calculated data, as obtained by Bounds, Petelenz, and Siebrand.[39] Graphical representation of the dependence $U_{CT}(r) = E_G^{Opt} - E_{CT}(r) = f(r_{CT}^{-1})$ for the experimental and calculated $U_{CT}(r)$ values ought to yield a straight line with a slope $-e^2/4\pi\bar{\epsilon}\epsilon_0 r_{CT}$. However, as shown in Fig. 4 in Ref. 318, there appears a wide spread in the calculated values, and deviations sometimes reach a magnitude of 0.1–0.2 eV.

Bounds et al.[42] refined the calculations of the energy spectrum of CT-state parameters in Ac and also performed similar calculations for Tc and Pc. In addition to accounting for the interaction term between the CP-state charges and the self-consistent field of induced dipoles ΔP_{CT}, they also considered the interaction term

TABLE 4.5. *CP-State Energy Parameters in an Anthracene Crystal, as Dependent on Separation Distance r_{CP} Calculated in Submolecular Approximation*[42] [a]

Coordinates of negative charge in the lattice*	r_{CP} (Å)	ΔP_{CP} (eV)	W_c (eV)	W_Q (eV)	U_{CP} (eV)	E_{CP} (eV)	ϵ_{eff}
(1/2,1/2,0)	5.24	1.59	−2.55	0.102	−0.86	3.24	2.97
(1/2,−1/2,0)	5.24	1.59	−2.55	0.102	−0.86	3.24	2.97
(0,1,0)	6.04	1.46	−2.19	−0.004	−0.735	3.365	2.99
(1,0,0)	8.56	1.14	−1.65	0.050	−0.46	3.64	3.55
(1,1,0)	10.48	0.89	−1.36	−0.003	−0.47	3.63	2.87
(1,−1,0)	10.48	0.87	−1.34	0.035	−0.43	3.67	3.09
(0,0,1)	11.18	0.96	−1.38	−0.055	−0.48	3.62	2.88
(0,2,0)	12.07	0.79	−1.16	0.002	−0.37	3.73	3.13
(3/2,3/2,0)	15.72	0.61	−0.91	0.005	−0.29	3.81	3.13
(3/2,−3/2,0)	15.72	0.61	−0.91	0.005	−0.29	3.81	3.13
(2,0,0)	17.12	0.58	−0.84	0.006	−0.25	3.85	3.36
(2,2,0)	20.95	0.47	−0.69	−0.001	−0.21	3.89	3.20
(2,−2,0)	20.95	0.47	−0.68	0.005	−0.21	3.89	3.20
(0,0,2)	22.37	0.43	−0.65	−0.005	−0.23	3.87	2.89

[a]*Note*: the asterisk (*) denotes positive charge positioned in the (0,0,0) site of the lattice.

between the charges with the constant quadrupole moment W_{Q0} of the molecules. In particular, this accounts for the change in quadrupole moment $\Delta\theta_{AB}$ of the molecular ion, due to addition of an excess charge. The change in the quadrupole moment $\Delta\theta_{AB}$ is larger than the value of the quadrupole moment θ_{AB} of a neutral molecule and it increases sharply with increase in the number of benzene rings in the polyacene series. As shown in Ref. 42, the contribution of the nuclei to the total quadrupole moment value remains unchanged, the main contribution to $\Delta\theta_{AB}$ being due to the electronic part caused by the additional charge. Both cation and the anion show approximately the same $\Delta\theta_{AB}$ value, being only of opposite signs.

The method of calculating the quadrupole interaction term W_{Q0} is given in Section 3.5.5. The authors of Ref. 42 employ submolecular approach (see Section 3.5.4) for calculating the $\Delta P_{CP}(r)$ terms, as well as for estimating $\Delta\theta_{AB}$ and, accordingly, $W_{Q0}(r)$. The Coulomb interaction term $W_C(r)$ is also calculated in a submolecular approximation, with account for charge distribution over the benzene rings of the molecular ion (polaron).

The calculated values of the energy parameters of CP-states in an Ac crystal, as obtained according to the above-mentioned improved method, are presented in Table 4.5. As may be seen, accounting for the term W_{Q0} is particularly essential for the CP-state of the nearest neighboring molecules at $r_{CP} = 5.24$ Å where its value reaches 0.1 eV.

It should also be noted that the term W_c of Coulombic interaction, as obtained in submolecular approximation, when the charge is distributed over the benzene rings of the ion, has a lower value than in the case of point charge approximation. Thus, in Ac, we have a W_c value of −2.75 eV for the nearest neighboring molecules in point charge approximation while the corresponding value equals −2.55 eV in

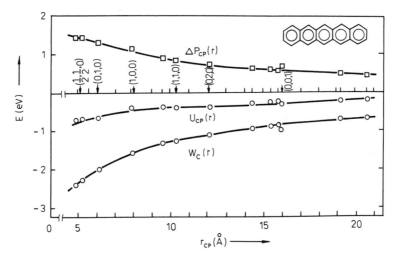

FIGURE 4.14. Dependences of calculated CP-state energy parameters $\Delta P_{CP}(r)$, $W_C(r)$, and $U_{CP}(r)$ on separation distance r_{CP} in a pentacene crystal according to Ref. 42, as determined by refined method of submolecular approximation.

submolecular approximation (see Table 4.5). It is of interest to remark, at this point, that calculations of the Coulombic term, applying the model of the charge distributed over the molecular ion, as in Ref. 74, the value of $W_c = -2.50$ eV has been obtained for the given CP-state. This shows that the submolecular approximation, as proposed by Bounds and Munn[38] (see Section 3.5.4), yields a satisfactory picture of charge distribution over the molecular ion.

Figure 4.14 demonstrates the calculated dependences of CP-state parameters $\Delta P_{CP}(r)$, $U_{CP}(r)$, and $W_c(r)$ obtained in the framework of the refined approximation.[42]

On the whole, the refined calculations of CP-states, according to Ref. 42, appear to be more reliable and describe more adequately the physical situation (see Fig. 4.15) than the initial calculations in which rougher approximations were applied (see Chapter 3). As may be seen from Fig. 4.15, the experimental and theoretical $E_{CT}(r)$ values are in fair agreement. This circumstance supports, on the one hand, the legitimacy of determining the optical energy gap through linear extrapolation, and the feasibility of the method of calculation,[42] on the other one.

The last column of Table 4.5 presents the values of the "local" effective dielectric constant, as obtained from the ratio $\epsilon_{eff} = W_c/U_{CP}$. It characterizes the degree of local polarizational screening of the given CP-state (see Fig. 4.14). ϵ_{eff} can be seen to show a maximum value for the deepest CP-state along the **a** axis[100] and minimum value along the **c** axis. With increasing charge separation distance r_{CP} the value of ϵ_{eff} evens itself out and approaches the mean value of $\bar{\epsilon} = 3.2$ for Ac.

The authors of Ref. 42 apply the microelectrostatic approximation, extended for large values of charge separation r_{CP}. They use this approach for proving the

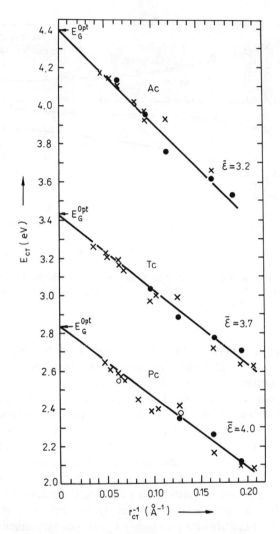

FIGURE 4.15. Experimental and calculated CT-state energy values E_{CT} as dependent on charge transfer distance r in $E_{CT} = f(r_{CT}^{-1})$ coordinates for anthracene (Ac), tetracene (Tc), and pentacene (Pc) crystals. The experimental points (●) have been determined by means of electromodulated absorption spectra,[317,318] but points (○) for pentacene have been determined from photoconductivity spectra at low temperature;[203] the calculated points (×) have been obtained by improved Fourier transformation method in submolecular approximation.[42] The straight lines represent Coulombic approximation according to formula (4.27) at $\bar{\epsilon} = 3.2$ (Ac), 3.7 (Tc), and 4.0 (Pc).[351]

adequacy of the phenomenological equation (4.23), as well as of the full correspondence between the microelectrostatistic and the macroscopic (continuum) approximation for sufficiently large r_{CP} values. This correspondence is most complete if, instead of the mean isotropic dielectric constant $\bar{\epsilon}$ one employs the tensor ϵ of the macroscopic dielectric permeability, having the main components ϵ_i. For a definite crystallographic direction $\mathbf{r} = (\lambda,\mu,\nu)$ with fixed values of the sines of the direction angles λ, μ, and ν we then have the following expression for U_{CP}:[42]

$$U_{CP} = E_C(r) + \Delta P_{CP}(r) = -\frac{e^2}{4\pi\epsilon_{\lambda,\mu,\nu}\epsilon_o r}. \qquad (4.28)$$

It may be seen that if we replace the virtual effective dielectric constant $\epsilon_{\lambda,\mu,\nu}$, which macroscopically accounts for the anisotropy of the lattice, by the mean isotropic dielectric constant $\bar{\epsilon}$, then Eqs. (4.23) and (4.28), connecting the microscopic and the macroscopic approximations, are completely *equivalent*. Expression (4.28) also supports the Coulombic nature of the $U_{CP}(r)$ dependence [and, correspondingly the $E_{CP}(r)$ dependence in *all* directions of the lattice].

As shown in Ref. 42, the value of $\epsilon_{\lambda,\mu,\nu}$ is directly connected with the main components ϵ_i of the dielectric tensor ϵ by the formula

$$\epsilon_{\lambda,\mu,\nu} = (\epsilon_2\epsilon_3\lambda^2 + \epsilon_3\epsilon_1\mu^2 + \epsilon_1\epsilon_2\nu^2)^{1/2}. \qquad (4.29)$$

A comparison between the calculated $E_{CT}(r)$ values[42] with experimental ones[318] in $E_{CT}(r) = f(r_{CT}^{-1})$ coordinates yields for Tc and Pc just as fair an agreement as for Ac (see Fig. 4.15). These results confirm once more the adequacy of the method for determining E_G^{Opt}, as well as the reliability of the refined calculations of Bounds *et al.*[42]

It might be worthwhile, in conclusion, to compare calculated and experimental values of the terms E_{CP}^1 and U_{CP}^1 of the CP-states of the nearest neighboring molecules $(0,0,0) \leftrightarrow (1/2,1/2,0)$ in Ac, Tc, and Pc (see Table 4.6).

The data in Table 4.6 show that the calculations of parameters E_{CP}^1 and U_{CP}^1 have been most intensely performed, in various approximations, for an Ac crystal. The best agreement with the reliable experimental E_{CP}^1 value[318] has been achieved by Bounds *et al.*,[42] using the refined calculation method in terms of submolecular approximation. (It ought to be noted that the frequently quoted value of E_{CP}^1 for Ac $E_{CP}^1 = 3.45$ eV is only an experimental estimate[287] and yields only the correct order of magnitude of this value.)

Satisfactory agreement between calculated and experimental E_{CP}^1 values has also been observed for Tc; fair correspondence has been found in the case of Pc (see Table 4.6).

Figure 4.16 shows experimental and calculated bond energy U_{CP} of the deepest CP-state, as dependent on the number N of benzene rings in polyacene molecules. It can be seen that with increase in N there is an insignificant decrease of bond energy U_{CP}. This is understandable, since one may expect only a slight increase of the screening effect of Coulombic interaction between nearest neighboring

TABLE 4.6. E_{CP}^1 and U_{CP}^1 Parameter Values for the Deepest CP-State of the Nearest Neighboring Molecules (0,0,0) (1/2,1/2,0) in Polyacene Crystals

Crystal	r_{CP} (Å)	Method of determination	E_{CP}^1 (eV)	U_{CP}^1 (eV)	E_G^{Ad} (eV)
Anthracene	5.4	Calculated by the SCPF method in point charge approximation[332,342]	3.35	0.75	4.10
		Calculated by Fourier transformation in point charge approximation[39]	3.32	0.78	—
		Calculated by Fourier transformation in submolecular approximation[39]	3.13	0.97	—
		Refined method of submolecular approximation[42]	3.24	0.86	—
		Experimental method of optical CT-transitions[317]	3.24	0.86	—
		Experimental estimate after Ref. 287	3.45	0.65	—
Tetracene	5.12	Refined method of submolecular approximation[42]	2.41	0.72	3.13
		Experimental method of optical CT-transitions[318]	2.34	0.79	—
Pentacene	5.15	Refined method of submolecular approximation[42]	1.76	0.71	2.47
		Experimental method of optical CT-transitions[318]	1.74	0.73	—

molecules due to insignificant increase in electronic polarization energy for polyacene series (see Fig. 3.8).

Kurik and Piryatinski[203] have studied intrinsic photoconductivity spectra in thin Pc layers at 4.2 K (see Fig. 4.17). These spectra revealed a fine structure in the near-threshold region of photogeneration at $h\nu > 2.4$ eV. The declining part of the near-threshold spectral dependence $I_{ph} = f(h\nu)$ clearly shows separate peaks of narrow bands. The intensity of these peaks increases with electrical field strength. It

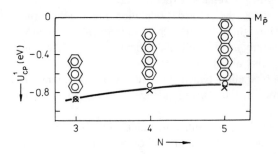

FIGURE 4.16. Experimental $(\times)^{318}$ and calculated $(\bigcirc)^{42}$ bond energy values U_{CP}^1 for deepest CP-states $(0,0,0) \leftrightarrow (1/2,1/2,0)$ in anthracene, tetracene, and pentacene crystals, as dependent on the number of benzene rings (N) in the molecule.

is therefore reasonable to assume that the appearance of these bands is caused by direct photoionization from the optical CP-transition levels. In this case the peak at $h\nu = 2.54$ eV may be identified as the CT-transition $(0,0,0) \to (0,0,1)$ with a value of $r_{CT} = 16.01$ A, i.e., as a transition along the **c** axis of Pc crystallites. The peaks at $h\nu = 2.70$, 2.85, 3.00, and 3.16 eV, positioned at equal energy distances of $\Delta E = 0.155$ eV, may be considered as possible vibronic satellites of the lowest CT-transition at $h\nu = 2.54$ eV, having the vibronic quantum number $\nu = 0$. Since in the work[203] the electric field $\mathcal{E}$ was directed at right angles to the **ab**-plane of the oriented pentacene crystallites in a "sandwich"-type cell, one may naturally expect that just this transition must possess the highest intensity, unlike the results of electromodulated Pc spectra measurements in Ref. 317. In the latter work a surface cell was used, and CT transitions were observed only in the **ab** face of the crystallites. As regards the peak at $h\nu = 2.41$ eV, it might be attributed to a CT-transition of type $(0,0,0) \to (1,0,0)$ with $r_{CT} = 7.90$ A along the **a**-axis of the crystallites. These E_{CT} values are plotted in $E_{CT} = f(r_{CT}^{-1})$ in Fig. 4.15 for Pc. The figure shows that the E_{CT} values of the given transitions in Pc, as obtained from the spectral dependence of the photocurrent $I_{ph} = f(h\nu)$ at a low temperature (see Fig. 4.17), are in good agreement both with experimental E_{CT} values determined by electro-modulation methods in Ref. 317, as well as with calculated data in Ref. 42, and are satisfactorily approximated by a Coulombic straight line according to expression (4.27). The fact of direct manifestation of optical CT-transitions in the low-temperature spectrum of photoconductivity in Pc is of considerable interest from the physical point of view. This means that at least part of photogenerated charge carriers emerge, after direct ionization of CT-states, as nonrelaxed free carriers beyond the Onsager radius. This assumption is supported by experimental and computer-simulated studies (see Refs. 355 and 356) which demonstrate that at low temperatures and sufficiently high electric fields the carriers, moving along the field direction, do not thermalize but leave the Coulombic well as nonthermalized free carriers (see Chapter 7). As an additional confirmation to this may serve the fact that according to the authors of Ref. 203 the observed low-temperature photoconductivity spectral curve (Fig. 4.17) cannot be approximated by the Onsager formula.

Thus, with the background of relaxed carriers, above the adiabatic energy gap E_G^{Ad} there appear the CT-transition peaks of nonrelaxed carriers which converge to the optical bandgap E_G^{Opt} (see Fig. 4.15).

As shown in Section 4.2, in polyacenes the optical CT-transitions are practically completely obscured by intensive bands of intramolecular (Frenkel) transitions (see, e.g., Fig. 4.6) and may be detected only by electromodulation techniques. However, in crystals of some polar molecules the CT-bands may be shifted toward the long-wave region of the spectrum beyond the band of the first singlet transition, and they can be directly detected in optical absorption spectra. Thus, Takase and Kotani[379] observed weak absorption bands at $\lambda = 415$ and 430 nm in phenothiazine crystals, situated beyond the long-wave tail of the fundamental absorption band. The intensity of these bands grows with increase in electric field strength, and they can be identified as bands of intermolecular CT-transition.

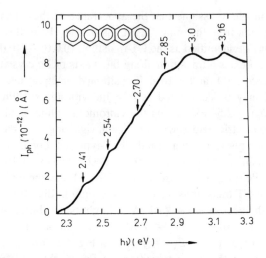

FIGURE 4.17. Spectral dependence of photoconductivity $I_{ph} = f(h\nu)$ in thin-layer pentacene samples in the near-threshold region of intrinsic photogeneration at 4.2 K according to Ref. 203. Arrows indicate fine structure in the spectrum, due to ionization from assumed optical CT-transition levels.

4.5. ENERGY STRUCTURE OF POLARON STATES IN AN IDEAL (PERFECT) CRYSTAL. A SYNOPSIS

Analysis of experimental data and calculation, as discussed above, shows that ionized states in OMC, i.e., conduction levels of charge carriers in the energy level diagram of an ideal (perfect) crystal are determined by the self-energy value of polaron-type quasiparticles, namely, electronic and MPs.

The energy levels of vibrationally *nonrelaxed* electronic polaron states in the crystal E_P^+ and E_P^- are separated by the optical energy gap E_G^{Opt}, while the conduction levels of relaxed states of the MPs M_P^+ and M_P^- are separated by the corresponding adiabatic energy E_G^{Ad}.

At present, the E_G^{Opt} and E_G^{Ad} values have been determined experimentally with sufficient reliability only for polyacene crystals — Ac, Tc, and Pc. For these crystals refined self-consistent calculations of effective electronic polarization energies have also been performed. In the latter, the asymmetry of electronic polarization of positive and negative charges, due to the asymmetrical interaction term between the charge and the quadrupole moments of the molecules, has been taken into account. On the basis of these calculations and their comparison with experimental results of the determination of energy parameters of the molecules and the crystal, refined and reevaluated energy level diagrams of ionized states of Ac, Tc, and Pc crystals have been obtained. Generalized experimental data and results of refined calculations for polyacene crystals yielded the energy spectrum of Coulombic field-bonded pairs of electronic and molecular polarons. It has been shown that in optical CT-transitions nonrelaxed electron-polaron pairs are formed which produce, after vibrational relaxation, Coulombic field-bonded geminate CP states.

4.6. ENERGY STRUCTURE OF LOCAL STATES IN REAL (IMPERFECT) OMC

We have discussed above the energy structure of electronic states in ideal, perfect organic crystals.

However, an ideal crystal with perfect lattice structure is an abstraction. A real crystal possesses different kinds of structural defects as well as contain impurities, e.g., different types of guest molecules in the host lattice. It has been shown that these defects create in OMC local electronic states: trapping states (traps) and scattering states (antitraps) both for charge carriers and excitons.

The local states in OMC, their nature, distribution characteristics, and methods of control have been discussed in some detail in a number of monographs,[288,332,351] review articles,[15,204] and conference proceedings.[333,335] Since these electronic states are of *polarization* origin and should be described in terms of many-electron interaction, we shall present here a brief analytic overview, mainly emphasizing the role of electronic polarization in the local state formation in OMC. For more detailed studies we refer the reader to the above-mentioned sources.

4.6.1. Local States of Structural Origin

Only in an ideal crystal, having perfect periodicity of structure and strictly equal unit cell parameters throughout the lattice, are effective polarization energies of P_{eff}^- and P_{eff}^+ also strictly fixed and have discrete values for a given crystal, and, consequently, the conduction levels M_P^- and M_P^+ in the energy diagram also have discrete values (see Figs. 4.8–4.10).

However, in every real molecular crystal there are local deviations from ideal geometry, for instance, in the regions of deformation around structural defects. It may be shown that structural irregularities and lattice defects create local states for electronic polarons.[332,333,335] A detailed phenomenological model describing the formation and nature of local electronic states of structural origin in OMC was proposed by Silinsh[330] and, independently, by Sworakowski.[374] According to this model, local charge carrier trapping states with quasicontinuous energy spectra are due to the local electronic polarization variations for charge carriers located in the regions of structural irregularities of the crystal.

It may be shown[330,332,333] that every deviation of intermolecular distance Δr_i from its mean value $\bar{r}_i$ ($\Delta r_i = r_i - \bar{r}_i$; where r_i is the distance between the molecule with localized charge carrier and the ith molecule) produces a corresponding change in the carriers' effective electronic polarization energy:

$$\Delta P_i = P_i(\bar{r}_i + \Delta r_i) - P(\bar{r}_i).$$

The total value of ΔP, taking into account only the contribution of the main polarization term of the charge-induced dipole $(q-d)$ interaction is given by the formula:[330,332]

$$\Delta P = 2e^2\alpha \sum_{i=1}^{N-1} (\Delta r_i/\bar{r}_i^5). \tag{4.30}$$

The formation of local electronic states in a real OMC, caused by structural defects of the lattice, has been illustrated, according to the polarization model,[330,332,333] in Fig. 4.18.

Thus, in the regions of lattice compression local states with increased polarization energy ($\Delta P < 0$) are formed which act as charge carrier traps in the energy gap E_G^{Ad}. On the other hand, in dilated lattice regions local states with diminished polarization energy ($\Delta P > 0$) are created above the negative charge carrier conduction level M_P^- and below the positive charge carrier conduction level M_P^+ which act as energetically unfavorable antitraps for charge carriers (Fig. 4.18). One should mention that in this approach it is assumed that the possible change in MP energy $(E_b)_{eff}$ in the defect region is negligible and that the local state formation is practically determined only by the change in electronic polarization energy ΔP_{eff} (see Ref. 335).

As can be seen from formula (3.55), in case of $(q-d)$ interaction, the polarization term, as well as the ΔP term in (4.30) is not dependent on the charge carrier's sign. That means that the same structural defect may act as a trapping state for both electrons and holes and the corresponding local electronic states are situated symmetrically in the energy diagram (see Fig. 4.18).

4.6.1.A. The Gaussian Distribution Model of Local States in OMC

The random distribution of structural irregularities in a real crystal leads to a statistical dispersion of electronic polarization energy. It can be shown that the most feasible approximation of energy spectra for such local states of structural origin is the Gaussian distribution model[272,330,332,333] (see for details Ref. 322, pp. 141 and 271). Two basic kinds of Gaussian distribution, namely, $G_e(E)$ and $G_g(E)$ (see Fig. 4.18) may be introduced.[272]

The shallow, quasicontinuous local trapping states, with $P < 0$ at the conduction level edge, and their symmetrical counterpart, "antitrapping" states, with $P > 0$ may be approximated by $G_e(E)$ distribution, centered symmetrically around the conduction levels M_P^- and M_P^+ with $E_t = 0$ and $E_t = E_G^{Ad}$, respectively. The $G_e(E)$-type of distribution may be caused, e.g., by edge dislocations, having both structural deformation counterparts, namely, a compressed lattice region above and a dilated lattice region below the dislocation line (Figs. 4.18 and 4.19).

On the other hand, trapping states, located in the energy gap, may be approximated by $G_g(E)$ type of distribution centered at E_t below the conduction level M_P^- (Fig. 4.18). This type of distribution may be formed in compressed lattice regions

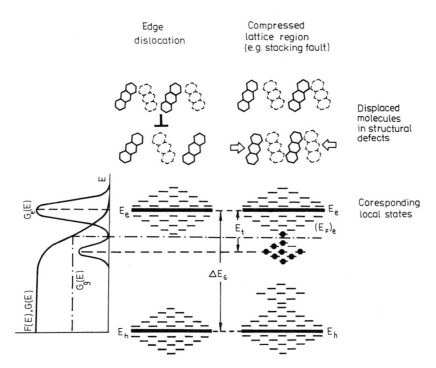

FIGURE 4.18. Schematic energy diagram for a molecular crystal with Gaussian distributions of local states of structural origin.[335] (In this figure former notations, see Ref. 335, are used. According to more up-dated notation, used in this book, $E_e = M_P^-$, $E_h = M_P^+$, and ΔE_G in this case is equal to E_G^{Ad}.)

of various more complex extended structural defects, such as stacking fault ribbons, dislocation aggregations, second-phase inclusions, etc.

The Gaussian model thus make it possible to consider the integral profile of local state energy distribution within the energy gap E_G^{Ad} as a superposition of separate Gaussian distributions.

Both kinds of Gaussian distribution may be described by the formula:[272,332]

$$N(E) = \frac{(N_t)_i}{(2\pi)^{1/2}(\sigma_p)_i} \exp\left[-\frac{(E-E_t)_i^2}{2(\sigma_p)_i^2} \right], \tag{4.31}$$

where $(N_t)_i$ is the total density of local states of the corresponding distribution; $(\sigma_p)_i$ is the distribution parameter, $(E_t)_i$ is the position of the Gaussian distribution peak inside the energy gap (Fig. 4.18).

The Gaussian approach is physically self-consistent in the framework of a generalized phenomenological ionized state energy model for real OMC, developed in Refs. 332 and 351, according to which the energy of conduction levels, as well as that of the local states of structural origin, are determined by the electronic polarization of the crystal by charge carriers.

Such a generalized Gaussian model includes in the common description limiting cases of local state distribution. Thus, at the limit, when the parameter $\sigma \to 0$, the $G_e(E)$ distribution gives an idealized picture of discrete conduction levels M_P^- and M_P^+ for a perfect crystal (see Figs. 4.8–4.10). On the other hand, the $G_g(E)$ type of distribution at the limit, when $\sigma \to 0$, describes a discrete trapping state at E_t, which may be considered as a δ-function. In our opinion, this model may serve as a phenomenological and conceptual basis for future development of a comprehensive polaron theory of local electronic states in OMC.

4.6.1.B. Formation of Local States for Charge Carriers in Structural Defects in OMC

A molecule which has been displaced by Δr_i from its initial equilibrium state r_i actually forms a local electronic state with respect to a regular lattice and may possess an altered self-energy value for a charge carrier localized on the given molecule.[330,332,333,335] Thus, the energy spectra of local states are directly correlated with the *microtopography* of molecules in the structural defects. This isomorphic correlation may be presented quantitatively, in zero-order approximation, as:[15]

$$\Delta P/P = 4 \sum_{i=1}^{N-1} \Delta r_i / r_i. \tag{4.32}$$

Thus, if the coordinates and orientation of all displaced molecules are known, then the complete energy spectrum of local states can be calculated.

For the calculation of the configuration of molecules in structural defects the atom-atom potential (AAP) method can be employed (see Refs. 184 and 283 and references therein) (see also Chapter 1, Section 1A). Recently, the traditional AAP method has been supplemented with the molecular dynamics approach.[130,275] It has been shown that both methods give identical results.

The first successful application of the AAP method for calculation of the equilibrium postrelaxation configuration of molecules in a stacking fault ribbon, bounded by partial dislocations in Ac and in substituted Ac crystals, was demonstrated by Silinsh and Jurgis[343] and Ramdas et al.[293] (see for details Ref. 332, p. 188).

Later Mokichev and Pachomov[243] employed the AAP method for calculating the equilibrium configuration of molecules for an edge dislocation of (001)[010] type in the Nph crystal (see Fig. 4.19).

Recently Kojima and coworkers[130,191a,275] have calculated the equilibrium molecular configuration both for edge and screw dislocations in an Ac crystal (Figs. 4.20 and 4.21). In these extended and refined calculations more than 1000 molecules around the dislocation core have been included.

Achievements in high-resolution electron microscopy have opened the way for direct observation of molecules in the regions of structural defects.

Thus, Kobayashi and coworkers (see Refs. 187 and 190 and references therein) have obtained superb electron micrographs of molecular configurations in the regions of extended structural defects, such as stacking faults, edge dislocations, grain boundaries, inclusions, etc., in crystals of Me phthalocyanines. These micrographs confirm the plausibility of the models proposed for description of such defects in OMC.[332] Further advancements in this field[187] are still more promising and one may hope that in the very near future it will be possible to obtain microphotographic maps of molecular configuration in defect structures, suitable for checking the calculation data, as well as for direct calculation of energy spectra of local trapping states.

Let us first briefly discuss calculations of local state energy spectra in extended structural defects. The local state energy spectra have been calculated for such extended structural defects as stacking faults[343] as well as edge and screw dislocations in Ac-type crystals.[130,243,275]

In Ref. 343 linear and angular displacements of molecules, relaxed to the equilibrium configuration, were calculated by the AAP method for stacking fault ribbons bounded by partial dislocations $1/2(110)$ and $1/2(1\bar{1}0)$ in the ab plane of an anthracene crystal. The corresponding calculations of local state energy spectra by the SCPF method demonstrated that in the compressed regions, inside the stacking fault, a set of trapping states with E_t of the order of $0.1-0.2$ eV are formed.

Similar approach was used by the authors of Ref. 243 in case of edge dislocations. First, they calculated the equilibrium configuration of molecules for an edge dislocations of (001) [010] type in a Nph crystal (Fig. 4.19). After that the authors evaluated by the SCPF method (see Section 3.5.2) the corresponding ΔP values for every molecule in the dislocation core. As can be seen from Fig. 4.19 in the compression region, above the dislocation line, trapping states with increased polarization energy ($\Delta P < 0$) are formed. On the other hand, in the dilated lattice region, below the dislocation line, scattering antitraps with decreased P values ($\Delta P > 0$) emerge.

Recently Kojima and coworkers[130] calculated by SCPF method the local state energies for 1062 displaced molecules around an edge dislocation in an Ac crystal (see Fig. 4.20).

Both calculations of local state energy spectra of displaced molecules around the edge dislocation in naphthalene and anthracene crystals[130,243] are in good agreement. The estimated value of the deepest trapping states $E_t \approx 0.1$ eV just above the dislocation line are confirmed by other independent estimates and experimental observations.

Kojima and coworkers[130] have calculated also the energy spectra of local states around a screw dislocation in an anthracene crystal (see Fig. 4.21).

As can be seen from Fig. 4.21, the spatial distribution of traps and antitraps around the screw dislocation is quite different from that of edge dislocation. In some directions one observes an alternative sequence of traps and antitraps while in other directions lines of traps and antitraps are formed. Such a line of traps creates a quasi-one-dimensional potential "valley" along which the trapped carrier is free to move.

On the whole these calculations confirm the phenomenological model of local state formation in OMC displayed in Fig. 4.18. In general, every kind of local lattice compression should lead to formation of corresponding local trapping states, e.g., compression regions on grain boundaries, predimer configurations of molecules in Ac-type crystals, inclusions of another, higher-density phase in the matrix of the parent crystal, etc. (see for details Ref. 332).

These conclusions may be extended to other types of organic solids. For example, crystalline domains in an amorphous matrix in polymers or LB layers may act as trapping centers for charge carriers. Local states are also formed in the vicinity of such point defects as lattice vacancies.

Lattice vacancies are dominant point defects in OMC: their thermal equilibrium density at 300 K has been estimated to be above 10^{15} cm^{-3} in Ac-type crystals.[332]

As a carrier approaches a vacancy, the absolute value of P decreases ($\Delta P > 0$) according to Eq. (3.55), due to removal of the polarizability contribution of the molecule formerly occupying the vacant site. This means that the vacancy would act as a scattering center. Earlier calculations of ΔP in the vicinity of a lattice vacancy in Ac[342] showed that around the vacancy a repulsive scattering potential of polarization origin $P(r)$ is formed for carriers of both signs. Later Eisenstein and Munn[100] also included in their treatment the charge-quadrupole interaction term (3.5.25) (see Section 3.5.5). Since the sign of the term W_{q-Q_0} is dependent both on the sign of the charge carrier and the quadrupole moment Θ, the ΔP may have positive, as well as negative value. Indeed, the calculation results of Ref. 100 demonstrated that in the vicinity of a lattice vacancy a complicated potential hypersurface is formed including both trapping ($\Delta P < 0$) and scattering ($\Delta P > 0$) states (Fig. 4.22). The calculations predicted also an asymmetry of hole and electron trapping (see Fig. 4.22). This trapping asymmetry was actually observed experimentally in Pc crystals[352] and thus the existence of a new kind of structural traps of quadrupolar origin was confirmed.

As we see, charge carrier-quadrupole interaction produces a certain kind of asymmetry both in energy structure diagrams (see Figs. 4.8–4.10) and trapping phenomena.

Local states appear at the crystal surface and metal–crystal interface. It has been shown that local states of polarization origin are also formed at near-surface molecules of the crystal.[332,333] If the charge carrier approaches the free surface, its polarization energy P decreases, as can be seen from expression (3.5.5) for W_{q-d}. The terms of the sum in (3.5.5) disappear for molecules which are replaced by vacuum outside the boundaries of the surface. As a result, a near-surface potential barrier of polarization origin

$$\varphi_p(r) = |P_v| - |P_S(r)|$$

is created for charge carriers, where P_v and P_S are the polarization energies inside

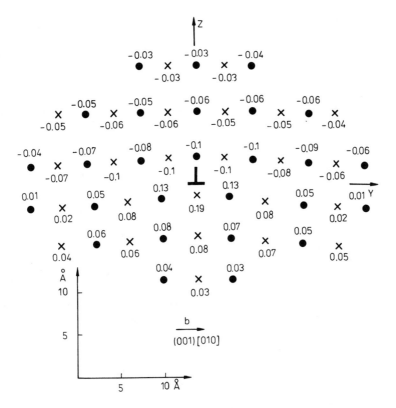

FIGURE 4.19. Calculated equilibrium configuration of molecules in the core of an *edge* dislocation of (001) [010] type in a naphthalene crystal and the corresponding changes of electronic polarization energy ΔP (in eV) for charge carriers localized on the given molecules according to Ref. 243.

the bulk and in the near-surface regions respectively. Calculations of the P_v and $P_S(r)$, performed by the SCPF method,[342] yield polarization barrier values up to 0.5 eV, e.g., for the top layer of the **ac** surface of the Ac crystal with penetration depth ca. 12–15 Å.

The surface polarization barrier, as predicted in Ref. 342, has been observed experimentally in Ac by Salaneck,[304] applying angle-dependent photoelectron emission measurements. On the contrary, if the charge carrier approaches the crystal surface interface with metal, its polarization energy P increases due to interaction with the free electrons of the metal. The estimated increase ΔP may be ca. 0.5 eV or more for Pc crystals,[204] the penetration depth of this interaction may be about 50 Å.

Munn has recently proposed a theory of electrical properties of molecular crystal surfaces.[255a] The author discusses crystal surface electrical fields, surface polarization energies, charge carrier energetics, and nonlinear optics.

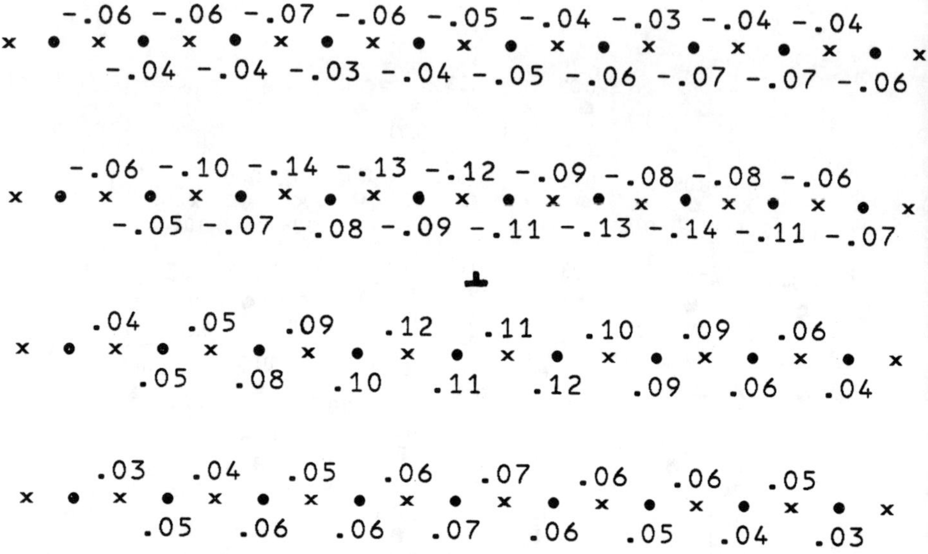

```
 -.06  -.06  -.07  -.06  -.05  -.04  -.03  -.04  -.04
x  •  x  •  x  •  x  •  x  •  x  •  x  •  x  •  x  •  x
  -.04  -.04  -.03  -.04  -.05  -.06  -.07  -.07  -.06


 -.06  -.10  -.14  -.13  -.12  -.09  -.08  -.08  -.06
x  •  x  •  x  •  x  •  x  •  x  •  x  •  x  •  x  •  x
  -.05  -.07  -.08  -.09  -.11  -.13  -.14  -.11  -.07

                          ⊥

   .04   .05   .09   .12   .11   .10   .09   .06
x  •  x  •  x  •  x  •  x  •  x  •  x  •  x  •  x  •  x
   .05   .08   .10   .11   .12   .09   .06   .04


   .03   .04   .05   .06   .07   .06   .06   .05
x  •  x  •  x  •  x  •  x  •  x  •  x  •  x  •  x  •  x
   .05   .06   .06   .07   .06   .05   .04   .03
```

FIGURE 4.20. Calculated equilibrium configuration of molecules in the core of an *edge* dislocation in an anthracene crystal and the corresponding changes of electronic polarization energy ΔP (in eV) according to Refs. 130 and 191a.

4.6.1.C. Local States of Structural Origin for Excitons: Isomorphism of the Energy Spectra of Local States for Charge Carriers and Excitons

It has been shown that one and the same structural defect of a molecular crystal can form local trapping states both for charge carriers and for excitons.[204,332] As already shown in Chapter 2, exciton self-energy is determined by the topology (arrangement) of the molecules in the crystal in a way similar to charge carriers' self-energy. It should be emphasized that the interaction forces are similar in their physical nature, namely, both are of electronic polarization origin (see Chapter 1, Section 1B). However, the r-dependences of the total interaction energy are different in both cases. For a charge carrier the interaction energy decreases as $1/r^4$ while for an exciton as $1/r^6$.

Due to r-dependence the term D [as well as the less important term I_{mn}, see formula (2.1.1)] will be unequivocally set only for a perfect crystal, and only in this case will the exciton energy levels be strictly fixed (see Fig. 2.1) and equal for the whole crystal. If, however, there are local deviations from the ideal geometry in the crystal, namely, in the regions of deformation around structural defects, then the exciton self-energy also assumes corresponding local values in complete analogy with charge carriers. In regions of lattice compression the value of interaction term D will be increased, thus forming local exciton trapping states and, vice versa, in regions of lattice dilation D will be diminished creating local exciton antitrapping states. Thus, as we see, the same structural defect can, indeed, form local states both

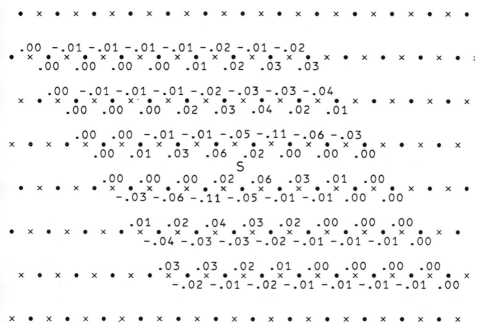

FIGURE 4.21. Calculated equilibrium configuration of molecules in the core of a screw dislocation in an anthracene crystal and the corresponding changes of electronic polarization energy ΔP (in eV) according to Refs. 130 and 191a.

for charge carriers and excitons. However, the energy of the corresponding terms P and D will differ owing to different r-dependences and parameters used in formulas (2.1.1) and (3.55). Common in both cases is the pronounced dependences of these terms on lattice geometry and their high sensitivity to changes in this geometry.

The change ΔD of the parameter D caused by displacement of molecules in a structural defect may be evaluated according to the following expression:[204]

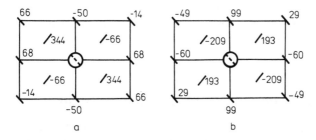

FIGURE 4.22. Calculated changes of polarization energy ΔP (in meV) due to a vacancy at (000) in the **ab** plane of a pentacene crystal for various positions of an excess charge carrier located in the vicinity of the vacancy: (a) for holes (b) for electrons.

$$\Delta D = 6A^* \sum_{i \neq k} \Delta r_{ik}/\bar{r}_{ik}^7.$$ (4.33)

These local states for excitons may be described by a Gaussian distribution formula, analogous to the formula[272] for charge carriers:

$$N(\Delta D) = \frac{N_s}{(2\pi)^{1/2}\sigma_{\mathrm{D}}} \exp\left[-\frac{\Delta D^2}{2\sigma_{\mathrm{D}}^2}\right].$$ (4.34)

It has been shown that the ratio of the corresponding dispersion parameters equal:[15]

$$\sigma_P/\sigma_{\mathrm{D}} = 2P/3D.$$ (4.35)

Thus, for $P = 1.5$ eV and $D = 0.3$ eV, it follows that $\sigma_P/\sigma_{\mathrm{D}} \approx 3$. That means the same structural defect forms about three times wider local state distribution bell width for charge carriers, as compared with that of excitons.

The corresponding trap depth for charge carriers E_t and excitons E_s are determined by the respective total terms of polarization energy changes in structural defect, namely, ΔP and ΔD, evaluated according to formulas (4.30) and (4.33):

$$E_t = \Delta P$$ (4.36)

and

$$E_s = \Delta D.$$ (4.37)

The ratio of E_t/E_s is of the order of 2–5 and increases with increasing lattice compression.[332] Consequently, the same lattice defect form two to five times deeper traps for charge carriers than for singlet excitons.

The principle of isomorphism of local state energy spectra for charge carriers and excitons is illustrated in Fig. 4.23.[204,333] It demonstrates the most important property of local states in OMC, i.e., that the same structural defect of compressed lattice region produces *isomorphic* energy spectra for charge carriers and excitons which differ only by the corresponding E_t and E_s, as well as by σ_P and σ_D values.

The energy and dispersion parameters for both cases are *quantitatively* correlated; therefore if one determines these parameters for exciton trapping (E_s and σ_S) then the corresponding parameters for carrier trapping (E_t and σ_P) can be evaluated, and vice versa.[204]

4.6.2. Local States of Chemical Impurity Origin

Guest (impurity) molecules in a host molecular crystal form local states for charge carriers and excitons. The energy levels of these states can be described in the framework of the polarization model.[332] Figure 4.24 taken from an early paper by Karl[153] illustrates formation of local trapping and antitrapping states for charge

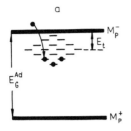 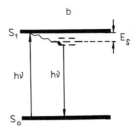

FIGURE 4.23. Schematic diagram of local trapping states of structural origin for charge carriers (a) and excitons (b) according to Ref. 204. This diagram illustrates the principle of energy spectra isomorphism of local trapping states for charge carriers and excitons.

carriers formed by some guest molecules in an anthracene host. These trapping states are often called "chemical traps" to distinguish them from structural ones.

As may be seen from Fig. 4.24, the energy levels of the guest molecules are shifted with respect to the conduction levels (E_e, E_h) in the many-electron interaction energy diagram of the host crystal.

In zero-order approximation the local states formed by guest molecules in a host lattice may be regarded as discrete, determined only by molecular parameters of the guest and host molecules. Thus, local states for negative charge carriers (electrons), formed by guest molecule E_{guest}, are determined by the difference between the electron affinity of guest $(A_G)_{\text{guest}}$ and host $(A_G)_{\text{host}}$ molecules, respectively:

$$E^e_{\text{guest}} = (A_G)_{\text{guest}} - (A_G)_{\text{host}}. \tag{4.38}$$

If $E^e_{\text{guest}} > 0$ traps and, vice versa, if $E^e_{\text{guest}} < 0$ antitraps are formed.

On the other hand, local states for positive charge carriers (holes), formed by guest molecule E^h_{guest}, are determined by the difference between the ionization energy of guest $(I_G)_{\text{guest}}$ and host $(I_G)_{\text{host}}$ molecules, respectively:

$$E^h_{\text{guest}} = (I_G)_{\text{guest}} - (I_G)_{\text{host}}. \tag{4.39}$$

In this case, if $E^h_{\text{guest}} < 0$ traps, and if $E^h_{\text{guest}} > 0$-antitraps are formed.

According to formulas (4.38) and (4.39) Tc, in an Ac host lattice forms traps for both electrons and holes (see Fig. 4.24) and, vice versa, Ac in a Tc host forms antitraps for both electrons and holes. Typical electron acceptors (e.g., anthraquinone) creates only trapping states for electrons, while typical donors create trapping states for holes (see Fig. 4.24). As may be seen from Fig. 4.24, formulas (4.38) and (4.39) may serve as a rough approximation. Actually, a guest molecule produces some distortion of surrounding host lattice. Therefore, the polarization energy of the carrier, localized on the guest molecule $(P_{\text{eff}})_{\text{guest}}$ may differ from that of the host molecule $(P_{\text{eff}})_{\text{host}}$, namely, $(P_{\text{eff}})_{\text{guest}} \neq (P_{\text{guest}})_{\text{host}}$, and one should introduce a correction term of $(\Delta P_{\text{eff}})_{\text{guest}}$.

One should also mention that the local state energy levels of guest molecules may not be discrete owing to slight difference between distortions surrounding various

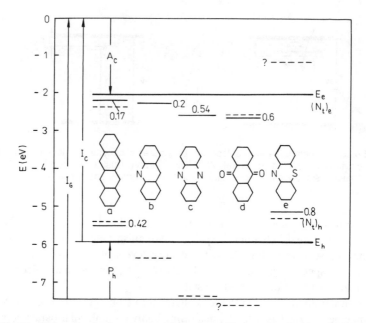

FIGURE 4.24. Charge carrier traps and antitraps formed by guest molecules: (a) tetracene, (b) acridine, (c) phenazine, (d) antraquinone, and (e) phenothiazine in anthracene host crystal according to Ref. 153; solid lines, experimental trapping levels; dashed lines, levels estimated according to formulas (4.38) and (4.39).

guest molecules, thus, the energy levels may exhibit narrow Gaussian distribution with $\sigma_p > 0$. On the other hand, displaced host molecules, surrounding the guest one, should form *local* states of structural origin. Thus the impurity states (chemical traps) are closely related to the structural ones.

If the guest molecule possesses a permanent dipole moment $\mathbf{d}_0$ shallow, so-called dipole traps can be formed in the vicinity of the guest molecule.[150,278]

If such molecules are introduced in a nonpolar host, e.g., Ac, a set of local traps are formed due to charge-permanent dipole $(q - d_0)$ interaction:

$$W_{q-d_0} = e\mathbf{d}_0\mathbf{r}_g/4\pi\epsilon\epsilon_0 r_g^3.$$

The authors of Refs. 150 and 278 have demonstrated that the depth E_t of these dipole traps is directly proportional to the dipole moment $\mathbf{d}_0$ of the guest molecule. For $\mathbf{d}_0$ values from 3 to $4D$ the corresponding E_t value is ca. 0.1 eV. Thus, the density and energy spectra of these trapping states may be controlled by introducing the guest molecules with an appropriate $\mathbf{d}_0$ value.

Ostapenko et al.[150,278] and Sworakowski[376] have recently carried out more refined calculations of the energy spectra of the dipole traps. In further studies the authors of Ref. 151 have also observed formation of dipole clusters consisting of two or three guest molecules, and performed calculations of local state energy spectra of such clusters. Sworakowski[376a] has recently calculated the depth E_t and cross sections of local trapping states caused by dipolar defects in highly anisotropic polydiacetylene crystals. The author demonstrates that these parameters of dipolar trap states are linearly dependent on the magnitude of the dipole moment.

4.6.3. Experimental Methods of Local State Studies

Among the variety of experimental methods, the more appropriate ones for studying the local states of Gaussian distribution [their energy spectra $(E_t)_i$, total density $(N_t^0)_i$, and distribution parameters $(\sigma_P)_i$] are the methods of space-charge-limited current (SCLC) and thermally activated spectroscopy techniques, the method of thermally stimulated current (TSC) being preferable.

Detailed descriptions of these and other methods are given elsewhere (see, e.g., Refs. 15, 262, 270, 272, 273, 288, 332). We shall give here only a brief comparative characterization of them.[335]

In the case of Gaussian traps the current-voltage (CV) characteristics exhibit, under the SCLC conditions of monopolar carrier injection, S-shaped dependence of $j = f(U)$ in lgU, lgj coordinates for trap-filled limit voltages $U > U_{TFL}$. If several Gaussian trap distributions are trap-filled, a steplike CV-characteristic emerges (Fig. 4.25).

A phenomenological SCLC theory of Gaussian traps[272] yields analytical expressions suitable for approximation of experimental $j = f(U)$ characteristics. The validity range and criteria of such an approximation are given in Refs. 272 and 332.

As a result of approximation one obtains the main parameters of trap distribution N_t^0, σ and also the product μN_e, where μ is the drift mobility of carriers and N_e is the density of states on the conduction level. In the framework of the SCLC method one should also investigate the thickness dependence $j = f(L)$, and the $j = f(1/T)$ dependence to estimate the $(E_t)_i$ values. The conventional SCLC method is most appropriate for studying the shallow trap distributions. However, as an integral method, it provides relatively low resolution. The differential method of SCLC (DM-SCLC)[273] yields directly the distribution profiles $h(E)$ and parameters E_t, N_t^0, and σ (see Fig. 4.25). The parameter μN_e serves in this case as an input parameter for the approximation. A recently developed method of thermally modulated SCLC (TM-SCLC)[270] is based on careful measurements of the conductivity activation energy E_a dependence $E_a = f(U)$ (Ref. 262) (see Fig. 4.25). The method yields the parameters E_t, N_t, and σ and provides also a higher resolution of energy spectra. The SCLC methods described may be regarded as mutually complementary ones.

The TSC method can be used as an independent approach for checking the results of the SCLC techniques. This method also yields the parameters E_t, N_t, and σ, as well as the distribution profiles $h(E)$ (Fig. 4.25) and the trapping cross section q.

Only a complex application of all these methods can provide the required plausibility of trap control in OMC.[262,335]

Apart from these experimental techniques some additional indirect methods of trap control may be recommended, such as thermally stimulated luminescence,[150,278,288,332] defect luminescence[204,332] (based on the isomorphism of exciton and carrier energy spectra), thermally stimulated depolarization technique,[288,332] etc.

The possible influence of spatially nonuniform trap distribution on the SCLC characteristics have already been discussed in Ref. 332, p. 331.

Recently some new methods have been developed, which allow to estimate the spatial trap distribution profiles, such as the method of photo-EMF,[271] as well as photo- and electrorefraction techniques.[114]

Finally, one should emphasize that the experimentally determined parameter N_t^0 (total density of traps) actually corresponds to the density of the deepest trapping sites in the domains of local states.[335] If a charge carrier enters a local state domain, it quickly relaxes to the deepest trapping site, i.e., to the molecule at the end of the extra plane in an edge dislocation core (see Figs. 4.19 and 4.20), to the molecule adjacent to the vacancy (see Fig. 4.22), or to the middle of a stacking fault ribbon, and so on.

Thus, the carrier is being detrapped from this deepest trapping site (or from energetically nearby sites due to the Boltzmann distribution factor). Hence, the value of N_t^0 actually corresponds to the number of local state domains, but not to the total number of local states N_t [see formula (4.31)]. Consequently, $N_t \gg N_t^0$ since every domain contains hundreds of local states (see Figs. 4.19 and 4.20). Also the observed Gaussian distribution of local trapping state energy spectra is, presumably, formed by statistical overlaps of the E values of different types of trapping state domains. On the other hand, every domain may have its own Gaussian distribution of local states [see Eq. (4.31)]. One should keep these factors in mind when interpreting the experimental results.

For description of such extended local trapping state domains Kalinovski, Godlewski, and Mondalski[151a,151b] have proposed so-called macrotrap model. These authors apply a simplified approach according to which local state domain is considered as a macrotrap of spherical symmetry consisted of local microtraps with energy (E) distributed in space (r) so that $E = (3kT/\sigma)\ln(r_o/r)$, where σ is a characteristic parameter of exponential energy distribution function, and r_o is the radius of a macrotrap.

The Gaussian distribution model [Eq. (4.31)] has been used extensively in recent years for description of local state energy spectra in disordered organic solids, e.g., molecularly doped polymers and related disordered materials (see review articles by Baessler[15a] and by Borsenberger, Magin, Van der Auweraer, and De Schryver[35c]).

4.7. ENERGY STRUCTURE OF ELECTRONIC STATES IN LANGMUIR–BLODGETT (LB) MULTILAYERS

In concluding this chapter on electronic states in typical OMC, let us make now a brief excursion into the realm of a new field of contemporary solid-state physics and discuss the specific energy structure of electronic states in a LB molecular multilayer assembly.

Typical characteristics	Measurable dependences	Parameters obtained
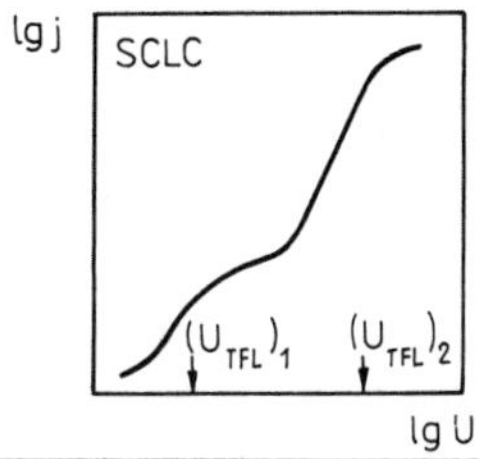 SCLC	Basic $j = f(U)$ Supplementary: $j = f(L)$ $j = f(1/T)$ Input parameters: $h(E)$; E_t	N_t^o G μN_e
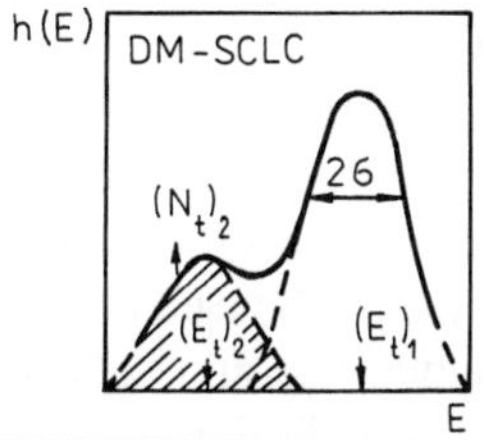 DM-SCLC	Basic: $j = f(U)$ Input parameters: μN_e	$h(E)$ E_t N_t^o G
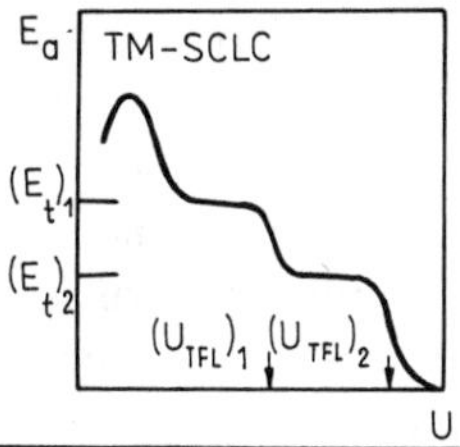 TM-SCLC	Basic: $j = f(1/T)$ Supplementary: $j = f(U)$	E_t N_t^o G
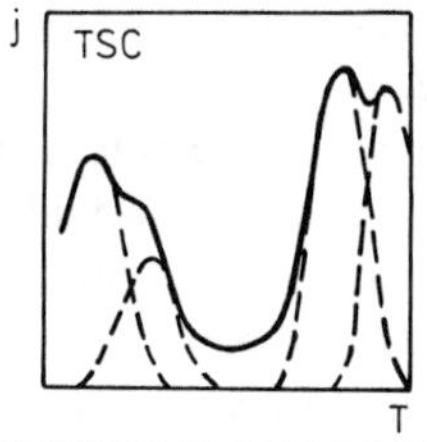 TSC	Basic: $j = f(T)$ Supplementary: $j = f(U)$	E_t N_t^o G ν(frequency factor) q (trapping cross-section)

FIGURE 4.25. Typical characteristics of experimental methods used for trapping state studies: SCLC, conventional method of space-charge limited current; DM-SCLC, differential method of SCLC; TM-SCLC, thermally modulated SCLC; TSC, thermally stimulated current technique according to Ref. 335.

It should be emphasized that LB multilayers may serve as excellent model systems of highly organized molecular assemblies. If an LB film consists of a conjugated amphiphilic compound like phthalocyanine with aliphatic "tails," it may be envisaged as a multilayer of two-dimensional sheets of semiconducting molecules separated by isolating aliphatic spacers. Energetically such a superlattice forms a sequence of quantum-wells — an organic equivalent of AlGaAs structures currently fabricated by molecular beam epitaxy techniques.[353]

We shall present here the main results of energy structure studies in vanadyl phthalocyanine LB multilayers according to Ref. 353.

An amphiphilic surface active compound — tetra (octadecylaminosulphanyl) vanadyl phthalocyanine (VOPhc) with four aliphatic "tails" (see Fig. 4.26a) was deposited on a polished hydrophobized silicon substrate of p-type by standard LB technology.[5,392] The LB multilayer assemblies of Y-type were obtained with a different number of layers ($n = 6$–50) (Fig. 4.26b). The top metallic electrode of aluminum (Al) or bismuth (Bi) was deposited by vacuum evaporation technique.

The parameters of the energy structure of LB multilayer assembly were determined by combined intrinsic photoconductivity and photoelectron emission threshold methods[381,400] (see also Ref. 332, p. 120).

Structural studies of the VOPhc LB films show that the thickness l of a monolayer is ca. 21 Å;[280,392] the flat "heads" of the VOPhc molecules are oriented parallel[280] or almost ($\angle \sim 13°$) parallel[392] to the plane of the substrate. The aliphatic chains are tilted at mean angles of ca. 40° (Ref. 392) (see Fig. 4.26b). Such molecular structure of a periodic multilayer superlattice determines its energy structure (Fig. 4.26c). The VOPhc LB multilayer assembly forms a sequence of quantum wells. These potential wells are created by the bilayer of semiconductive VOPhc molecules of ~ 10 Å thickness. There emerge two symmetrical wells — one for electrons with affinity $A_c^e = 2.6$ eV, the other for holes with affinity $A_c^h = 4.3$ eV. The position of the conduction level of holes E_h in the energy diagram is determined by the ionization energy I_c of the VOPhc molecules in the layer $I_c = 4.7$ eV.[400] On the other hand, the conduction level of electrons E_e is determined by the electron affinity of the VOPhc molecules $A_c^e = 2.6$ eV. The conduction levels E_h and E_e are separated by an energy gap of $E_G = 2.1$ eV (Ref. 381) — a typical value for phthalocyanines (see Ref. 288).

The bilayer of aliphatic "chains" ($l = 32$ Å) separates the potential wells of the bilayer of VOPhc "heads." This dielectric bilayer possesses a wide energy gap, $E_G = 9.0$ eV, which is determined by the first ionization potential of the hydrocarbon chains[400] (their electron affinity being practically equal to zero).

Thus, by choosing appropriate polyconjugated molecules as a "head" of amphiphilic compounds, it is possible to predetermine the energy gap E_G and the thickness of quantum wells in the process of designing LB multilayers. On the other hand, the periodic sequence of dielectric aliphatic hydrocarbon chains ($E_G = 9$ eV) provides extremely high dielectric breakdown strength of over $\mathcal{E} \geq 10^7$ V/cm.

The Fermi levels of the metallic electrodes (Si, Bi, or Al) are positioned close to the hole conduction level E_h (see Fig. 4.26c), thus creating favorable conditions for effective monopolar hole injection in SCLC measurements.[5]

The dark conductivity measurements of the VOPhc LB films in a SCLC regime under conditions of monopolar injection of holes yield nonlinear, S-shaped CVC characteristics which indicate that the transport of holes is trap-controlled and that these carrier traps are of Gaussian distribution (see Sections 4.6.1 and 4.6.3). Using appropriate approximation formulas[332] one can obtain from these CV characteristics

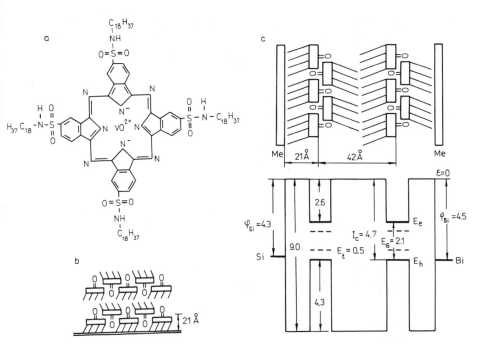

FIGURE 4.26. Molecular and energy structure of LB multilayer assembly of vanadyl phthalo-cyanine according to Ref. 353: (a) molecular structure of amphiphilic tetra(octadecylaminosul-phanyl)vanadyl phthalocyanine (VOPhc); (b) schematic structure of LB multilayer; (c) schematic diagram of the energy structure of electronic states of VOPhc LB multilayer assembly.

the mean trap density N_t^0 and the Gaussian distribution parameter σ. On the other hand, the mean depth E_t of these traps was determined by the TM-SCLC method (see Section 4.6.3).

These measurements demonstrate that at the vicinity of the Si(p) substrate, traps with mean depth $E_t = 0.3 \pm 0.1$ eV and $N_t^0 = 10^{17} - 10^{18}$ cm^{-3} dominate, while in the vicinity of the top evaporated electrode, the trap depth equals 0.5 ± 0.1 eV with $N_t^0 = 10^{18} - 10^{19}$ cm^{-3}. It is interesting to mention that the energy spectra of trapping states in the LB films exhibit more discrete, almost monoenergetic character in comparison with those of vacuum evaporated films in which one usually observes a whole set of trapping states of almost quasicontinuous distribution (cf. Ref. 262). Presumably, this phenomenon is determined by the more regular structure of LB multilayers. It has been suggested[353] that the observed trapping states are of structural origin.

Structural investigations of VOPhc LB films demonstrate that inside the separate layers an amorphous structure dominates with inclusion of dispersive crystalline regions.[392] However, one should emphasize that, contrary to this finding, in the direction perpendicular to the layers and the substrate, strict quasicrystalline periodicity is observed.

According to the polarization model (see Section 4.6.1), the crystallite domains, with more closely packed molecules, form charge carrier trapping states of structural origin with increased polarization energy ΔP, equal to the mean trap depth E_t. On the other hand, the mean density of these traps N_t^0 presumably equals the density of crystalline domains in the LB films. As has already been shown, the structural traps are bipolar and may serve as trapping centers for both electrons and holes. Therefore the corresponding trapping levels E_t are positioned symmetrically in the energy gap (see Fig. 4.26c).

The possible charge carrier transport mechanisms in LB multilayers will be discussed briefly in Chapter 7.

4.8. CONCLUSIONS ON ELECTRONIC STATES IN REAL (DEFECTIVE) ORGANIC CRYSTALS

Summing up the last part of this chapter on local states of structural and impurity origin in OMC, we wish to emphasize once more that both the self-energy of charge carriers and excitons in a perfect crystal and energy levels of local states in a real (imperfect) one are determined by the same physical phenomenon — electronic polarization of the surrounding lattice.

The polarization model is phenomenologically and conceptually based on direct interrelation between the topology of molecules at structural defects and the corresponding energy spectra of local states.

At present, refined computer simulation approaches have been developed for studying both counterparts — molecular configuration and local state energy spectra at structural defects. In turn, polarization-type interaction, which is similar in nature also determines the isomorphism of local state energy spectra of charge carriers and excitons.

At present, a number of reliable independent experimental techniques of local state characterization have been developed and a rich data pool collected which unequivically confirms the validity of the Gaussian distribution model of local states. The local states studied in OMC have revealed a number of new and exciting physical properties which render promising possibilities for future application of organic crystals in molecular electronics.

CHAPTER 5

Exciton and Charge Carrier Transport Phenomena

What we observe is not Nature itself,
but Nature exposed to our method of questioning.

WERNER HEISENBERG

. . . but Nature is built that the asking of one question
automatically excludes us from asking at the same time,
in the same connection, the complementary question.

JOHN ARCHIBALD WHEELER

This chapter is devoted to exciton and charge carrier transport theories in OMC.

Recent achievements in the field of transport theories, based on the models of the generalized stochastic Liouville equation and generalized master equations (GMEs), will be discussed in detail, as well as improvements and refinements of these models and their relations to some earlier transport theories.

These contemporary generalized approaches allow one to describe both the coherent (bandlike) and diffusive motion of excitons and charge carriers in the framework of a single unified model, as well as time- and temperature-dependent dynamics of transition from one mode of transport to the other one.

Trapping phenomena and finite-life-time effects for excitons are also discussed briefly.

Finally, some problems of application of the Fokker-Planck equations for the description of the behavior of hot, nonthermalized charge carriers are also considered (see Chapters 6 and 7).

5.1. SOME CHARACTERISTIC FEATURES OF TRANSPORT PHENOMENA IN ORGANIC CRYSTALS

The dynamics of exciton and charge carrier transport strongly depends on the properties of the crystal and on the interaction of these quasiparticles with the surrounding lattice. In organic crystals the following characteristic features play the more decisive role:

1. Strong localization and low mobility of excitons and charge carriers due to the molecular nature of the crystal (see Chapter 1A).

2. Relatively strong coupling to intramolecular vibrations, especially to those that lead to vibrations of charged groups.

3. Relatively strong coupling to intermolecular acoustic or optical vibrations and librations.

For electrons, an additional typical feature is a relatively strong coupling to Frenkel excitons.

The statement of strong intermolecular coupling in item (3) above might seem surprising. First, let us specify what we mean by "relatively strong." By that we usually understand strong as compared to the unrenormalized bandwidths δE, as well as band gaps E_G (as obtained from band structure calculations for electrons and rigid molecular calculations for excitons). Probably item (2) is not surprising since the strong coupling means here (as generally accepted) the necessity of working with vibronic states. Strong coupling to libration and intermolecular modes of vibrations is, however, not so usual. To justify and explain our approach, we should mention that, in principle, when describing electronic states of molecules in a crystal, one may work either in terms of adiabatic basis (i.e., basis of states moving together with individual molecules) or of rigid basis (corresponding to fixed positions of nuclei). The statement in item (3), as formulated above, actually corresponds to the rigid basis approach. The strong coupling to libration or intermolecular vibrations, however, surprising as it might seem, is then a matter of necessity, as it is just this coupling that makes the electrons or holes, forming molecular polarons or excitons forming excitonic polarons, move together with the librating or vibrating molecules. On the other hand, when working in terms of adiabatic basis, the electrons or excitons in the corresponding states are (by definition) moving together with the librating or mutually vibrating molecules. Hence, in this case the coupling to optical or acoustical intermolecular vibrations and librations would be necessary to incorporate the actually existing but presumably weak effect of influence of changes of intermolecular distances or of mutual orientations on internal states of a given molecule in its own center-of-mass system. Therefore for adiabatic basis such a coupling would certainly be weak.

For quantitative reasons, the adiabatic method of approach seems to be undoubtedly better, from the physical point of view, since the moving basis corresponds well to reality. On the other hand, if we are going to introduce and work with overlaps between electronic states of neighboring molecules (to include a possibility of electrons to move), a nonorthogonality problem for electron states localized on neighboring molecules appears to lead to technical complications when we formally introduce, e.g., the second quantization formalism. Moreover, working in the adiabatic basis practically means, in the treatment of vibrations, to introduce the adiabatic approximation. Nonadiabaticity effects cannot be well treated by a simple perturbation series, since the corresponding energy denominators may be (in a macroscopic system) arbitrarily small. Anyway, many people still believe that the adiabatic approximation well describes the physics of many electronic phenomena (including transport) in solids. However, the opposite is probably true since rigorous corrections to the adiabatic approximation involve divergences, and therefore the validity of this approximation is questionable (see Chapter 2 for more about this subject). Thus, if we are technically able to avoid lowest-order arguments (in the above coupling to vibrations), it is preferable to use the rigid basis formalism. We will follow this approach, although, in such a way, we risk involvement in possible technical problems. These are, in addition to using technical means appropriate for the strong coupling (partial summation of perturbation series to infinity, etc.), mainly problems connected with the necessity of working with highly excited states. As an illustration, let us realize that, e.g., any unexcited but shifted localized molecular electronic state may be expanded into a series as a linear combination of rigid (nonshifted) but arbitrarily excited states. For transport theories, this then means introduction of an additional trace over excited states of given molecule. This is easily possible but additional technical problems are not outweighed by anything new in physics of transport phenomena. Thus, for our purposes here, we shall avoid higher excited molecular (as well as band) states as far as possible.

In connection with this, a few words are worth mentioning here about reflection of the strength of the coupling in the values of coupling constants. In practical calculations, one usually resorts to models with the coupling to phonons linear in the nuclear displacements, i.e., to linear phonon creation and annihilation operators. For instance, with planar molecules and intramolecular out-of-plane vibrational modes, corresponding coefficients (coupling constants) of the linear coupling are then simply zero (irrespective of the strength of the coupling in reality) because of symmetry arguments. So in similar cases, one should rather preserve a sufficiently general form of coupling (including linear as well as higher-order terms) to obtain correct physical conclusions.

5.2. COHERENT AND DIFFUSIVE MOTION OF EXCITONS AND CHARGE CARRIERS

Dynamics of quantum quasiparticles — excitons and charge carriers — depend in organic molecular crystals on two (seemingly contradictory) tendencies: that of delocalization (owing mainly to resonance or hopping integrals in the Hamiltonian) and that of localization (mostly due to the interaction of quasiparticles with the surrounding lattice). Thus, in transport phenomena the dual nature of the quantum quasiparticles emerges: the wave and corpuscular giving rise to two corresponding modes of propagation; namely, coherent (wavelike) and diffusive. Generally speaking, localization and delocalization in the above sense do not yet necessarily mean the possibility (or impossibility) to find the quasiparticle on a specified molecule (or in its vicinity) in the crystal. Nor do they mean the necessity of using the localized or extended basis of states when working with the density matrix. Actually, regarding the possibility of localizing the quasiparticle in space, it strongly depends on the physical situation described by the density matrix ρ.

Historically both approaches, the delocalized and localized representation of quasiparticle transport, were developed independently and were initially regarded as mutually exclusive and incompatible limiting cases.

The delocalized treatment of quasiparticles was developed for the description of molecular Frenkel excitons and introduced in the physics of organic crystals by Davydov, Agranovich, and others (see Refs. 7, 8, 75, 84, 85, 288, and 351). According to this treatment, a molecular exciton is regarded as an extended collective excitation of the crystal which propagates as a quantum mechanical plane-wave, characterized by a wave vector **k** (see Section 1B.2.12). This representation prevailed mainly due to good theoretical foundations and elaborate mathematical formalism. Experimental evidence supporting the picture of a delocalized exciton at low temperatures led to its general recognition as one of basic notions in the physics of organic solids.[84,85] However, this approach may be strictly justified only for a rigid (i.e., nonvibrating) molecular crystal, a condition which is fulfilled only at low temperature range.

With increasing temperature (typically above 50–60 K in case of singlet excitons in Ac-type crystals) this picture breaks since scattering on lattice phonons becomes so strong that the wave vector **k** cannot be taken any more as a good quantum number. The exciton actually starts to behave as a corpuscle diffusing in the lattice like a Brownian particle and, thus, the localized description becomes more appropriate in the higher temperature range.

A qualitatively similar transition with growing temperature from coherent, wave-like motion to diffusive may be observed also in case of charge carriers. Above 80 K the carrier's mean free path $\bar{l}$ in aromatic crystals already approaches the value of lattice constant $\bar{l} \approx a_0$, the scattering takes place at every lattice site and the picture of localized charge carrier becomes valid (see Section 7.2.3).

For the description of a localized exciton or charge carrier phenomenological equations of diffusive motion are used and corresponding parameters (diffusion coefficient, mean diffusion length, mean free path, etc.) are introduced. At present there is overwhelming experimental evidence supporting the localized behavior of quasiparticles (excitons and charge carriers) at high temperatures.

Actually, the exciton localization effect was already predicted by Frenkel (see Ref. 7, p. 79), and discussed later by Trilifaj,[391] Agranovich,[7] and others (see also Refs. 288 and 351).

Besides temperature, *time* is another factor determining the transition of an exciton or charge carrier from delocalized to a localized state. At short time intervals, just after excitation, the quasiparticle behaves in a coherent wavelike fashion. However, after some critical time interval Δt (also dependent on temperature) the quasiparticle, due to fast scattering on lattice phonons, becomes localized and further behaves like a corpuscle subjected to diffusive mode of motion.

Now we shall try to introduce some basic notions of coherent and diffusive motion of excitons and charge carriers.

For tutorial purposes, let us assume that we have a one-dimensional crystal (a chain) composed of single molecules with one molecule per elementary cell. Let us have just one exciton or charge carrier in the system. Designating by $P_n(t)$ the probability of finding the exciton (or carrier) at the nth site, the usual Pauli master equations[281] reads

$$\frac{\partial}{\partial t} P_m(t) = \sum_{n(\neq m)} [W_{mn}P_n(t) - W_{nm}P_m(t)], \quad m, n = 0, \pm 1, \ldots .$$

(5.2.1)

Here W_{mn} is the rate (probability per second) of the $n \rightarrow m$ transitions. Due to periodicity, W_{mn} depends on the distance $|m - n|a$ (where a is the lattice constant). Let us choose the initial condition

$$P_m(0) = \delta_{m0},$$

(5.2.2)

which means that initially the exciton is located at the origin. From Eqs. (5.1.2) and (5.2.2) the mean-square displacement

$$\Delta(t) = \sum_m (ma)^2 P_m(t)$$

(5.2.3)

may be shown to fulfil the following equation

$$\frac{\partial}{\partial t} \Delta(t) = \sum_{mn} (ma)^2 [W_{mn}P_n(t) - W_{nm}P_m(t)]$$

$$= a^2 \sum_n \sum_{m(\neq n)} (m-n)^2 W_{mn}P_n(t) + 2a^2 \sum_n nP_n(t) \sum_{m(\neq n)} (m-n)W_{mn}.$$

(5.2.4)

Because of the above assumptions, $\sum_{m(\neq n)}(m-n)^2 W_{mn}$ is independent of n and $\sum_{m(\neq n)}(m-n)W_{mn}$ equals zero. Since $\sum_n P_n(t) = 1$, Eq. (5.2.4) yields

$$\Delta(t) = ta^2 \sum_{m(\neq n)} (m-n)^2 W_{mn}.$$

(5.2.5)

Thus, we have obtained a linear time-dependence of the mean-square displacement $\Delta(t) \sim t$ which is typical of the diffusive (hopping, incoherent) regime.

On the other hand, it is well known that diffusion cannot take place without sufficiently strong scattering. If we deal with the rigid chain (i.e., without vibrations), one should rather start from the wave (Schrödinger) equation

$$i\hbar \frac{\partial}{\partial t} c_m(t) = \sum_n J_{mn}c_n(t),$$

(5.2.6)

where $|\psi(t)\rangle = \sum_n c_n(t)|\phi_n\rangle$ (with $|\phi_n\rangle$ being centered as site n) is the wave function at time t and $J_{mn} = \langle \phi_m|H|\phi_n\rangle$ (depending just on the distance $a|m-n|$)

are diagonal (for $m = n$) and off-diagonal (for $m \neq n$) elements of the Hamiltonian H. The solution with the initial condition

$$c_m(0) = \delta_{m0} \tag{5.2.7}$$

reads

$$c_m(t) = \frac{1}{N} \sum_k e^{i(kma - \epsilon(k)t/\hbar)} \tag{5.2.8}$$

where

$$\epsilon(k) = \sum_n J_{mn} e^{ika(m-n)} \tag{5.2.9a}$$

are the stationary plane-wave eigenstate energies with

$$k = \frac{2\pi}{a} \frac{n}{N}, \quad n = 0, \pm 1, \ldots \pm \left[\frac{N}{2}\right] \tag{5.2.9b}$$

(or $n = 0, 1, 2, \ldots, N-1$). The limit $N \to \infty$ should be taken.

Thus, the probability of finding the exciton at site m equals

$$P_m(t) = |c_m(t)|^2 \to_{N \to \infty} \left| \frac{a}{2\pi} \int_{-\pi/a}^{\pi/a} e^{i(kma - \epsilon(k)t/\hbar)} dk \right|^2 \tag{5.2.10}$$

and the mean square displacement

$$\begin{aligned}
\Delta(t) = \sum_m a^2 m^2 P_m(t) &= \frac{a^2}{4\pi^2} \sum_m \left| \int_{-\pi/a}^{\pi/a} e^{-i\epsilon(k)t/\hbar} \frac{\partial}{\partial k} e^{ikma} dk \right|^2 \\
&= \frac{a^2 t^2}{(2\pi\hbar)^2} \sum_m \left| \int_{-\pi/a}^{\pi/a} \frac{d\epsilon(k)}{dk} e^{i(kma - \epsilon(k)t/\hbar)} dk \right|^2 \\
&\approx_{N \to +\infty} t^2/\hbar^2 \sum_m \left| \frac{1}{N} \sum_k \frac{d\epsilon(k)}{dk} e^{i(kma - \epsilon(k)t/\hbar)} \right|^2 \\
&= \frac{t^2}{\hbar^2} \frac{1}{N} \sum_{k, k'} \frac{d\epsilon(k)}{dk} \frac{d\epsilon(k')}{dk'} e^{i(\epsilon(k) - \epsilon(k'))t/\hbar} \frac{1}{N} \sum_m e^{-i(k-k')ma} \\
&= \frac{t^2}{\hbar^2} \frac{1}{N} \sum_k \left(\frac{d\epsilon(k)}{dk} \right)^2 = \frac{t^2 a^2}{\hbar^2} \sum_m J_{nm}^2 (n-m)^2.
\end{aligned} \tag{5.2.11}$$

Thus, as a result, we have obtained a quadratic time-dependence of the mean-square displacement $\Delta(t) \sim t^2$ which is typical of the coherent (wavelike) regime.

These two results illustrate two limiting cases of the exciton and charge carrier motion: the first the diffusive one, in the presence of sufficiently strong scattering on lattice vibrations, the second the coherent one for a *rigid* periodic lattice, corresponding to the low-temperature situation.

One should indicate here that the difference between Eq. (5.2.5) $[\Delta(t) \sim t]$ and Eq. (5.2.11) $[\Delta(t) \sim t^2]$ is due to different kinds of description (master equation for the probabilities and Schrödinger equation for the probability amplitude). The difference is not due to the strength of interaction, and on the model level, dependencies (5.2.5) and (5.2.11) are incompatible.

On the other hand we shall demonstrate later that, in reality, $\Delta(t)$ behaves as in Eq. (5.2.11) [i.e. $\Delta(t) \sim t^2$] for sufficiently short t intervals while at times greater than a critical time, $\Delta(t)$ starts to be a linear function of t $[\Delta(t) \sim t$, as in Eq. (5.2.5)]. Thus, the behavior of excitons (and, similarly, charge carriers) changes over from the coherent regime of motion to an incoherent one with increasing time. Apparently, the critical time of such a transition also depends on the temperature region at which it occurs, and, presumably, also on the nature of the quasiparticle (e.g., it is different for singlet and triplet excitons). Possible reasons for contingent absence of such a transition observed in direct experiments might be connected with either insufficient time resolution or with other side effects (finite lifetime of excitons, electron capture, etc.) not included here.

However, there is overwhelming indirect evidence in favor of such transition.

In the following sections of this chapter we present a detailed description of generalized contemporary transport models and theories which permit one to overcome the apparently contradictory character of earlier theories and allow one to consider the coherent (wavelike) and *diffusive* motion of excitons and charge carriers in organic crystals in the framework of a unified model including both forms of quasiparticle motion in a complementary picture. And, most importantly, these models, as we shall see later, allow one to describe, in a unified approach, the dynamics of time- and temperature-dependent transition of quasi-particles from coherent (delocalized) to diffusive (localized) modes of transport.

5.3. STOCHASTIC LIOUVILLE EQUATION MODEL

This approach was first suggested in early works by Lax[211] and Primas.[291] Later, especially in connection with parameterization and applications, the approach was developed and refined by Haken and Strobl.[126,128] A good review of physical ideas connected with the approach up to 1981 as well as an introduction to the problem may be found in Ref. 296. The main idea of the approach is to describe a single exciton (possibly also a carrier) by the single-exciton density matrix $\rho_{mn}(t)$ (with indices designating the sites) assuming that the influence of vibrations (forming a genuine thermodynamic bath or reservoir) may be substituted by an external stochastic potential field with prescribed statistical properties. In this section, we will treat a one-, two-, or three-dimensional model molecular crystal, not necessarily periodic, which includes as special cases not only perfect periodic crystals but also dimers, chains, etc.

In Section 5.4 we introduce the generalized master equation (GME) theory of the excitation (or carrier) transfer. It is therefore helpful to start from the same formalism. First, let us realize that we have a Hilbert space with the scalar product of two wave functions $\langle \varphi | \psi \rangle$ having the following properties

$$\langle \varphi | \lambda_1 \psi_1 + \lambda_2 \psi_2 \rangle = \lambda_1 \langle \varphi | \psi_1 \rangle + \lambda_2 \langle \varphi | \psi_2 \rangle$$

$$\langle \varphi | \psi \rangle^* = \langle \psi | \varphi \rangle,$$

$$\langle \varphi | \varphi \rangle \geq 0, \quad \langle \varphi | \varphi \rangle = 0 \Leftrightarrow | \varphi \rangle = 0. \tag{5.3.1}$$

With wave function as functions of coordinates, such a scalar product may be introduced as $\int \varphi^*(\mathbf{r}) \psi(\mathbf{r}) d^3\mathbf{r}$. In addition, we have operators acting on such wave functions which fulfill the Schrödinger equation

$$i\hbar \frac{\partial}{\partial t} | \psi(t) \rangle = H(t) | \psi(t) \rangle, \tag{5.3.2}$$

where $H(t)$ is the Hamilton operator. When working with the operator of the density matrix $\rho(t)$, expression (5.3.2) implies the Liouville equation

$$i \frac{\partial}{\partial t} \rho(t) = \frac{1}{\hbar} [H(t), \rho(t)] \equiv L(t)\rho(t), \tag{5.3.3a}$$

where

$$\rho(t) = \sum_j | \psi_j(t) \rangle p_j \langle \psi_j(t) |,$$

$$p_j \geq 0, \quad \sum_j p_j = 1. \tag{5.3.3b}$$

Here $L(t) \ldots = (1/\hbar)[H(t), \ldots]$ is the Liouville superoperator. By superoperator, we mean in general a prescription making a new operator from an old one. One can also introduce a scalar product of operators [e.g., $(A|B) = Tr(A^\dagger B)$] with following properties

$$(A | \lambda_1 B_1 + \lambda_2 B_2) = \lambda_1 (A|B_1) + \lambda_2 (A|B_2),$$

$$(A|B)^* = (B|A),$$

$$(A|A) \geq 0, \quad (A|A) = 0 \Leftrightarrow A = 0. \tag{5.3.4}$$

Notice that Eq. (5.3.4) forms a full analogy to Eq. (5.3.1). Thus, the operators then form a new Hilbert space called the Liouville space and superoperators [such as $L(t)$ in (5.3.3a)] are operators in the Liouville space.

If D is an arbitrary linear projection superoperator (i.e., having the idempotency property $D^2 = D$), the same applies also to $(1-D)(= (1-D)^2)$. One can then rewrite (5.3.3a) as

$$i \frac{\partial}{\partial t} D\rho(t) = DL(t)D\rho(t) + DL(t)(1-D)\rho(t), \tag{5.3.5a}$$

$$i \frac{\partial}{\partial t} (1-D)\rho(t) = (1-D)L(t)D\rho(t) + (1-D)L(t)(1-D)\rho(t). \tag{5.3.5b}$$

Then express $(1-D)\rho(t)$ from (5.3.5b) and introduce in (5.3.5a). This yields the Nakajima-Zwanzig[265,413-415] identity

$$\frac{\partial}{\partial t} D\rho(t) = -iDL(t)D\rho(t) - \int_{t_0}^{t} DL(t)\exp_{\leftarrow} \left(-i(1-D) \int_{\tau_1}^{t} L(\tau_2)d\tau_2 \right)$$

$$\times (1-D)L(\tau_1)D\rho(\tau_1)d\tau_1 - iDL(t)\exp_{\leftarrow} \left(-i \int_{t_0}^{t} (1-D)L(\tau)d\tau \right)$$

$$\times (1-D)\rho(t_0), \tag{5.3.6}$$

where $\exp_{\leftarrow} \ldots$ means the time-ordered exponential (the arrow indicates the direction of increasing time). For example

$$\exp_{\leftarrow} \int_{t_0}^{t} A(\tau)d\tau = 1 + \frac{1}{1!} \int_{t_0}^{t} A(\tau)d\tau + \frac{1}{2!} \int_{t_0}^{t} \left[\int_{t_0}^{\tau_1} A(\tau_1)A(\tau_2)d\tau_2 \right.$$

$$\left. + \int_{\tau_1}^{t} A(\tau_2)A(\tau_1)d\tau_2 \right] d\tau_1 + \ldots \tag{5.3.7a}$$

as compared to the usual exponential

$$\exp \int_{t_0}^{t} A(\tau)d\tau = 1 + \frac{1}{1!} \int_{t_0}^{t} A(\tau)d\tau + \frac{1}{2!} \int_{t_0}^{t} d\tau_1 \int_{t_0}^{t} d\tau_2 A(\tau_1)A(\tau_2) + \ldots . \tag{5.3.7b}$$

For the stochastic Liouville equation model, we have

$$H(t) = H_0 + V(t) \tag{5.3.8}$$

where

$$H_0 = \sum_{m \neq n} J_{mn}|m\rangle\langle n| + \sum_n \epsilon_n|n\rangle\langle n| \equiv \sum_{m \neq n} J_{mn}a_m^\dagger a_n + \sum_n \epsilon_n a_n^\dagger a_n$$

$$(5.3.9)$$

$(|n\rangle \equiv |\varphi_n\rangle)$ is the Hamiltonian of the periodic rigid crystal in the orthogonal tight-binding (Wannier) representation and $V(t)$ is the potential of the above mentioned real stochastic field. First, we rewrite Eq. (5.3.3a) in the interaction (Dirac) picture as

$$i\frac{\partial}{\partial t}\tilde{\rho}(t) = \exp(iL_0 t)\mathcal{L}(t)\exp(-iL_0 t)\tilde{\rho}(t) \equiv \tilde{\mathcal{L}}(t)\tilde{\rho}(t), \quad (5.3.10a)$$

$$\tilde{\rho}(t) = \exp(iH_0 t/\hbar)\rho(t)\exp(-iH_0 t/\hbar) \equiv \exp(iL_0 t)\rho(t),$$

$$L_0 \ldots = \frac{1}{\hbar}[H_0, \ldots], \quad \mathcal{L}(t) \ldots \equiv \frac{1}{\hbar}[V(t), \ldots]. \quad (5.3.10b)$$

Then [in the same way as Eq. (5.3.6) follows from (5.3.3a)], we obtain a closed identity for $D\tilde{\rho}(t)$ of the same form as Eq. (5.3.6) with substitutions

$$\rho(t) \rightarrow \tilde{\rho}(t), \quad L(t) \rightarrow \tilde{\mathcal{L}}(t). \quad (5.3.11)$$

Finally, for D, we choose the stochastic averaging, i.e.,

$$D \ldots = \langle \ldots \rangle \quad (5.3.12)$$

where $\langle \ldots \rangle$ means averaging over all realizations of the stochastic potential $V(t)$ with prescribed properties. From this it follows that whenever we refer to the stochastic Liouville equation model, we mostly omit the sign of averaging $\langle \ldots \rangle$ assuming tacitly that this averaging is always performed.

Let us now assume that $V(t)$ is a Gaussian stochastic process, i.e.,

$$\langle V(t_1) \ldots V(t_{2k-1}) \rangle = 0,$$

$$\langle V(t_1) \ldots V(t_{2k}) \rangle = \sum_{l=2}^{2k} \langle V(t_1)V(t_l) \rangle$$

$$\times \langle V(t_2) \ldots V(t_{l-1})V(t_{l+1}) \ldots V(t_{2k}) \rangle,$$

$$k = 1, 2 \ldots, \quad (5.3.13)$$

which means that any correlation function is either zero or is determined by the pair correlation function $\langle V(t)V(t') \rangle$. Further, let $V(t)$ be a δ-correlated (white noise) Markov process, i.e.,

$$\langle V_{mn}(t_1)V_{pq}(t_2) \rangle = \delta(t_1-t_2)2\Lambda_{mnpq}. \quad (5.3.14)$$

Then, returning to the Schrödinger picture, we obtain [provided that the initial condition $\rho(t_0)$ is not stochastic, i.e., $\rho(t_o) = \langle \rho(t_o) \rangle$]:

$$\frac{\partial}{\partial t} \rho(t) = -\frac{i}{\hbar} [H_0, \rho(t)] - \sum_{mnpq} \Lambda_{mnpq}[a_m^\dagger a_n, [a_p^\dagger a_q, \rho(t)]]. \quad (5.3.15)$$

In the special case, when we use the Haken-Strobl-Reineker[126,128,296] parameterization, i.e.,

$$\Lambda_{mnpq} = \gamma_{mn} \delta_{mq} \delta_{np} + \bar{\gamma}_{mn} \delta_{mp} \delta_{nq}(1 - \delta_{mn}),$$

$$\gamma_{mn} = \gamma_{nm}, \quad \bar{\gamma}_{mn} = \bar{\gamma}_{nm}, \quad (5.3.16)$$

we obtain

$$\frac{\partial}{\partial t} \rho_{mn}(t) = -\frac{i}{\hbar} ([H_0, \rho(t)])_{mn} + \delta_{mn} \sum_p [2\gamma_{mp}\rho_{pp}(t) - 2\gamma_{pm}\rho_{mm}(t)] - (1$$

$$- \delta_{mn})[2\Gamma_{mn}\rho_{mn}(t) - 2\bar{\gamma}_{mn}\rho_{mn}], \quad (5.3.17a)$$

$$2\Gamma_{mn} = \sum_r [\gamma_{rm} + \gamma_{rn}] = 2\Gamma_{nm}. \quad (5.3.17b)$$

Equation (5.3.17a) is the basic one for the so-called Haken-Strobl-Reineker model.

In case of periodicity, Γ_{mn} becomes independent of m and n, i.e. ($\Gamma_{mn} = \Gamma$); in non-periodic systems (finite chains, etc.), the m, n-dependence survives. One should realize that Eq. (5.3.16) does not yield all the elements Λ_{mnpq} which might be nonzero. However, we shall return to this point later on.

For illustration, let us take $J_{pq} = 0$ in Eq. (5.3.17a). Then, for $m = n$ in Eq. (5.3.9), we get the Pauli master equation[281]

$$\frac{\partial}{\partial t} P_m(t) = \sum_{n(\neq m)} [2\gamma_{mn}P_n(t) - 2\gamma_{nm}P_m(t)],$$

$$P_m(t) = \rho_{mm}(t). \quad (5.3.18)$$

Identifying W_{mn} with $2\gamma_{mn}$, Eq. (5.3.18) reduces to Eq. (5.2.1). On the other hand, for $m \neq n$, Eq. (5.3.17a) yields

$$\frac{\partial}{\partial t} \rho_{mn}(t) = -\frac{i}{\hbar} (\epsilon_m - \epsilon_n)\rho_{mn}(t) - 2\Gamma_{mn}\rho_{mn}(t) + 2\bar{\gamma}_{mn}\rho_{nm}(t). \quad (5.3.19)$$

Thus, diagonal and off-diagonal elements of ρ develop independently. Therefore, in general with parameterization according to Eq. (5.3.16), nonzero elements J_{pq} in Eq. (5.3.9) are the only reason for a contingent coupling of the diagonal and the off-diagonal elements of ρ. We also return later to this point.

One can verify that at $t \to \infty$, Eq. (5.3.17a) yields (even for nonperiodic systems) an asymptotic solution

$$\rho_{mn}(t \to \infty) = \delta_{mn}/N. \tag{5.3.20}$$

Here, N is the number of sites accessible. However, for any system this solution is valid only at infinite temperature. Thus, irrespective of possible nonequal ϵ_m and nonperiodic J_{mn}, the site populations become equal and the site-off-diagonal elements always decay to zero. This is a well-known deficiency of all stochastic models. In the next section, we attempt to seek an improvement.

An interesting feature of Eq. (5.3.17a) may be seen in case of dimers. For simplicity, let us take $\bar{\gamma}_{12} = \bar{\gamma}_{21} = 0$. With $J_{12} = J_{21} = 0$ (i.e., for $|1\rangle$ and $|2\rangle$ being eigenstates of H_0), Eq. (5.3.11) yields

$$\rho_{12}(t) = \rho_{21}(t)^* = S(t)\exp(-i(\epsilon_1 - \epsilon_2)t/\hbar),$$

$$dS(t)/dt = -1/T_2 S(t),$$

$$1/T_2 = 2\Gamma_{12} = \gamma_{11} + \gamma_{22} + 2\gamma_{12}. \tag{5.3.21}$$

On the other hand, from Eq. (5.3.18) follows:

$$\frac{d}{dt}[P_1(t) - P_2(t)] = -\frac{1}{T_1}[P_1(t) - P_2(t)], \quad \frac{1}{T_1} = 4\gamma_{12}, \tag{5.3.22}$$

where T_2 and T_1 are transversal and longitudinal relaxation times, respectively.

Thus, the off-diagonal γ_{mn} $(m \neq n)$ contribute both to transfer, i.e., site (or longitudinal) relaxation in Eq. (5.3.22), and to dephasing (i.e., transverse relaxation) in Eq. (5.3.21), while the diagonal γ_{mm} contribute just to dephasing. From definition (5.3.16) it is easy to show that $\gamma_{11} = \Lambda_{1111} \geq 0$, $\gamma_{12} = \gamma_{21} = \Lambda_{1221} \geq 0$. Thus Eqs. (5.3.21) and (5.3.22) yield a standard relation between the longitudinal (T_1) and transverse relaxation (T_2) times

$$2T_1 \geq T_2, \tag{5.3.23}$$

which is a standard and well-known result always stemming from the lowest-order theories. In our case, because of Eqs. (5.3.13) and (5.3.14), only the lowest-(second)-order (in V) terms contribute to the right-hand side of Eq. (5.3.15). So, for more general (colored-noise) stochastic models, higher-order terms become important, and Eq. (5.3.23) may become invalid (see Refs. 297 and 323). In other words, for some real situations or for more general models, the phase coherence may persist even after populations have relaxed.

For infinite, completely periodic lattices, Eq. (5.3.17a) can be solved. For initially fully localized excitation, i.e., for

$$\rho_{mn}(0) = \delta_{m0}\,\delta_{n0}, \tag{5.3.24}$$

one obtains the following dependence for the mean square displacement $\Delta(t)$ on a periodic chain with one molecule per elementary cell:[296]

$$\Delta(t) = \sum_m a^2 m^2 [2\gamma_{m0} + J_{m0}^2/(\Gamma + \bar{\gamma}_{m0})]t + \frac{1}{2} \sum_m a^2 m^2 J_{m0}^2 (\Gamma + \bar{\gamma}_{m0})^{-2}$$

$$\times (\exp(-2(\Gamma + \bar{\gamma}_{m0})t) - 1). \tag{5.3.25}$$

The result (or more general results for more general initial conditions (see Ref. 296) is important for the interpretation of, e.g., photoconductivity measurements in OMCs. One should notice that for small t

$$\Delta(t) \approx \sum_m a^2 m^2 [2\gamma_{m0} t + J_{m0}^2 t^2], \quad \text{for } t \ll \Gamma^{-1}, \tag{5.3.26}$$

while, for long t

$$\Delta(t) \approx \sum_m a^2 m^2 [2\gamma_{m0} + J_{m0}^2/(\Gamma + \bar{\gamma}_{m0})]t - \frac{1}{2} \sum_m a^2 m^2 J_{m0}^2/(\Gamma + \bar{\gamma}_{m0})^2,$$

$$t \gg \text{Max}(\Gamma + \bar{\gamma}_{m0}). \tag{5.3.27}$$

The linear term of $\Delta(t)$ in t in Eq. (5.3.26) when $t \to 0$ is an artifact of the model owing to the unphysical assumption of the δ-correlation in time [see Eq. (5.3.14)]. It will be shown below that actually in real situations always

$$\Delta(t) \sim t^2, \quad \text{if } t \to 0. \tag{5.3.28}$$

At long times, if one introduces the diffusion constant in the chain as

$$D = \lim_{t \to \infty} \frac{1}{2t} \Delta(t), \tag{5.3.29}$$

Eq. (5.3.27) yields

$$D = \sum_m a^2 m^2 \left[\gamma_{m0} + \frac{1}{2} J_{m0}^2/(\Gamma + \bar{\gamma}_{m0}) \right]. \tag{5.3.30}$$

One can easily see in Eq. (5.3.30) how the coherent and incoherent channels of transport combine to yield diffusion constant D corresponding to reality. Eq. (5.3.30) remains meaningful in the purely incoherent case ($J_{mn} = 0$). Formally, it is also meaningful in the purely coherent case ($\gamma_{mn} = \bar{\gamma}_{mn} = 0$) but the result then has nothing to do with reality since Eq. (5.3.29) is (when $\gamma_{mn} = \bar{\gamma}_{mn} = 0$) divergent $[\lim_{\gamma, \bar{\gamma} \to 0} \lim_{t \to \infty} 1/2t\Delta(t)$ is finite while $\lim_{t \to \infty} \lim_{\gamma, \bar{\gamma} \to 0} 1/2t\Delta(t) = +\infty]$.

In addition to Eqs. (5.3.26) and (5.3.20), there is at least one other disadvantage of the stochastic Liouville equation model with the Haken-Strobl-Reineker parameterization [Eq. (5.3.16)]. For simplicity, let us assume a one-dimensional periodic chain and introduce (for an initially homogeneous distribution in space) a distribution function in the k-space

$$f(k, t) = \sum_n \exp(-ika(m-n))\rho_{mn}(t). \qquad (5.3.31)$$

With

$$\bar{\gamma}(k) = \sum_{n(\neq m)} \bar{\gamma}_{mn} \exp(-ika(m-n)), \qquad (5.3.32)$$

Eq. (5.3.17a) yields the Boltzmann-like equation

$$\frac{\partial}{\partial t} f(k, t) = -2\Gamma[f(k, t) - 1/N] + \frac{2}{N} \sum_{k'} \bar{\gamma}(k-k')f(k',t). \qquad (5.3.33)$$

Since

$$\sum_{k'} \bar{\gamma}(k-k') = 0, \qquad (5.3.34)$$

[compare with Eq. (5.3.32)], it gives

$$f(k, t \to \infty) = 1/N \qquad (5.3.35)$$

which corresponds to infinite temperature. One may check that this conclusion is not connected with the symmetry properties of Eq. (5.3.16). Below we show that one can generalize the stochastic Liouville equation model replacing the classical field $V(t)$ (leading only to induced transitions) by a quantum field (yielding both induced and spontaneous transitions). Thus, the symmetry relations $\gamma_{mn} = \gamma_{nm}$ become invalid, being (in the lowest order in J_{mn}) replaced by the detailed balance conditions. In the Haken-Strobl-Reineker parameterization, however, Eq. (5.3.17a) [i.e., also Eq. (5.3.35)] remains valid. Thus, at least in systems of equivalent sites, one should refrain from this parameterization at finite temperatures including more parameters than in Eq. (5.3.17a).

These results demonstrate the necessity to extend and generalize the model.

5.4. GENERALIZED MASTER EQUATIONS AND GENERALIZED STOCHASTIC LIOUVILLE EQUATION MODEL

The aim of the present section is to generalize the Stochastic Liouville equation model to make it applicable for finite temperatures and non-periodic systems. In molecular crystals, this feature would make it possible to include in the description impurities, trapping centers, etc., and the dynamics of excitations in the vicinity of these imperfections.

Although we can start from time-convolutionless (time-independent) GMEs (see, e.g., Ref. 69), we will prefer here to follow Ref. 59. As in Eq. (5.3.8), we assume the following total Hamiltonian

$$H = H_0 + V. \tag{5.4.1a}$$

Here V is, however, the interaction Hamiltonian of a given exciton with a quantum field (e.g., harmonic phonon field) forming a reservoir while, in contradistinction to Eq. (5.3.9), H_0 includes the Hamiltonian of this field H_R, i.e.,

$$H_0 = \sum_{m \neq n} J_{mn} a_m^{\dagger} a_n + \sum_n \epsilon_n a_n^{\dagger} a_n + H_R \equiv H_e + H_R. \tag{5.4.1b}$$

In a similar way as in Eq. (5.3.10), we introduce the interaction (Dirac) picture, i.e., Eq. (5.3.3a) reduces to

$$i \frac{\partial}{\partial t} \tilde{\rho}_T(t) = \exp(iL_0(t-t_0))\mathcal{L} \exp(-iL_0(t-t_0))\tilde{\rho}_T(t) \equiv \tilde{\mathcal{L}}(t)\tilde{\rho}_T(t),$$

$$\tilde{\rho}_T(t) = \exp(iH_0(t-t_0)/\hbar)\rho_T(t)\exp(-iH_0(t-t_0)/\hbar) \equiv \exp(iL_0(t-t_0))\rho_T(t), \tag{5.4.2a}$$

$$L_0 \ldots = \frac{1}{\hbar}[H_0, \ldots], \mathcal{L} \ldots \equiv \frac{1}{\hbar}[V, \ldots]. \tag{5.4.2b}$$

Here $\rho_T(t)$ is the total density matrix of the exciton and the field (reservoir). Then [as from Eq. (5.3.3) via Eqs. (5.3.5a, b) to Eq. (5.3.6)], one derives, starting from Eq. (5.4.2a), the following equation:

$$\frac{\partial}{\partial t} D\tilde{\rho}_T(t) = -iD\tilde{\mathcal{L}}(t)D\tilde{\rho}_T(t) - \int_{t_0}^t D\tilde{\mathcal{L}}(t)\exp\left[-i(1-D)\int_{\tau}^t \tilde{\mathcal{L}}(\tau_1)d\tau_1\right]$$

$$\times (1-D)\tilde{\mathcal{L}}(\tau)D\tilde{\rho}_T(\tau)d\tau - iD\tilde{\mathcal{L}}(t)\exp\left[-i\int_{t_0}^t (1-D)\tilde{\mathcal{L}}(\tau)d\tau\right]$$

$$\times (1-D)\rho_T(t_0). \tag{5.4.3}$$

Compared to the previous section, however, we have now not a stochastic but rather a quantum field over states of which one has to take the average. So, the projector $D(= D^2)$ must be taken in another way.

We choose the Argyres and Kelley[13] form

$$(D \ldots)_{m\mu, n\nu} = \rho_{\mu\nu}^R (Tr_R \ldots)_{mn} = \rho_{\mu\nu}^R \sum_{\lambda} (\ldots)_{m\lambda, n\lambda} \tag{5.4.4}$$

where $\rho^R_{\mu\nu}$ is arbitrary except for

$$\sum_{\nu} \rho^R_{\nu\nu} \equiv Tr_R \rho^R = 1 \tag{5.4.5}$$

(otherwise $D \neq D^2$). In Eqs. (5.4.4), (5.4.5) and below, Greek indices designate (arbitrarily chosen) states of the reservoir. On the other hand, as we assume just one exciton in the molecular condensate, the Latin indices (denoting general states of the system in question) designate its location (e.g., the molecule on which the exciton is located). Taking then Tr_R of (5.4.3) yields the GME for the exciton density matrix in the interaction picture $\tilde{\rho}(r)$, since

$$(D\tilde{\rho}_T(t))_{m\mu,n\nu} = \rho^R_{\mu\nu} \sum_{\lambda} (\tilde{\rho}_T(t))_{m\lambda,n\lambda} = \rho^R_{\mu\nu}(Tr_R\tilde{\rho}_T(t))_{mn} \equiv \rho^R_{\mu\nu}\tilde{\rho}_{mn}(t). \tag{5.4.6}$$

The exciton density matrix $\rho(t)$ in the Schrödinger picture can then be obtained as

$$\rho_{mn}(t) = \sum_{\lambda} (\rho_T)_{m\lambda,\,n\lambda}(t)$$

$$= \sum_{\lambda} (\exp(iH_0(t-t_0)/\hbar)\tilde{\rho}_T(t)\exp(-iH_0(t-t_0)/\hbar))_{m\lambda,n\lambda}$$

$$= (\exp(iH_e(t-t_0)/\hbar)\tilde{\rho}(t)\exp(-iH_e(t-t_0)/\hbar))_{mn}. \tag{5.4.7}$$

Before transforming back to the Schrödinger picture, let us specify our initial conditions. Let us assume that the exciton is created so quickly (by, e.g., a very short light pulse) during a very short time interval just preceding t_o that the reservoir (e.g., phonons) not taking part in the creation act of the exciton does not manage to react to its presence at $t = t_o$. Thus, the exciton and reservoir remain initially statistically independent, i.e.,

$$(\rho_T(t_0))_{m\mu,\,n\nu} = \rho_{mn}(t_0)(\rho_R)_{\mu\nu}, \tag{5.4.8}$$

where ρ_R is the initial density matrix of the reservoir (i.e., the quantum field in question). The same arguments may be repeated as long as we do not treat the exciton but a charge carrier (electron) which is sufficiently quickly generated or injected into the system.

Notice that so far we have not specified $\rho^R_{\mu\nu}$ entering the problem via our projector D in Eq. (5.4.4). Let us therefore choose $\rho^R_{\mu\nu}$ to be equal to the initial reservoir density matrix $(\rho_R)_{\mu\nu}$, i.e.,

$$\rho^R_{\mu\nu} = (\rho_R)_{\mu\nu}. \tag{5.4.9}$$

Then

$$(D\rho(t_0))_{m\mu,\,n\nu} = (\rho_R)_{\mu\nu}\rho_{mn}(t_0) = (\rho(t_0))_{m\mu,\,n\nu} \qquad (5.4.10)$$

i.e., $(1-D)\rho(t_o) = 0$ and the entire last term on the right-hand side of Eq. (5.4.3) (the initial condition term) disappears. From now on, we shall always work with the initial condition (5.4.8) and with our chosen equation (5.4.9), i.e., henceforth there will be therefore no initial condition term in (5.4.3).

Let us now say a few words about the first term on the right-hand side of Eq. (5.4.3). It equals

$$-i(D\tilde{L}(t)D\tilde{\rho}_T(t))_{m\mu,n\nu} = -i\rho_{\mu\nu}^R(Tr_R(\tilde{L}(t)(\rho^R\otimes\tilde{\rho}(t))))_{mn}$$

$$= -i\rho_{\mu\nu}^R(Tr_R(\exp(iL_0(t-t_0))\mathcal{L}[(\exp(-iL_R(t-t_0))\rho^R)$$

$$\otimes\rho(t)]))_{mn}. \qquad (5.4.11)$$

However, it disappears when, e.g., Eq. (5.4.9) is used provided that the initial density matrix of the reservoir is canonical

$$\rho_R = \exp(-\beta H_R)/Tr_R(\exp(-\beta H_R)), \qquad (5.4.12)$$

H_R is the Hamiltonian of harmonic vibrations and V is linear in creation $(b_k^\dagger)$ and annihilation (b_k) operators of phonons. In other cases, it yields just a weak temperature-dependent correction to the first term on the right-hand side of Eq. (5.4.17) below. We ignore this assuming Eqs. (5.4.9) and (5.4.12) and V linear in exciton creation and annihilation operators.

Taking Tr_R in Eq. (5.4.3) and using (5.4.2b), one arrives at

$$\frac{\partial}{\partial t}\tilde{\rho}_{mn}(t) = \sum_{pq}\int_{t_0}^{t}\tilde{w}_{mn,pq}(t,\tau)\tilde{\rho}_{pq}(\tau)d\tau \qquad (5.4.13)$$

with

$$\tilde{w}_{mn,\,pq}(t,\,\tau) \approx -\sum_{\mu\pi\lambda}(\tilde{L}(t)\tilde{L}(\tau))_{m\mu,n\mu,p\pi,q\lambda}(\rho_R)_{\pi\lambda} \qquad (5.4.14)$$

being the (second order in $\tilde{L}$) memory function describing the weight with which the state [at time τ, i.e., $\tilde{\rho}(\tau)$ in Eq. (5.4.13)] determines the rate of change of the state $(\partial\tilde{\rho}(t)/\partial t)$ at time $t \geq \tau$. Later, in this as well as the next section, we return to the problem of using the second-order approximation which might lead to some objections in case of realistic, in particular, strong coupling. In Eqs. (5.4.13), (5.4.14), we have used a standard notation for matrix elements of linear superoperators with four (double) indices determined by the rule

$$(\mathcal{B}A)_{ab} = \sum_{cd}\mathcal{B}_{abcd}A_{cd}, \qquad a,b,\ldots = m\mu,n\nu,\ldots \qquad (5.4.15)$$

($\mathcal{B}$ is an arbitrary linear superoperator while A is an operator). The second-order (Born's) approximation used in Eq. (5.4.14) is not necessary but it is helpful. It becomes exact (even for the finite coupling strength) in the limit of fast reservoir when just τ in the vicinity of t contributes to the exp $\leftarrow$ ... term in the second term on the right-hand side of Eq. (5.4.3) and it becomes practically unity.

Now, assume that the reservoir is actually fast as compared to the dynamics of the quasiparticle (exciton, charge carrier). This assumption is inherent in the standard stochastic Liouville equation model [see Eq. (5.3.14)]. This means that dephasing in Eq. (5.4.14) leads (after taking the limit of infinite reservoir) to such a fast decay of $\tilde{w}_{mn,pq}$ with increasing $t-\tau$ that $\tilde{w}_{mn,pq}(t,\tau)$ becomes already zero without practically any change of $\tilde{\rho}_{pq}(\tau)$ as compared to $\tilde{\rho}_{pq}(t)$. Hence

$$\frac{\partial}{\partial t}\tilde{\rho}_{mn}(t) \approx \sum_{pq}\int_{-\infty}^{t}\tilde{w}_{mn,pq}(t,\tau)d\tau\tilde{\rho}_{pq}(t), \tag{5.4.16}$$

which gives [using Eq. (5.4.2b)]

$$\frac{\partial}{\partial t}\rho_{mn}(t) = -\frac{i}{\hbar}[H_e,\rho(t)]_{mn}+\sum_{pq}W_{mnpq}\rho_{pq}(t), \tag{5.4.17}$$

with

$$W_{mnpq} = -\int_{-\infty}^{t}\sum_{\mu\pi\lambda}(\exp(-iL_0(t-t_0))\bar{L}(t)\bar{L}(\tau)\exp(iL_0(t-t_0)))_{m\mu,n\mu,p\pi,q\lambda}$$

$$\times\rho_{\pi,\lambda}^{R}d\tau = -\int_{-\infty}^{0}d\tau\sum_{\mu\pi\lambda}(L\bar{L}(\tau))_{m\mu,n\mu,p\pi,q\lambda}\rho_{\pi\lambda}^{R}. \tag{5.4.18}$$

Equation (5.4.16) is the so-called Markov approximation to Eq. (5.4.13). Markovian (i.e., local in time) Eqs. (5.4.17) are our final set of equations forming the generalized stochastic Liouville equation model (GSLEM). In general,

$$\sum_{m}W_{mmpq} = 0, \tag{5.4.19}$$

which is a property which should be kept for any parameterization. As may be seen from Eq. (5.4.17), it ensures that the sum $\sum_m\rho_{mm}(t)$ (being the probability of finding the quasiparticle anywhere in the system) is time-independent, which means the conservation of the number of quasiparticles.

Let us now introduce real coefficients

$$\gamma_{mn} = \frac{1}{2\hbar^2}\int_{-\infty}^{0}\sum_{\mu\nu}[V_{m\mu,n\nu}V_{n\nu,m\mu}(\tau)+V_{n\nu,m\mu}V_{m\mu,n\nu}(\tau)]p_\nu\,d\tau. \tag{5.4.20}$$

(Here and below, we assume

$$(\rho_R)_{\mu\nu} = p_\mu \delta_{\mu\nu} \qquad (5.4.21)$$

which is true if ρ_R is canonical [see Eq. (5.4.12)] and the Greek indices denote eigenstates of H_R.) It can then be verified that $2\gamma_{mn}$ is the mean reservoir-assisted lowest-order-golden-rule hopping rate $n \to m$. In the limit $J_{rs} \to 0$, we get the detailed balance condition:[59]

$$\gamma_{mn} = \exp(-\beta(\epsilon_m - \epsilon_n))\gamma_{mn}, \qquad (5.4.22)$$

where β is the reciprocal temperature in energy units first appearing in Eq. (5.4.12). Then, for $m \neq n$,

$$W_{mmnn} = 2\gamma_{mn}, \qquad (5.4.23)$$

and from Eq. (5.4.19)

$$W_{mmmm} = -2 \sum_{n(\neq m)} \gamma_{mn}. \qquad (5.4.24)$$

Further, for $m \neq n$ (omitting renormalization corrections and some small terms disappearing when $J_{rs} \to 0$) we have

$$W_{mnmn} \approx -\sum_r (\gamma_{rm} + \gamma_{rn}). \qquad (5.4.25)$$

Introducing

$$\bar{\gamma}_{mn} = \bar{\gamma}_{nm}^* = \frac{1}{2\hbar^2} \int_{-\infty}^0 \sum_{\mu\nu} [V_{m\mu,n\nu} V_{m\nu,n\mu}(\tau) + V_{m\nu,n\mu} V_{m\mu,n\nu}(\tau)] p_\nu d\tau,$$

$$\qquad (5.4.26)$$

we find for $m \neq n$

$$W_{mnnm} = 2\bar{\gamma}_{mn}. \qquad (5.4.27)$$

If all other elements W_{mnpq} are omitted, we obtain from Eq. (5.4.17) a set of equations (5.3.17a, b) where, however [instead of Eq. (5.3.16)], $\bar{\gamma}_{mn} = \bar{\gamma}_{nm}^*$ and γ_{mn} (instead of being equal to γ_{nm}) fulfils Eq. (5.4.22) when $J_{rs} \to 0$. With these two changes, Eqs. (5.3.17a, b) then form the GSLEM in the Haken-Strobl-Reineker parameterization. It is not difficult to see that for periodic systems and a real basis of localized states, the generalized stochastic Liouville equation model mathematically may be reduced to the standard stochastic Liouville equation model although the physics underlying them is completely different.

From that, one can see that [irrespective of finite temperature $T = (k_B\beta)^{-1}$ in Eq. (5.4.12)], one again obtains Eq. (5.3.35) in GSLEM within Haken-Strobl-Reineker parameterization. This is so because (as mentioned above) the only coupling of diagonal and off-diagonal matrix elements of ρ is via $-i/\hbar([H_e, \rho(t)])_{mn}$ in Eq. (5.3.17a) which disappears when $t \to +\infty$ for systems of equivalent sites. For illustration, let us take, e.g., a symmetric dimer with $\epsilon_1 = \epsilon_2 = 0$, i.e.,

$$H_e = J[|1\rangle\langle 2| + |2\rangle\langle 1|] = E_+|+\rangle\langle +| + E_-|-\rangle\langle -|,$$

$$E_\pm = \pm J,$$

$$|\pm\rangle = \frac{1}{\sqrt{2}}[|1\rangle \pm |2\rangle]. \tag{5.4.28}$$

Equations (5.3.17a) of the GSLEM, within the Haken-Strobl-Reineker parameterization approach, yield

$$\frac{\partial}{\partial t}\rho_{11}(t) = -\frac{i}{\hbar}J[\rho_{21}(t) - \rho_{12}(t)] + 2[\gamma_{12}\rho_{22}(t) - \gamma_{21}\rho_{11}(t)], \tag{5.4.29a}$$

$$\frac{\partial}{\partial t}\rho_{12}(t) = -\frac{i}{\hbar}J[\rho_{22}(t) - \rho_{11}(t)] - 2\Gamma_{12}\rho_{12}(t) + 2\bar{\gamma}_{12}\rho_{21} \tag{5.4.29b}$$

and similarly with the interchange $1 \leftrightarrow 2$. Hence (owing to $\epsilon_1 = \epsilon_2$, i.e., as a consequence of symmetry) $\rho_{jj}(t) \to 1/2$, $j = 1,2$ irrespective of the initial conditions, the first term on the right-hand side of (5.4.29b) disappears, i.e., $\rho_{ij}(t) \to 0$, $t \to +\infty$, $i \neq j$. Thus, the first term on the right-hand side of (5.4.29a) also drops out. Therefore

$$\rho_{11}(+\infty) = \rho_{22}(+\infty) = 1/2,$$

$$\rho_{12}(+\infty) = \rho_{21}(+\infty) = 0 \tag{5.4.30}$$

and

$$\rho_{++}(t) = 1 - \rho_{--}(t) = \frac{1}{2}[\rho_{11}(t) + \rho_{22}(t) + \rho_{12}(t) + \rho_{21}(t)] \to \frac{1}{2},$$

$$\rho_{+-} = (\rho_{-+}(t))^* = \frac{1}{2}[\rho_{11}(t) - \rho_{22}(t) + \rho_{21}(t) - \rho_{12}(t)] \to 0 \tag{5.4.31}$$

irrespective of the finite temperature and finite difference $E_+ - E_- = 2J$. One can, however, easily see that outside the Haken-Strobl-Reineker parameterization [i.e., using Eq. (5.4.17)] one correctly obtains a solution that leads (with $t \to +\infty$) to different populations of $|+\rangle$ and $|-\rangle$ states. Keeping the term

$$W_{1222} = W_{2111} = W^*_{1211} = W^*_{2122} = W \tag{5.4.32}$$

[W being real and equality in Eq. (5.4.32) is due to the symmetry] in addition to the Haken-Strobl-Reineker parameters, Eqs. (5.4.29a, b) are replaced by ($\gamma \equiv \gamma_{12} = \gamma_{21}$, $\Gamma \equiv \Gamma_{12} = \Gamma_{21}$, $\bar{\gamma} \equiv \bar{\gamma}_{12} = \bar{\gamma}_{21}$)

$$\frac{\partial}{\partial t}\rho_{11}(t) = -\frac{i}{\hbar}J[\rho_{21}(t)-\rho_{12}(t)]+2\gamma[\rho_{22}(t)-\rho_{11}(t)],$$

$$\frac{\partial}{\partial t}\rho_{12}(t) = -\frac{i}{\hbar}J[\rho_{22}(t)-\rho_{11}(t)]-2\Gamma\rho_{12}(t)+2\bar{\gamma}\rho_{21}(t)+W[\rho_{11}(t)+\rho_{22}(t)].$$

$$(5.4.33)$$

Thus

$$\rho_{11}(+\infty) = \rho_{22}(+\infty) = 1/2,$$

$$\rho_{12}(+\infty) = \rho_{21}(+\infty) = W/(2\Gamma-2\bar{\gamma}) \qquad (5.4.34)$$

i.e.,

$$\rho_{++}(+\infty)/\rho_{--}(+\infty) = 1+2W/(\Gamma-\bar{\gamma}-W) \qquad (5.4.35)$$

becomes different from unity as it should be in reality. Thus, for systems of equivalent sites and finite temperature, the Haken-Strobl-Reineker parameterization in GSLEM should be extended.

5.5. APPLICABILITY OF THE STOCHASTIC LIOUVILLE EQUATION AND GENERALIZED STOCHASTIC LIOUVILLE EQUATION MODELS

First, we discuss our approximations to either Eq. (5.3.15) or Eq. (5.4.17). In both cases, we have employed the idea of the fast (as compared to the exciton or carrier dynamics) reservoir. This approach automatically preserved just the lowest (second)-order terms in the time integrals in Eqs. (5.3.6) and (5.4.3). However, this condition is hardly true for singlet excitons and charge carriers unless strong polaron effects (band narrowing) take place. (Notice that the bandwidth $\delta E \approx 0.01$ eV corresponds to the coherent hopping time τ_h to the nearest neighbor molecules in the lattice $\tau_h \approx 10^{-14}$ sec. This is by several orders of magnitude shorter than typical reciprocal phonon frequencies $< t_d$ = decay time of correlation functions, see Eq. (5.4.14).) Thus the Stochastic Liouville equation model as well as the GSLEM theories may be applied without change or further justification just to the triplet exciton migration. For singlet excitons and charge carriers, fortunately one can use other arguments based on the second-order time-convolutionless GMEs. One then obtains again Eq. (5.4.17) except that within a short initial time interval (up to t_d), coefficients W_{mnpq} might be time-dependent.[69]

Anyway, in case of a not-very-fast-reservoir, a problem of justification of the Born (second-order) approximation appears, especially due to the above-mentioned strong coupling to phonons in the rigid, or crude basis of quasiparticle states. One should notice that the introduction of the Born approximation is by no means absolutely necessary and is being used here just for technical reasons. As far as the real quasiparticle dynamics [determined by the exciton or electron density matrix in the interaction picture $\bar{\rho}_{mn}(t)$] is slower than decay of all the infinite order memory functions $\bar{w}_{mn,\,pq}(t)$, one can still arrive at Eq. (5.4.17) with only minor generalizations in Eq. (5.4.18) and ensuing formulas. [The latter assumption may even be canceled when one is working, as in Ref. 69, with the infinite-order time-convolutionless (time-independent) GMEs leading to Eq. (5.4.17) with, however, infinite-order coefficients W_{mnpq}. As in the second-order approximation, the results are that these coefficients are time-dependent.] However, for the sake of simplicity, we will discuss here the traditional second-order (Born) approximation.

With the Born approximation, the situation is, fortunately, not so bad as it might seem at first sight. As already mentioned, the strong coupling to intramolecular vibrational modes leads to vibronic (small polaron) states. Working with this vibronic basis automatically includes most of this coupling. The strong coupling to intermolecular, libration, and lattice vibration modes (inherent to the rigid basis, in contrast to the adiabatic basis) is, however, necessary to "force" the quasiparticle to move with vibrating or librating molecules. A shifted or swung (adiabatic) state of a molecule (molecular ion) can be well expressed using higher excited states of the same rigid, i.e., unshifted or unswung molecule — that is why higher excited but rigid molecular states become involved via this coupling. Thus, the whole theory should be rebuilt anew with only the change that the Latin indices denote not only the molecules but also the (ground or excited) states on these (rigid) molecules. Such a theory has not been presented so far. A hope remains that on returning at a proper stage to the adiabatic basis, the standard weak coupling situation might be restored in which the standard second-order (Born) approximation is acceptable. This hope is supported by the observation that one might from the very beginning work in the adiabatic representation (in which the coupling to the intermolecular, libration, and lattice vibration modes is undoubtedly weak), provided that the implicit dependences of the exciton or electron creation and annihilation operators on the nuclear displacement operators is properly kept (otherwise, one would return to the adiabatic approximation). Then, one should get structurally analogous equations as Eq. (5.4.17) except for some renormalizations in the first term on the right-hand side. That is why these equations are, for qualitative studies at least, sometimes used even in situations when some of the assumptions used in deriving them are not very well fulfilled.

On the other hand, one should not be apprehensive by the second-order approximation used here. For illustration, let us, for example consider an asymmetrical molecular dimer using the Haken-Strobl-Reineker parameterization. For simplicity, let us take $\bar{\gamma}_{12} = \bar{\gamma}_{21} = 0$, $J_{12} = J_{21} = J$, and $\Gamma_{12} = \Gamma_{21} = \Gamma$. The Eq. (5.3.17a) then yields

$$\frac{\partial}{\partial t}\rho_{12}(t) = -\frac{i}{\hbar}(\epsilon_1-\epsilon_2)\rho_{12}(t)-\frac{i}{\hbar}J[\rho_{22}(t)-\rho_{11}(t)]-2\Gamma\rho_{12}(t).$$

$$(5.5.1)$$

So

$$\rho_{12}(t) = \exp(-i(\epsilon_1 - \epsilon_2 - 2i\hbar\Gamma)t/\hbar)\rho_{12}(0) - \frac{i}{\hbar} J \int_0^t \exp(-i(\epsilon_1 - \epsilon_2 - 2i\hbar\Gamma)$$

$$\times (t-\tau)/\hbar)[\rho_{22}(\tau) - \rho_{11}(\tau)]d\tau \qquad (5.5.2)$$

which may be introduced back to Eq. (5.3.17a) for, e.g., $m = n = 1$. We then obtain

$$\frac{\partial}{\partial t}\rho_{11}(t) = 2[\gamma_{12}\rho_{22}(t) - \gamma_{21}\rho_{11}(t)] + \frac{i}{\hbar} J \exp(-2\Gamma t)[\exp(-i(\epsilon_1 - \epsilon_2)t/\hbar)$$

$$\times \rho_{12}(0) + C.C.] + 2\frac{J^2}{\hbar^2}\int_0^t \exp(-2\Gamma(t-\tau))$$

$$\cdot \cos((t-\tau)(\epsilon_1 - \epsilon_2)/\hbar) \cdot [\rho_{22}(\tau) - \rho_{11}(\tau)]d\tau, \qquad (5.5.3)$$

and a similar expression for $\partial \rho_{22}(t)/\partial t$. Realizing that $\Gamma \sim g^2$ where g is the coupling constant to the reservoir, we see that Eq. (5.3.17a) [and, in general, Eq. (5.4.17)] is equivalent to a system of coupled integrodifferential equations for probabilities $P_j(t) = \rho_{jj}(t)$ with the memory kernel partially summed up to infinity. This feature supports our view that Eq. (5.4.17) may remain meaningful (at least for a qualitative discussion) even when the coupling to vibrations is not very weak.

Practical calculations based on the stochastic Liouville equation model or GSLEM have been performed many times. For earlier references see Ref. 296. With inclusion of life-time effects for excitons (see below), the method has been applied even to such complicated systems as, e.g., antenna systems of photosynthetic units.[269,377] For nonperiodic systems, some illustrative results may be found in Refs. 59 and 69. In general, such calculations are computationally difficult as for N sites, one has to solve a system of N^2 differential equations for N^2 quantities except for cases when symmetry considerations may be applied.

5.6. GENERALIZED MASTER EQUATION THEORIES WITH DIAGONALIZING PROJECTOR

Though one can use diagonalizing (in sites, i.e., molecular indices) projection superoperators in combination with time-local (convolutionless) identities, historically most of theories of the exciton or charge carrier propagation in molecular crystals and aggregates resulted from the application of the diagonalizing projectors to the usual Nakajima-Zwanzig identity; see Eq. (5.3.6). Let us take as the simplest (and most widely used) diagonalizing projector, the Peier[282] projector

$$(DA)_{m\mu, n\nu} \equiv \sum_{a\alpha, b\beta} D_{m\mu, n\nu, a\alpha, b\beta} A_{a\alpha, b\beta} = \rho_{\mu\nu}^R \delta_{mn} \sum_{\lambda} A_{m\lambda, m\lambda}$$

$$\equiv \rho^R_{\mu\nu}\delta_{mn}(Tr_R A)_{mn} \tag{5.6.1}$$

i.e.,

$$D_{m\mu, n\nu, a\alpha, b\beta} = \rho^R_{\mu\nu}\delta_{mn}\,\delta_{ma}\,\delta_{nb}\,\delta_{\alpha\beta} \tag{5.6.2}$$

with any $\rho^R_{\mu\nu}$ fulfilling the condition (5.4.5). Because $(D\rho_T(t))_{m\mu, n\nu}$ $= \rho^R_{\mu\nu}\rho_{mm}(t)\delta_{mn} \equiv \rho^R_{\mu\nu}\delta_{mn}P_m(t)$, application of Eq. (5.6.1) in Eq. (5.3.6) yields the GME in the form of a closed set of integrodifferential equations for probabilities $P_m(t)$ of finding the exciton (or charge carrier) at individual molecules. Below we see yet another (so-called partitioning) projector of the diagonalizing type but here, we adhere to Eq. (5.6.1). From Eqs. (5.6.1) and (5.3.6) we obtain

$$\frac{\partial}{\partial t}P_m(t) = \sum_n \int_{t_0}^t w_{mn}(t-\tau)P_n(\tau)d\tau + I_m(t-t_0), \tag{5.6.3a}$$

$$w_{mn}(t) = -\sum_{\mu\nu\lambda} [L \exp(-i(1-D)Lt) \cdot (1-D)L]_{m\mu, m\mu, n\nu, n\lambda}\,\rho^R_{\nu\lambda}, \tag{5.6.3b}$$

$$I_m(t-t_0) = -i\sum_{\mu} [L \exp(-i(1-D)L(t-t_0)) \cdot (1-D)\rho(t_0)]_{m\mu, m\mu}. \tag{5.6.3c}$$

Here we have already assumed that the Liouville superoperator in the Schrödinger picture is (owing to lack of external time-dependent fields) time-independent. In addition to that, we have employed the identity $DLD = 0$.

A theory of this type for excitons was first developed in detail by Kenkre starting from the first paper by Kenkre and Knox,[171] although some earlier papers also existed. The approach based on Eqs. (5.6.3a–c) is often called the Kenkre-Knox theory. A nice presentation of it with references (up to 1982) may be found in Ref. 169.

Because

$$L_{m\mu, n\nu, a\alpha, b\beta} = \frac{1}{\hbar} [H_{m\mu, a\alpha}\,\delta_{n\nu, b\beta} - H_{b\beta, n\nu}\,\delta_{m\mu, a\alpha}], \tag{5.6.4}$$

it gives

$$\sum_{m\mu} L_{m\mu, m\mu, a\alpha, b\beta} = 0. \tag{5.6.5}$$

Therefore

$$\sum_m w_{mn}(t-\tau) = 0, \tag{5.6.6a}$$

$$\sum_m I_m(t - t_0) = 0 \tag{5.6.6b}$$

and, from Eq. (5.6.3a)

$$\partial/\partial t \sum_m P_m(t) = 0. \tag{5.6.7}$$

Thus, the total probability of finding the exciton (or charge carrier) anywhere in the lattice is constant (finite-lifetime effects are not included). From Eq. (5.6.6a), one can obtain the diagonal memory functions w_{mm} via the off-diagonal ones

$$w_{mm}(t - \tau) = - \sum_{n(\neq m)} w_{nm}(t - \tau) \tag{5.6.8}$$

i.e., Eq. (5.6.3a) may be rewritten in another form as follows

$$\frac{\partial}{\partial t} P_m(t) = \sum_{n(\neq m)} \int_{t_0}^t [w_{mn}(t - \tau) P_n(\tau) - w_{nm}(t - \tau) P_m(\tau)] d\tau + I_m(t - t_0). \tag{5.6.9}$$

Equation (5.6.9) reminds one of the Pauli master (balance) equations. See below for a more direct correspondence.

First, one should emphasize that our memory functions $w_{nm}(t)$ introduced in Eq. (5.6.3b) are in general different from those ($w_{nm, pq}$) introduced in Eq. (5.4.14). Their decay with increasing time argument is (in addition to a possible small nonexponential damping in infinite systems, even for zero coupling to vibrations) determined by dephasing processes in the crystal lattice (including those inside individual molecules). Thus, one cannot expect appreciable damping of the memory functions on time intervals $< \omega_{\text{Max}}^{-1}$ = reciprocal maximum vibrational frequency (intramolecular modes included). In other words, for a time interval much shorter than a typical reciprocal vibrational frequency, the lattice is practically rigid, i.e., no effect of vibrations suppressing the coherent character of the exciton (or carrier) motion is to be expected. Since opposite results are sometimes reported, we return to this point later.

Now, let us pay some attention to the initial condition term $I_m(t - t_0)$. To get its reliable form, one would probably have to sum up its perturbation expansion (in powers of coupling to vibrations) at least partially to infinity. That is why one usually assumes that the initial condition again [as in Eq. (5.4.8)] corresponds to the initially statistically independent exciton (or charge carrier) and reservoir (vibrations including intramolecular ones). If, in addition to Eq. (5.4.8), one assumes that the exciton (or carrier) density matrix is initially site-diagonal, i.e.,

$$\rho_{mn}(t_o) = \delta_{mn} P_m(t_o) \tag{5.6.10}$$

in Eq. (5.4.8), $(1-D\rho)(t_o) = 0$ with $\dot{D}$ given as in Eq. (5.6.1) [provided that $\rho_{\mu\nu}^R$ in (5.6.1) equals to $(\rho_R)_{\mu\nu}$ in Eq. (5.4.8) as assumed henceforth in this section]. Thus, $I_m(t-t_o) = 0$. For the sake of simplicity, we will assume this situation here.

Applying these assumptions, two important facts follow from Eq. (5.6.9) using $I_m = 0$. For very short t, $(t \to t_o+)$, $\partial P_m(t)/\partial t \to 0$. Thus, unlike, e.g., the Pauli master equations, $P_m(t)$ starts to develop as

$$P_m(t) \approx P_m(t_o) + \text{const.}(t-t_o)^2, \quad t \to t_o+. \tag{5.6.11}$$

The reader should be warned, however, that this quadratic regime may appear at much shorter time intervals than, e.g., 1 ps. Further, assuming the exciton (or carrier) initially localized at a molecule located at the origin, i.e., $P_m(t_o) = \delta_{m0}$, from Eq. (5.6.3) in a three-dimensional molecular crystal we have

$$\frac{\partial}{\partial t} \Delta(t) \equiv \frac{\partial}{\partial t} \sum_m \mathbf{r}_m^2 P_m(t) = \sum_{mn} \int_{t_o}^t \mathbf{r}_m^2 w_{mn}(t-\tau) P_n(\tau) d\tau$$

$$= \int_{t_o}^t \sum_{mn} [(\mathbf{r}_m - \mathbf{r}_n)^2 + 2\mathbf{r}_n(\mathbf{r}_m - \mathbf{r}_n) + \mathbf{r}_n^2] w_{mn}(t-\tau) P_n(\tau) d\tau.$$

$$\tag{5.6.12}$$

(Here $\mathbf{r}_m$ is position of the mth molecule.) Using Eq. (5.6.6a) and the fact that in a perfect crystal (owing to periodicity)

$$\sum_m (\mathbf{r}_m - \mathbf{r}_n) w_{mn}(t-\tau) = 0 \tag{5.6.13}$$

and $\sum_m (\mathbf{r}_m - \mathbf{r}_n)^2 w_{mn}(t-\tau)$ is independent of n, we have from Eq. (5.6.12) for the mean square exciton displacement $\Delta(t)$:

$$\frac{\partial}{\partial t} \Delta(t) = \int_0^{t-t_o} \sum_m (\mathbf{r}_m - \mathbf{r}_n)^2 w_{mn}(\tau) d\tau. \tag{5.6.14}$$

So for very small time intervals, $\Delta(t)$ behaves as $(t-t_o)^2$ but, for very long times, the situation is not so simple.

Let us first take the (rather academic, but trivial) case of zero coupling to the reservoir (vibrations). Then, for a linear chain with just the nearest-neighbor transfer (hopping or resonance) integral J

$$w_{mn}(\tau) = \frac{1}{\tau} \frac{d}{d\tau} J_{|m-n|}^2(2J\tau/\hbar), \tag{5.6.15}$$

where $J_m(x)$ is the Bessel function. This exact result was first obtained by Sokolov[364] (from the known solution of the Schrödinger equation in this case) and

Reineker and Kühne.[200,298] Then $w_{mn}(\tau)$ itself is integrable to infinity as indicated in Eq. (5.6.14) but the sum over sites

$$\sum_m (\mathbf{r}_m - \mathbf{r}_n)^2 w_{mn}(\tau) = 2 \sum_{m=1}^{+\infty} a^2 m^2 \frac{1}{\tau} \frac{d}{d\tau} J_m^2(2J\tau/\hbar) = 4a^2 J^2/\hbar^2$$

(5.6.16)

is not integrable to infinity at all. In this case, therefore, $\Delta(t) = 2a^2 t^2 J^2/\hbar^2$ for any t [in full correspondence with Eq. (5.2.11)], the linear dependence $\Delta(t) \sim t$ never appears and one cannot correspondingly introduce the diffusion constant. This behavior is, however, typical of just the coherent regime. For arbitrarily small coupling to any realistic reservoir with dispersion, not only $w_{mn}(\tau)$ but also the sum $\sum_m (\mathbf{r}_m - \mathbf{r}_n)^2 w_{mn}(\tau)$ becomes integrable to infinity, i.e., $\Delta(t) \sim t$ for sufficiently large t and the diffusion constant D

$$D = \frac{1}{2d} \lim_{t \to +\infty} \frac{\Delta(t)}{t}$$

(5.6.17)

can be introduced (here $d = 1$, 2, or 3 is the dimension).

Let us assume that the exciton (or charge carrier) is slow as compared to the rate of decay of the memory functions $w_{mn}(t)$. (In reality, this assumption is usually fulfilled well only just for triplet excitons.) This should be connected with smallness of the hopping (transfer on resonance) integrals J_{rs} in our Hamiltonian H_0 in Eq. (5.3.9), i.e., with that of the unrenormalized band width. So the problem immediately turns to which is at least the lowest-order perturbation (in powers of J_{rs}) result for $w_{mn}(\tau)$ and which are the corresponding higher-order corrections. Before treating this problem, notice that for the above case of a slow dynamics, Eq. (5.6.9) (with $I_m = 0$) may be approximated well by the Pauli master equations (provided that t is not very small)

$$\frac{\partial}{\partial t} P_m(t) \approx \sum_{n(\neq m)} [W_{mn} P_n(t) - W_{nm} P_m(t)],$$

(5.6.18a)

$$W_{mn} = \int_0^{+\infty} w_{mn}(\tau) d\tau.$$

(5.6.18b)

Equations (5.6.18a, b) are the Markov approximation to Eq. (5.6.9) which may, for the slow dynamics, be applicable for $t \gg t_D$ (= the decay time of the memory functions). On the other hand, it may in principle happen that such an approximation becomes meaningless when integrals in Eq. (5.6.18b) become zero.[70] Moreover, it was shown that even if this is not the case, the condition $t \gg t_D$ (i.e., the macroscopic time scale) itself is in some cases (of not very slow dynamics) still insufficient for justifying Eqs. (5.6.18a, b), in contrast to the so-called Balescu-Swenson theorem (see Ref. 169) stating the opposite. Nevertheless, even in such a situation, it may happen that for $t \gg t_D$, some effective Pauli master equation like (5.6.18a) would apply but the corresponding transition (hopping) rates (W_{mn}^{eff}) do not necessarily fulfil Eq. (5.6.18b).[19,70] Thus, W_{mn}^{eff} are then not necessarily connected with the lowest-order Golden rule as will be shown below. If we are not interested in determining the transition rates from first principles, this fact is hardly relevant. This observation then makes it meaningful to model the exciton (or carrier) dynamics by the Pauli master (-like) equations provided that the real transport is (on a given time-scale) sufficiently incoherent.

Now, let us return to the case of small resonance (hopping) integrals J_{rs}. Assume for simplicity the Hamiltonian

$$H = H_e + H_{\text{ph}} + H_{e-\text{ph}},$$

$$H_{\text{ph}} = \sum_k \hbar \omega_k b_k^\dagger b_k,$$

$$H_{e-\text{ph}} = \frac{1}{\sqrt{N}} \sum_{n,k} g_k \exp(i\mathbf{k}\mathbf{r}_n) \hbar \omega_k a_n^\dagger a_n (b_k + b_{-k}^\dagger), \qquad (5.6.19)$$

i.e., conditions of harmonic vibrations and local (in site indices) linear (in the lattice point displacements) exciton (or electron)-phonon coupling. With H_e being given in Eq. (5.4.1), we have

$$H = H_0 + \mathcal{H},$$

$$H_0 = \sum_r \epsilon_r a_r^\dagger a_r + H_{\text{ph}} + H_{e-\text{ph}},$$

$$\mathcal{H} = \sum_{r \neq s} J_{rs} a_r^\dagger a_s. \qquad (5.6.20)$$

Introducing

$$L_0 \ldots = \frac{1}{\hbar} [H_0, \ldots],$$

$$\mathcal{L} = \frac{1}{\hbar} [\mathcal{H}, \ldots] \qquad (5.6.21)$$

one can verify that, assuming $\rho_{\alpha\beta}^R = (\rho_R)_{\alpha\beta} = \delta_{\alpha\beta} p_\beta$,

$$DL_o = L_0 D = 0 \qquad (5.6.22)$$

which (in addition to $DLD = 0$) turns Eq. (5.6.3b) to the form

$$w_{mn}(t) = -\sum_{\mu\nu} [DL \exp(-i(1-D)Lt)(1-D)LD]_{m\mu,\,m\mu,\,n\nu,\,n\nu}P_\nu$$

$$= -\sum_{\mu\nu} [\mathcal{L} \exp(-iL_0 t)\mathcal{L}]_{m\mu,m\mu,n\nu,n\nu}P_\nu + O(J^3)$$

$$= \frac{1}{\hbar} \sum_{\mu\nu\alpha\beta} [\mathcal{H}_{m\mu,n\alpha}\mathcal{H}_{n\nu,m\beta}(\exp(-iL_0 t))_{n\alpha,m\mu,n\nu,m\beta}$$

$$+ \mathcal{H}_{n\beta,m\mu}\mathcal{H}_{m\alpha,n\nu}$$

$$\times(\exp(-iL_0 t))_{m\mu,n\beta,m\alpha,n\nu}]P_\nu + O(J^3), \quad m \neq n. \qquad (5.6.23)$$

This formula still cannot be made more explicit as long as $|m\mu\rangle$ are eigenstates of $\sum_r \epsilon_r a_r^\dagger a_r + H_{\mathrm{ph}} = H_0|_{g_{\mathbf{k}}=0}$, i.e.,

$$|m\mu\rangle = \prod_{\mathbf{k}} \left[\frac{1}{\sqrt{\mu_{\mathbf{k}}!}} (b_{\mathbf{k}}^\dagger)^{\mu_{\mathbf{k}}} \right] a_m^\dagger |vac\rangle,$$

$$\mu_{\mathbf{k}} = 0,1,2, \dots . \qquad (5.6.24)$$

In this case, the initial condition (5.4.8) with (5.6.10) means that the lattice is initially unrelaxed around the exciton (or carrier) standing in the lattice. The memory functions can be calculated only approximately; for a system of equivalent sites (molecules), expansion in powers of J_{rs} is not necessary and one obtains roughly

$$w_{mn}(t) \approx w_{mn}^{\mathrm{coh}}(t)\exp(-\alpha t). \qquad (5.6.25)$$

Here $w_{mn}^{\mathrm{coh}}(t)$ are the *coherent* memory functions, i.e., those for the given system but without any coupling to phonons ($g_{\mathbf{k}} = 0$).[62]

Validity of Eq. (5.6.25) is limited only to weak coupling to phonons. This seems to be of no use for the molecular crystals where the strong coupling to intramolecular vibrations leads to formation of nearly small polaron (in the solid-state language) states. Thus, usually, instead of Eq. (5.6.24), one takes

$$|m\mu\rangle = \prod_k \left[\frac{1}{\sqrt{\mu_{\mathbf{k}}!}} \left(b_{\mathbf{k}}^\dagger + \frac{1}{\sqrt{N}} g_{\mathbf{k}} \exp(i\mathbf{k}\mathbf{r}_m) \right)^{\mu_{\mathbf{k}}} \right]$$

$$\times \exp\left(\frac{1}{\sqrt{N}} \sum_{\mathbf{k}} g_{\mathbf{k}} \exp(i\mathbf{k}\mathbf{r}_m)(b_{\mathbf{k}} - b_{-\mathbf{k}}^\dagger) \right) a_m^\dagger |vac\rangle,$$

$$\mu_{\mathbf{k}} = 0,1,2 \dots . \qquad (5.6.26)$$

In such a case the initial condition (5.4.8) with (5.6.10) mean that the lattice is initially relaxed around the standing particle. Because states $|m\mu\rangle$ in Eq. (5.6.26) are eigenstates of H_0,

$$[\exp(-iL_0 t)]_{a\alpha, b\beta, c\gamma, d\delta} = \delta_{a\alpha, c\gamma}\delta_{b\beta, d\delta}\exp(-i(E_{a\alpha}-E_{b\beta})t/\hbar)$$
(5.6.27)

where

$$E_{m\mu} = \epsilon_m + \sum_k \hbar\omega_k\mu_k - \frac{1}{N}\sum_k \hbar\omega_k|g_k|^2$$
(5.6.28)

is the eigenenergy corresponding to $|m\mu\rangle$ in Eq. (5.6.26). Thus, to the lowest order in J_{rs}, Eq. (5.6.23) yields

$$w_{mn}(t) \approx \frac{2}{\hbar^2}\sum_{\mu\nu} |\mathcal{H}_{m\mu,n\nu}|^2\cos((E_{m\mu}-E_{n\nu})t/\hbar) = \frac{2J_{mn}^2}{\hbar^2}\,\mathrm{Re}\,\exp[h_{mn}(t)$$

$$-h_{mn}(0)],$$

$$h_{mn}(t) = \frac{2}{N}\sum_k |g_k|^2(1-\cos(\mathbf{k}(\mathbf{r}_m-\mathbf{r}_n)))\cdot[n_B(\hbar\omega_k)\exp(i\omega_k t)$$

$$+[1+n_B(\hbar\omega_k)]\exp(-i\omega_k t)],$$

$$n_B(z) = [\exp(\beta z)-1]^{-1}.$$
(5.6.29)

This is the result first obtained by Kenkre and Rahman.[169,175] [Here, one should mention that the second-order approximation like Eq. (5.6.9) is far less exact than the second-order approximation used above in connection with projector (5.4.4) — compare Eq. (5.6.29) with, e.g., Eq. (5.5.3) and ensuing discussion.] Before discussing physical consequences of Eq. (5.6.29), let us remember that, according to this formula, $w_{mn}(t)$ does not (even after the infinite reservoir limit) decay to zero when $t \to +\infty$. This would mean that, owing to (5.6.14), the exciton (or carrier) transport remains partly coherent at arbitrarily long times. This contradicts not only the stochastic Liouville equation or GSLEM results, but does not correspond even to a simple physical reasoning about role of dephasing and relaxation processes. Fortunately, these striking conclusions are due to the second-order approximation used in Eqs. (5.6.23) and (5.6.29). For equivalent sites ($\epsilon_m = 0$) and only the nearest neighbor hopping integral J, this may be shown explicitly taking a partial sum to the infinite order[22,60] which gives

$$w_{mn}(t) \approx \exp(-2\Gamma t)w_{mn}^{\mathrm{coh}}(t)|_{J_{mn}\to\tilde{J}_{mn}} + 2\frac{\tilde{J}_{mn}^2}{\hbar^2}[\mathrm{Re}\,\exp(h_{mn}(t))-1],$$
(5.6.30a)

where m, n is for the nearest neighbors and

$$\tilde{J}_{mn} = \begin{cases} \tilde{J} = J \exp[-h_{mn}(0)/2] & \text{for nearest neighboring molecules} \\ 0 & \text{otherwise} \end{cases}$$

$$2\Gamma = 2z(\tilde{J}/\hbar)\tau_0,$$

$$\tau_0 = \int_0^{+\infty} [\text{Re} \exp(h_{mn}(t)) - 1]. \tag{5.6.30b}$$

Here z is the coordination number (number of nearest neighbors to a given site). Thus the structure of the memory function is of a two-channel form: the first term on the right-hand side of Eq. (5.6.30a) corresponds to a quasicoherent propagation with small exponential damping, while the second one is due to phonon-assisted processes. We are still uncertain whether the connection between the integral of the second term in Eq. (5.6.30a) and the damping constant 2Γ of the first term [see Eq. (5.6.30b)] means anything physically deeper or not. Notice that in the second order in J, Eq. (5.6.30) reduces to (5.6.29) but $w_{mn}(t) \to 0$, as $t \to +\infty$ in Eq. (5.6.30), in contrast to Eq. (5.6.29).

5.7. RELATION BETWEEN THE KENKRE-KNOX AND THE FÖRSTER-DEXTER THEORIES

In 1974, Kenkre and Knox[171] introduced a turning point in activity in the exciton transport problem. The main reason was that it suggested a possibility of determining the memory functions (difficult to calculate theoretically; see above) from independent optical experiments. This was done specially for the case typical of solutions of molecules (in a liquid nonabsorbing in the spectral range of absorption of these molecules) or for the case of molecular crystals. The basic assumptions are:

1. One can write J_{rs}, i.e., transfer (hopping or resonance) exciton integrals as matrix elements of just the dipole-dipole interaction modified by the surrounding medium via the "solvent" refraction index n.

2. Intramolecular vibration modes are not appreciably influenced by the "solvent" effects.

3. For the memory functions, one can use the second-order formula (5.6.29) with **k** denoting mostly intramolecular vibrations.

Similarity of the matrix elements leads then to the following approximate formula

$$w_{mn}(t) \approx \frac{3[\ln 10]^2 \hbar^4 c^4}{4\pi^2 n^4 N' |\mathbf{r}_m - \mathbf{r}_n|^6} \int_{\Delta W = -\infty}^{+\infty} d(\Delta W/\hbar) \cos(\Delta W \cdot t/\hbar)$$

$$\cdot \int_0^{+\infty} dW \frac{\bar{A}(W - \Delta W/2)\bar{\epsilon}(W + \Delta W/2)}{(W - \Delta W/2)^3 (W + \Delta W/2)}. \tag{5.7.1}$$

Here c is the velocity of light, $N' = 6.02 \times 10^{20}$ is the total number of molecules per milimol, $\bar{A}(W)$ represents the fluorescence intensity at energy W, $[\bar{A}(W)dW$ is the number of quanta emitted per unit time per molecule in the energy range $dW]$ and $\epsilon(W)$ is the extinction coefficient representing the absorption spectrum. Using Levshin's law of mirror symmetry, Eq. (5.7.1) may also be turned to

$$w_{mn}(t) = \frac{3[\ln 10]^2 \hbar c^2}{4\pi^2 n^2 (N')^2 |\mathbf{r}_m - \mathbf{r}_n|^6} \int_{-\infty}^{+\infty} d(\Delta W/\hbar) \cos(\Delta W \cdot t/\hbar)$$

$$\cdot \int_0^{+\infty} dW \frac{\bar{\epsilon}(W + \Delta W/2)\bar{\epsilon}(2W_0 - W + \Delta W/2)}{(W + \Delta W/2)(2W_0 - W + \Delta W/2)} . \qquad (5.7.2)$$

For a check, one may try to substitute Eq. (5.7.1) in (5.6.18b) in an attempt to find the corresponding Markovian transfer rates. It is really remarkable that it yields

$$W_{mn} = \int_0^{+\infty} w_{mn}(t)dt = \frac{3[\ln 10]^2 \hbar c^2}{4\pi^2 n^2 (N')^2 |\mathbf{r}_m - \mathbf{r}_n|^6} \int_0^{+\infty} dW \frac{\bar{\epsilon}(W)\bar{\epsilon}(2W_0 - W)}{W(2W_0 - W)} . \qquad (5.7.3)$$

The resulting Eq. (5.7.3) is actually the original famous Förster formula for the transition rate of excitons between neighboring molecules, being proportional to the sixth power of the reciprocal distance $|\mathbf{r}_m - \mathbf{r}_n|$ between the molecules (see Refs. 90 and 108). This coincidence, certainly not fortuitous, encouraged many people to try to use Eqs. (5.7.1) and (5.7.2) to calculate the memory functions from the optical measurement data thus avoiding complicated first principle calculations (see Refs. 171 and 172). The results were surprising and encouraging for standard Markovian modeling of the excitation transfer based on the Pauli master equations (5.6.18a). Namely, decay times t_D of memory functions were found to be less or of the order of 10^{-13} s. In view of the above analysis of the role of dephasing, this result seems to be hardly acceptable in general so that some further analysis was needed. Nedbal,[268] using approximate (but partially summed to the infinite order) formula for the memory functions by Sokolov and Hizhnyakov,[365] calculated and compared both sides of Eq. (5.7.1). In the second order in the dipole-dipole coupling, he was able to verify Eq. (5.7.1) as either an asymptotic formula for $|\mathbf{r}_m - \mathbf{r}_n| \gg a$, the lattice constant, or for interaction with just intramolecular vibrations without any dispersion. Such a treatment is, however, not very convincing, since in this order, one obtains an infinitely narrow zero-phonon line in absorption as well as in emission and, correspondingly, nondecaying memory functions [as shown in connection with Eq. (5.6.29)]. Going to higher orders meant first of all to generalize the Lax[210] formula for optical absorption and emission spectra to include the natural broadening of optical lines. In case of sufficiently fast dephasing [decay of $h_{mn}(t)$ to zero when $t \to +\infty$ in Eq. (5.6.29)], the Kenkre-Knox relation (5.7.1, 5.7.2) was found valid (as an asymptotic formula for $|\mathbf{r}_m - \mathbf{r}_n| \gg a$ or for dispersionless intramolecular vibrations), provided that the quasicoherent channel [the first term on the right-hand side of Eq. (5.6.30a)] is negligible.[268] Unfortunately, this is not true in general. Thus, the Kenkre-Knox relation (5.7.1), however stimulating at first sight, may be taken only as a qualitative method of a rough estimation [in case of a completely negligible quasicoherent channel in Eq. (5.6.30a)] of the phonon-assisted (i.e., the second) term in the memory function (5.6.30a). Even for this term, the method applies only for long-distance transfers as far as the dispersion of the vibration (intramolecular, libration, optical, and acoustic) modes are kept finite. One should realize that experimental spectra may become smeared owing to a finite resolution which may make the right-hand sides of Eqs. (5.7.2) and (5.7.3) decay more rapidly than they should. Thus, the previous estimate $t_D \gtrsim 10^{-11}-10^{-12}$ s remains realistic.

5.8. RELATION TO THE CONTINUOUS-TIME-RANDOM-WALK (CTRW) METHOD

In the coherent regime, an exciton or charge carrier propagates as a plane wave, keeping phase coherence at arbitrarily large distances in space as well as on the time axis. Such a plane-wave (or Bloch-like wave in realistic molecular crystals) may be easily expressed via localized states (e.g., the Wannier states in case of the single-particle approximation). Thus, even in the coherent case, one might use the language of localized states, writing the GME for probabilities of finding the exciton (or carrier) at individual molecules. The coherent character of the motion (i.e., preserving the above-mentioned phase coherence) is then reflected in a pronounced *memory* with faint or even fully lacking decay of $\Sigma_m(\mathbf{r}_m-\mathbf{r}_n)^2 w_{mn}(t)$ [but not necessarily that of individual $w_{mn}(t)$] as a function of t [see Eq. (5.6.16) with (5.6.14)]. Thus, it is physically incomprehensible (and in the light of the above discussion physically unacceptable) that the coherent (or quasicoherent) transport of excitons or charge carriers could be well described as a succession of mutually uncorrelated hops with some distribution of pausing or waiting times between two succeeding hops. Formally, the exact proof of equivalence of GME and the latter (CTRW) description, however, exists (see Refs. 169, 173, and 244). Let us, therefore, have a look at it in more detail (we follow Ref. 61).

Using the retarded Fourier-Laplace transformation

$$f(t)\Theta(t) = \int_{-\infty}^{+\infty} f^{\omega+i\delta} \exp(-i(\omega+i\delta)t) \frac{d\omega}{2\pi}, \qquad (5.8.1a)$$

$$f^{\omega+i\delta} = \int_0^{+\infty} f(t)\exp(i(\omega+i\delta)t)dt \qquad (5.8.1b)$$

[$\Theta(t)$ being the Heaviside step function = 0, 1/2 or 1 for $t < 0$, $t = 0$ or $t > 0$, respectively], one can easily reduce Eq. (5.6.9) with $I_m = 0$ to

$$P_m^z = \frac{i}{z}\left[1 - \frac{\Sigma_{r(\neq m)} w_{rm}^z}{-iz + \Sigma_{r(\neq m)} w_{rm}^z}\right]P_m(t_0)\exp(izt_0)$$

$$+ \sum_{n(\neq m)} \frac{w_{mn}^z}{-iz + \Sigma_{r(\neq m)} w_{rm}^z} P_n^z,$$

$$z = \omega + i\delta \tag{5.8.2}$$

with w_{mn}^z and P_m^z being the transforms of $w_{mn}(t)$ and $P_m(t)$, respectively. Denoting

$$\psi_m^z = 1 + \frac{iz}{-iz + \Sigma_{r(\neq m)} w_{rm}^z}, \tag{5.8.3}$$

$$Q_{mn}^z = \frac{w_{mn}^z}{-iz + \Sigma_{r(\neq m)} w_{rm}^z} \tag{5.8.4}$$

and using the reciprocal transformation (5.8.1a), one turns Eq. (5.8.2) to

$$P_m(t) = \left[1 - \int_0^{t-t_0} \psi_m(\tau)d\tau\right]P_m(t_0) + \sum_{n(\neq m)} \int_0^{t-t_0} Q_{mn}(\tau)P_n(t-\tau)d\tau. \tag{5.8.5}$$

Let us interpret $\psi_m(\tau)d\tau$ as a probability [i.e., $\psi_m(\tau)$ is the probability density] that there has been a hop from site m to anywhere else with the time delay from interval $(\tau, \tau+d\tau)$ after the time at which the quasiparticle (exciton or charge carrier) has certainly been found at m; let $Q_{mn}(\tau)$ be the same quantity for hop $n \to m$. Then Eq. (5.8.5) is the total balance condition serving as a basis for the CTRW method. It is clear that if Eq. (5.8.5) were (in the above way) properly interpreted, this would be an indispensable basis for modeling the exciton or carrier dynamics.

Though formally correct, Eq. (5.8.5) cannot in general be interpreted as discussed above. Objections against this interpretation are as follows:

From the above interpretation, one easily deduces that

$$\sum_{m(\neq n)} Q_{mn}(\tau) = \psi_n(\tau) \tag{5.8.6a}$$

i.e.,

$$\sum_{m(\neq n)} Q_{mn}^z = \psi_n^z. \tag{5.8.6b}$$

From Eqs. (5.8.3) and (5.8.4) we see that Eq (5.8.6b) is true just when $\Sigma_{r(\neq m)} w_{rm}^z$ [i.e., also $\Sigma_{r(\neq m)} w_{rm}(t)$] is independent of m. Unfortunately, this is in general not true; it applies only for systems of equivalent sites (e.g., perfect molecular crystals made of identical molecules in physically equivalent positions). In this case (but not in general), both sides of Eq. (5.8.6a) become independent of n so that from Eq. (5.8.5), we obtain

$$\sum_m P_m(t) = 1 \qquad (5.8.7)$$

so far as $\Sigma_m P_m(t') = 1$ for any $t' \in (t_o, t)$ (number-of-quasiparticles conservation law). So, if at all, CTRW methods based on Eq. (5.8.5) are justified for systems of equivalent sites only.

The second objection against the above interpretation [i.e. CTRW methods based on Eq. (5.8.5)] is that both $\psi_m(t)$ and $Q_{mn}(t)$ must be non-negative. Namely this is the point which excludes any applicability of CTRW methods based on Eq. (5.8.5) (owing to its improper interpretation) for the coherent (or quasicoherent) regime, in accordance with the introductory discussion. The problem is that the inequalities

$$\psi_m(t) \geqslant 0, \qquad Q_{mn}(t) \geqslant 0 \qquad (5.8.8)$$

do not apply for every t in the coherent or quasicoherent regimes. For example, for a symmetric dimer with the memory function

$$w_{12}(t) = w_{21}(t) = 2 \frac{J^2}{\hbar^2} \exp(-2\Gamma t) \qquad (5.8.9)$$

[see, e.g., (5.5.3) for $\gamma_{12} = \gamma_{21} = 0$ and $\rho_{12}(0) = \rho_{21}(0) = 0$]

$$\psi_1(t) = \psi_2(t) = Q_{12}(t) = Q_{21}(t) = (J_0^2/\sqrt{J_0^2 - \Gamma^2})\exp(-\Gamma t)\sin(t\sqrt{J_0^2 - \Gamma^2}) \qquad (5.8.10a)$$

for $|J_0| > \Gamma$, or

$$\psi_1(t) = \psi_2(t) = Q_{12}(t) = Q_{21}(t) = (J_0^2/(\sqrt{\Gamma^2 - J_0^2})\exp(-\Gamma t)\sinh(t\sqrt{\Gamma^2 - J_0^2}) \qquad (5.8.10b)$$

for $|J_0| < \Gamma$, with $J_0 = J\sqrt{2}/\hbar$. Thus Eq. (5.8.8) does not apply for $\Gamma < |J_0|$. Other examples where Eq. (5.8.8) does not apply may be found in Ref. 61. Thus, CTRW methods are not in general equivalent to GME and are well applicable [as far as being based on Eq. (5.8.5) with the above interpretation] for systems of equivalent sites and, simultaneously, sufficiently incoherent transfer only, in contrast to the usual opinion.

5.9. DIFFUSIVITY OF EXCITONS AND CHARGE CARRIERS IN MOLECULAR CRYSTALS IN TERMS OF GME

Here we want to show some general results of the GME for the exciton and charge carrier diffusivity. Using the Einstein relations, these results can be then converted into those for the carrier mobility. We want to show here, for a model periodic (and, for simplicity, cubic) infinite molecular crystal with one molecule per elementary cell, that, as a result of cooperation and interplay of the coherent and incoherent mechanisms, the diffusivity is roughly

$$D \approx D_{\text{band}} + D_{\text{hop}}. \tag{5.9.1}$$

Here D_{band} and D_{hop} are the band and hopping contributions (both of them, however, being influenced by the presence of both mechanisms), respectively. Since D_{band} and D_{hop} have conspicuously different temperature dependence, Eq. (5.9.1) may help in the determination of the prevailing character of the exciton or charge carrier motion using experimentally determined temperature dependence of D. In this section, we mainly follow Refs. 64 and 66.

Let us start from the Hamiltonian

$$H = H_0 + \mathcal{H},$$

$$H_0 = \sum_{k} \hbar \omega_k b_k^\dagger b_k + \frac{1}{\sqrt{N}} \sum_{nk} g_k \exp(i\mathbf{k}\mathbf{r}_n) \hbar \omega_k a_n^\dagger a_n (b_k + b_{-k}^\dagger),$$

$$\mathcal{H} = \sum_{m \neq n} J_{mn} a_m^\dagger a_n + \frac{1}{\sqrt{N}} \sum_{m \neq n} \sum_{k} g_k^{mn} \hbar \omega_k a_m^\dagger a_n (b_k + b_{-k}^\dagger). \tag{5.9.2}$$

So we have included in $\mathcal{H}$ not only the site- (or molecule-) off-diagonal part of the exciton (or carrier) Hamiltonian but also the site-off-diagonal part of the linear exciton- (or carrier-) phonon coupling ignored so far. Here, again, the reader should be warned that for charge carriers no problems in deriving Eq. (5.9.2) exist (see Section 1B of Chapter 1), while for excitons, the second term in $\mathcal{H}$ exists just in the adiabatic basis. Since in condensed matter, the adiabatic approximation intimately connected with the adiabatic basis is probably incorrect (see Chapter 2), everything here (and in Ref. 66) that is connected with this term in $\mathcal{H}$ (site off-diagonal or nonlocal exciton-phonon coupling) should be taken with care in case of excitons. As for the site-diagonal part of the quasiparticle Hamiltonian with site-energies (i.e., $\sum_m \epsilon a_m^\dagger a_m$), this may be ignored because of the possibility of choosing zero-value of energy, i.e., $\epsilon = 0$.

Eigenstates of H_0 read as in Eq. (5.6.26); corresponding energies are according to Eq. (5.6.28) with $\epsilon_m = \epsilon = 0$.

Definition of the diffusion constant [see Eq. (5.6.17)] together with Eq. (5.6.14), gives

$$D = -\frac{1}{2d} \lim_{\omega + i\delta \to 0} \lim_{\mathbf{Q} \to 0} \frac{\partial^2}{\partial \mathbf{Q}^2} w(\mathbf{Q}, \omega + i\delta) \tag{5.9.3}$$

where

$$w(\mathbf{Q},\omega+i\delta) = \sum_n \exp(-i\mathbf{Q}(\mathbf{r}_m-\mathbf{r}_n)) \int_0^{+\infty} w_{mn}(t)\exp(i(\omega+i\delta)t)dt \tag{5.9.4}$$

is the space- (as well as time-) Fourier-transformed memory function. Using its definition, one obtains

$$w(\mathbf{Q},\omega+i\delta) = \frac{i}{\hbar N} \sum_{\mathbf{q}} \sum_{\pi\mu} [\mathcal{H}_{\mu\pi}(\mathbf{q}) - \mathcal{H}_{\mu\pi}(\mathbf{q}+\mathbf{Q})]S_{\pi\mu}(\mathbf{q}+\mathbf{Q},\mathbf{q};\omega+i\delta) \tag{5.9.5}$$

where $\mathcal{H}_{\mu\pi}$ and $S_{\mu\pi}(\mathbf{q}+\mathbf{Q},\mathbf{q};\omega+i\delta)$ are introduced by

$$\mathcal{H}_{p\pi,r\rho} = \exp\left[-i\sum_{\mathbf{q}} \mathbf{q}(\mathbf{r}_p\,\pi_{\mathbf{q}} - \mathbf{r}_r\,\rho_{\mathbf{q}})\right] \frac{1}{N} \sum_{\mathbf{k}} \mathcal{H}_{\pi,\rho}(\mathbf{k})\exp(i\mathbf{k}(\mathbf{r}_p-\mathbf{r}_r)) \tag{5.9.6}$$

and

$$S_{p\pi,\,r\rho;n}(\omega+i\delta)$$

$$\equiv \delta_{p\pi,\,r\rho}\,\delta_{pn} p_\pi - i\sum_{\nu} \int_0^{+\infty} dt[\exp(-i(1-D)Lt)]_{p\pi,\,r\rho,\,n\nu,\,n\nu}p_\nu$$

$$\times \exp(i(\omega+i\delta)t)$$

$$= \frac{1}{N^2} \sum_{\mathbf{k}_1\mathbf{k}_2} S_{\pi\rho}(\mathbf{k}_1,\mathbf{k}_2;\omega+i\delta)\exp(i\mathbf{k}_1(\mathbf{r}_p-\mathbf{r}_n) - i\mathbf{k}_2(\mathbf{r}_r-\mathbf{r}_n)). \tag{5.9.7}$$

Here, the projector D is given as in Eq. (5.6.1) with $\rho^R_{\mu\nu} = (\rho_R)_{\mu\nu} = \delta_{\mu\nu}, p_\nu$. [As in Section 5.6, here we assume the initial conditions such that the quasiparticle is initially localized at the origin and the initial reservoir density matrix is ρ_R, i.e., $(1-D)\rho(t_o) = 0$; thus, there is no initial condition term here.]

Detailed analysis shows that one should distinguish in Eq. (5.9.5) the role of the "phonon-diagonal" contributions ($\sim S_{\pi\pi}$) from that of the "phonon-off-diagonal" terms ($\sim S_{\pi\mu}$ for $\pi \neq \mu$).[64] The diagonal quantity $S_{\pi\pi}$ can be shown to fulfil an equation of the same type as the Boltzmann equation for the band transport in solids. Its treatment is quite complicated because of the inelastic character of the phonon scattering. If one can, irrespective of that, introduce the relaxation time approximation, this equation can be solved and the result yields according to Eqs. (5.9.3) and (5.9.5):

$$D_{\text{band}} \approx \frac{1}{d} \sum_{\mathbf{q}} \sum_{\pi} \tau_\pi(\mathbf{q})|v_{\pi\pi}(\mathbf{q})|^2\rho^{eq}_{\pi\pi}(\mathbf{q}), \tag{5.9.8}$$

where $\rho_{\pi\pi}^{eq}$ is the Fourier transformed (in space) diagonal (in phonon indices) element of the equilibrium exciton (carrier) density matrix

$$v_{\rho\pi}(\mathbf{q}) = \frac{1}{\hbar} \frac{\partial \mathcal{H}_{\rho\pi}(\mathbf{q})}{\partial \mathbf{q}} \qquad (5.9.9)$$

[i.e. $|v_{\pi\pi}(\mathbf{q})|^2$ is the squared Fourier transformed (in space) diagonal (in phonon indices) matrix element of velocity] and $\tau_\pi(\mathbf{q})$ is the above-mentioned relaxation time. Clearly, because $\mathcal{H}_{\pi\pi}(\mathbf{q})$ strongly depends on π, one can easily recognize the influence of the effective polaron band narrowing (with increasing temperature) on dominating diagonal elements of the exciton- or carrier-polaron velocity. This fact then supports a tendency of a sharp increase of the D_{band} with decreasing temperature; the main reason for this tendency is, however, a sharp increase of $\tau_\pi(\mathbf{q})$.[66] This behavior as well as the structural similarity supports our interpretation of Eq. (5.9.8) as the band contribution to the diffusivity.

The off-diagonal terms $S_{\pi\mu}(\pi \neq \mu)$ in Eq. (5.9.5) yield in Eq. (5.9.3)

$$D_{\text{hop}} \approx \frac{\pi\hbar}{d} \sum_{\pi \neq \lambda} \sum_{\mathbf{q}} |v_{\pi\lambda}(\mathbf{q})|^2 \frac{1}{2} [\rho_{\pi\pi}^{eq}(\mathbf{q}) + \rho_{\lambda\lambda}^{eq}(\mathbf{q})]$$

$$\cdot \delta[-E_\pi + E_\lambda - \mathcal{H}_{\pi\pi}(\mathbf{q}) + \mathcal{H}_{\lambda\lambda}(\mathbf{q}) | \gamma_{\pi\lambda}(\mathbf{q})] \qquad (5.9.10)$$

where

$$\delta[x|\epsilon] = \frac{\hbar}{\pi} \frac{\text{Im } \epsilon}{(x + \hbar \text{ Re } \epsilon)^2 + (\hbar \text{ Im } \epsilon)^2} \to \delta(x), \quad \epsilon \to 0 \qquad (5.9.11)$$

is a broadened δ-function yielding [in Eq. (5.9.10)] an approximate form of the energy conservation law. Energies E_μ in Eq. (5.9.10) are those in Eq. (5.6.28) with all $\epsilon_m = 0$; $\mathcal{H}_{\mu\mu}(\mathbf{q})$ then determines the bandwidth (corresponding to the phonon population μ). For the broadening parameters $\gamma_{\pi\rho}(\mathbf{q})$ (being connected with the quasiparticle lifetime in plane-wave states) we refer the reader to Ref. 64. One can find that D_{hop} in Eq. (5.9.10) is little temperature-dependent or increases with increasing temperature (even possibly in an activated manner) in a limited temperature interval. As an example, let us mention that for the nearest neighbor interaction and a Gaussian density of phonon states around ω_0, one obtains[66,329]

$$D_{\text{hop}} \approx \text{const} \cdot \sinh(\beta\hbar\omega_0/2)\exp[-2g^2 \tanh(\beta\hbar\omega_0/4)] \qquad (5.9.12)$$

where g is an average coupling constant. Details of the temperature dependence for both contributions in $\mathcal{H}$ in Eq. (5.9.2) may be found in Ref. 66. Correction terms to Eq. (5.9.1) are practically negligible and may be identified (if at all) in just a limited temperature interval in the vicinity of the critical temperature at which the D_{band} (dominating at low T and strongly decreasing with increasing T) starts to be outweighed (with increasing temperature) by D_{hop}. This type of behavior corresponds to both experimental data and other theoretical calculations (see, e.g., Refs.

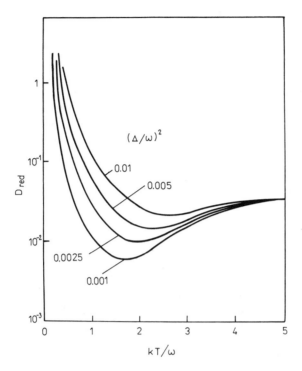

FIGURE 5.1. Calculated curves of reduced diffusion coefficient D_{red} of excitons in a molecular crystal as dependent on parameter kT/ω (where ω is the phonon frequency) according to Ref. 257; $D_{red} = D/(2a^2J^2/\Delta)$, where J and Δ are the width of exciton and phonon bands, respectively; a is the lattice parameter. Calculations have been performed for different values of the ratio $(\Delta/\omega)^2$ at the value of parameter $\gamma^2 = 0.01$, where γ is a function of the ratio g/ω, g being the exciton-phonon coupling constant.

106, 329, and 371 and Fig. 5.1). Before comparing with experiment, however, one should realize that here we have treated the cubic model crystal. If the crystal symmetry is highly anisotropic, one can easily obtain (in a limited temperature interval), e.g., prevailing band diffusion (D_{band}) in one direction and, simultaneously, prevailing hopping contribution (D_{hop}) in another one. Because of too many uncertain parameters in the problem and several other approximations not mentioned here (like omission of the quadratic exciton- or carrier-phonon coupling), one still cannot insist on the quantitative agreement of theory with experiment.

5.10. GME AND PARTITIONING PROJECTORS; TRAPPING, SINKS, AND FINITE-LIFETIME EFFECTS

In principle, one can use the GME theories discussed in Sections 5.4 and 5.6 even to nonperiodic problems though the lack of symmetry causes some technical complications. In this way, one can also add one further state of the system, in particular, the state without the exciton or a deep trapping state, to the states used above numbered by the Latin index denoting location of the exciton (or carrier) in the lattice. Hence, using the formalism developed so far, one can easily model finite-lifetime as well as trapping effects. Unfortunately, this method brings additional technical complications; so it is not commonly used.

Instead of that, it is popular to work with the original system (as if there were no finite-lifetime or trapping effects) and to include these effects by additional terms to the corresponding form of GME. For example, Eq. (5.6.9) then reads

$$\frac{\partial}{\partial t} P_m(t) = \sum_{n(\neq m)} \int_{t_o}^{t} [w_{mn}(t-\tau)P_n(\tau) - w_{nm}(t-\tau)P_m(\tau)]d\tau + I_m(t-t_o)$$

$$-\frac{1}{\tau_o} P_o(t)\delta_{m0} \qquad (5.10.1)$$

assuming that site zero is either connected with a trap to which the exciton (or carrier) located at this site can escape from the system with the escape rate τ_o^{-1}, or this is a state with a lowered exciton lifetime (τ_o). The channel via which the exciton or carrier may thus escape from the system, may be called sink and Eq. (5.10.1) is then the so-called sink model. It is very easy to generalize Eq. (5.10.1) to a greater number of sinks which may be connected with any site (molecule) in the system, with possibly different sink parameters τ_o^{-1}. One should realize that, in accordance with the physics of such processes, Eq. (5.10.1) implies

$$\frac{\partial}{\partial t} \sum_m P_m(t) = -\frac{1}{\tau_o} P_o(t) \leqslant 0 \qquad (5.10.2)$$

as far as $\tau_o > 0$ and $P_m(t) \geqslant 0$. So, the total probability of finding the exciton (or carrier) in the system is not only nonconserving but it should always monotonously decrease. This should be the basic property of the sink model which might be well accepted in many situations. However, according to Eq. (5.10.2), the total probability of finding the exciton (or charge carrier) in the system goes to zero when $t \to +\infty$ which contradicts (at any finite temperature) the equilibrium statistical physics (although at the room temperature, the probability of finding a thermally activated exciton is exceedingly small). Before going to another (and the most relevant) objection, let us mention that in calculating experimental integral quantities like, e.g., the quantum yield, the sink model has been found to be very useful (see, e.g., Ref. 169 and the literature quoted therein). The serious problem is, however, that the sink model yields probabilities which are, in the vicinity of the coherent regime, not necessarily non-negative for any t.[68] [Thus, the inequality (5.10.2) may be also disturbed.]

To understand the problem physically, let us mention that such a simple inclusion of the sink effect as in Eq. (5.10.1), with the memory functions being unperturbed by the sink, is equivalent, in the GSLEM (see Section 5.4), to inclusion (adding) the term $-\delta_{m0}\delta_{n0}P_o(t)/\tau_o$ to the right-hand side of Eq. (5.3.17a). [Following then the procedure leading to Eq. (5.5.3) gives, as compared to Eq. (5.5.3) having the form of (5.6.9), just the additional term $-\delta_{m0}P_o(t)/\tau_o$ as in Eq. (5.10.1).] So we would get no influence of the sink on the off-diagonal elements of the density matrix — the effect which is known to be incorrect from conversion of the Schrödinger equation with a non-hermitian ("optical") potential (describing decay of particles) to the effective Liouville equation. This shows that proper inclusion of the sink effect should be connected, at least, by adding the term

$$-\frac{1}{2\tau_o}[\delta_{m0} + \delta_{n0}]\rho_{mn} \tag{5.10.3}$$

to the right-hand side of Eq. (5.3.17a). (Here, we consider the situation when the sink is connected only with site 0; possible and the most direct generalization to any greater number of sites connected with such sinks means taking a sum of analogous terms.) Numerical studies then confirm that the probabilities $P_m(t) = \rho_{mm}(t)$ of finding the exciton (or carrier) at individual sites then remain non-negative.[59,69] Excluding the off-diagonal elements [in the same way as in Eq. (5.5.3)] shows that inclusion of the sink effect to the GME for probabilities only [like Eq. (5.10.1)] must consist of two steps:

1. Inclusion of the additional sink term as in Eq. (5.10.1)

2. Changing the memory functions in the vicinity of the sink[68]

This second effect (suppression of the memory functions) then leads to an interesting behavior of the quasiparticle (exciton, carrier) that it tries to avoid (at least partly) the sites (molecules) connected to the sink (i.e., places where it could be annihilated or captured) as far as the sink parameter $\gamma_o = 1/(2\tau_o)$ exceeds a certain critical value (so-called "fear-of-death" effect). Increasing the sink parameter γ_o to infinity then dynamically splits off completely those molecules that are connected to the sink. On the other hand, any change of the memory functions in the vicinity of the sink is very unpleasant from the technical point of view (by disturbing, e.g., the periodicity). Thus, rigorous studies of the sink effect are connected rather with adding a term (5.10.3) to the right-hand side of Eq. (5.3.17a) derived in terms of the GSLEM.

Rigorous inclusion of the sink effect which would not lead to the above contradiction with the equilibrium statistical physics at finite temperatures is possible using so-called partitioning projectors. Except for basic ideas (see Ref. 58) no detailed theory has yet been published. The task becomes relatively easy, provided that the transitions to the sink as well as back coexist and are only incoherent. For illustration, let us assume that the system is represented by just one molecule and the transitions to the sink and back may be represented (in the way analogous to the GSLEM as above) by the Haken-Strobl-Reineker parameters $\gamma_{1s} \equiv \gamma_{1 \leftarrow s}$ and $\gamma_{s1} \equiv \gamma_{s \leftarrow 1}(\neq \gamma_{1s})$. Then the sink analogy of Eq. (5.3.17a) reads

$$\frac{\partial}{\partial t} \rho_{11}(t) = -2\gamma_{s1}\rho_{11}(t) + 4\gamma_{s1}\gamma_{1s} \int_0^t \exp(-2\gamma_{1s}(t-\tau))\rho_{11}(\tau)d\tau,$$

$$\rho_{11}(0) = 1. \tag{5.10.4}$$

If the back-transitions from the sink are forbidden, $\gamma_{1s} = 0$ and Eq. (5.10.4) reduces to a special case of Eq. (5.10.1), yielding $\rho_{11}(t) \to 0$, $t \to +\infty$. For arbitrarily small but positive γ_{1s}, the second term on the right-hand side of Eq. (5.10.4) is negligible for $t \ll \gamma_{1s}^{-1}$; for other cases, one must resort to a general solution. Equation (5.10.4) yields

$$\rho_{11}(t) = [\gamma_{1s} + \gamma_{s1} \exp(-2(\gamma_{1s} + \gamma_{s1})t)]/(\gamma_{1s} + \gamma_{s1}) \to \frac{\gamma_{1s}}{\gamma_{1s} + \gamma_{s1}}, \quad t \to +\infty$$
$$\tag{5.10.5}$$

which is not in a contradiction with (and, for γ_{1s} and γ_{s1} fulfilling the detailed balance condition, it is even in a full correspondence with) the equilibrium statistical physics at any finite temperature. It is worth mentioning that Eq. (5.10.4) can be also rewritten in the GME form

$$\frac{\partial}{\partial t} \rho_{11}(t) = \int_0^t w(t-\tau)\rho_{11}(\tau)d\tau, \tag{5.10.6a}$$

$$w(t-\tau) = -2\gamma_{s1}\delta(t-\tau-0+) + 4\gamma_{1s}\gamma_{s1}\exp(-2\gamma_{1s}(t-\tau)). \tag{5.10.6b}$$

The fact that $\rho_{11}(t)$ tends to $\rho_{11}(+\infty) \neq 0$ is connected with the fact that our memory function $w(t)$ is singular in the sense

$$\int_0^{+\infty} w(t)dt + \infty = 0. \tag{5.10.7}$$

This property then makes illusory any attempt to describe realistic sink effects at $T \neq 0$ by a Markovian approximation to GME. Very small values of γ_{1s} (as compared to γ_{s1}) then lead to a problem with the long-time memory. That is why one usually limits the sink effect treatments to either a sufficiently incoherent regime [using then, e.g., Eq. (5.10.1)] or to any regime [using Eq. (5.3.17a) with the above mentioned correction (5.10.3)], assuming in both cases that transitions from the sink back to the system are practically negligible.

5.11. COHERENCE AND QUANTUM YIELD OF EXCITON TRANSFER

Quantum yield of exciton transfer is one of experimentally accessible quantities in molecular crystals which are relatively simple to treat theoretically. It is introduced as the ratio of the number of photons collected from the guest (radiating) molecules (exciton acceptors) to that of photons absorbed by the host molecules (exciton donors). The greatest advantage for theoreticians is that this is an integral quantity (the site occupation probabilities, if they enter at intermediate stages, are under time integrals).

Most of the exciton quantum yield and trapping theory based on GME is due to Kenkre, usually using the simplest version of the sink model like Eq. (5.10.1). In addition to Ref. 169, basic ideas may be found in, e.g., Refs. 174, 177, and 178. For application to experiments, mostly in molecular crystals, and for some typical experiments related to the exciton diffusion and annihilation in molecular crystals see Refs. 168 and 176. For influence of initial conditions on the quantum yield in sensitized luminescence see Ref. 361. Overall agreement with experiment is still rather qualitative but fairly good.

5.12. METHODS OF SOLUTION OF TIME-CONVOLUTION GME

Except for a few academic cases (with mostly idealized or in the absence of influence of the thermodynamic reservoir) where GME (of any type) might be analytically soluble, one must usually resort to approximation.

Solution of the convolution integro-differential equations [like Eqs. (5.4.16) or (5.6.9)] using, in the time domain, step-by-step integrations becomes very soon (with increasing time) impractical. A better method is to use the Laplace-Fourier transformation applying a theorem asserting that the Laplace (Fourier) transform of a convolution is a product of the corresponding transforms. Thus, in the Laplace or Fourier variables, the time integrodifferential equations reduce to a set of algebraic equations which can be readily solved. The problem (usually soluble by numerical methods which might become inaccurate with increasing time) is, however, the inverse Laplace or Fourier transformation.

The third and very useful (though little known) method is that one based on expansion of the memory functions into generalized Bessel functions. The trick is that convolution as well as derivatives of such functions are simply and analytically expressable in terms of the same functions so that the time-convolution GME turns to an algebraic set of equations for expansion coefficients. Details may be found in, e.g., Refs. 359 and 360.

Instead of solving the time-convolution GME, we have found it much more useful (for at least finite systems of molecules) to solve the set of differential equations of the type (5.4.17) (with possible inclusion of the sink effects — see Section 5.10) or, in the Haken-Strobl-Reineker parameterization (but still within GSLEM) [see Eq. (5.3.17a)]. This idea is physically based on the observation that (on excluding the site-off-diagonal elements of ρ) the problem is mathematically equivalent to that of the integrodifferential equations but with an appreciably increased accuracy as compared to the simple expansion of the memories in Eq. (5.6.9) [see, e.g., Eq. (5.5.3)].

Systems like Eqs. (5.4.17) [or (5.3.17a)] may be solved using standard methods for systems of differential equations. Another (and, in our opinion, more useful) method is to express the solution to, e.g., Eq. (5.4.17) in the matrix form as

$$\rho(t) = \exp(A(t-t_o))\rho(t_o), \tag{5.12.1}$$

where A is the matrix of the right-hand side of Eq. (5.4.17), i.e.,

$$A_{mnpq} = W_{mnpq} - i/\hbar[(H_e)_{mp}\,\delta_{nq} - (H_e)_{qn}\,\delta_{mp}]. \tag{5.12.2}$$

The matrix exponential

$$B \equiv \exp(A.\Delta t) = (\exp(A^{\Delta t}/_n))^n \tag{5.12.3}$$

may be then calculated numerically for such a large n that

$$\|A^{\Delta t}/_n\| < 1. \tag{5.12.4}$$

This approach then permits a polynomial expansion of the exponential on the right-hand side of Eq. (5.12.3). Having then calculated B for a given time step Δt, we may then use the recurrent relation

$$\rho(t_o + (m+1)\Delta t) = B\rho(t_o + m\Delta t), \quad m = 0, 1, 2, \ldots, \tag{5.12.5}$$

to obtain $\rho(t)$ in a sufficiently dense manifold of points. The interested reader is referred to Ref. 243a.

5.13. GROVER AND SILBEY AND MUNN AND SILBEY THEORIES

The theory of Grover and Silbey[121] is one of the earlier microscopic and, simultaneously, advanced theories of the exciton migration in molecular crystals. Correspondence of this approach with the stochastic Liouville equation model and the theory by Kenkre and Knox has been discussed in Refs. 165 and 166. Some more attention will be devoted to this approach here.

First, one should notice that Grover and Silbey involved the canonical transformation to the small (exciton or carrier) polaron picture. Their quantity

$$\mathscr{G}_{nm}(t) = \langle 0|A_0\langle A_n^\dagger(t)A_m(t)\rangle A_0^\dagger|0\rangle \tag{5.13.1a}$$

[here $|0\rangle$ denotes the electronic as well as vibrational ground state, $A_n^\dagger$ creates the excitonic (or carrier) polaron at the nth molecule, while $\langle\ldots\rangle$ denotes a thermal phonon average, see Ref. 121] is, according to general opinion, directly related to our exciton density matrix $\rho_{mn}(t)$ in the nonpolaronic basis. Using the notation of Ref. 121 and the fact that, with their S, $\exp(-S|0\rangle = |0\rangle$, one obtains

$$G_{nm}(t) \approx Tr(A_0^\dagger |0\rangle\langle 0| A_0 \exp(i\tilde{\mathcal{H}}t/\hbar) A_n^\dagger A_m \exp(-i\tilde{\mathcal{H}}t/\hbar))$$

$$= Tr(\exp(-S)a_0^\dagger |0\rangle\langle 0| a_0 \exp(S)\exp(-S)\exp(i\mathcal{H}t/\hbar)a_n^\dagger$$

$$\cdot\ a_m \exp(-i\mathcal{H}t/\hbar)\exp(S))$$

$$= Tr(\rho(0)\exp(i\mathcal{H}t/\hbar)a_n^\dagger a_m \exp(-i\mathcal{H}t/\hbar)) = \rho_{mn}(t). \quad (5.13.1b)$$

Grover and Silbey, however, use the time development of the "dressed" exciton operators given by the original Hamiltonian $\mathcal{H}$ {Eq. (13) of Ref. 121 instead of the transformed Hamiltonian $\tilde{\mathcal{H}}$ [Eq. (13) of Ref. 121]} as formally used in (5.13.1b). Hence, they do not in fact apply the polaron canonical transformation though they work with dressed exciton operators. Thus, our equality (5.13.1b) does not apply, i.e., in particular $G_{nm}(t) \neq \rho_{mn}(t)$ for $m \neq n$. In other words, one should stress that instead of the usual unclothed (nonpolaronic, i.e., bare) exciton density matrix $\rho_{mn}(t)$, Grover and Silbey work with the dressed (i.e., polaron) density matrix $G_{nm}(t)$. This partly explains all formal differences between Eqs. (5.3.17a) and (5.13.2) below. On the other hand, though one usually assumes that "the species which is migrating (at least on a long time scale) through the crystal is the clothed exciton,"[121] description of its dynamics in terms of the unclothed (non-polaronic, i.e., bare) states [i.e., via $\rho_{mn}(t)$ as in previous sections] is equally good unless approximations become involved. This is because $G_{mm}(t) = \rho_{mm}(t)$ has the meaning of probability of finding the (clothed, i.e., dressed, as well as unclothed, i.e., bare) exciton at site m.

The second reason for formal differences, though possibly of minor importance, are different initial conditions. Grover and Silbey use the initial condition with an exciton polaron standing initially in an otherwise relaxed lattice at, e.g., site zero. This, however, should have little impact on diffusivity results as far as the clothing process of initially unrelaxed exciton takes place within $\sim 10^{-12}$ s.[121]

The equations at which Grover and Silbey arrive (after some manipulations and approximations, in particular the second-order approximation in the resonance hopping integrals renormalized by the canonical transformation to the small polaron picture) read for the perfect infinite molecular crystal

$$\frac{\partial}{\partial t} G_{nm}(t) \approx \frac{i}{\hbar} \sum_p [\tilde{J}_{n-p} G_{pm}(t) - \tilde{J}_{m-p} G_{np}(t)]$$

$$-\frac{1}{\hbar^2} \sum_{pr} [\tilde{J}_{r-p} \tilde{J}_{p-n} \gamma_{rppn}(t) G_{rm}(t) + \tilde{J}_{p-r} \tilde{J}_{m-p} \gamma_{prmp}(t) G_{nr}(t)$$

$$-\tilde{J}_{m-r} \tilde{J}_{p-n} \gamma_{mrpn}(t) G_{pr}(t) - \tilde{J}_{p-n} \tilde{J}_{m-r} \gamma_{pnmr}(t) G_{pr}(t)].$$

$$(5.13.2)$$

Here $\tilde{J}_{mn} = \tilde{J}_{m-n}$ is given as in Eq. (5.6.30a) and we explicitly use that, owing to the symmetry of the model, $J_{mn} = J_{m-n} = J_{n-m} = J_{nm}$. Finally, in Eq. (5.13.2) and in terms of the above notation

$$\gamma_{mnpq}(t) = \int_0^t \exp\left[\frac{1}{N}\sum_k |g_k|^2[(\exp(i\mathbf{kr}_m)-\exp(i\mathbf{kr}_n))(\exp(-i\mathbf{kr}_p)\right.$$

$$-\exp(-i\mathbf{kr}_q)) \cdot n_B(\hbar\omega_k)\exp(i\omega_k t)+(\exp(-i\mathbf{kr}_m)-\exp(-i\mathbf{kr}_n))$$

$$\left.\times(\exp(i\mathbf{kr}_p)-\exp(i\mathbf{kr}_q)) \cdot [1+n_B(\hbar\omega_k)]\exp(-i\omega_k t)]\right]d\tau.$$

$$(5.13.3)$$

In the narrow phonon band limit, i.e., in terms of Einstein's model of dispersionless phonons, which is a good approximation for optical intramolecular modes in molecular crystals, $\gamma_{mnpq} = 0$ except for the following cases:

1. Both $m = q$ and $n = p$, $\gamma_{mnnm}(t) = \gamma_1(t)$.
2. Either $m = q$ or $n = p$, $\gamma_{mnpm}(t) = \gamma_2(t)$.
3. Either $m = p$ or $n = q$, $\gamma_{mnmq}(t) = \gamma_3(t)$.
4. Both $n = p$ and $m = q$, $\gamma_{mnmn}(t) = \gamma_4(t)$.

With dominating $\gamma_1(t) = \gamma_4(t)$, (i.e., neglecting γ_2 and γ_3) it yields, with just the nearest-neighbor resonance integral J in one dimension

$$\frac{\partial}{\partial t}\mathcal{G}_{nm}(t) \approx \frac{i}{\hbar}\tilde{J}[\mathcal{G}_{n+1,m}(t)+\mathcal{G}_{n-1,m}(t)-\mathcal{G}_{n,m+1}(t)-\mathcal{G}_{n,m-1}(t)]$$

$$-2\frac{\tilde{J}^2}{\hbar^2}\gamma_1(t)[2\mathcal{G}_{nm}(t)-(\mathcal{G}_{n+1,m+1}(t)+\mathcal{G}_{n-1,m-1}(t))\delta_{nm}$$

$$-(\delta_{n,m+1}+\delta_{n,m-1})\mathcal{G}_{mn}(t)].\qquad(5.13.4)$$

For instance, at $T = 0$, $\gamma_1 = 0$ so that

$$\mathcal{G}_{nm}(t)|_{T=0} \approx I_n(2i\tilde{J}t/\hbar)I_m(-2i\tilde{J}t/\hbar),\qquad(5.13.5)$$

which corresponds to the coherent motion of the small exciton (or charge carrier) polaron.[235] Here I_m is the modified Bessel function. At $T \neq 0$ and long times, one obtains the linear time dependence ($\sim t$) of the mean square displacement $\Delta(t)$; the diffusion constant (5.6.17) (for $d = 1$) then reads

$$D = a^2\left(\frac{2}{3}\tilde{J}^2/\alpha+2\alpha\right),$$

with

$$\alpha = 2\tilde{J}^2\gamma_1(+\infty)/\hbar^2\qquad(5.13.6)$$

which again corresponds to two channels of the exciton motion. For other cases, we refer the reader to Ref. 121.

Let us return to Eq. (5.13.4) and let us try to compare it with Eq. (5.3.17a). Clearly, the first term in the square brackets on the right-hand side of Eq. (5.13.4) corresponds to the first term $-i/\hbar [H_0, \rho(t)]_{mn}$ in Eq. (5.3.17a) except that Eq. (5.13.4) contains $\tilde{J}$ instead of J. Similarly, the structure of the remaining terms in Eq. (5.13.4) corresponds to (5.3.17a) as if it were at $t \to +\infty$

$$\gamma_{n,\,n\pm 1} \approx \tilde{J}^2 \gamma_1 (+\infty)/\hbar^2,$$

$$\gamma_{n,\,n} \approx 0. \qquad (5.13.7)$$

These differences are certainly due to different schemes used; Eq. (5.3.17a) (and a similar equation for the GSLEM) is based on expansion in terms of the exciton-phonon coupling, i.e., is good when this coupling is weak enough. On the contrary, Eqs. (5.13.2) and (5.13.4) are based on expansion in powers of the clothed (i.e., renormalized owing to the exciton-phonon coupling) exciton resonance integrals (or exciton band width) so that the Grover and Silbey approach is good when this coupling is rather strong. Therefore the stochastic Liouville equation (or GSLEM) and the Grover and Silbey theory are two complementary approaches which are very difficult to compare because of different areas of applicability and quantities involved.

In Ref. 62a, an interpolation scheme was suggested between the Grover and Silbey approach and GSLEM illustrating that there is in fact no contradiction between these methods, and all the seeming differences are just due to different formal schemes used.

It is worth mentioning that the Grover and Silbey theory can be constructed also using the projection formalism leading to GME; for this case, the Argyres and Kelley projector[13] performing averaging over the phonon bath should be used (see Appendix B of Ref. 121). Comments about encountered technical problems (negative probabilities) resulting from application of the diagonalizing projector (see Peier [282]) are, however, likely to be based on a misunderstanding. The theory by Grover and Silbey is of the second order in the hopping (resonance) integrals, but, on excluding the off-diagonal elements $\mathcal{G}_{mn}$, $(m \neq n)$, it becomes [in the same way as for the analogous situation with coupling to the reservoir in Eq. (5.5.3)] partially summed up to the infinite order. Thus, any theory of the second order for only diagonal elements $\rho_{mm}(t) = \mathcal{G}_{mm}(t)$ resulting from application of the Peier[282] projector and the second-order approximation (as in Ref. 175) is worth an incomparable amount. So the negative probabilities mentioned in Ref. 121 must be ascribed to the appreciably worse (though formally of the same order) approximation. This must be so because no rigorous treatment can yield the negative probabilities $P_n(t)$ as far as it is an exact (or sufficiently exact) consequence of the Liouville (i.e., also of the Schrödinger) equation always leading to $P_n(t) \geq 0$.

From this point of view, accepting that the Grover and Silbey theory exceeds any second order (in hopping integrals) GME theory for probabilities $P_m(t)$, one might expect comparable results with those theories based on GME [see Eq. (5.6.9)] which have the memory kernels expanded in powers of Jrs and resummed partially to the infinite order [see Eqs. (5.6.30a, b)].[22,60] While still no detailed analysis of this correspondence exists, one can easily reveal at least related quantities and terminology. For example, in Eqs. (5.13.3) and (5.13.4), $\gamma_4(t) = \int_0^t [\exp(h_{mn}(\tau)) - 1]d\tau$ with $h_{mn}(t)$ from Eq. (5.6.29), i.e., Re $\gamma_1(+\infty) \approx$ Re $\gamma_4(+\infty) = \tau_o$ in Eq. (5.6.30b). Since the partially infinite-order GME theory[22,60] with Eq. (5.6.9) and memory kernels Eq. (5.6.30a, b) is expected to be at least analogous to the Grover and Silbey theory.[121] Both these approaches are believed to be applicable, at least when the migrating species (e.g., a "dressed" exciton, i.e., excitonic polaron) moves almost incoherently. Finally, due to the fact that for periodic crystals a correspondence with the CTRW approach may be well established for the GME theory (except at the vicinity of the coherent regime), one can expect equivalent results stemming from all these approaches under the above conditions. For such a comparative study of different approaches see, e.g., Ref. 378. Another comment regards the fact that, though the Grover and Silbey theory has been constructed for excitons, it may be readily translated to the language of charge carriers in molecular crystals. Then a close correspondence may be found with the old and detailed theory of small polarons by Firsov and collaborators (see, e.g., Ref. 106), going back to early sixties[208,209] and even with the small polaron theories by Holstein,[136] Yamashita and Kurosawa,[416] and Sewell.[324]

The method of canonical transformation extensively used by the Russian school[106] and reminding of the paper by Grover with Silbey[121] proved to be well applicable to the exciton and charge carrier migration in molecular crystals in a later series of papers by Silbey and Munn.[258,259,329] In the first paper of the series, only the site-diagonal linear coupling to phonons was used. Instead of calculating the diffusivity (and, via the Einstein relations, mobility) from the definition, a formula

$$D = \langle\langle v_{\mathbf{k}}^2 \Gamma_{\mathbf{kk}}^{-1} + \gamma_{\mathbf{kk}} \rangle\rangle \qquad (5.13.8)$$

for the diffusivity D was used. This result (derived originally by Yarkony and Silbey[417]) is not exact in the band limit (see Ref. 65), but may serve here as a very good and in general as an at least qualitatively correct interpolation formula involving band velocities $v_{\mathbf{k}}$, reciprocal scattering rate (i.e., the quasiparticle lifetime) $\Gamma_{\mathbf{kk}}^{-1}$ out of the plane wave state, and the hopping rate term $\gamma_{\mathbf{kk}}$. In Eq. (5.13.8), $\langle\langle \ldots \rangle\rangle$ means statistical averaging over the band states. Though the authors showed the applicability of this treatment without the above canonical approximation in the weak coupling limit (standard band-limit result $D \sim T^{-1}$ being obtained with scattering on acoustic phonons with the deformation potential coupling at low T), this transformation was used in general. For various relations between relevant parameters (exciton or charge carrier bandwidth B, typical phonon frequency ω_0, strength of the coupling to phonons g, and temperature in energy units $k_B T$), D was calculated from Eq. (5.13.8) for the phonon bandwidth $\Delta \ll \hbar\omega_0$. It was numerically very well verified (see Ref. 329). We observe that:

1. For any ratio $B/(\hbar\omega_0)$ and any g, D decreases roughly as T^{-m}, $m > 0$ with increasing T at low temperatures.

2. For $g^2 \ll 1$, no qualitative change of this temperature dependence appears at higher T.

3. For $g^2 \approx 0.3$ and $B/(\hbar\omega_0) \approx 1$, a plateau of almost-temperature-independent D appears at higher $T(k_B T > \hbar w_0)$.

4. For $g^2 \approx 1$, this plateau may be found in a broader interval of $B/(\hbar\omega_0)$ with a possible dip for $B/(\hbar\omega_0) \gtrsim 1$ developing between the low- and high-temperature regions.

5. For $g^2 \gg 1$, this dip (local minimum) develops further to yield a sharp increase followed by saturation for $k_B T \gtrsim \hbar\omega_0$; for $k_B T \ll \hbar\omega_0$, the previous sharp decrease with increasing T survives, but the breaking point T_c becomes shifted to lower temperatures.

Realizing that, in realistic anisotropic molecular crystals, two different regimes may apply at the same temperature in two different crystallographic directions (B may differ in different directions), this behavior is usually found to correspond very well to experiment.

In Refs. 258 and 259, an attempt was made to extend the theory to the nonlocal (i.e., site-off-diagonal) linear coupling to phonons. Here, as in the above material, the reader should be warned that for excitons, the nonlocal coupling appears in the adiabatic basis intimately connected with the adiabatic approximation. As shown above, the latest results by Wagner and coworkers[401–404] (see Chapter 2) indicate inapplicability of this approximation in at least some dynamic problems because of divergent corrections to adiabatic results. So for excitons, the results of Refs. 258 and 259 should be taken with care.

For charge carriers, however, Refs. 258 and 259 form a relevant contribution to the transport theory. In Ref. 258, the standard canonical (small polaron) transformation was generalized to treat the nonlocal coupling. Inclusion of the nonlocal coupling was found to further distort the band shape (leading possibly to new minima) and to further suppress (as compared to the local coupling case) the electron and hole bandwidths. Using this formalism, the transport was then treated in Ref. 259. It was found that the nonlocal coupling further suppresses the band contribution (to the diffusivity and mobility) and increases the hopping contribution. The main effect is the increase of magnitude but suppressing of the temperature dependence of the diffusivity. Relation to a sometimes observed temperature-insensitive mobility in molecular crystals (for a simple theory see also, e.g., Ref. 371) is, however, still rather uncertain (see also Chapter 7).

Numerical study of the above canonical transformation to solve self-consistently the problem of the small polaron states with nonzero but not very strong nonlocal carrier-phonon coupling appeared recently in Ref. 410a. Polaron energy shifts much larger and Debye-Waller factors much smaller than originally assumed were found. No direct consequence for transport has, however, been deduced so far.

In this connection, we should stress that so far we have treated only the linear coupling to phonons, for both excitons and charge carriers. In some cases (e.g., in case of out-of-plane vibrational modes in planelike molecules), the linear coupling disappears for reasons of symmetry, i.e., the quadratic coupling becomes inevitable. Such a coupling is sometimes inevitable in other cases, too. This may be, e.g., the case when local excitation (for of the exciton-phonon coupling) or localization of a charge carrier on a given molecule (for the electron- or hole-phonon coupling) leads to a change in frequency of the relevant (e.g., intramolecular) vibrational modes. In general, this coupling is known to enhance the tendency to localization. When it is strong, it might yield bound states which might complicate the transport theory appreciably. Relatively little is known about the real effect of this coupling on transport. The weak coupling theory (which may be well constructed in a way analogous to Refs. 258, 259, and 417 — see, e.g., Ref. 257) yields that the transport is again hopping-like at high temperatures with a gentle temperature dependence at low T while sharp decrease of the quasicoherent diffusivity with increasing T occurs[256,257] (see Fig. 5.1).

5.14. SOME OTHER METHODS AND TOPICS

We have discussed above GME theories based on the convolution-type (i.e. non-Markovian) Nakajima and Zwanzig identity [Eq. (5.3.6)] or its modifications. In practical calculations, this often forces one to use the Markov approximation since otherwise, the time-convolution equations are difficult to solve. In this connection, one should realize that there are also GME that directly, without approximations, lead to systems of convolutionless differential equations with time-dependent coefficients. A corresponding theory for excitons and charge carriers in molecular systems is, however, not well developed. Basic ideas and illustrations for applications may be found, e.g., in Ref. 69.

There is a practical semiempirical method how to avoid GMEs of any type. It is based on propagators $\Psi_{mn}(t)$ (i.e., probabilities that the exciton or carrier, being initially at site n, will be found at site m at $t \geq 0$). These propagators are usually simple to calculate in the limiting cases of the fully coherent as well as fully incoherent regimes. In the intermediate regime, a simple interpolation may be quite operational as shown by Skala and Kenkre.[362]

From the above theories, one can see that, at sufficiently high temperatures, the motion of excitons and charge carriers can be usually well described as a *diffusive migration* reminicent of the Brownian motion. Under such conditions, for modeling purposes, one may thus apply the usual textbook methods like the Langevin or Fokker-Planck equations. For thermalized or almost thermalized excitons and carriers and high enough temperatures, these methods thus need no further justification. In practice, on the other hand, they are phenomenologically used successfully even for hot, unthermalized carriers[344,348,355,356] (see Chapters 6 and 7). So this problem deserves some further attention.

5.15. HOT UNTHERMALIZED CHARGE CARRIERS AND THE FOKKER-PLANCK EQUATION

Currently, one derives the Fokker–Planck equation from the Bachelier-Smoluchowski-Chapman-Kolmogorov[244] chain equation for conditioned probabilities (or probability densities) assuming for the particle (the charge carrier henceforth) the validity of the Langevin equation. The above chain equation is, however, Markovian and the electron (or the molecular polaron) is regarded as a classical particle behaving in accordance with Markovian kinetics. In addition to that, in case of applications (see, e.g., Ref. 61) one needs this equation for a combined distribution function in the six-dimensional phase space of coordinates and momenta (velocities), i.e., the Wigner distribution function. Introduction of this distribution function must be done with some care, since coordinates and momenta are noncommuting variables. Designating this distribution function as $W(\mathbf{r},\mathbf{v},t)$, the Fokker–Planck equation (generalized to include an external electric field $\vec{\mathcal{E}}$ owing to, e.g., the Coulomb field of the parent positive ion acting on an escaping electron during photogeneration of the electron-hole pair) reads [see Eq. (6.41)]:

$$\frac{\partial}{\partial t} W + \mathbf{v}\,\frac{\partial W}{\partial \mathbf{r}} + \frac{e\vec{\mathcal{E}}}{m_{\mathrm{eff}}}\,\frac{\partial W}{\partial \mathbf{v}} = \beta\left[\frac{\partial}{\partial \mathbf{v}}\,(\mathbf{v}W) + \frac{k_B T}{m_{\mathrm{eff}}}\,\Delta_{\mathbf{v}}\,W\right]. \qquad (5.15.1)$$

Here, m_{eff} is the effective mass of the charge carrier (polaron) appearing (together with the friction constant β) in the starting Langevin equation

$$\frac{d\mathbf{v}}{dt} = -\beta\mathbf{v} + \frac{e\vec{\mathcal{E}}}{m_{\mathrm{eff}}} + \mathbf{A}(t). \qquad (5.15.2)$$

In Eq. (5.15.2) $\mathbf{A}(t)$ is the stochastic force per unit mass (owing to the charge carrier scattering with its surroundings) whose equilibrium correlation function $\Phi_{ij}(s) = \langle A_i(t+s)A_j(s)\rangle$, with assumed very strong decay around $s \approx 0$, must fulfill equation

$$\int \Phi_{ij}\,ds = \delta_{ij}2\beta\,\frac{k_B T}{m_{\mathrm{eff}}}. \qquad (5.15.3)$$

Requirement (5.15.3) is necessary for the equipartition theorem to be compatible with Eq. (5.15.2). The friction constant β entering the thermalized electron mobility

$$\mu = \frac{e}{m_{\mathrm{eff}}\,\beta} \qquad (5.15.4)$$

as well as the equilibrium diffusion constant (diffusivity)

$$D = \frac{k_B T}{e}\,\mu = \frac{k_B T}{m_{\mathrm{eff}}\,\beta} \qquad (5.15.5)$$

then enters Eq. (5.15.1) via Eqs. (5.15.2) and (5.15.3).

Although the situation is more favorable for the thermalized electrons, neither Eq. (5.15.2) nor Eq. (5.15.3), nor the equipartition theorem, nor the Markov character of the kinetics are well established for hot unthermalized electrons in molecular crystals. Thus, one must resort to more intricate arguments.

First, one should mention that Eq. (5.15.1) is sometimes called the generalized Fokker-Planck (or Fokker-Planck-Krammer) equation to indicate the **r**-dependence of W and the spatially inhomogeneous character of the problem owing to the external electric (or any other type of) force (see Section 6.3.1). There is a possibility of deriving some other type of generalized Fokker-Planck equation, e.g., generalization connected with inclusion of the memory effects.[119] One should emphasize that inclusion of such memory effects, e.g., in the case of photogeneration of free carriers in molecular crystals, is likely to be indispensable. The reason is that the originally hot unthermalized electron becomes thermalized on a distance of $r_{th} < r_c$ (see Section 6.2.2) owing to multiple losses of its kinetic energy due to inelastic scattering on lattice vibrations. However, during this thermalization stage, the mean square deviation (displacement variance) $\langle (r - \langle r \rangle)^2 \rangle (t)$ remains small as compared to $\langle r \rangle^2(t)$, i.e., the squared mean distance (the thermalization length) from the parent ion.[344,355,356] Thus, during the thermalization stage and irrespective of the multiple scattering, the electron (or molecular polaron) keeps going almost in the same direction (see Section 6.3.3 and Fig. 6.29). This, in the language of kinetic theories, is a reflection of the memory but, on the other hand, it is clearly incompatible with the idea of uncorrelated scattering events according to Eq. (5.15.2) except that the scattering is on light species unable to appreciably change the electron momentum in any individual scattering event.

Such behavior is possible only in the case of a heavy particle scattered on light ones, e.g., lattice phonons.

It should be emphasized that computer simulation of the charge carrier thermalization process in the framework of a modified Sano-Mozumder's model,[344] based on Fokker-Planck equations (see Section 6.3.3), requires introduction of heavy polaron-type particle (molecular polaron) at the very initial phase of the thermalization.

Thus according to this approach it is assumed that the relaxation of the hot charge carrier, i.e., "dressing" up the electron in a polarization cloud, occours in the time domain of the first few scattering events. The created heavy molecular polaron then moves in a quasicoherent fashion; its mean trajectory and velocity actually reflect the presence of some kind of memory (see Figs 6.27 and 6.29).

A microscopic theory of such interplay of electron (or exciton) relaxation and transfer has been recently developed by Čápek.[59a] The author applies a time-convolutionless GME approach and demonstrates an interplay of polaron cloud formation, i.e., relaxation and gradual loss of coherence of carrier propagation with increasing time as the hot carrier "cools" down.

The crucial question for the characterization of such "coupled" relaxation and transport process is how fast the molecular polaron is formed in comparison with mean thermalization time τ_{th} of the charge carrier (see Sections 6.2.6 and 6.2.7).

For estimation of the "cooling down" time $t_{rel}^{hot-cold}$ of the hot carrier Čápek[59a] has derived the following formula:

$$t_{rel}^{hot-cold} \gtrsim \hbar / \Delta W_{ph}^{opt}, \qquad (5.15.6)$$

where ΔW_{ph}^{opt} is the bandwidth of the most relevant intermolecular optical vibrational modes (optical phonons). A typical range of optical phonon energies in polyacene crystals is from 50 to 125 cm^{-1} (see Section 1A.4.2). Thus the corresponding bandwidth ΔW_{ph}^{opt} range is from 0.006 to 0.015 eV.

As a result we obtain the following most probable range of $t_{rel}^{hot-cold} = (0.44-1.1) \times 10^{-13}$ s.

According to Ref. 59a, $t_{rel}^{hot-cold}$ characterizes the relaxation time after which the molecular polaron starts to behave incoherently, i.e., losses its memory. However, the time of onset of genuine diffusive behaviour t_{diff} of the carrier (which may be assumed to be equal to the mean thermalization time τ_{th}) may be greater than the left side of inequality [Eq. (5.15.6)], namely

$$t_{diff} \approx \tau_{th} \gtrsim t_{rel}^{hot-cold} \qquad (5.15.7)$$

Computer simulation of the charge carrier thermalization processes[344,355] yields the following values of the mean thermalization time: $\tau_{th} = 2.6 \times 10^{-13}$ s in Nph and $\tau_{th} = 8.5 \times 10^{-13}$ s in Pc crystals (see Figs. 6.27 and 7.14). As one may see, these τ_{th} values are of the same order of magnitude as those estimated according to formulas (5.15.6) and (5.15.7).

On the other hand, the estimated range of the MP formation τ_v is of one or even two orders of magnitude shorter ($\tau_v = 1.7 \times 10^{-15} - 1.1 \times 10^{-14}$ s) than τ_{th} (see Table 3.7 and Fig. 3.9).

These data show that the polaron formation can actually take place already during the first moments of thermalization, i.e., after only a few scattering events.

5.16. PHENOMENOLOGICAL APPROACHES

Molecular crystals are good examples of solids where propagation of at least three types of species is possible: positive charges (holes), negative charges (electrons), and neutral quasiparticles (excitons). As already mentioned above, most of the transport theories consider both excitons and charge carriers. Experiment seems to be qualitatively well compatible with theory for excitons (see, e.g., Ref. 288); better-than-qualitative agreement seems to be rather a matter of future investigations for reasons connected with both theory (unclear input parameters, lack of reliable computer simulations) and experiment (limited possibility of using electric fields as probes, masking of tiny effects by the finite exciton lifetime, etc.). On the other hand, for charge carriers the situation is recently slowly becoming more lucid as compared with that described in standard textbooks and monographs.[152,288] Starting from the mid-eighties, ultrapure samples of some organic crystals (Nph, perylene) have been prepared and used for measurements of time-of-flight charge carrier mobilities down to 4.2 K; also hot charge carriers have been observed at low temperatures with drift velocities greater than those of equilibrium carriers[406,407] (see Chapter 7). This changing situation is partly reflected in a recent review by Movaghar[250,251] and is also accompanied by progress in computer simulation.[355,356] Since the contemporary state of experimental breakthrough in carrier transport phenomena will be discussed in Chapter 7, let us only give here some warnings one has to keep in mind when comparing such advanced experiments with so diverse types of theory.

The question of possible coexistence of the band and hopping character of the carrier motion in molecular crystals has been in theory positively solved in the sense that (at least) the low-field mobility is really an approximate sum of the bandlike and hopping-like terms [see Eqs. (5.9.1) with Eqs. (5.9.8), (5.9.10), and (5.9.12)]. Here, one should realize, however, that neither the isolated band nor the isolated hopping terms are well defined, since both theory and experiment know only their sum. On the other hand, e.g., experimentally observed decreasing values of the mobility with increasing temperature of the type $\mu(T) \sim T^{-n}$ do not yet necessarily mean that the transport is certainly bandlike, as it is sometimes assumed (see Chapter 7). On a semiphenomenological level, the simplest reason for this dependence might be a strong temperature dependence of the effective polaron mass yielding approximately the above power-law contribution to μ outside the band regime approach (see Refs. 344, 355, and 356). On a microscopic level, the reason is that statistically the carrier may mostly behave as a hopping particle, but that the band contribution to the mobility prevails as it is relatively large, being weighed by its own specific factors. Finally, the hopping contribution [as Eq. (5.9.12)] is, in principle, not necessarily thermally activated. Thus, possible contradictions between the $\mu(T) \sim T^{-n}$, $n > 0$ dependence, and, e.g., the value of the mean free path comparable with the intermolecular distances are not necessarily irreconcilable.

Here, another word of caution is necessary. The notion of the mean free path l_o is well defined in connection with a given theory of band conduction. For bare (unclothed) and polaronic carriers, however, this quantity may yield completely different results. A similar observation can be made for, e.g., a carrier in a rigid perfect periodic crystal. In the single-particle approximation, the Bloch wave has an infinite mean free path, while for the plane wave, a naively introduced l_o may easily become comparable with the lattice constant (see Chapter 3).

Similarly, the question is often posed on grounds of experimental results, whether the migrating species in the molecular crystal is either a band or a hopping small polaron carrier. Such a question is, in fact, badly posed since the small polaron can also propagate coherently, i.e., as a wave [compare, e.g., Eq. (5.13.5)], though the central particle (electron) moves in a zigzag (i.e., hopping-like) manner (this motion being superimposed over the motion of the polaron as a whole), owing to continuous absorption and emission of virtual phonons. Thus, for the same physical situation, both the coherent and hopping description of the conduction process might be correct.

In general, one should realize that the wavelike motion of both the unclothed carrier and the polaron can be described in terms of both polaronic and nonpolaronic states and vice versa since this means using another basis in the corresponding Hilbert space. Thus, the genuine character of the migrating species should be estimated from the relative magnitude of the off-diagonal matrix elements (with respect to the diagonal ones) in the corresponding representation. Normally, this is, however, an academic problem having little to do with experiment. So at the moment of posing the question about the true character of the migrating species, one should realize how little can be said for a given experiment. Though we tend to say that under normal conditions (e.g., room temperature), the carrier motion in molecular crystals reminds us of the hopping (though not necessarily uncorrelated) motion of the nearly small molecular polaron, a warning is worth mentioning here, namely,

1. No good answer exists to questions that are not rigorously posed.

2. Any physical situation has more than one mathematical description.

In particular, one should not forget that, from the point of view of nuclear displacements, the above-mentioned nearly small molecular polaron around the charge carriers is in fact never nearly small. From the point of view of deformation of (mostly π-) electron shells of the neighboring molecules, it encompasses as much as 10^4 molecules or even more. Still, interpretation of the transport phenomena in terms of these nearly small molecular polarons is well justified (at least as one possible description) provided that one does not simultaneously forget about mechanisms (the long-range electron-electron coupling, i.e., also the electron-exciton coupling) leading to formation of the electronic polaron. The role of these mechanisms in the transport phenomena is almost uninvestigated. Currently it is believed that they should not cause any qualitative change. Accepting this and keeping all the above facts in mind, at the qualitative level we do not see any contradiction, but rather we see agreement between sufficiently rigorous theory and experiment on transport properties in molecular crystals. We give below detailed comparison (see Chapters 6 and 7) in terms of phenomenological approaches.

CHAPTER 6

Charge Carrier Photogeneration and Separation Mechanisms in OMC

When it comes to atoms, language can be used only as in poetry. The poet, too, is not nearly so concerned with describing facts as with creating images

NIELS BOHR

In this chapter we shall discuss models and mechanisms of charge carrier photogeneration and separation in OMC. Due to specific features of electronic and crystallographic structure of OMC and weak intermolecular interaction forces (see Section 1A of Chapter 1) charge carrier photogeneration and separation phenomena are highly complicated and proceed via multistep processes. At the near-threshold region of intrinsic photogeneration there may emerge two competitive photogeneration mechanisms, namely, autoionization (AI) of excited molecular states and direct optical charge transfer on nearest neighbor molecules. At higher excitation energies the AI mechanism becomes dominant and appears in two possible modes, namely, as direct ionization of photoexcited neutral molecular states, and autoionization from fixed electronic levels after intramolecular relaxation. At higher excitation energies the quantum efficiency of photogeneration are also affected by competitive transitions from lower-lying valence states.

The great variety of photophysical processes in OMC compels one to introduce rather sophisticated phenomenological models for their description.

We shall discuss in some detail these models and the experimental data on which they are based. Computer-simulated studies of charge carrier photogeneration and separation will also be described. It will be demonstrated that both experimental and simulation data may be interpreted in terms of a MP approach (see Chapters 3 and 4).

254

6.1. BASIC PHOTOGENERATION MECHANISMS AT THE NEAR-THRESHOLD SPECTRAL REGION

Photogeneration phenomena have been most thoroughly studied in polyacene crystals. Anthracene especially has served for several decades as a model compound of OMC. These studies have demonstrated that photogeneration phenomena in organic molecular crystals exhibit a number of peculiarities the interpretation of which required novel approach. Owing to the complex nature of the problem the explanation of the photogeneration phenomena in OMC and proposed models and mechanisms have been a subject of controversy for over two decades.

In the late sixties Pope and Geacintov demonstrated[115,116,286] that the behavior of intrinsic photocurrent spectra dependences in Ac-type crystals unequivocally supports the AI mechanism of carrier photogeneration. They showed that the conventional mechanism of band-to-band transitions, widely used for traditional semiconductors, is, in this case, not valid.

Thus, the nonapplicability of simple band theory approach for the description of the electronic processes in OMC, widely discussed in the previous chapters, naturally extends also to photogeneration phenomena.

According to the AI model, proposed in Refs. 115, 116, and 286 ionization is considered as a configurational interaction between an excited discrete neutral molecular state ψ_s^* and a continuum of ionized states in the crystal φ_{E_C} in the spirit of Fano's[105] theoretical treatment of AI processes in the gaseous phase.

A further theoretical basis for the AI mechanism in organic crystals was provided by Jortner,[148] similar in spirit to the intramolecular vibrational relaxation theories. According to this approach, in addition to the configurationally interacting terms of the excited molecular state ψ_s^* (a Frenkel exciton state) and a continuum of ionized states of the crystal φ_{E_C}, an intramolecular relaxation term $\sum_i C_i \chi_i$ is introduced:

$$\psi_E = a_E \psi_s^* + b_E \int \varphi_{E_C} dE + \sum_i C_i \chi_i, \qquad (6.1)$$

where a_E, b_E, and C_i are numerical coefficients that characterize the contribution of each term in the ionization process. The third term forms a competitive relaxation channel which actually lowers the quantum efficiency of AI. Equation (6.1) illustrates the molecular and solid-state aspects of the photogeneration process (see Fig. 1.2) and shows that the molecular properties may be prevailing over the solid-state ones in OMC.

The AI mechanism has been discussed in some detail both from the phenomenological and the microscopic point of view in the monograph (Ref. 288, p. 473 and 497).

However, detailed studies of the temperature dependences of intrinsic carrier photogeneration in Ac crystals, performed by Braun and co-workers[24,73,160] showed that the initial AI model should be significantly modified. The existence of photogeneration activation energy E_a^{ph} and its spectral dependences $E_a^{ph}(h\nu)$ demonstrated that after the act of AI the hot electron thermalizes inside the Coulombic well

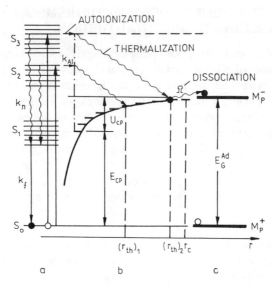

FIGURE 6.1. Schematic diagram of multistep photogeneration model in organic molecular crystals (from Refs. 332, 349, and 351]. (a) Neutral ground (S_0) and excited (S_1, S_2, S_3) electronic states of molecules in the crystal; rate constants: k_{AI}, of direct autoionization, k_n, nonradiative and k_f, radiative intramolecular relaxation. (b) Charge pair (CP) states: E_{CP}, energy of CP state; U_{CP}, Coulombic binding energy of CP state; Ω, dissociation probability of CP state; r_{th}, mean thermalization length of hot charge carrier; r_c, critical Coulombic radius. (c) Ionized state levels of free charge carriers; E_G^{Ad}, adiabatic energy gap; M_P^+, M_P^-, conductivity levels of relaxed charge carriers (molecular polarons).

of the parent ion forming bonded CP states for the dissociation of which additional activation energy is necessary. As a result, for the description of photogeneration phenomena in OMC a multistep photogeneration model has been developed.[73,332,344,349,351] According to this model (see Fig. 6.1) photogeneration of free charge carriers in OMC proceeds through various intermediate stages.

The first stage is photon absorption and formation of a neutral Frenkel exciton state (Fig. 6.1). In the second stage, AI of the excited molecular state takes place, with quantum efficiency $\Phi_0(h\nu)$, creating a positively charged parent ion and a hot electron with excess kinetic energy E_k. In the third stage, the hot quasifree electron rapidly loses its initial excess energy, due to fast inelastic scattering, and becomes thermalized at a mean thermalization length r_{th} inside the critical Coulombic radius r_c ($r_{th} < r_c$), thus forming an intermediate CP state of bound charge carriers (see Chapter 4). At the final stage of photogeneration free carriers are created via dissociation of the CP state the probability of which can be described in the framework of a modified[73,279,344,351] Onsager model.[277]

At the initial phase of AI the motion of the hot electron may be coherent. However, its coherence of motion is very rapidly lost, owing to inelastic scattering on lattice vibrations, on acoustic and optical lattice phonons, the scattering events taking place practically at every lattice site (see Section 7.2). Thus, the electron becomes strongly localized and further energy relaxation presumably occurs in an incoherent motion by hopping from site to site until it finally thermalizes at the distance r_{th}.

The mean thermalization length r_{th} is further used as an input parameter in the Onsager model. After the thermalization the CP state can either dissociate, forming free charge carriers, or alternatively, undergo geminate recombination. The Onsager theory allows to determine the probability of these processes at given temperature T and electric field strength $\mathcal{E}$.

Initially Onsager treated rigorously the Brownian motion of a pair of charged particles in a Coulombic field within the framework of diffusion model. Later, the Onsager theory was used extensively by radiation chemists (see e.g., Refs. 140 and 141). Braun et al.[24,73] applied this theory to intrinsic photoconductivity in Ac single crystals to gain an understanding of the observed T and $\mathcal{E}$ dependences of the free charge carrier photogeneration quantum yield. The modified Onsager model[73] allowed these authors to describe the final dissociation stage of photogeneration (see Fig. 6.1) in terms of thermal dissociation of the CP states as the result of Brownian diffusion of the charge carriers under the influence of mutual Coulombic and external electric fields.

According to the Onsager model, the activation energy for the probability of charge carrier geminate escape, i.e., activation energy of intrinsic photogeneration E_a^{ph} is determined by the Coulombic binding energy U_{CP} of the CP state[73,349] (see Fig. 6.1)

$$E_a^{ph} = -U_{CP} = \frac{e^2}{4\pi\epsilon\epsilon_0 r_{th}}. \qquad (6.2)$$

Thus, measuring the spectral dependence of photogeneration activation energy $E_a^{ph}(h\nu)$ in a crystal, it is possible to evaluate the spectral dependence of the so defined mean thermalization length $r_{th}(h\nu)$:

$$r_{th}(h\nu) = \frac{e^2}{4\pi\epsilon\epsilon_0 E_a^{ph}(h\nu)}. \qquad (6.3)$$

In earlier work (see, e.g., Refs. 24 and 73) it was assumed that the mean thermalization length r_{th} depends only on $h\nu$, i.e., $r_{th} = r_{th}(h\nu)$, according to Eq. (6.3), and therefore the distribution function $g(r)$ of thermalized electrons is isotropic, possessing spherical symmetry. In their initial use of the Onsager theory for interpretation of photogeneration stages in Ac, Batt et al.[24] used a delta function to describe the distribution $g(r)$ of the mean thermalization length r_{th} of the hot ejected (autoionized) electron

$$g(r) = \frac{\delta(r - r_{th})}{4\pi r_{th}^2}. \qquad (6.4)$$

Later Chance and Braun[73] introduced Gaussian distribution of $g(r)$ as being more realistic

$$g(r) = \frac{1}{(2\pi)^{1/2}\sigma} \exp\left[-\frac{(r-r_{th})^2}{2\sigma^2}\right],$$ (6.5)

where σ is the dispersion parameter of the Gaussian distribution.

The Gaussian distribution, however, is a two-parameter curve. Since experiment yields only one parameter, namely, r_{th} (see Section 6.2.2), it is usually assumed that the distribution parameter σ is small, and delta function approximation may be used. However, the σ value can be estimated in a synthetic experiment applying computer simulation technique (see Ref. 344 and Section 6.3).

Later it was shown by Silinsh et al., both experimentally[349] and by computer simulation[344] (see also Ref. 351) that the mean thermalization length r_{th} of the carrier is dependent not only on $h\nu$ (see Fig. 6.1), but also on electric field $\mathcal{E}$ and temperature T, namely, $r_{th} = r_{th}(h\nu, \mathcal{E}, T)$ (see Sections 6.2.6, 6.2.7, and 6.3.3).

In the framework of the multistep AI model (see Fig. 6.1), the quantum efficiency of free charge carrier photogeneration Φ is a function of photon energy $h\nu$, electric field $\mathcal{E}$, and temperature T, i.e., $\Phi = \Phi(h\nu, \mathcal{E}, T)$ and may be presented as a product of two probabilities: CP state photogeneration quantum yield $\Phi_0(h\nu)$ and CP state dissociation efficiency Ω

$$\Phi(h\nu, \mathcal{E}, T) = \Phi_0(h\nu)\Omega[r_{th}(h\nu, \mathcal{E}, T), \mathcal{E}, T].$$ (6.6)

Thus, the parameters $\mathcal{E}$ and T influence both the thermalization and dissociation stages of photogeneration in OMC. Therefore the isotropic Onsager approach should be substituted by a modified Onsager model in which the field-induced anisotropic distribution of thermalized carriers is taken into account[344,349,351] (see Sect. 6.2.7).

According to this modified approach the dissociation efficiency $\Omega(r_{th}, \mathcal{E}, T)$ is calculated by the Onsager formula for given values of an r_{th} anisotropic distribution with $h\nu$, $\mathcal{E}$, and T as parameters:[349,351]

$$\Omega_i[r_{th}(h\nu, \mathcal{E}, T), \mathcal{E}, T] = \frac{1}{2\gamma r_{th}} \exp(-r_c/r_{th}) \sum_{m=0}^{\infty} \frac{(r_c/r_{th})^m}{m!}$$

$$\sum_{n=0}^{\infty} [1-\exp(-2\gamma r_{th})] \sum_{k=0}^{m+n} \frac{(2\gamma r_{th})^k}{k!},$$ (6.7)

where r_c is the critical Coulombic radius of thermal dissociation of the CP state

$$r_c = \frac{e^2}{4\pi\epsilon\epsilon_0 kT}$$ (6.8)

and γ is the parameter of the $\mathcal{E}$-dependent correction term

$$\gamma = e\mathcal{E}/2kT.$$ (6.9)

It should be mentioned that the anisotropic distribution of the r_{th} in the electric field cannot be obtained directly from experiment but should be calculated for the given field strength $\mathcal{E}$ and the angle θ between the initial velocity vector $\mathbf{v}_0$ of the ejected electron and the electric field $\mathcal{E}$ (see Section 6.3.3).

Anisotropic distribution of the r_{th} may be affected also by diffusional and dielectric anisotropy in Ac-type crystals (see Refs. 251a and 284a).

On the whole, the validity of the Onsager model for the description of photogeneration quantum efficiency $\Phi(h\nu, \mathcal{E}, T)$ in Ac single crystals has been convincingly proved by a number of investigators (see e.g., Refs. 17, 72, 73, 160, and 222). The Onsager model also appears to have general validity in describing photogeneration processes in other organic solids; e.g., such as Tc and Pc vacuum-deposited layers,[349] polyvinylcarbazol,[234] polydiacetylene single crystals,[219] as well as amorphous selenium.[279]

It should, however, be emphasized here that for higher electric fields ($\mathcal{E} \geq 10^4$ V/cm) and longer thermalization distances ($r_{th} \geq 40$ Å) the anisotropic approach (6.7) provides better agreement with experimental data than the isotropic one (see Section 6.3.3).

Rackovski and Scher[292a,292b,316a] have proposed a molecular theory of photogeneration based on time-dependent random walk (RW) process involving charge transfer (CT) steps between molecular sites (see Section 5.8). The authors use the RW process to calculate the dissociation efficiency $\eta(\mathcal{E})$ as a function of applied electric field $\mathcal{E}$. In the proposed theory the Onsager approach is extended, especially for small charge carrier separation distance r_o and large electric fields $\mathcal{E}$; molecular parameters and crystal structure are taken into account.

The term $\Phi_0(h\nu)$ in Eq. (6.6) is determined by the first two photogeneration stages: excitation and autoionization (see Fig. 6.1). The $\Phi_0 = \Phi_0(h\nu)$ dependence may be described by the following ratio (Ref. 332, p. 115):

$$\Phi_0^{AI}(h\nu) = \frac{k_{AI}(h\nu)}{k_{AI}(h\nu) + \Sigma_i k_i(h\nu)}, \qquad (6.10)$$

where k_{AI} is the autoionization rate constant and the k_i are rate constants of different competitive intramolecular (nonradiative and radiative) relaxation channels.

The relation (6.10) actually determines the character of $\Phi(h\nu, \mathcal{E}, T)$ and $E_a^{ph}(h\nu)$ dependences over a wide spectral range of photoexcitation.

In the near-threshold spectral region (e.g., in the region from 3.9 to 4.5 eV in case of Ac crystals; see Figs. 6.5 and 6.8) $E_a^{ph}(h\nu)$ dependence monotonously decreases, and the corresponding $r_{th}(h\nu)$ dependence increases with increasing photon energy $h\nu$ in accordance with the kinetic diagram shown in Fig. 6.1. However, the diagram in Fig. 6.1 gives a correct description of photogeneration processes only at the near-threshold region of intrinsic photoconductivity.

At higher excitation energies there may appear spectral regions in the range of which the E_e^{ph} and the corresponding r_{th} values remain constant, independent of photon energy $h\nu$ (see Fig. 6.10). This is due to AI from fixed electronic levels after intramolecular vibronic relaxation of the photoexcited molecule. This effect and its

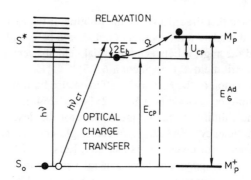

FIGURE 6.2. Schematic diagram of photogeneration of free charge carriers in OMC via direct optical CT-transitions.[351] E_b, the energy of formation of molecular polaron (M_p) in the relaxation process of CT excited state. Ω, the probability of CP state dissociation according to Onsager mechanism.

implications will be discussed later (see Section 6.2.5).

On the whole, the multistep autoionization model is widely recognized as the dominant mechanism of photogeneration in OMC (see Refs. 73, 288, 332, and 351 and references therein).

In the near-threshold spectral region an alternative photogeneration mechanism, proposed in the early eighties by Siebrand et al.,[39-42] may be relevant. According to this approach the CP states can be created by direct optical CT transitions (see Fig. 6.2).

The existence of the anticipated direct optical CT transitions was confirmed experimentally by Sebastian et al. in Tc and Pc,[317] and later also in an Ac crystals[318] (for details see Chapter 4).

However, it has been shown that the direct optical CP-state generation is actually also a two-step process. After optical CT transition relaxation of the created CP state takes place[318,344,349,351] (see Fig. 6.2).

In the relaxation process a MP is formed (see Chapters 3 and 4). The relaxation energy of both charge carriers $2E_b$ causes further increase of their separation distance. This process may be described in terms of the thermalization phenomenon.[318] Thus, the intermediate thermalization stage has become a common part of both photogeneration models.[344,351]

Also the final stage of the CP state dissociation is similar in both cases independent of the pathway of the CP state generation (by AI or direct optical CT-transition, see Figs. 6.1 and 6.2). Therefore Eq. (6.6) may be regarded as universal and valid for both photogeneration mechanisms.

However, as will be shown later (see Section 6.2.4), the efficiency of CT-transitions may be dependent both on photon energy $h\nu$ and electric field $\mathcal{E}$. Therefore, in this case, the term $\Phi_0(h\nu)$ should be substituted by the term $\Phi_0(h\nu, \mathcal{E})$. The complementary character of both photogeneration mechanisms (Figs. 6.1 and 6.2) in the near-threshold spectral region will be shown in Sections 6.2.4 and 6.3.4.

6.2. EXPERIMENTAL INVESTIGATION OF PHOTOGENERATION PROCESSES IN SOME MODEL OMC

Intrinsic photogeneration processes have been most extensively studied in Ac crystals, the most popular model compound of OMC for the last two decades. As previously mentioned Batt et al.[24] were first to demonstrate that for intrinsic photogeneration of free charge carriers an experimentally detectable activation energy E_a^{ph} is needed. These results showed that photogeneration in Ac crystals actually proceeds via an intermediate stage of bounded charge pair (CP) state and its subsequent thermal dissociation (see Fig. 6.1). Spectral dependences of $E_a^{ph}(h\nu)$ in Ac obtained by Batt et al.[24] were later refined in meticulous experimental studies by Braun and coworkers.[73,160]

These $E_a^{ph}(h\nu)$[24,73,160] as well as described field dependence studies[24,72,222] allows one to describe, in terms of the Onsager model, the *final* dissociation stage of charge carrier separation (see Fig. 6.1), accessible to direct experimental observation. On the other hand, data on intermediate stages, including the primary ionization act, thermalization of hot electron and formation of geminate CP states, were scarce and inadequate, and their interpretation controversial (see, e.g., Refs. 17, 24, 39–41, 73, 160, 222, and 288). Silinsh and coworkers[332,349,337] have extended the photogeneration studies to Tc and Pc crystals. Due to much higher photosensitivity of Tc and Pc as compared to Ac, it was possible to obtain new details of the whole photogeneration process, and to gain an insight into the mechanisms of the intermediate stages, especially the thermalization process.[349] Since these data in Ref. 349 have been obtained mainly on Pc crystals, we shall give in the next section a brief description of the experimental methods used.

6.2.1. Description of Experimental Methods

The Pc crystals were purified by vacuum sublimation techniques[10] (gradient sublimation and subsequent zone sublimation of twenty passes).

For photoconductivity measurements thin-layer samples were prepared by vacuum evaporation of Pc sandwiched between Au (bottom) and Al (top) semitransparent electrodes on quartz substrate at $p = 10^{-5}-10^{-6}$ Torr and average evaporation rate 5–15 Å/s (see Fig. 6.3c). The thickness L of the layers varied from 0.8 to 2 μm; the substrate temperature was ~ 300 K.

Optical spectra and electron micrographs confirmed that the deposited Pc layers consist of planar, closely packed crystallites the ab plane of which is oriented predominantly parallel to the substrate surface, the average crystallite dimensions being 0.1–0.15 μm.[349]

The thin-layer sample was illuminated through the semitransparent bottom Au electrode at negative potential (see Fig. 6.3c). The given electrode configuration was blocking for electron as well as hole injection up to field intensities $\mathcal{E} = (2-3)\times10^5$ V/cm.

Transient photocurrent was measured in the quasiuniform light absorption regime

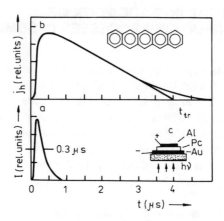

FIGURE 6.3. Transient photocurrent measurements in thin layers of Pc. (a) Xenon lamp light pulse used for photoexcitation. (b) Corresponding transient hole current pulse $j_n(t)$ in Pc ($\tau = 0.3$ μs, $R = 1$ kΩ). (c) Schematic configuration of the thin-layer sample Au/Pc/Al on a quartz substrate.

in the sample, when $1/\alpha \approx L$. Light pulses from a xenon arc lamp with peak half-width of ~ 0.3 μs (Fig. 6.3a) via a monochromator were used for studying the quantum efficiency of transient photoconductivity $\eta(h\nu, \mathcal{E})$ per absorbed photon as dependent on $h\nu$ and $\mathcal{E}$ in the near-threshold region of intrinsic photogeneration of Pc ($h\nu = 2.2$–3.4 eV) in the temperature range 140–300 K. A typical transient pulse obtained is shown in Fig. 6.3b. The transient pulse in this case is provided by the flow of majority carriers, i.e., holes, the lifetime of holes $\tau_h = 10^{-5}$–10^{-4} s being of three to four orders greater than that of electrons (τ_e being $\sim 10^{-8}$ s) (see Sections 6.2.3 and 6.2.8).

The photogenerated charge q per pulse was estimated as $q = \int_0^{t_{tr}} i(t)dt$, where t_{tr} is the extrapolated transit time (see Fig. 6.3b). From the temperature dependence $q = q(1/T)$ with $h\nu$ and $\mathcal{E}$ as parameters, the photogeneration activation energy dependences $E_a^{ph}(h\nu, \mathcal{E})$ were determined, as well as the transient photoconductivity quantum efficiency $\eta(h\nu, \mathcal{E})$ per absorbed photons. Measurements were performed in a nitrogen cryostat with an automatic temperature regulation providing ± 0.02 K stability. The accuracy of the E_a^{ph} determination was $\pm(0.010$ to $0.015)$ eV.

For the investigation of the fast photocurrent component of the minority carriers (electrons) 20-ns laser pulses with $h\nu = 2.34$ eV from a neodymium-yttrium-aluminum-garnet (Nd-YAG) laser were used (see Sections 6.2.8 and Fig. 6.17).

Owing to available information about the hole trap energy spectra in the Pc samples (see, Ref. 332) it was possible to evaluate their effect on the transient photocurrent. The trapping data analyses showed that the tail part of the pulse behind t_{tr} (see Fig. 6.3b) contributes, in our case, only ca. 5–10% of the total transit charge and may be easily taken into account.

Methods of stationary photoconductivity quantum efficiency $\eta(h\nu, \mathcal{E}, T)$ measurements are described in some detail in Section 6.2.3.

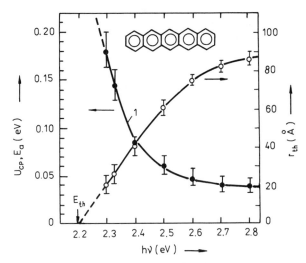

FIGURE 6.4. Experimental zero field thermalization length $r_{th}(h\nu)$ (open circles) and corresponding $U_{CP}(h\nu)$ (filled circles) spectral dependences in Pc. Curve (1) — calculated according to Eq. (6.1) $U_{CP}(h\nu)$ spectral dependence. E_{th}, the extrapolated intrinsic photoconductivity threshold value.[349] Measurements have been performed in transient photocurrent regime.

6.2.2. Determination of the Mean Thermalization Length r_{th} in Polyacene Crystals

According to Eq. (6.3) the spectral dependence of the mean thermalization length $r_{th}(h\nu)$ in the near-threshold region can be evaluated by measuring the corresponding spectral dependence of the photogeneration activation energy $E_a^{ph}(h\nu)$. This evaluation is valid in terms of the Onsager model according to which the E_a^{ph} value is determined by the Coulombic binding energy U_{CP} of a CP state [see formula (6.2) and Fig. 6.1].

However, this assumption is actually true only for low field intensities, e.g., $\mathcal{E} \leqslant 10^3$ V/cm. First, at higher electric fields the effective Coulombic binding energy U_{CP} is described according to the $\mathcal{E}$ -dependent term of the Onsager formula (6.7) and, correspondingly, the experimentally observed value of E_a^{ph} is also decreased. Second, the value of r_{th} can be increased by electric field $\mathcal{E}$ while a hot electron is cascading down to thermal energies, since $r_{th} = r_{th}(\mathcal{E})$ at electric fields $\mathcal{E} \geqslant 10^4$ V/cm (see Section 6.2.7).

To obtain field-independent r_{th} values the authors of Ref. 349 extrapolated the experimental $r_{th} = f(\mathcal{E})$ dependences, obtained at different photon energies $h\nu$ in the threshold spectral region of Pc to $\mathcal{E} \rightarrow 0$. The extrapolation yields the spectral dependence of the zero-field thermalization length $r_{th}(h\nu)$ (Fig. 6.4, open circles). From the $r_{th} = f(h\nu)$ dependence (Fig. 6.4) it is easy to estimate, according to Eq. (6.2) the corresponding $U_{CP}(h\nu)$ dependence, i.e., the zero-field depth of the corresponding CP states (Fig. 6.4, filled circles; see Fig. 6.1).

In order to prove the validity of the zero-field $r_{th}(h\nu)$ values (Fig. 6.4), obtained by extrapolation, the authors of Ref. 349 used these values to calculate the temperature dependence of the CP-state dissociation efficiency $\Omega(r_{th}, \mathcal{E}, T)$ at low electric fields ($\mathcal{E} < 10^3$ V/cm), according to the Onsager formula (6.7). Subsequently, from the $\Omega(r_{th}, T) = f(1/T)$ dependences they determined the corresponding activation energies $E_a^{ph} = E_a^{ph}(h\nu)$ which should be equal to the $U_{CP} = U_{CP}(h\nu)$ values. As can be seen from Fig. 6.4, the calculated $E_a^{ph}(h\nu)$ curve (1) at $\mathcal{E} < 10^3$ V/cm lies within the error bars of $U_{CP}(h\nu)$ values determined directly from the extrapolated zero-field $r_{th}(h\nu)$ values.

As may be seen from Fig. 6.4, the $r_{th}(h\nu)$ curve yields a sublinear dependence on photon energy $h\nu$ and reaches the asymptotic value of r_{th} at $h\nu \approx 2.8$ eV. However, as we will see later, in Section 6.2.5, the $r_{th}(h\nu)$ curve at higher excitation energies yields a more complicated structure with "plateau" regions where r_{th} becomes independent of $h\nu$, i.e., $r_{th} = $ const (see Fig. 6.10). The cause of such behavior is discussed in Section 6.2.5.

The parabolic-type $r_{th}(h\nu)$ dependence in the near-threshold region follows from the basic ideas of the ballistic model (see Fig. 6.1). In this case one can expect that the r_{th} value would be proportional to the initial velocity v_0 of the ejected hot electron. Consequently, the r_{th} dependence on kinetic energy E_k of the electron is expected to be

$$r_{th} \sim v_0 \sim E_k^{1/2} = (h\nu - E_G^{Ad})^{1/2}. \tag{6.11}$$

The dependence (6.11) has been confirmed by computer simulation studies of thermalization processes in Pc crystals (see Section 6.3.3).

The extrapolation of the $r_{th}(h\nu)$ curve in Fig. 6.4 gives the spectral threshold value $E_{th} = 2.2$ eV of intrinsic photogeneration in Pc crystals which is in good agreement with the E_{th} value obtained from photoconductivity quantum efficiency spectral dependences $\eta(h\nu)$ (see Chapter 4).

Kato and Braun[160] have measured the $E_a^{ph}(h\nu)$ dependence in stationary regime in the near-threshold region (3.9–4.4 eV) of an Ac crystal (Fig. 6.5, curve 1) and obtained, according to formula (6.3) the corresponding $r_{th}(h\nu)$ dependence (Fig. 6.5, curve 2). Since the $E_a^{ph}(h\nu)$ measurements have been performed at low electric field values ($\mathcal{E} \leq 5 \times 10^3$ V/cm) the obtained $r_{th}(h\nu)$ magnitudes should be pretty close to the zero-field value, or may be only slightly overestimated.

As may be seen from Fig. 6.5, the $r_{th}(h\nu)$ dependence also demonstrates the sublinear dependence on excitation energy $h\nu$. The extrapolated E_{th} value $E_{th} = 3.87$ eV is close to the E_{th} values determined in Ac from $\eta(h\nu)$ dependences (see Chapter 4).

Zhukov et al.[411] have measured the $E_a^{ph}(h\nu)$ dependence in Pc single crystals in stationary photoconductivity regime, both in the intrinsic AI spectral region ($h\nu > E_{th} = 2.25$ eV) and in the region of extrinsic excitonic charge carrier generation ($h\nu < E_{th}$) (Fig. 6.6, open circles). As may be seen, in the AI region the E_a^{ph} value increases as the excitation energy $h\nu$ approaches the threshold E_{th}. For comparison, Fig. 6.6 also shows the $E_a^{ph}(h\nu)$ and $r_{th}(h\nu)$ dependences obtained in

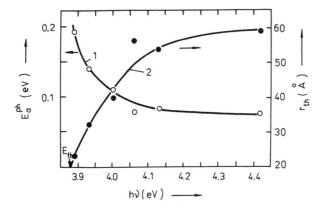

FIGURE 6.5. Experimental $E_a^{ph}(h\nu)$ dependence in an Ac crystal (1) and the corresponding $r_{th}(h\nu)$ curve (2) according to Ref. 160.

transient photocurrent regime[349] (filled circles). The $E_a^{ph}(h\nu)$ curve, obtained in Ref. 411 from stationary photoconductivity T dependences, should be regarded as a rough estimate, since this method does not exclude the possible influence of the carrier mobility temperature dependences $\mu(T)$. However, as may be seen from Fig. 6.6, both approaches give the same threshold value of intrinsic photogeneration, namely $E_{th} = 2.25 \pm 0.05$ eV. It is interesting to notice, that, in the spectral region of extrinsic photogeneration at $h\nu < E_{th}$, both the $E_a^{ph}(h\nu)$ and the $r_{th}(h\nu)$ dependences have a constant value, independent of the photon energy $h\nu$.

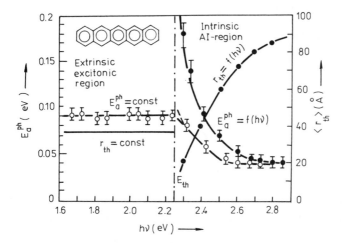

FIGURE 6.6. Experimental $E_a^{ph}(h\nu)$ dependence according to Ref. 411 (open circles) and $E_a^{ph}(h\nu)$ and $r_{th}(h\nu)$ dependences (filled circles) according to Ref. 349 in Pc crystals. The measurements of Ref. 349 performed in transient but those of Ref. 411 in stationary photocurrent regimes.

As has been shown in Ref. 204, the charge carrier generation in this subthreshold region presumably proceeds via a so-called excitonic photogeneration mechanism the essence of which is the following. In the $hv < E_G^{Ad}$ spectral region the absorbed light creates singlet S_1 excitons

$$S_0 + hv \rightarrow S_1^*. \tag{6.12a}$$

These excitons, during their lifetime, migrate in the crystal and may interact with the quinone-type impurity Q centers, present in the Pc (see Chapter 4). In the interaction process the S_1 exciton annihilates, the electron being trapped at the Q center while the hole obtains the relaxation energy and thermalizes in the Coulombic field of the quinone-type anion Q^- at the mean thermalization distance r_{th} creating a bound CP state. Subsequently this CP state may dissociate according to the Onsager mechanism

$$S_1^* + Q \rightarrow S_0 + Q^- \ldots h^+ \xrightarrow{kT, \, \mathcal{E}} S_0 + Q^- + h^+. \tag{6.12b}$$

As we see, this extrinsic photogeneration mechanism is also a multistep one, similar in nature to the intrinsic AI mechanism (see Fig. 6.1). It contains all the intermediate stages described in Section 6.1: ejection of the hot charge carrier (in this case a hole), its thermalization, formation of bound CP-state and thermal dissociation of this state. The main difference lies only in the primary photoexcitation stage [Eq. 6.12a] which yields constant values of $E_a^{ph}(hv)$ and $r_{th}(hv)$ independent in the given spectral range of the photon energy hv (see Fig. 6.6).

6.2.3. Approximation of Photoconductivity Quantum Efficiency $\eta(hv)$ Curves

The spectral dependence of the photoconductivity quantum efficiency $\eta(hv)$ in a stationary photocurrent regime can be determined by the following formula:[351]

$$\eta(hv) = \frac{j_{ph}(hv)}{ek(hv)I(hv)[1 - \exp(-\alpha(hv)L]} \text{ (electrons/phot.)}, \tag{6.13}$$

where $j_{ph}(hv)$ is the spectral dependence of the density of photocurrent; I is the light intensity at hv photons/cm^{2}s; $k(hv)$ is the spectral coefficient of the absorption of light in the semitransparent electrode; $\alpha(hv)$ is the coefficient of light absorption, cm^{-1}; L is the thickness of the sample.

It may be shown that, in first approximation, the stationary photoconductivity quantum efficiency $\eta(hv)$ curves in the threshold spectral region can be described by the Onsager formula (6.7), using a single adjusting parameter b at $\mathcal{E}, T = $ const,

$$\eta(hv, \mathcal{E}, T) = b\Omega[r_{th}(hv), \mathcal{E}, T]. \tag{6.14}$$

Figure 6.7 illustrates the approximation of experimental $\eta(hv)$ dependences in Pc by $\Omega[r_{th}(hv)]$ curves calculated according to Eq. (6.7) using zero-field $r_{th}(hv)$

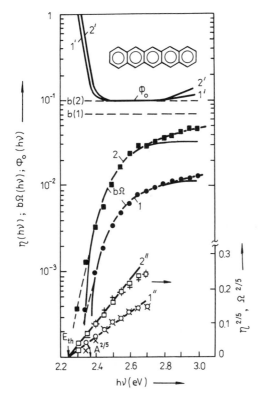

FIGURE 6.7. Experimental spectral dependences of photoconductivity quantum efficiency $\eta(h\nu)$ in Pc sample ($L = 1.5~\mu m$); ● at $\mathcal{E} = 1 \times 10^4$ V/cm and $T = 205$ K; ■ at $\mathcal{E} = 4 \times 10^4$ V/cm and $T = 205$ K; (1) and (2) calculated $b\Omega(h\nu)$ and (1') and (2') experimental $\Phi_0(h\nu)$ curves; (1") and (2") $\eta(h\nu)$ curves (○ and □) in $(\eta^{2/5}, h\nu)$ coordinates and $\Omega(h\nu)$ curves (× and +) in $(\Omega^{2/5}, h\nu)$ coordinates, respectively.

values at given $\mathcal{E}$ and T. The parameter b determines the vertical shift of the $\Omega(h\nu)$ curve in a semilogarithmic scale which should be adjusted to provide optimal fitting.

As may be seen from Fig. 6.7, the best fitting between the $\Omega(h\nu)$ curves (1,2) and the experimental $\eta(h\nu)$ dependences can be reached at $b(1) = 0.07$ and $b(2) = 0.1$.

The approximation results confirm that the $\eta(h\nu)$ threshold dependence is actually determined by the dissociation efficiency $\Omega[r_{th}(h\nu)]$ of the intermediate CP states. The empirical parameter b can be presented as a product of photogenerated geminate pair yield Φ_0 and carrier transit coefficient G

$$b = \Phi_0 G(\mathcal{E}). \tag{6.15}$$

In general, the coefficient G is determined by the ratio τ/t_{tr} where τ is the carrier lifetime and t_{tr} the transit time $t_{tr} = L/\mu \mathcal{E}$. In our case of blocking electrodes, the maximum value of G cannot exceed unity and may be described by the empirical Hecht's formula

$$G = 1 - \exp[-\tau/t_{tr}] = 1 - \exp[-\tau\mu \mathcal{E}/L]. \qquad (6.16)$$

To estimate the G value the authors of Ref. 349 investigated the $b(\mathcal{E})$ dependences. It was found that at $\mathcal{E} \geqslant 10^4$ V/cm $b(\mathcal{E})$ approaches saturation, confirming that the G value approaches unity. Under such experimental conditions both ratios: τ_h/t_{th} and S_h/L (where τ is the hole lifetime and S_h is the hole transit range $S_h = \mu\tau_h \mathcal{E}$)— become greater than unity, and $G \to 1$ (see Table 6 1). Under these conditions practically all photogenerated holes may be collected on the negative electrode. The estimated actual CP yield Φ_0 in Pc samples equals 0.1–0.7 CP states per absorbed photon.

As may be seen from Fig. 6.7 at $h\nu \geqslant 2.7$ eV the calculated curves deviate from the experimental points. This deviation increases with growing field strength $\mathcal{E}$. As will be shown in Section 6.3.3, the deviation is caused by the dependence of r_{th} on electric field $\mathcal{E}$, i.e., $r_{th} = r_{th}(\mathcal{E})$. This effect may be taken into account using a modified anisotropic Onsager approach, as it is demonstrated in computer simulation studies [see Section 6.3.3, formula (6.53), and Fig. 6.23]. The possible cause of deviation of the calculated $\Omega(h\nu)$ curves from the experimental points near the very threshold at $h\nu \leqslant 2.4$ cV will be discussed in Section 6.2.4.

On the whole, the approximation results of quantum efficiency $\eta(h\nu)$ curves by the Onsager formula (6.7) confirm quantitatively the phenomenological photogeneration model, proposed earlier by Silinsh (see Ref. 332, p. 123), according to which the quantum yield of photogeneration in the threshold spectral region in OMC is determined by the dissociation efficiency $\Omega(r_{th}, \mathcal{E}, T)$ of the intermediate CP states in terms of the Onsager theory.

Owing to the quantitative correlation between both dependences, i.e., $\Omega(r_{th}, \mathcal{E}, T)$ and $\eta(h\nu, \mathcal{E}, T)$ it is possible to also adopt a reverse procedure, i.e., to find the $r_{th}(h\nu)$ dependence from the $\eta(h\nu, \mathcal{E}, T)$ one. This approach is illustrated in Section 6.3.3 (see Fig. 6.21).

As has been shown by us and other authors (see Ref. 332, p. 122 for references), the $\eta(h\nu)$ dependence in a number of molecular crystals can be satisfactorily approximated in the threshold region by the following empirical formula: $\eta(h\nu) \sim (h\nu - E_{th})^n$ where $n \sim 5/2$ [see Chapter 4, Eq. (4.2)].

As indicated in Section 4.1, linear approximation of threshold dependence $\eta(h\nu)$ in $(\eta^{2/5}, h\nu)$ coordinates gives extrapolated limiting value of photon energy $h\nu$ at which there still is possibility of having free charge carrier intrinsic photogeneration. As in the case of often linearized extrapolations of threshold functions, the given approximation causes systematic deviation of a purely methodological nature. As can be seen from Figs. 4.3 and 4.7 the extrapolated E_{th} lie below the adiabatic energy gap E_G^{Ad} values in Pc crystals, i.e.,

$$E_{th} < E_G^{Ad}.$$

TABLE 6.1. Some Typical Values of the Empirical Parameter b, Hole Transit Coefficient G, Geminate CP Yield Φ_0, Mean Lifetime of holes τ_h, hole transit time t_{tr}, and hole transit range S_h in Pc Layers at Different Temperature T and Electric Field $\mathcal{E}$ Values[a]

Sample	T (K)	$\mathcal{E}$ (10^4 V/cm)	b	G	Φ_0 (CP states per photon)	τ_h (10^{-5} s)	t_{th} (10^{-5} s)	S_h (10^{-4} cm)
No. 19	205	1	0.22	0.39	0.56	1.7	3.5	0.7
$L = 1.4 \times 10^{-4}$ cm	205	4	0.48	0.86	0.56	1.7	0.9	2.7
	183	1	0.24	0.39	0.61	1.7	3.5	0.7
	293	1	0.36	0.52	0.69	1.7	2.3	1.0
No. 23	205	1	0.07	0.70	0.10	4.5	3.8	1.8
$L = 1.5 \times 10^4$ cm	205	4	0.10	0.99	0.10	4.5	0.9	7.2
	293	1	0.23	0.84	0.27	4.5	2.5	2.7
No. 34	294	1	0.20	0.99	0.20	14	2.8	8
$L = 1.65 \times 10^{-4}$ cm	294	1.6	0.20	1.00	0.20	14	1.7	13
	294	4.4	0.20	1.00	0.20	14	0.6	37
	294	8	0.20	1.00	0.20	14	0.3	67

[a]Silinsh E. A., Kolesnikov V. A., Muzikante I. J., and Balode D. R.: Phys. Status. Solidi (B) **113**, 379 (1982)

Due to the established proportionally of $\eta(h\nu)$ and $\Omega(h\nu)$, it can be shown that the empirical threshold function is actually determined by the CP state dissociation efficiency according to Eq. (6.7) and it can be written as

$$A(\mathcal{E},T)(h\nu-E_{th})^{5/2} = \eta(h\nu,\mathcal{E},T) = b\Omega[r_{th}(h\nu),\mathcal{E},T]. \quad (6.17)$$

Thus, the function $A(h\nu-E_{th})^{5/2}$ is actually an empirical approximation of the Onsager formula (6.7) in the threshold region. Approximation of the $\Omega[r_{th}(h\nu)]$ dependences by the threshold function with $\mathcal{E}$ and T as parameters shows that the coefficient A in Eq. (6.17) can be expressed as

$$A(\mathcal{E},T) = a\,\mathcal{E}\,\exp(\gamma T), \quad (6.18)$$

where a and γ are empirical constants; (see also Chapter 4).

Coefficient $A^{2/5}$ gives the slope of the linear dependence $\eta(h\nu)$ or $b\Omega(h\nu)$ in $\eta^{2/5},h\nu$ and $\Omega^{2/5},h\nu$ coordinates, used for the determination of the intrinsic photoconductivity threshold E_{th} by linear extrapolation (see Fig. 6.7). As can be seen from Fig. 6.7, $\eta^{2/5}(h\nu)$ and $\Omega^{2/5}(h\nu)$ dependences agree and yield equal threshold values E_{th} for Pc, namely, 2.24 eV. The approximation of $\eta(h\nu)$ dependences by (6.17) yields for Pc: $E_{th} = (2.24\pm0.05)$ eV (mean value for 34 samples) and for Tc: $E_{th} = (2.90\pm0.05)$ eV (mean value for four samples).

Silinsh et al.[349] approximated by the Onsager formula (6.7) the experimental $\eta(h\nu)$ dependence of an Ac crystal, obtained by Kato and Braun.[160] In the approximation procedure, the results of which are shown in Fig. 6.8, the $r_{th}(h\nu)$ values from Fig. 6.5 were used.

As may be seen from Fig. 6.8, in this case the $\eta(h\nu)$ dependence grows faster than the corresponding $\Omega(h\nu)$ curve and their fitting is poor beyond $h\nu \geqslant 4.1$ eV.

The possible cause of this deviation will be discussed in the next section. As it may be seen from Fig. 6.8 the value of Φ_0 (curve 3) increases with growing photon energy $h\nu$; its mean value is rather small and does not exceed 10^{-3} CP states per photon.

6.2.4. The CP State Population Pathways. Competitive and Complementary Channels

The CP state yield Φ_0 can be regarded as constant only as a first approximation. Actually, as can be seen from Fig. 6.7, $\Phi_0(h\nu)$ shows a steady increase in the proximity of the threshold, at $h\nu \leqslant 2.4$ eV, and a tendency of slight increase at $h\nu \geqslant 2.4$ eV in Pc. It is noteworthy that, according to Ref. 160, the $\Phi_0(h\nu)$ dependence in Ac crystals is similar in character (see Fig. 6.8).

The observed $\Phi_0(h\nu)$ dependences may be interpreted in terms of competitive pathways of CP state population and energy dissipation. Our experimental data as well as data reported by others strongly support the view that in Ac-type crystals in the spectral region ~ 0.2 eV above the threshold the dominant pathway of CP-state formation is the autoionization channel (see Fig. 6.1) and Φ_0 in this region practically equals autoionization yield Φ_0^{AI}.

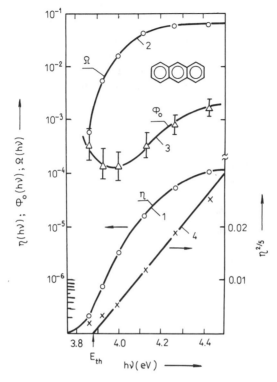

FIGURE 6.8. Experimental $\eta(h\nu)$ dependence in Ac crystal according to Kato and Braun[160] (1) calculated in Ref. 349 $\Omega(h\nu)$ curve using $r_{th}(h\nu)$ data from Ref. 160; (2) $\Phi_0(h\nu)$ dependence after Ref. 160; (3) $\eta(h\nu)$ curve in $(\eta^{2/5}, h\nu)$ coordinates yielding $E_{th} = (3.88\pm0.05)$ eV for Ac.[4]

The AI concept, introduced by Pope *et al.*[115,116,286] for the description of the primary ionization act in molecular crystals, forms a natural counterpart to the ballistic model (Fig. 6.1).

The main competitive channel to AI is the intramolecular relaxation according to a generalized equation (6.1) which may be expressed as a ratio of AI k_{AI} and intramolecular relaxation k_i rate constants [see formula (6.10)]. This ratio determines the value of the AI quantum yield Φ_0^{AI} and its spectral dependence $\Phi_0^{AI}(h\nu)$.

It has been suggested[73] that the Φ_0^{AI} value may increase for higher excited electronic states. This idea has been supported by Kato and Braun[160] who showed that the increase of $\Phi_0(h\nu)$ in Ac crystals at $h\nu > 4.1$ eV (see Fig. 6.8) correlates with the increase in the contribution of absorption in the S_3 band, in comparison with the total absorption. A slight increase in $\Phi_0(h\nu)$ in Pc at $h\nu \geqslant 2.7$ eV (see Fig. 6.7) may be of similar origin.

The value of Φ_0^{AI} allows one to estimate, according to Eq. (6.10), the contribution of the AI channel in comparison with other competitive channels. Thus, in case of Tc and Pc, with Φ_0^{AI} values of 0.1–0.8 electrons per photon, $k_{AI} \approx \sum_i k_i$. On the other hand, in the case of Ac, with a Φ_0 value of $\sim 10^{-3}$ electrons/photon (see Fig. 6.8) $\sum_i k_i \gg k_{AI}$ and almost all excitation energy is dissipated via intramolecular relaxation and subsequent fluorescence channels.

The intramolecular relaxation channel is especially critical in the near-threshold region. In this case the first step of intramolecular relaxation ends at the singlet S_1 state which lies below the intrinsic photogeneration threshold ($E_{th} > E_s$). Therefore such relaxation process may considerably reduce the AI quantum efficiency Φ_0^{AI}. The situation is more complicated for higher excited states (see Section 6.2.5).

The data of Table 6.1 also illustrate the observed tendency of increasing autoionization efficiency Φ_0^{AI} with temperature. However, this increase of Φ_0^{AI} by a factor of 1.2–2 in the 200–300 K range influences only slightly the total Φ dependence which is mainly determined by the exponential terms of (6.7).

The observed increase in $\Phi_0(h\nu)$ value at $h\nu \lesssim 2.4$ eV in Pc (Fig. 6.7) supports the idea that in the near-threshold region CP states may be populated via optical CT transitions which might become the dominant channel for near-neighbor CP state population.

Another explanation may be that at small r_{th} values the Onsager formula [Eq. (6.7)] may become invalid as indicated in Ref. 274. However, corrections to Eq. (6.7) introduced by Noolandi and Hong[274] for small r_{th} values are too small in the given $\mathcal{E}$ range to explain the observed $\eta(h\nu)$ deviations from $\Omega(h\nu)$ curve.

The charge carrier photogeneration mechanism via direct CT-transitions (Fig. 6.2) is based on theoretical predictions[39–42] which were later confirmed experimentally in Ac, Tc, and Pc crystals[317,318] (see Section 6.1). We have demonstrated[344,349,351] that this alternative photogeneration mechanism is complementary to the AI mechanism and actually is responsible for the increase in $\Phi_0(h\nu)$ value at the very photogeneration threshold (see Figs. 6.7 and 6.8).

The optical CT-transition energies should not be identified with the energies E_{CP} of the corresponding final CP states formed. After the very fast electronic transitions, with typical time scale $\tau_e = 10^{-16}-10^{-15}$ s, a much slower vibronic relaxation process ($\tau_v = 10^{-14}-10^{-13}$ s) takes place resulting in the formation of a molecular polaron state (see Chapters 3 and 4). Therefore the energy of the CP state E_{CP} equals, according to Fig. 6.2:

$$E_{CP} = h\nu_{CT}-E_b, \qquad (6.19)$$

where E_b is the energy of the formation of a MP state, and equals ca. 0.15 eV for Ac, Tc, and Pc crystals (see Chapter 3).

The surprisingly high oscillator strength ($f = 0.01$) observed in Ref. 317 for CT transitions in Tc and Pc layers may be interpreted in terms of the Bounds and Siebrand theory,[39,40] according to which the CT transition may derive the oscillator strength from coupling to neutral Frenkel exciton states. However, the analysis of electroabsorption data reported in Ref. 317 shows that the CT transition intensity α_{CT} decreases exponentially with increasing transition distance r_{CT}

$$\alpha_{CT}(r) = \alpha_{CT}^1 \exp\left[-\frac{r_{CT}-r_{CT}^1}{\rho_0}\right], \qquad (6.20)$$

where α_{CT}^1 and r_{CT}^1 are the CT absorption coefficient and transition distance to the nearest-neighbor molecule, respectively. The authors of Ref. 349 used Eq. (6.20) to approximate the $\alpha_{CT}(r)$ dependences reported in Ref. 317 for Tc and Pc. The least-squares fit yields the following attenuation length values for Pc: $\rho_0 = 2.50$ Å ($\gamma^2 = 0.96$) using the upper limits of reported α values and $\rho_0 = 1.10$ Å ($\gamma^2 = 0.98$) for the lower limits of α.

The corresponding $\Phi_0^{CT}(r)$ curve should have a similar functional dependence

$$\Phi_0^{CT}(r) = \Phi_0^{CT^1}(r)\exp\left[\frac{r_{CT}-r_{CT}^1}{\rho_0}\right]. \qquad (6.21)$$

Figure 6.9 shows that the $\Phi_0(r)$ curves actually represent a superposition of efficiencies of practically constant autoionization Φ_0^{AI} and of CT-transitions $\Phi_0^{CT}(r)$ exponentially decreasing with r. Approximation of a number of $\Phi_0^{CT}(r)$ curves, according to Eq. (6.21), yields an average ρ_0 value $\rho_0 = (5.2 \pm 1.2)$ Å. Both approximations, taking into account errors in the $r_{th}(h\nu)$ determination, are in satisfactory agreement and support the model of two independent CP population pathways in the threshold region of Pc. For Ac the increase in $\Phi_0(h\nu)$ near the threshold (see Fig. 6.8) may also indicate the presence of CT transitions in the region anticipated by the authors of Ref. 39 and later confirmed experimentally by Sebastian et al.[318]

These results confirm the concept of a complementary character of both AI and CT-transition models of photogeneration in the near-threshold spectral region of intrinsic photogeneration in polyacene crystals. The additive nature of both complementary photogeneration mechanisms may be described by the following empirical formula which can be derived from experimental data shown in Fig. 6.9:

$$\Phi_0(h\nu) = \Phi_0^{AI}(h\nu)+\Phi_0^{CT} = \Phi_0^{AI}(h\nu)+\Phi_0^{CT^1}\exp\left[\frac{r_{CT}-r_{CT}^1}{\rho_0}\right]. \qquad (6.22)$$

Second, the values of the parameter ρ_0 in the exponential dependences of $\alpha_{CT}(r)$ and $\Phi_0^{CT}(r)$ [see formulas (6.20) and (6.21) and Fig. 6.9] actually confirm the fact, that the initial separation length of both charge carriers increases in the relaxation process, i.e., $r_{CP} > r_{CT}$, as it was anticipated by Baessler et al.[318] This effect has been independently confirmed by computer simulation of charge carrier separation processes in Pc crystals.[344] Thus, e.g., the initial charge separation distance r_{CT}^1 for optical CT-transition to the nearest-neighbor molecule in Pc increases in the relaxation process from $r_{CT}^1 = 7.9$ Å to $r_o \approx 20$ Å (see Fig. 6.31). One can also see the postrelaxation distances of similar order of magnitude in Fig. 6.9.

Popovic and Sharp[289] have reported remarkable insensitivity of quantum efficiency to $h\nu$ and the absence of a distinct threshold in α-metal-free phthalocyanine (H$_2$Phc). Noolandi and Hong[274] explain this effect in terms of an alternative model according to which a photoexcited molecule first dissipates its energy via intramo-

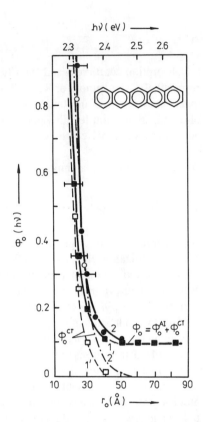

FIGURE 6.9. Typical experimental $\Phi_0(r_0, h\nu)$ curves in Pc (sample #23, $L = 1.5$ μm): (1) at $\mathcal{E} = 1\times10^4$ V/cm and $T = 205$ K; (2) at $\mathcal{E} = 4\times10^4$ V/cm and $T = 205$ K; (1') and (2') corresponding $\Phi_0^{CT}(r_0, h\nu)$ dependences obtained according to formula (6.21) yielding $\rho_0 = 4.56$ Å ($\gamma^2 = 1.0$) for curve 1' and $\rho_0 = 6.08$ Å ($\gamma^2 = 0.98$) for curve 2'.[349]

lecular conversion and may ionize only from a long-lived S_1 state, forming a CP-state with a constant separation length r_0 of the order of the intermolecular spacing. In this case, too, there may be another independent pathway of CP-state population via optical CT transition, which has been detected in CuPhc and H_2Phc layers by Blinov and Kirichenko,[35] using an electroabsorption technique. Since both CP state population channels yield presumably only deep-lying CP states, the quantum efficiency Φ of Phc is small, of the order of 10^{-4}–10^{-3} electron per photon.[289]

According to the initial model of autoionization proposed by Pope *et al.*[115,116,286] and later refined by Jortner[148] the authors assumed the process of autoionization as configurational interaction between the excited neutral molecular state ψ_s^* and a continuum of ionized states φ_{E_c} [see Eq. (6.1)]. It was supposed, in terms of the model, that the wave function φ_{E_c} actually describes a delocalized quasifree electron in the conductivity band E_c of the crystal. However, as we have seen in Chapter

3, devoted to polarization phenomena, charge carriers in OMC are strongly localized (with exception of very low temperatures) and move in the lattice in form of
polaron-type quasiparticles. Thus, as seen in Section 6.1, an ejected hot electron
actually becomes scattered at every lattice site, the free path of it being of the order
of a lattice constant, namely, $l_o \approx a_o$ (see Section 7.2.3).

One may therefore suggest that the ionization process in OMC can be described
in terms of Mulliken's theory used for the description of CT-complexes[252] (see also
Ref. 351, p. 25 and references therein).

In terms of this approach, the generalized wave equations, describing nonpolar
and polar components of a weak CT-complex, may be extended for description of
similar state in one-component systems like OMC.[351] In this case the ionized state
φ_E in the process of optical CT-transition may be presented as a mixture (superposition) of an excited neutral molecular state $\psi_s(M^*)$ and a state formed by the
CT-transition $\psi_i(M_1^+ + M_2^-)$, where ψ_i describes a charge pair bounded by
Coulombic interaction. It is important to emphasize, that in terms of such an
approach charge carriers are regarded as localized on molecules M_1 and M_2. Such
a mixed state may be described, in the spirit of Mulliken's approach, by a wave
function ψ_E constructed as a linear combination of the neutral molecular Frenkel
state $\psi_s(M^*)$ and of the ionized CT-state $\psi_i(M_1^+ + M_2^-)$:

$$\psi_E = a_E \psi_s(M^*) + b_E \psi_i(M_1^+ + M_2^-) + \sum_i c_i \chi_i, \qquad (6.23)$$

where the term $\sum_i c_i \chi_i$ describes the intermolecular relaxation channel [see Eq.
(6.1)]. Equations (6.23) and (6.1) are similar in the physical sense, the difference
lies in the fact that Eq. (6.23) takes into account the charge carrier localization
phenomenon.

Equation (6.23) allows us to describe the possible electric field $\mathcal{E}$ influence on the
ionization efficiency. In terms of such an approach, it may be assumed that in strong
electric fields the contribution of the ionic component increases, caused by the
electric field dependence of the coefficient b_E, namely, $b_E = b_E(\mathcal{E})$.

Recently Braun[43] (see also Ref. 319) discussed the influence of electric field $\mathcal{E}$ on
the rate constant of CT-state dissociation as an alternative photogeneration mechanism, without the intermediate thermalization stage, applicable to the CT-complexes. It is interesting to mention that in this case, Braun used another Onsager's
equation derived already in 1934[276] for the description of ion-pair dissociation in
weak electrolytes.

6.2.5. Some General Characteristics of the Autoionization Mechanism

Up to now we have mainly discussed the AI mechanisms at the near-threshold
spectral region of intrinsic photogeneration in OMC, as schematically presented in
Fig. 6.1. Recently Silinsh and Inokuchi[341] analyzed spectral $\eta(h\nu)$ and $E_a^{ph}(h\nu)$

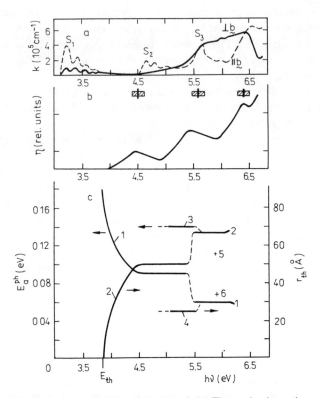

FIGURE 6.10. Spectral characteristics of Ac crystal. (a) Electronic absorption spectra parallel (∥) and perpendicular (⊥) to the **b** crystal axis. (b) Relative photocurrent $\eta(h\nu)$ electrons/photon (unpolarized light) (after Ref. 116). (c) Spectral dependence of photogeneration activation energy $E_a^{ph}(h\nu)$: (1) according to Refs. 73 and 160; (3) according to Refs. 161 and 162; (5) E_a^{ph} value at $h\nu = 5.9$ eV after Refs. 161 and 162 and mean thermalization length $r_{th}(h\nu)$: (2) according to Refs. 73 and 160; (4) according to Refs. 161 and 162; (6) r_{th} value at $h\nu = 5.9$ eV after Refs. 161 and 162.

characteristics of photoconductivity in polyacenes over a wide spectral range and have discussed some general aspects and the intrinsic nature of the AI mechanism.

As mentioned before, in the near-threshold region (e.g., in the region from 3.9 to 4.5 eV in case of Ac crystals, see Fig. 6.10c) the E_a^{ph} value monotonously decreases and the corresponding r_{th} value increases with increasing photon energy $h\nu$ in accordance with the kinetic diagram shown in Fig. 6.1.

However, at $h\nu \geqslant 4.5$ eV the behavior of the $E_a^{ph}(h\nu)$ dependence drastically changes remaining constant at the value of $E_a^{ph} = 0.09$ eV up to $\approx 5.2-5.4$ eV. On further increase of excitation energy E_a^{ph} drops down to another plateau at $h\nu \approx 5.4$ eV with $E_a^{ph} = 0.068$ eV. This means that, according to Eq. (6.3) we have in these spectral "plateau" regions a constant thermalization length of $r_{th} = 50$ Å and $r_{th} = 67$ Å, respectively (see Fig. 6.10c).

Chance and Braun suggested[73] that this unusual behavior of $E_a^{ph}(h\nu)$ and $r_{th}(h\nu)$ dependences may be connected with intramolecular relaxation, after excitation, to specific molecular states, having a higher autoionization probability. Such an assumption gives a plausible explanation to the constant value of both E_a^{ph} and r_{th} in these spectral regions.

According to the kinetic diagram in Fig. 6.1, one might expect that the quantum efficiency of photoconductivity $\Phi(h\nu)$ should also increase with growing excitation energy $h\nu$. However, such an increase is observed only at the near-threshold spectral region. At higher excitation energies the $\Phi(h\nu)$ dependences exhibit distinct structure with marked maxima and minima, as already demonstrated in earlier photoconductivity studies of Ac and Tc crystals,[116] and later in those of Pc crystals[349] (see Ref. 288, p. 472 and Fig. 6.10b). Pope and co-workers[115,288] showed that the structure of $\Phi(h\nu)$ dependences, i.e., the position of its maxima, may be directly connected with the appearance of the excitation of lower-lying valence levels.

Thus, the main problem of the photogeneration mechanism is reduced to the question of how to estimate the possible contribution of transitions from low-lying ground (valence) levels and relaxation processes of excited molecular states to the behavior of the main intrinsic photoconductivity spectral characteristics, viz $E_a^{ph}(h\nu)$ and $\Phi(h\nu)$. The controversy of the problem has been vividly formulated by Pope and Swenberg (Ref. 288, p. 494). Namely, that the unusual behavior of these characteristics "may be due to the existence of specific autoionizing states and/or to the excitation of low-lying valence levels".

The authors of Ref. 341 have demonstrated that the structure of $\Phi(h\nu)$ dependences, i.e., the sequence of maxima on the $\Phi(h\nu)$ curve, is actually caused by the excitation of low-lying valence levels. On the other hand, they have shown that the "plateaus" on the $E_a^{ph}(h\nu)$ dependences are unequivocally determined by intramolecular relaxation of the excited state to specific, quasimetastable AI states. Consequently, the authors of Ref. 341 suggest that the $\Phi(h\nu)$ and $E_a^{ph}(h\nu)$ dependences allow distinguishing between these two processes, thus changing the ambivalent conjunction "and/or" to the definite "and".

They also analyze the kinetics of direct autoionization of excited states and autoionization after relaxation to specific metastable states, as well as factors determining the contribution of these two competitive processes in photogeneration phenomena.

6.2.5.A. Spectral Dependences of Intrinsic Photogeneration Quantum Efficiency. The Role of Transitions from Low-lying Valence Levels

It may be shown[341] that the spectral dependence of photogeneration quantum efficiency $\Phi = \Phi(h\nu)$ (at given $\mathcal{E}$ and T), its structure, and its nonmonotonous behavior is determined by excitation of lower-lying valence levels S_{-n}.

It should be mentioned that in polyacenes the filled valence levels and excited

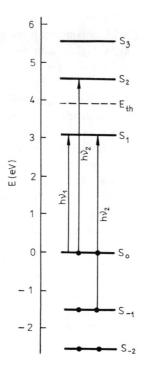

FIGURE 6.11. Schematic diagram of molecular orbital energy levels of valence and excited electronic states in an anthracene crystal. E_{th} — threshold energy of intrinsic photogeneration.[341]

state levels are positioned symmetrically below and above the site energies α due to resonance splitting: $E_i = \alpha + m_i\beta$.[133,288] This predicted symmetry of energy indices has been confirmed in the energy level diagram, e.g., for Ac (Fig. 6.11), based on optical and electron spectroscopy data (see references in Ref. 288).

In the near-threshold region at $h\nu > E_{th}$ excitation takes place from the S_0 ground level. In this spectral range the r_{th} value increases with growing photon energy $r_{th} = r_{th}(h\nu)$ (see Fig. 6.1), causing, according to Eq. (6.7), a corresponding increase of the photogeneration efficiency $\Phi(h\nu)$. The $\Phi(h\nu)$ dependence can be obtained from experimentally measured spectral dependence of intrinsic photocurrent quantum efficiency $\eta(h\nu)$, since $\eta(h\nu)$ is directly proportional to $\Phi(h\nu)$ viz. $\eta(h\nu) = b\Phi(h\nu)$, where b is the transfer coefficient of generated carriers to the collecting electrodes (see Section 6.2.3). The $\eta(h\nu)$ characteristics have been carefully measured for Ac,[116,160] Tc,[116] and Pc[349] crystals (see Fig. 6.10b and Ref. 288, p. 472).

However, the $\eta(h\nu)$ dependences preserve a rising character only in the threshold region inside the range $E_{th} < h\nu_1 < h\nu_2$. The $h\nu_2$ is the critical value of excitation energy at which the transition from level S_{-1} to S_1 becomes possible and a competitive excitation channel from the first lower-lying valence level S_{-1}

switches on. Since the excited level S_1 lies below the threshold E_{th} these transitions do not yield charge carriers. This causes a decrease in the $\Phi(h\nu)$ value and formation of a maximum on the $\eta(h\nu)$ curve (see Fig. 6.10b).

As can be seen from Fig. 6.10b, the first maximum on the $\eta(h\nu)$ curve actually corresponds to the reported peak of density of states in the valence band, as estimated from the kinetic energy distribution of photoemitted electrons (Ref. 288, p. 472) (see shaded region at $\approx$ 4.5 eV in Fig. 6.10a).

The decrease in the $\Phi(h\nu)$ value at $h\nu > h\nu_2$ can be estimated quantitatively in the following way.

It can be assumed that in the spectral region $E_{th} < h\nu < h\nu_2$ the CP-state photogeneration efficiency $\Phi_0(h\nu)$ remains constant and determines the total photogeneration efficiency according to Eq. (6.6). At $h\nu > h\nu_2$ the $\Phi(h\nu)$ value is decreased by the contribution of "futile" $S_{-1} \rightarrow S_1$ transitions

$$\Phi(h\nu, \mathcal{E}, T) = (1-k)\Phi_0\,\Omega(r_{th}, \mathcal{E}, T) \qquad (6.24)$$

where k is a coefficient characterizing the relative contribution of $S_{-1} \rightarrow S_1$ transitions.

On further increase of the excitation energy, other lower-lying levels $(S_{-2}, S_{-3}, \dots)$ switch on, and corresponding maxima in the $\eta(h\nu)$ curve appear in good agreement with the energy sequences of the density of respective valence levels (see Figs. 6.10a and 6.10b). Even better agreement is obtained in the case of $\eta(h\nu)$ curves for Tc crystals (see Ref. 288, p. 272). There are other critical turning points in the discussed photogeneration process. As the excitation energy increases, at the values of $h\nu \geqslant \Delta E = E_{th} - E_{S_{-n}}$ transitions from lower-lying valence levels begin to contribute to the photogeneration yield.

To make a rough evaluation of the impact of the appearance of these additional photogenerative channels, let us consider a possible increase in the $\Phi(h\nu)$ value comparing the yield of photogeneration transitions from S_0 and S_{-1} at $h\nu \geqslant E_{th} - E_{S_{-1}}$. Let us make a simplified analysis based on Eqs. (6.6) and (6.7). At field strength values $\mathcal{E} \leqslant 10^4$ V/cm the field-dependent correction term in Eq. (6.7) is small and may be neglected[344] (see also Ref. 288, p. 490). It has been shown that at such field values the thermalization length r_{th} is also practically field-independent,[344,349,351] and an isotropic Onsager approach can be applied. In such a case the dissociation efficiency Ω becomes only $r_{th}(h\nu)$ and T-dependent, and its value is practically determined only by the first term of Eq. (6.7). One should notice that in this term there is high prevalence of the exponential r_{th} dependence while the sum may be regarded as a slightly temperature-dependent constant $A(T)$. In such a simplified approach the CP state dissociation efficiency Ω may be presented in the following way:

$$\Omega[r_{th}(h\nu), T] = A(T)\exp[-r_c/r_{th}] = A(T)\exp[-e^2/4\pi\epsilon\epsilon_0 kTr_{th}]. \qquad (6.25)$$

As a result, the quantum efficiency $\Phi(h\nu, T)$ of both above mentioned transitions from S_0 and S_{-1} levels equals:

$$\Phi(h\nu, T) = (1-k)\Phi_0 A_1(T)\exp[-r_c/(r_{th})_1] + k\Phi_0 A_2(T)\exp[-r_c/(r_{th})_2], \tag{6.26}$$

where the first term describes the contribution of transitions from the S_0 level, and the second term transitions from the S_{-1} level.

Since the r_{th} value is $h\nu$-dependent (see Fig. 6.1), obviously $(r_{th})_1 > (r_{th})_2$, and consequently

$$\exp[-r_c/(r_{th})_1] \gg \exp[-r_c/(r_{th})_2].$$

This means that in the autoionization and subsequent thermalization processes two kinds of CP-states are formed, a group of shallow CP states with high dissociation probability and a group of deep-lying CP states with negligible dissociation yields (see Fig. 6.1). Therefore the total efficiency is essentially determined by the first term:

$$\Phi(h\nu, T) \approx (1-k)\Phi_0 A_1(T) \times \exp[-e^2/4\pi\epsilon\epsilon_0 kT(r_{th})_1]. \tag{6.27}$$

However, one should mention that, in the spectral region at which the excitation from the S_0 level reaches the plateau region with $(r_{th})_1 = $ const, the relative contribution of the second term increases with growing $h\nu$. Nevertheless, due to the exponential dependence the second term will remain much smaller than the first one. This may influence the shape of the $\eta(h\nu)$ curve but not substantially affect its main structure.

We may therefore conclude that the sequence of peaks on the $\eta(h\nu)$ curve in reality reflects the energy structure of low-lying valence levels.

6.2.5.B. Spectral Dependences of Photogeneration Activation Energy $E_a^{ph}(h\nu)$. The Role of Intramolecular Relaxation Processes

It may be shown[341] that the steplike character of $E_a^{ph}(h\nu)$ dependences (see Fig. 6.10c) is unequivocally determined by the kinetic processes in the upper, excited state "niche," namely, by the competition between the channels of direct photoionization (direct AI) and autoionization from specific AI states after the intramolecular relaxation stage and not by optical transitions from different low-lying valence levels, i.e., by the ground state "niche."

The E_a^{ph} value is usually determined from the temperature dependence of $\eta(h\nu)$ [or $\Phi(h\nu)$] in semilog coordinates $(\ln \Phi, 1/T)$. From Eqs. (6.6) and (6.25) it follows that

$$\ln \Phi(h\nu) = -\frac{e^2}{4\pi\epsilon\epsilon_0 kr_{th}(h\nu)} \times 1/T + \ln \Phi_0(h\nu) + \ln A. \tag{6.28}$$

Characteristic $\ln \Phi = f(1/T)$ dependences are shown schematically in Fig. 6.12.

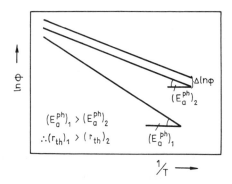

FIGURE 6.12. Typical temperature dependences of photogeneration quantum efficiency Φ in semilog coordinates ($\ln \Phi, 1/T$) used for determination of photogeneration activation energies E_a^{ph} in different spectral regions.[341]

As can be seen from Eq. (6.28) and Fig. 6.12 the E_a^{ph} value is equal, in ($\ln \Phi, 1/T$) coordinates, to the slope $e^2/4\pi\epsilon\epsilon_0 k r_{\text{th}}$ and is actually determined by the mean thermalization length r_{th} at given photon energy $h\nu$ [see Eqs. (6.2) and (6.3)]. Thus, the slope of the $\Phi(h\nu, T)$ in ($\ln \Phi, 1/T$) coordinates (see Fig. 6.12) is practically determined only by the first, T-dependent term in Eq. (6.28), while the $h\nu$-dependent changes of Φ_0, due to switching-on the lower-lying valence states [see (6.24) and (6.26)], yield only parallel shift of the linear $\ln \Phi = f(1/T)$ dependency. In other words, the determined E_a^{ph} values are insensitive to $\Phi_0(h\nu)$ changes: $\Delta\Phi_0(h\nu)$.

This approach can be proved by analyzing carefully measured $\ln \Phi = f(1/T)$ dependence in Ac crystals in Ref. 73 (see also Ref. 288, p. 491). The changing slopes of these straight lines actually yield the $E_a^{\text{ph}}(h\nu)$ dependence. However, in the "plateau" spectral range of $E_a^{\text{ph}}(h\nu)$ (from 5.4 to 6.0 eV) where $\Phi(h\nu)$ reveals a considerable change, one observes only a parallel shift of $\ln \Phi = f(1/T)$ dependence.

Let us now consider, in terms of this approach, the steplike character of the $E_a^{\text{ph}}(h\nu)$ dependence, obtained, for Ac crystals, by Braun and coworkers[73,160] (see Fig. 6.10c, curve 1), and the corresponding $r_{\text{th}}(h\nu)$ dependence (curve 2). In the threshold spectral region from $E_{\text{th}} = 3.9$ eV up to ≈ 4.5 eV, i.e., in the energy interval $E_{\text{th}} \leqslant h\nu \leqslant E_{S_2}$, $E_a^{\text{ph}}(h\nu)$ decreases and $r_{\text{th}}(h\nu)$ increases with growing photon energy $h\nu$ in accordance with the direct AI model, as shown in Fig. 6.1. In contrast to this behavior, in the energy range from 4.5 to ≈ 5.4 eV, i.e., in the interval $E_{S_2} \leqslant h\nu \leqslant E_{S_3}$, the $E_a^{\text{ph}}(h\nu)$ value remains constant as well as the corresponding $r_{\text{th}}(h\nu)$ value, i.e., $r_{\text{th}}(h\nu) = \text{const} = 50$ Å. This means that in this energy range the AI dominates after intramolecular relaxation to a specific AI level at ≈ 4.5 eV which apparently corresponds to the 0-0 vibronic band of the S_2 electronic state. Most probably, in this spectral range the rate constant of intramolecular relaxation k_n is greater than the rate constant of direct AI k_{AI}, i.e., $k_n > k_{\text{AI}}$, and the initial vibronically excited S_2 state rapidly relaxes to the 0-0 level from which AI takes place, providing a constant r_{th} value.

As the excitation energy $h\nu$ passes over the 0-0 transition level of the S_3 state at ≈ 5.5 eV a new "plateau" of $E_a^{ph}(h\nu)$ dependence emerges yielding a constant $r_{th}(h\nu)$ value $r_{th} \approx 67$ Å. These data demonstrate that in this case the 0-0 vibronic levels of the S_2 and S_3 electronic states serve as specific postrelaxation AI levels.

Similar $E_a^{ph}(h\nu)$ dependences in the threshold spectral region have been obtained for Pc crystals.[349,411] In this case also in the near-threshold region, at $h\nu > 2.25$ eV, the $E_a^{ph}(h\nu)$ dependence decreases and $r_{th}(h\nu)$ increases with growing $h\nu$ in accordance with the model of direct AI.

However, at $h\nu > 2.8$ eV, as the excitation energy $h\nu$ approaches the S_2 band, a "plateau" for both dependences appears (see also Ref. 351, p. 252).

It is important to mention that Kimura and coworkers[134,308] have observed a similar kind of competitive photoionization for various electronically excited states of molecules in the gas phase. Using a laser photoelectron spectroscopy technique combined with multiphoton resonance ionization method, the authors have demonstrated that in the case of Nph or triethylamine molecules, when excited in the S_2 spectral range, photoelectrons can be ejected either directly from the initial excited state, or from vibronically relaxed states. A different kind of competitive photoionization of organic molecules has been reported by Hager, Smith, and Wallace.[124] The authors have measured the two-color threshold photoionization spectra of gaseous jet-cooled aniline and observed both direct photoionization and vibrationally selective autoionizing Rydberg structures in these spectra. A theoretical interpretation of this phenomenon has been given by Lin, Boeglin, and Lin.[218]

The existence of two competitive autoionization channels in aromatic crystals has been convincingly demonstrated by Zagrubskii, Petrov, and Lovcjus.[410] These authors measured the energy spectra of photoemitted electrons from Tc crystals and observed two distinct kinds of electron energy distribution peaks: "moving" (dependent on photon energy) and fixed (independent of photon energy). These experimental data can serve as an additional proof of the presence of the above described two competitive autoionization channels of photoexcited molecules in the crystal: one of direct autoionization of the excited states, and one of autoionization from a fixed electronic level after intramolecular vibronic relaxation.

The most unambiguous results about the possible mechanism of charge carrier photogeneration through excited molecular states in aromatic crystals have been recently reported by Kotani and coworkers.[161,162,195,197] These authors used a straightforward two-color two-step excitation technique which completely excludes the participation of lower-lying valence states in the photogeneration processes. According to this technique, in the first step the molecules of the crystal are excited by a laser pulse to S_1 state: $S_0 + h\nu_1 \rightarrow S_1$. In the second step the S_1 exciton state is excited by a tuned dye-laser pulse: $S_1 + h\nu_2 \rightarrow S_n$. The ionization current of the S_n state ($S_n \rightarrow h^+ + e^-$) is detected by the transient carrier time-of-flight method. The authors measured the spectral dependence of the photoionization yield of S_n states in a p-terphenyl crystal which shows a distinct "plateau" from 5.8 to 6.4 eV.

It indicates that the kinetic energy of the ejected electrons and their corresponding mean thermalization length r_{th} remains constant in this energy range. These results can be regarded as an unequivocal proof in favor of the existence of a specific autoionization state at 5.8 eV which, after relaxation from the initial, higher-lying excited S_n state, is autoionizing with a high efficiency of the order of ≈ 0.1 electron/photon. To avoid ambiguities the authors of Refs. 195 and 197 have also shown that the initial S_n state is in reality a neutral excited molecular state but not a CT-state.

Recently, Kotani and coworkers have applied the two-color two-step excitation technique for photoionization studies of Ac crystals.[161,162,197] The authors observed a constant value of $E_a^{ph} = 0.14$ eV and a corresponding constant value of $r_{th} = 25 \pm 5$ Å in the energy interval between 5.1 and 5.5 eV (see Fig. 6.10c, curves 3,4). Although the $E_a^{ph}(h\nu) = $ const and $r_{th}(h\nu) = $ const dependences, as measured by Kotani and coworkers,[161,162,197] cover a narrower energy range, they coincide with the "plateau" region of the dependences obtained earlier by Chance and Braun[73] (see Fig. 6.10c, curves 1,2). This coincidence proves, beyond any doubt, the assumption that the regions of $E_a^{ph}(h\nu) = $ const and $r_{th}(h\nu) = $ const (see Fig. 6.10c) correspond to AI from specific AI state after intramolecular relaxation.

One may notice that the absolute values of $E_a^{ph}(h\nu) = $ const and $r_{th}(h\nu) = $ const differ in the two cases by a factor of two. However, it should be mentioned that the absolute values of E_a^{ph} and r_{th} are rather sensitive to the conditions of the experiment (quality of crystal samples, accuracy of measurement, parameters and approximation formulas used, etc.).

The most important in this case is the fact that the $E_a^{ph}(h\nu)$ and $r_{th}(h\nu)$ dependences remain constant at the given energy range, thus proving the presence of postrelaxation autoionization.

Kotani et al.[161,162,197] also observed an increase of $r_{th}(h\nu)$ at $h\nu > 5.5$ eV which approximately corresponds to a steplike increase in the $r_{th}(h\nu)$ dependence of Ref. 73 (see Fig. 6.10c, curves 2 and 4). In an additional experiment the authors of Refs. 161 and 162 applied one-color two-step excitation at higher energies ($2h\nu = 2 \times 2.95 = 5.90$ eV). At this discrete value of the excitation energy they observed decreased $E_a^{th} = 0.11$ eV and increased $r_{th} = 34$ Å values (see Fig. 6.10c, points 5 and 6).

It is interesting to mention that the relative increase of r_{th} in this case is similar to that of Ref. 73, namely, $34/25 \approx 67/50$ (see Fig. 6.10c). This may indicate that the point 6 on Fig. 6.10c actually corresponds to the expected second "plateau" of $r_{th}(h\nu)$ at $h\nu > 5.5$ eV.

Thus, the results of Kotani and co-workers[161,162,195,197] confirm unequivocally that the step-like character of $E_a^{ph}(h\nu)$ and $r_{th}(h\nu)$ dependences is exclusively determined by kinetics of the relaxation processes in the upper, excited state "niche."

6.2.5.C. Kinetics of the Competitive Autoionization Processes[341]

If a molecule in the crystal is photoexcited to some higher S_n state $S_0 \rightarrow S_n$ ($v = i$), for this vibrationally excited electronic state there exist two possible pathways of energy dissipation (see Fig. 6.13a): a "horizontal" channel of direct autoionization from the level $v = i$ with a rate constant k_{AI}, and a "vertical" channel of intramolecular relaxation with a rate constant k_n. We will show that in case of $k_{AI} > k_n$, the direct autoionization will be prevalent, and, vice versa, if $k_n > k_{AI}$, the intramolecular relaxation process dominates.

The whole relaxation kinetics of internal conversion from one electronic state S_n to another S_m, i.e., $S_n \rightarrow S_m$, is determined by the probability of electron transition between the corresponding potential surfaces of these states. The rate constant k_m of such a transition is actually determined by the overlap integral $\int \chi_i^{(m)} \chi_o^{(n)}$ of the vibrational wavefunctions of the corresponding S_n and S_m electronic states, i.e., by their Franck-Condon factor[394] (see Fig. 6.13b).

If the overlap is small, as shown in Fig. 6.13b, and the integral $\int \chi_i^{(m)} \chi_o^{(n)} \approx 0$, the relaxation rate of the electron becomes retarded at the bottom of the potential surface of the $S_n(v = 0)$ state. If in this case the rate constant k_m for the transition from S_n to S_m state becomes smaller than the rate constant k'_{AI} of autoionization from the $S_n(v = 0)$ level, a new "horizontal" channel of postrelaxed AI opens. Thus, as can be seen from Fig. 6.13, the energy structure of excited states and the corresponding Franck-Condon factors actually determine which of the two AI pathways will emerge as the dominant one. Let us briefly discuss the kinetics of the two competitive AI mechanisms.

The steady-state population of the level $S_n(v = i)$ (see Fig. 6.13a) may be obtained from the following equation:

$$I = (k_{AI} + k_n)[S_n(v = i)], \qquad (6.29)$$

where I is the flow of photons with corresponding energy $h\nu$ photons/cm^{2}s; $[S_n(v = i)]$ is the population of the given level.

On the other hand, the population of the level $S_n(v = 0)$ can be determined from the equation

$$k_n[S_n(v = i)] = (k'_{AI} + k_m)[S_n(v = 0)]. \qquad (6.30)$$

Consequently

$$[S_n(v = i)] = I/(k_{AI} + k_n) \qquad (6.31)$$

and

$$[S_n(v = 0)] = \frac{k_n[S_n(v = i)]}{k'_{AI} + k_m} = \frac{k_n I}{(k'_{AI} + k_m)(k_{AI} + k_n)}. \qquad (6.32)$$

The corresponding autoionization quantum yields $\Phi_0(v = i)$ and $\Phi_0(v = 0)$ can be presented in the following way:

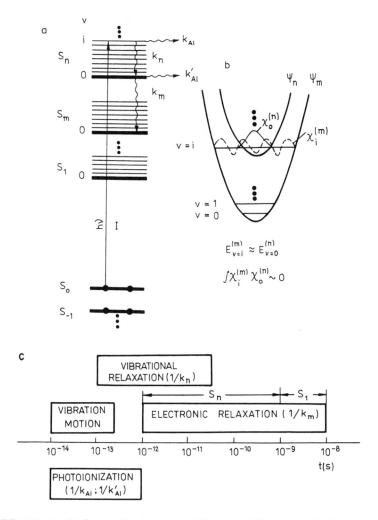

FIGURE 6.13. (a, b) Schematic diagrams of the competitive autoionization processes in OMC.[341] (a) Diagram illustrating the two competitive channels of autoionization from excited molecular S_n state. k_{AI}, rate constant of direct ionization; k_n, rate constant of vibrational relaxation of electronic S_n state; k'_{AI}, rate constant of autoionization after intramolecular relaxation; k_m, rate constant of internal conversion from S_n to S_m electronic state. (b) Schematic diagram of the potential surfaces S_n and S_m of electronic states, illustrating the Franck-Condon principle of overlaping vibrational wave functions $\chi_0^{(n)}$ and $\chi_0^{(m)}$ of the corresponding electronic S_n and S_m states.[341] (c) Characteristic time scales of photoionization and electronic and vibrational relaxation processes in molecules (after Ref. 107a).

1. For direct autoionization

$$\Phi_0(v = i) = \frac{k_{AI}[S_n(v = i)]}{I} = \frac{k_{AI}}{k_{AI}+k_n}. \tag{6.33}$$

It may be seen that Eq. (6.33) is a particular case of the more general equation (6.10). In this case, at $k_{AI} > k_n$, direct AI from the initial excited state $S_n(v = i)$ will dominate, and the mean thermalization length r_{th} will be dependent on photon energy hv (see Fig. 6.1):

$$r_{th} = r_{th}(hv). \tag{6.34}$$

2. Autoionization after intramolecular relaxation

$$\Phi_0(v = 0) = \frac{k'_{AI}[S_n(v = 0)]}{I} = \frac{k'_{AI}k_n}{(k'_{AI}+k_m)(k_{AI}+k_n)}. \tag{6.35}$$

Equation (6.35) is not convenient for analysis; for this purpose is better to use the ratio of both AI quantum yields

$$\frac{\Phi_0(v = 0)}{\Phi_0(v = i)} = \frac{k'_{AI}}{k_{AI}}\left[\frac{k_n}{k'_{AI}+k_m}\right]. \tag{6.36}$$

If the overlap of the vibrational wavefunction of the S_n and S_m states is small, as in Fig. 6.13b, and the Franck-Condon factor $\int \chi_i^{(m)}\chi_o^{(n)} \to 0$, one may expect that $k'_{AI} \gg k_m$. For such a case we obtain the following simple ratio of the two AI quantum yields:

$$\frac{\Phi_0(v = 0)}{\Phi_0(v = i)} \approx \frac{k_n}{k_{AI}}. \tag{6.37}$$

Consequently, if $k_n > k_{AI}$ (or even $k_n \gg k_{AI}$) AI after vibrational relaxation from a fixed, quasimetastable $S_n(v = 0)$ level dominates (or even completely prevails) over direct autoionization from the initially photoexcited $S_n(v = i)$ state. In this case r_{th} does not depend on photon energy, i.e.,

$$r_{th}(hv) = \text{const} \tag{6.38}$$

at least in the energy range of vibronic levels of the given electronic state S_n (see Fig. 6.13a). Thus, the prevalence of this AI mechanism yields the "plateau" regions of $r_{th}(hv)$ and $E_a^{ph}(hv)$ dependences (see Fig. 6.10c).

If $k_n \approx k_{AI}$, both AI mechanisms may coexist, as demonstrated, e.g., by the photoelectron emission spectra of Tc crystals.[410]

The discussed kinetics work for higher excited electronic states above the $S_2(v = 0)$ level. At the threshold spectral region, in the energy range $E_{th} < hv < E_{S_2}(v = 0)$, a specific situation arises. In this case only a direct AI channel emerges, since electrons, relaxed to $S_1(v = 0)$ state (which in aromatic crystals is below the intrinsic photogeneration threshold E_{th}, inside the energy gap), do not participate in the intrinsic photogeneration process. Thus, one should apply Eq. (6.33) for direct AI yield. If $k_n > k_{AI}$, the AI quantum yield is small. This is so in the case, e.g., of an Ac crystal, having a Φ_0 value of $\approx 5\times10^{-3}$ electrons/photon,[73] which shows that $k_n \gg k_{AI}$. In this case the nonradiative intramolecular relaxation to $S_1(v = 0)$ and subsequent fluorescence $S_1 \rightarrow S_0 + h\nu$ strongly dominate.

The condition of high autoionization efficiency is $k_{AI} \geq k_n$, which apparently holds for Pc and Tc crystals.[349] Thus, the ratio $k_{AI}/(k_{AI}+k_n)$ (6.33) may serve as a criterion for efficient organic photoconductors in the threshold spectral region of intrinsic photoconductivity.

In Fig. 6.13c the characteristic time scales of electronic processes in molecules[107a] gives a schematic illustration of the kinetics of ionization processes discussed above. At time scale $\tau = 10^{-14}$–10^{-13} s the AI rate constant τ_{AI} is greater than the vibrational relaxation one τ_n, i.e., $\tau_{AI} > \tau_n$, and direct autoionization prevails. At time scale $\tau = 10^{-13}$–10^{-12} s both processes overlap ($k_{AI} \leq k_n$), and the vibrational relaxation channel becomes competitive with the direct AI. On the other hand, electronic relaxation according Fig. 6.13c is always slower than AI, i.e., $k_m < k'_{AI}$ and consequently, in this time domain AI after intramolecular relaxation dominates.

Thus, it may be concluded that the structure of intrinsic photoconductivity $\eta(h\nu)$ or photogeneration $\Phi(h\nu)$ spectral dependences, measured over a wide energy range, is mainly due to excitation of low-lying valence levels (S_0, S_{-1}, S_{-2},... S_{-n}) and reflects the energy structure of these levels.

On the other hand, the step-like character of photogeneration activation energy $E_a^{ph}(h\nu)$ spectral dependence is determined by the electronic and vibronic energy structure of excited state levels (S_1, S_2,... S_n) and kinetics of competitive AI and intramolecular relaxation processes.

Consequently, these main photogeneration spectral characteristics allow us to identify the corresponding photogeneration mechanisms.

6.2.6. Temperature Dependence of the Mean Thermalization Length r_{th}

Silinsh et al.[349] studied the $r_{th}(T)$ dependence in Pc layers. The authors observed a distinct decrease of transient photoconductivity quantum efficiency Y above 200 K in the threshold region from 2.3 to 2.5 eV (Fig. 6.14a). It has been shown that this effect is caused by the dependence of r_{th} on T at higher temperatures presumably due to increasing localization of the charge carrier with growing temperature. The authors of Ref. 349 evaluated from the Y(T) curve (Fig. 6.14a) the corresponding $r_{th}(T)$ dependence (Fig. 6.14b) applying calculated zero-field $\Omega(r_{th})$ and experimental Y(r_{th}) values (taking into account also a slight dependence of Φ_0 on T). Figure 6.14b shows that r_{th} is practically T-independent at low temperatures but decreases steadily above 200 K. Similar $r_{th}(T)$ dependence has been observed by Lochner, Reimer, and Baessler,[219] in polydiacetylene crystals.

The r_{th} dependence on temperature can be evaluated also from $r_{th}(h\nu)$ curves obtained at different temperatures. In Section 6.3.3.A (see Fig. 6.21b) are shown, according to Ref. 344, two $r_{th}(h\nu)$ dependences at 204 and 293 K, respectively (curves 1 and 2). These $r_{th}(h\nu)$ curves have been obtained from corresponding

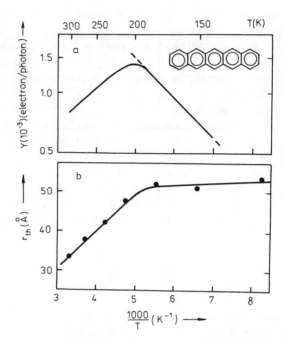

FIGURE 6.14. (a) Quantum efficiency of the transient photoconductivity Y for Pc layers as a function of $1/T$ at $h\nu = 2.4$ eV. (b) Corresponding T-dependence of the mean thermalization length $r_{th} = r_{th}(T)$.

$\eta(h\nu)$ dependences of stationary photoconductivity using for approximation formulas (6.7) and (6.14). As may be seen from Fig. 6.21b, the $r_{th}(h\nu)$ curve at 293 K lies ca. 10–15 Å lower than that at 204 K, i.e., the r_{th} value correspondingly decreases by $\Delta r_{th} = 15$ Å. This is in good agreement with refined transient photoconductivity measurements as well as with computer simulation results (see Fig. 6.21a). More detailed analysis of computer simulated $r_{th}(T)$ dependences will be given in Section 6.3.3.A.

6.2.7. Electric Field Dependence of the Mean Thermalization Length r_{th}

At higher electric fields ($\mathcal{E} \geqslant 10^4$ V/cm) and at sufficiently large r_{th} values ($r_{th} \geqslant 40$ Å) one should take into account the influence of $\mathcal{E}$ on the r_{th} value due to hot electron drift during the thermalization.[349,351] The magnitude of such a drift $r_{\mathcal{E}}$ can be roughly evaluated:

$$r_{\mathcal{E}} = \bar{\mu}\,\mathcal{E}\,\tau_{th}, \tag{6.39}$$

where $\bar{\mu}$ is the mean electron mobility during the thermalization process.

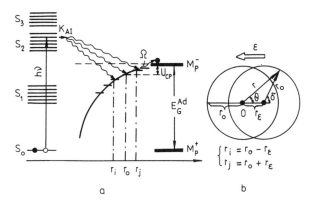

FIGURE 6.15. (a) Formation of spectrum of CP states caused by electron drift in electrical field during thermalization. (b) Schematic picture of field-induced drift of the sphere of thermalized electrons according to Ref. 349.

Due to the drift during the thermalization time τ_{th}, the initial sphere of thermalized electrons, with radius $r_0 = r_{th}$, is presumably shifted along the field by a value $r_{\mathcal{E}}$ (Fig. 6.15b). As a result, quasimonoenergetic CP levels should be replaced by a spectrum of distribution of CP states covering an energy interval from $U_{CP}(r_0 + r_{\mathcal{E}})$ to $U_{CP}(r_0 - r_{\mathcal{E}})$ (Fig. 6.15a). In this case, to calculate the total dissociation efficiency $\Omega(h\nu, T, \mathcal{E})$ of the whole set of anisotropic distribution of the formed CP-states, the isotropic Onsager formula[73] should be replaced by a modified anisotropic formula according to which the $\Omega(h\nu, T, \mathcal{E})$ is found as a sum of the dissociation efficiencies of all kth CP states:[344,349]

$$\Omega(h\nu, T, \mathcal{E}) = \sum_k \Omega_k[r_{th}(h\nu, T, \mathcal{E}), T, \mathcal{E}], \qquad (6.40)$$

where Ω_k is the Onsager formula (6.7).

The problem of anisotropy of CP state distribution has been formulated, in a less general way, already in Section 6.1 [see formula (6.7)].

Figure 6.16 illustrates the summation of the dissociation probabilities, according to the formula (6.40), of a set of CP-states formed as the result of the drift of electrons in an electric field $\mathcal{E}$ during the thermalization (see Fig. 6.15). In Fig. 6.16 the dependences $lg\Omega_k = f(1/T)$ are shown for eight CP-states of different depth U_{CP}, as well as the total dissociation efficiency dependence $lg\Omega = f(1/T)$ of the whole set of CP-states. These eight CP-states have the values of r_{th} from $r_{th} = r_o - r_{\mathcal{E}} = 15$ Å to $r_{th} = r_o + r_{\mathcal{E}} = 65$ Å, where $r_o = 40$ Å and $r_{\mathcal{E}} = 25$ Å.

As may be seen from Fig. 6.16 every CP-state possesses a characteristic energy of activation E_a equal to the depth U_{CP} of the corresponding CP-state (see Fig. 6.15). The dependence $lg\Omega = f(1/T)$ (filled circles), obtained by summing the values of Ω_k according to (6.40) yields the effective activation energy $(E_a)_{eff}$ which

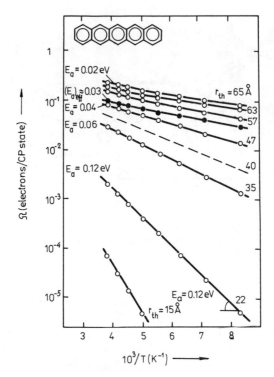

FIGURE 6.16. Calculated dissociation efficiency dependences $lg\Omega_k = f(1/T)$ for a set of eight CT-states (open circles) and the total dissociation efficiency $\Omega = f(1/T)$ of the whole set CT-states according to Eq. (6.40) (filled circles). Parameters used: $r_0 = 40$ Å, $r_{\mathcal{E}} = 25$ Å, $\mathcal{E} = 4 \times 10^4$ V/cm. Dashed line $\Omega = f(1/T)$ for $r_{\mathcal{E}} = 0$.

can be observed experimentally for the given set of CP-states. As may be seen from Fig. 6.16, the slightly temperature dependent curve $lg\Omega = f(1/T)$ (filled circles) is positioned over the $lg\Omega_\kappa = f(1/T)$ curve (dashed line), corresponding the CP state zero-shift $r_{\mathcal{E}} = 0$.

That means that drift of electrons during the thermalization (see Fig. 6.15) increases the total dissociation efficiency due to lowered value of effective activation energy $(E_a)_{eff}$ for generation of free charge carriers.

The authors of Ref. 349 have modified analytically the isotropic Onsager formula, taking into account the electric-field induced drift of electrons during the thermalization. The formula has been derived under assumption that the sphere of distribution of thermalized electrons $r_{th} = r_o$ is shifted symmetrically in the field $\mathcal{E}$, practically without distortion, by the distance $r_{\mathcal{E}}$ retaining its δ-function character of distribution (see Fig. 6.15b). However, computer simulation of the electron drift during the thermalization[344] demonstrates that the initial distribution sphere is strongly distorted at higher electric fields obtaining an elongated, ellipsoidal form with the axis of symmetry parallel to the electric field (see Section 6.3.3.D and Fig.

6.29). The applicability of the analytically derived anisotropic formula of Ref. 349 is eliminated to a relatively low electric field at which better agreement with experimental data is provided, as compared with the conventional isotropic Onsager formula.

To obtain more direct evidence on the $r_{th}(\mathcal{E})$ dependence, the authors of Ref. 349 investigated $\eta(\mathcal{E})$ characteristics with $h\nu$ as parameter in the threshold region of thin-layer Pc samples, having $G \approx 1$ [see (6.16)]. Then, using the established correlation between η and Ω [see (6.14)], these characteristics were approximated by formula (6.7).

As a result the effective, field-dependent thermalization length $(r_{th})_{eff} = f(\mathcal{E})$, for different values of photon energy $h\nu$, were obtained in which the influence of $\mathcal{E}$ via $\mathcal{E}$-dependent term of the Onsager formula (6.7) has been taken into account.

The $(r_{th})_{eff} = f(\mathcal{E})$ curves, as evaluated from experimental $\eta(\mathcal{E},h\nu)$ dependences, obtained in Pc layers at $T = 205$ K, are shown in Fig. 6.30 (solid circles) (see Section 6.3.3.D). These $(r_{th})_{eff} = f(\mathcal{E})$ curves have been compared with calculated maximum thermalization length $(r_{th})_{max}$ along the direction of the external electric field $\mathcal{E}$ at $T = 204$ K (empty circles). Both experimental and calculated $r_{th}(\mathcal{E})$ dependences (see Fig. 6.30) show that, at sufficiently high values of r_{th} ($r_{th} \geqslant 50$ Å) and kinetic energy E_k ($E_k = m_{eff}v_0^2/2 = (h\nu - E_G^{Ad}) - U_0 \geqslant 0.6$ eV), the thermalization time τ_{th} is long enough to provide a noticeable electron drift during thermalization. As can be seen from Fig. 6.30 the value of $(r_{th})_{eff}$ increases with increasing $\mathcal{E}$ and $h\nu$. These $(r_{th})_{eff}$ and $(r_{th})_{max}$ dependences give additional evidence supporting the ballistic thermalization model (see Fig. 6.1).

6.2.8. The Fast Photocurrent Pulse

Besides the hole transient pulse (Fig. 6.3b) the authors of Ref. 349 also observed a fast electron photocurrent pulse $j_e(t)$ (Fig. 6.17b) in Pc layers, which repeats the laser pulse of 20 ns duration at $h\nu = 2.34$ eV. The amplitude of j_e shows a linear dependence on $\mathcal{E}$ (for $\mathcal{E} \leqslant 7 \times 10^4$ V/cm) and light intensity I. The observed duration of j_e may most probably be determined by the lifetime τ_e of electrons which is estimated to be the order of $\tau_e = (vqN_t)^{-1} = 10^{-9} - 10^{-8}$ s for trapping in deep electron traps of quinone-type impurity origin ($N_t = 10^{16} - 10^{17}$ cm^{-3}) below 0.7 eV. The corresponding electron transit range $S_e = \mu\tau_e \mathcal{E}$ for $\mathcal{E} = 10^4$ V/cm is of the order of 100–1000 Å. Although these values significantly surpass the thermalization parameters τ_{th} and $r_{\mathcal{E}}$, we anticipate that polarization currents $j = dP/dt \approx Ner_{\mathcal{E}}/\tau_{th}$, caused by electron drift during thermalization, might give a considerable contribution to the formation of the fast photocurrent pulse. The genuine value of such polarization currents might be resolved by shorter ps laser pulses.

Similar fast photocurrent pulses have been observed by Hesse and Baessler[132] in amorphous Pc under different experimental conditions, having a rate constant of current decay of $\approx 10^7$ s^{-1}.

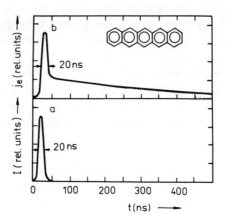

FIGURE 6.17. (a) Laser pulse used for photoexcitation at $h\nu = 2.34$ eV. (b) Fast transient current-pulse j_e in Pc ($\tau = 7$ ns; $R = 75$ Ω).[349]

6.2.9. Brief Survey of Electric Field-Dependent Electronic Processes in OMC

Analyzing in previous sections the experimental results of investigation of photo-generation mechanisms in OMC we found that the electric field may affect different electronic processes which are directly or indirectly connected with the photogeneration process or measurements of its parameters.

Figure 6.18 gives a brief visualized survey of the most important electronic processes, affected be the electric field, as well as the characteristic field range of their influence.

First, in order to determine adequately the quantum yield of photogeneration (quantum yield of CP state population Φ_0) it is necessary to take into account the electric field-dependence of the charge carrier transit coefficient G, i.e., $G = G(\mathcal{E})$ [see formulas (6.15) and (6.16)]. The carrier transit coefficient G characterizes quantitatively that part of photogenerated carriers which are collected on electrodes and determines the quantum efficiency of the photocurrent j_{ph}. The $G(\mathcal{E})$ dependence can be approximated by an empirical formula (6.16) and, as it was demonstrated in Section 6.2.3, in the case of thin-layer systems of OMC having the thickness of the order of $\sim \mu$m, that already at fields $\mathcal{E} \geqslant 10^4$ V/cm it is possible to collect on the electrodes practically all generated carriers (in the case of monopolar conductivity, the majority carriers), i.e., to realize the conditions $G \rightarrow 1$ (see Table 6.1).

The most important electronic process, directly affected by the electric field, is the $\mathcal{E}$-dependent CP-state dissociation efficiency. This process can be adequately described by the Onsager formula (6.7). In this case the influence of the electric field, due to the $\mathcal{E}$-dependent term of the Onsager formula, begins at $\mathcal{E} > 10^3$

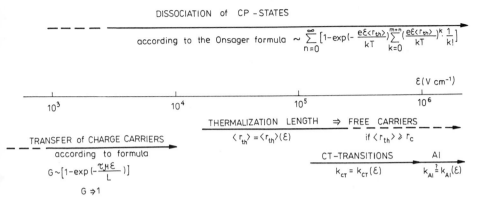

FIGURE 6.18. Schematic diagram of electric field-dependent electronic processes in OMC.[351]

V/cm. On the other hand, the S-shaped dependence of $\Omega(\mathcal{E})$ approaches the saturation value ($\Omega \rightarrow 1$) at $\mathcal{E} > 10^5$ V/cm.

As shown in Section 6.2.7, electric field can change the effective thermalization length $(r_{th})_{eff}$ due to the drift of the hot electron during thermalization.[349] This effect can be observed at field strength $\mathcal{E} > 10^4$ V/cm and for sufficiently great mean thermalization length $r_{th} \geq 40$ Å. For still higher electric field values ($\mathcal{E} > 10^5$ V/cm at room temperature) the situation arises when $r_{th} \geq r_c$ and the carrier moves out of the Coulombic field as a free thermalized carrier, or as a hot carrier with drift velocity v_d greater than the mean thermal velocity v_{th}, i.e., $v_d > v_{th}$ (see Section 7.1.3). In this electric field range the microscopic mobility μ_0 becomes field dependent, that is, $\mu_0 = \mu_0(\mathcal{E})$, a characteristic feature of a hot carrier generation regime (see Section 7.1.3).

Recently other possible mechanisms of the influence of electric field on the photogeneration of charge carriers have also been discussed. First, electric fields of the order of $\mathcal{E} > 10^5$ V/cm can most probably increase the rate constant k_{CT} of the direct optical CT-transitions and, consequently, affect the corresponding competitive photogeneration channel (see Section 6.2.4). At still higher electric fields $\mathcal{E} > 10^6$ V/cm one may expect direct influence of $\mathcal{E}$ on the AI rate constant k_{AI}. This influence may be considered in terms of mixing the neutral molecular Frenkel state with a localized charge transfer state via field-dependent coefficient $b_E = b_E(\mathcal{E})$ [see Eq. (6.23)].

Finally, as previously mentioned, Braun[43] discussed the possibility of electric field influence on the CT-state dissociation rate constant as an alternative photogeneration mechanism for CT-complexes. In this model, charge carrier generation proceeds without the intermediate stage of thermalization. It is interesting to notice that in this case the author of Ref. 43 uses as a starting model an earlier approach by Onsager, proposed in 1934[276] for the description of dissociation of ionic pairs in weak electrolyte solutions, instead of the famous 1938 Onsager formula.[277]

Owing to such variety of the possible effects of electric field on electronic processes of photogeneration and charge carrier transfer in OMC, the investigator should use highly sophisticated approaches in order to identify these electric field effects and determine their contribution and range of influence.

Finally, we should emphasize that the described electric field effects are characteristic only in the case of blocking electrodes. If the electrodes are capable of field injection of charge carriers, then one may additionally observe the influence of electric field both on charge carrier injection phenomena and on injection currents, or so-called space charge limited currents.[332]

6.3. COMPUTER SIMULATION OF THE PHOTOGENERATED CHARGE CARRIER SEPARATION PROCESSES IN OMC

As shown in Section 6.2, the intermediate processes of charge carrier separation, namely, the thermalization and dissociation stages, are of major importance in the photogeneration phenomena in OMC. Unfortunately, the details of these intermediate stages are not easily accessible to direct experimental observation.

In order to get some insight in these processes Silinsh et al.[344,348,354,355,356] used the technique of an artificial experiment, i.e., computer simulation studies of charge carrier separation, and transport phenomena.

In this aspect the most crucial process is a hot carrier's thermalization which is a constituent part of both intrinsic photogeneration mechanisms (see Section 6.1, Figs. 6.1 and 6.2) as well as extrinsic photogeneration (see Section 6.2.2, Fig. 6.6).

As previously mentioned, thermalization of a hot carrier in Ac-type crystals proceeds via fast scattering and, since the mean free path l_o of epithermal electrons is of the order of intermolecular distances, scattering events take place practically at every molecule (see Section 7.2.3 and Figs. 7.15 and 7.28). Consequently, the excess kinetic energy of the electron is lost very rapidly resulting in a thermalized Coulomb-field-bounded CP state (see Section 6.1, Fig. 6.1). Because of such strong localization, a hot electron with energy in the subexcitonic range may be regarded as a quasiclassical Brownian particle moving in the lattice by hopping from molecule to molecule (Ref. 351). Therefore the authors of Ref. 344 have found it appropriate to use an extended Sano-Mozumder (SM) model[306] for computer-simulated studies of thermalization processes in Ac-type crystals. This model has been initially applied to describe the thermalization of quasifree epithermal electrons in dielectric liquids. Pentacene has been chosen as an object of investigation in Ref. 344 since its photogeneration mechanisms, including thermalization processes, have been studied experimentally in some detail (see Section 6.2).

The initial aim of our simulation studies has been to find a physically reasonable explanation of the experimentally observed spectral, temperature and electric field dependences of the mean thermalization length $r_{th}(h\nu, T, \mathcal{E})$ of photogenerated charge carriers, and the impact of the $r_{th}(h\nu, T, \mathcal{E})$ dependences on the total dissociation efficiency (probability) $\Omega(r_{th}, T, \mathcal{E})$ and the corresponding quantum efficiency $\eta(h\nu, T, \mathcal{E})$ of intrinsic photoconductivity.

However, we have found that to attribute physical meaning to simulation results we had to introduce the concept of a heavy quasiparticle with a temperature-dependent effective mass m_{eff} instead of a "bare" quasifree electron. It has been shown that this quasiparticle may be described phenomenologically in terms of an adiabatic nearly-small *molecular polaron* (MP) model already described in some detail in Chapter 3 (see also Ref. 344). We shall also demonstrate that this polaron approach may be extended outside the initial scope of describing the charge carrier behavior during thermalization and may also be used for the description of transport properties of carriers after thermalization (see Section 6.3.4 and Chapter 7). The MP concept provides a physically viable solution to several controversial empirical facts which were difficult to explain by current theories. First, it gives a satisfactory answer to the longstanding question why the temperature dependence of mobility μ_0 in Ac-type crystals ($\mu_0 \sim T^{-n}$) is formally in accord with band-like motion (see Ref. 288, p. 338), although the carrier mean free path l_o is of the order of the lattice constant and, consequently, should be presumably represented in terms of a hopping model.

Second, it also helps to solve another existing controversial situation. Indeed, on the one hand, energy structure studies of ionized states in Ac-type crystals show that relaxed thermalized charge carriers are actually molecular polarons and the corresponding conductivity levels are, in fact, MP state levels M_P^+, M_P^- separated by an adiabatic energy gap E_G^{Ad} (see Chapter 4, Figs. 4.8–4.10). On the other hand, some of current transport theories, rejecting as inappropriate the small-lattice polaron approach, are compelled to use a quasifree charge-carrier concept in terms of a simple band model, shown to be invalid for Ac-type crystals.[288,332,351].

In our simulation approach the two main temperature-dependent variable parameters of the model, i.e., the effective mass $m_{\text{eff}}(T)$ of the carrier and its thermalization relaxation rate constant $\beta(T)$ have not been chosen arbitrarily, but are shown to follow empirical formulae consistent with polaron theory concepts. A comparison between calculated dependences and experimental data allowed to adjust slightly the variable parameters of the model, the initial values of which were chosen from purely physical considerations.[344]

6.3.1. Description of the Model

According to an extended SM approach, let us consider a "dressed" electron produced in a photoionization process with effective mass m_{eff} and excess kinetic energy E_k (not exceeding 1.3 eV, i.e., corresponding the subexcitonic range), having at zero time ($t = 0$) a radial velocity v_0 and an initial position r_o relative to the parent positive ion. We want to determine the time scale in which the velocity distribution of the hot electron approaches a thermal (Maxwellian) one, as well as the distribution of the position of thermalized electrons r_{th} at time $t = \tau_{\text{th}}$. Assuming that the Brownian motion of the electron can be represented as a Markovian process, one can derive a generalized Fokker-Planck equation[125,299a,306] for W, the distribution function for position r and velocity v of the electron:

$$\partial W/\partial t + \mathbf{v}\ \text{grad}_\mathbf{r}\ W + \mathbf{F}\ \text{grad}_\mathbf{r}\ W\ =\ \beta\ \text{div}_\mathbf{v}(W\mathbf{v}) + q\nabla_\mathbf{r}^2 W. \qquad (6.41)$$

[The Fokker–Planck equation is often used in modern synergetics since it describes both the stochastic and the deterministic processes in nature (see Refs. 125 and 299a)].

In Eq. (6.41), as applied for the description of thermalization in Pc crystals, $\mathbf{F}$ represents the Coulombic force of the parent positive charge plus the external electric field $\mathcal{E}$, and β is the friction constant which, in this case, describes the inelastic scattering of an electron with excess kinetic energy on lattice phonons and which equals the characteristic thermalization relaxation rate constant, namely, $\beta = 1/\tau_r$, where τ_r is the characteristic relaxation time. The term $q = \beta kT/m_{\text{eff}}$ describes the diffusive stochastic motion of the electron (see Section 5.15).

There is no explicit analytical solution of Eq. (6.41). However, a set of equations for the *mean* values can be derived, suitable for computer simulation of the problem. Using the SM formalism, the following set of equations for the mean values were obtained from Eq. (6.41)

$$d\langle x_i\rangle/dt = \langle v_i\rangle, \qquad (6.42a)$$

$$d\langle v_i\rangle/dt = F_i - \beta\langle v_i\rangle, \qquad (6.42b)$$

$$d\langle x_i x_j\rangle/dt = \langle v_i x_j\rangle + \langle v_j x_i\rangle, \qquad (6.42c)$$

$$d\langle v_i v_j\rangle/dt = \langle F_i v_j\rangle + \langle F_j v_i\rangle - 2\beta\langle v_i v_j\rangle + 2\delta_{ij}q, \qquad (6.42d)$$

$$d\langle v_i x_j\rangle/dt = \langle F_i x_j\rangle + \langle v_i v_j\rangle - \beta\langle v_i x_j\rangle \qquad (6.42e)$$

where $i, j = 1,2,3$, and

$$F_i = 1/m_{\text{eff}}\left[-\frac{e^2}{4\pi\epsilon_0}(x_i/r^3) - e\,\mathcal{E}\,\delta_{i3}\right], \qquad (6.43)$$

$i = 3$ defines the field direction.

The mean values in (6.42a)–(6.42e), containing force F, are approximated, as in the SM paper,[306] via the mean values of coordinates and correlation functions.

Equation (6.42b) represents the motion of the electron against the Coulombic field and its drift in the external electric field $\mathcal{E}$, as well as the motion of the electron against the friction force caused by inelastic scattering on lattice vibrations. Equations (6.42c)–(6.42e) are used for finding the correlation functions for the Brownian motion. Thus, in the extended model, in addition to the Coulombic field, the electron drift in the external field according to Eq. (6.43), is also taken into account. This was of major importance since we had to simulate experimentally observed electron drift during thermalization.[349] Cartesian coordinates were used in Eqs. (6.42) as more suitable for the non-symmetrical problem.

Finally, it is interesting to notice that the set of equations (6.42) actually incorporates both counterparts inherent to the generalized Fokker–Planck equation (6.41), i.e., necessity and chance.

Thus, Eq. (6.42b) describes the deterministic motion of the charge carrier in the force fields (Coulombic and external electronic fields, as well as "frictional" field of inelastic scattering). On other hand, Eqs. (6.42c)–(6.42e) take into account the influence of stochastic forces of Brownian motion which causes the statistical distributions of position and velocity of the carriers (see Figs. 6.26, 6.27, and 6.29). This is in accordance with the famous statement of Herman Haken: "Chance and necessity — reality needs both."[125] This combined approach has been so successfully used in modern physics in general and contemporary synergetics and related fields in particular. And that in spite of the celebrated warning by Albert Einstein, "God does not play dice." (See Ref. 54a, p. 343.)

6.3.2. Parameterization of the Model

There are thus two basic initial input quantities of the model which may formally be regarded as adjustable parameters of the problem, namely, m_{eff} and β. A feasible initial value of β was chosen on the base of the following considerations. Relaxation time $\tau_r = \beta^{-1}$ gives the characteristic interval during which the electron's "memory" of radial velocity is lost (see Ref. 125 and Fig. 6.27), which approximately equals the thermalization time τ_{th}, namely $\tau_r \approx \tau_{th}$ (see Ref. 306). The lower limit of β, determined by inelastic scattering by a one-phonon process, may be evaluated according to the following expression (see Ref. 186):

$$\beta^{-1} = \tau_{th} = E_k/h\nu_{ph}^2 = E_k/h\nu_{ph} \cdot 1/\nu_{ph} = n \cdot 1/\nu_{ph}, \qquad (6.44)$$

where E_k is the excess kinetic energy of the electron. ν_{ph} is the frequency of emitted phonons, and n is the number of inelastic scattering events. Assuming that scattering takes place on optical phonons of the lattice and using a typical value for the optical phonon energy $h\nu_{ph} = 120$ cm^{-1} = 0.015 eV for Ac-type crystals (see Ref. 332, p. 16) and $E_k = 1$ eV, one finds that $\tau_{th} \approx \tau_r \approx 2 \times 10^{-11}$ s, or $\beta = 0.5 \times 10^{11}$ s^{-1}. On the other hand, the upper limit of β is set by the characteristic hopping rate constant of a charge carrier in the **ab** plane of an Ac-type crystal, namely, $\beta_h \approx (2-5) \times 10^{13}$ s^{-1} (see Ref. 332, p. 36). In case of multiphonon scattering, which may be even more probable than the one-phonon scattering,[247] β should have some intermediate value within the limits $0.5 \times 10^{11} < \beta < 10^{13}$ s^{-1}. Consequently, $\beta = 10^{12}$ s^{-1} was used as the initial zero-approximation value of the parameter in the simulation procedure.

Another crucial simulation parameter was found to be the effective mass m_{eff} which determines the initial radial velocity v_0 of the electron. If one derives the value of v_0 from the kinetic energy term $mv_o^2/2$ for an excess value of E_k of, say, 1 eV, using the free electron mass $m = m_e$, then $v_0 = 5.9 \times 10^7$ cm s^{-1} is so high that the electron easily escapes out of the Coulombic capture radius $r_C = e^2/4\pi\epsilon\epsilon_0 kT$ during thermalization assuming a physically reasonable range of β values. To match the real situation and "keep" the electron inside the Coulombic radius after thermalization, in accordance with experimental data, v_0 should be considerably smaller, typically of the order of 10^5–10^6 cm s^{-1}. This means that one should introduce a corresponding value of m_{eff} in the excess kinetic energy term of the electron

$$E_k = (m_{\text{eff}} v_0^2)/2 = (h\nu - E_G^{\text{Ad}}) - U_0, \qquad (6.45)$$

where the difference between the photon energy $h\nu$ and the adiabatic energy gap E_G^{Ad} yields the excess kinetic energy gained by the electron during the ionization, and U_0 is the Coulombic potential energy at r_o, i.e., $U_0 = -e^2/4\pi\epsilon_0 r_o$ which provides an equivalent amount of kinetic energy to the electron at r_o. To give a good fit to experimental $r_{\text{th}}(h\nu)$ and photogeneration quantum efficiency $\eta(h\nu)$ dependences in the temperature range from 165 to 300 K (see Section 6.3.3) the assumption had to be made in the simulation procedure that the relative effective mass of the electron $m^* = m_{\text{eff}}/m_e$ increases exponentially with temperature T according to

$$m^*(T) = m_0^* \exp[T/T_0], \qquad (6.46)$$

where m_0^* was found to be equal to $m_0^* = 18.7 \pm 0.1$, and $T_0 = 66.4$ K. The $m^*(T)$ dependence is shown in a semilogarithmic plot in Fig. 6.19. As can be seen, the value of m^* equals $230 m_e$ at 165 K and increases to $1550 m_e$ at 293 K. Such values for the electron effective mass and its temperature dependences may be shown to be consistent with a nearly small molecular polaron concept (see Section 6.3.4). It should be mentioned that formula (6.46) was introduced in Ref. 344 empirically as an ad hoc assumption for a relatively high temperature region (165–300 K). A similar formula was proposed in early works of small lattice polaron theories (see Refs. 12, 145, 285, and 324). It has been derived in more general terms in Section 1B of Chapter 1 [see Eq. (1B.8.18)]. However, the exponential $m_{\text{eff}}(T)$ dependence is physically more reasonable for the low T region while for the high T region, as it will be shown in Chapter 7, the $m_{\text{eff}}(T)$ may follow a power-law dependence.

Adjustment of the initial zero-approximation value of β to provide a satisfactory fit to experimental $r_{\text{th}}(h\nu)$ and $\eta(h\nu)$ curves (see Section 6.3.3) yields a $\beta(T)$ dependence which decreases with temperature (see Fig. 6.19). It was found to be more convenient to use the friction coefficient $\beta_m(T)$ instead of $\beta(T)$:

$$\beta_m(T) = \beta(T) m_{\text{eff}}(T), \qquad (6.47)$$

which yields a linear temperature dependence (see Fig. 6.19), and can be described, for $T \geqslant 165$ K, by

$$\beta_m(T) = 6.0 \times 10^{-16} + 5.46 \times 10^{-18}(T - 165)s^{-1}kg. \qquad (6.48)$$

The reported experimental adiabatic energy-gap value $E_G^{\text{Ad}} = 2.25$ eV for Pc,[349] obtained by the photogeneration threshold method, was initially used in expression (6.45). It was found, however, in the course of the simulation procedure, that for the best fit with the experimental dependences of $\eta(h\nu)$ the value of E_G^{Ad} should be slightly adjusted to take into account a small increase in E_G^{Ad} with temperature

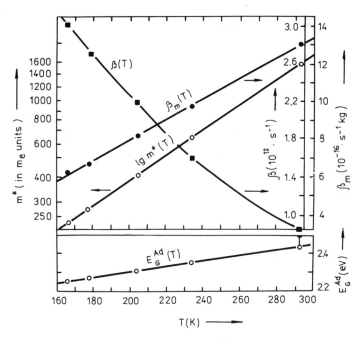

FIGURE 6.19. Temperature dependences of the thermalization relaxation rate constant $\beta(T)$ and the three adjustable parameters of the model: $m^*(T)$, relative effective mass of the electron; $\beta_m(T)$, friction coefficient; and $E_G^{Ad}(T)$, adiabatic energy gap for charge carriers in Pc crystals.[344]

according to the relationship (see Fig. 6.19):

$$E_G^{Ad}(T) = 2.25 + 1.42 \times 10^{-3}(T - 165) \text{ eV}, \qquad (6.49)$$

for $T \geqslant 165$ K.

(The methods for the determination of E_G^{Ad} have been described in Chapter 4).

Finally, we must choose the initial radial separation r_o of the electron from the geminate positive ion at $t = 0$ from where thermalization starts. Taking into account the nature of ionization processes in molecular crystals,[288,332,351] it seemed most appropriate to identify r_o with the distance between the ion and the nearest neighboring molecules, i.e., with the intermolecular spacing of the lattice. This is consistent with both ionization models (see Figs. 6.1 and 6.2). According to the AI model the first scattering event, due to $l_o \approx a$, should most probably occur at the nearest adjacent molecule. On the other hand, in terms of the optical CT-transition model, r_o may be identified with the transition distance r_{CT}, since during relaxation, the thermalization process obviously sets out from this position. Accordingly, we used for Pc crystals a value of r_o equal to the lattice parameters, namely, $r_o = 0.515, 0.60$, and 0.79 nm. Trial calculations showed, however, that the thermalization distance r_{th} is practically independent of these variations in r_o (see

Section 6.3.3.A). Hence, in further calculations we used a constant value of $r_o = 0.79$ nm as an averaged magnitude of interspatial distance around the positive ion and considered the parameter r_o as isotropic for any given space angle θ.

It can thus be seen that there are two main adjustable parameters of the model, namely, m^* and β_m, the temperature dependences of which are given by Eqs. (6.46) and (6.48).

6.3.3. Simulation Results. Comparison with Experiment

6.3.3.A. Thermalization distance r_{th} and thermalization time τ_{th}

Fig. 6.20 shows the variation of average total energy $\langle E(r) \rangle$, average kinetic energy $\langle E_k(r) \rangle$, and potential energy $U_{CP}(r)$ of a photogenerated hot electron as functions of the separation distance r from the positive ion during the thermalization process. The diagram in Fig. 6.20 represents the description of thermalization in the framework of the autoionization model of photogeneration in Pc crystals (see Section 6.1 and Fig. 6.1). The $\langle E(r) \rangle$ and $\langle E_k(r) \rangle$ dependences were calculated using the parameter values according to Eqs. (6.46), (6.48), and (6.49) (see Fig. 6.19) for two different values of photon energy, namely, $h\nu = 2.4$ and 2.7 eV at $T = 204$ K, $\mathcal{E} = 0$, and r_o equal to 0.79 nm. As can be seen, the corresponding thermalization distances are $r_{th} = 4.7$ and 8.0 nm, respectively. These data satisfactorily simulate the experimentally observed dependences of r_{th} on $h\nu$ at the threshold spectral region of intrinsic photogeneration in Pc crystals (see Fig. 6.21a). The r dependence of $\langle E_k \rangle$ is determined by

$$\langle E_k(r) \rangle = \langle E(r) \rangle + |U_{CP}(r)| \qquad (6.50)$$

while the initial excess kinetic energy at r_o is given by (see Fig. 6.20):

$$E_k(r_o) = h\nu - E_G^{Ad} + |U_{CP}(r_o)| \qquad (6.51)$$

Fig. 6.21a shows calculated $r_{th}(h\nu)$ dependences at zero electric field ($\mathcal{E} = 0$) for two different temperatures $T = 204$ (1) and 293 K (2). In Fig. 6.21a the reported[349] experimental zero-field $r_{th}(h\nu)$ dependences at $T = 200$ K are also shown, as well as experimental r_{th} values at $(h\nu) = 2.4$ eV for $T = 204$ and 293 K, demonstrating the decrease of r_{th} with increasing temperature in the threshold spectral region in Pc crystals (see Fig. 9 in Ref. 349). Fig. 6.21b shows $r_{th}(h\nu)$ curves for temperatures $T = 205$ and 293 K, respectively, obtained in Ref. 350 by approximating the experimental photogeneration quantum efficiency dependences $\eta(h\nu)$ according to the Onsager formula (for details see Section 6.2.3). As can be seen from Figs. 6.21a and 6.21b, calculated $r_{th}(h\nu)$ dependences for different values of T are in satisfactory agreement with experimental data and simulate the realistic thermalization picture in Pc crystals.

The authors of Ref. 150a have estimated temperature dependence of the mean thermalization length $r_{th} = r_{th}(T)$ for hot electrons ($E_k = 1.3$ eV) at zero field

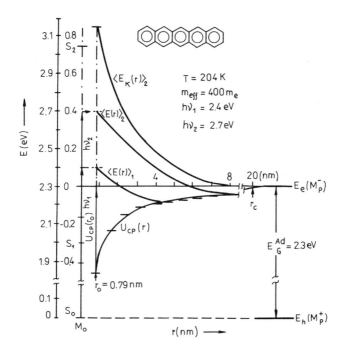

FIGURE 6.20. Energy diagram representing the thermalization process in the framework of the autoionization model of photogeneration in Pc crystals.[344] Curves show the variation of the average total energy $\langle E(r) \rangle$, the average kinetic energy $\langle E_k(r) \rangle$, and the potential energy $U_{CP}(r)$ of photogenerated hot electrons as functions of separation distance r from the positive ion during thermalization.

($\mathcal{E} = 0$) in the **a** and **c'** directions in a naphthalene crystal over a wide range of temperature (50–300 K). The simulation was performed in the framework of the modified SM model according to Eqs. (6.42a–e). The calculation results are shown in Fig. 6.21c (filled circles, the **a** direction; open circles, the **c'** direction).

It has been shown in Ref. 150a that the simulated $r_{th}(T)$ dependences may be approximated by the following formula (solid lines)

$$r_{th}(T) = C_1/[\ln(C_2 T - 1) - C_3], \qquad (6.52)$$

where C_1, C_2, and C_3 are temperature-independent empirical constants.

As may be seen, the r_{th} value shows strong nonlinear increase with lowering temperature; nevertheless, the carrier at $\mathcal{E} = 0$ still remains inside the Coulombic radius r_c ($r_c = 178$, 533, and 1066 Å for 300, 100, and 50 K, respectively). However, in the low T region the carrier can be "pulled out" from the Coulombic well at relatively low electric fields (see Chapter 7).

Figure 6.21c (curve 3) shows the calculated r_{th} dependence on the hot carrier's kinetic energy E_k at $T = 300$ K. As one might expect, the $r_{th}(E_k)$ dependence is sublinear according to formula (6.11).

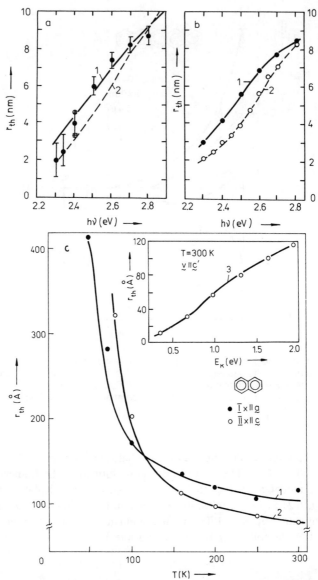

FIGURE 6.21. (a) Calculated zero-field ($\mathcal{E} = 0$)$r_{th} = r_{th}(h\nu)$ curves for Pc: (1) at $T = 204$ K ($m_{eff} = 400\ m_e$); (2) at $T = 293$ K ($m_{eff} = 1550\ m_e$) as compared with experimental data.[344] ●, experimental zero-field $r_{th}(h\nu)$ dependence at $T = 200$ K;[349] ⊗ and □, experimental values of r_{th} at $T = 204$ and 293 K, respectively.[349] (b) $r_{th}(h\nu)$ dependences obtained by approximation of experimental photogeneration quantum efficiency $\eta(h\nu)$ curves according to the Onsager formula at $T = 205$ K (1) and at $T = 293$ K (2).[350] (c) Calculated mean thermalization length r_{th} dependence on temperature $r_{th} = f(T)$ of hot electrons ($E_k = 1.3$ eV) at zero field ($\mathcal{E} = 0$) in the a direction (filled circles, 1) and c' direction (open circles, 2) in naphthalene crystal.[150a] Solid lines (1, 2), approximation to empirical formula (6.52) using the following coefficients C_1, C_2, C_3: 317.8; 0.043; 0.631 (curve 1) and 196.6; 0.024; 0.696 (curve 2). Curve 3, calculated r_{th} dependence on hot electron initial kinetic energy $r_{th} = f(E_k)$ in the c' direction of the crystal at $T = 300$ K and $\mathcal{E} = 0$.

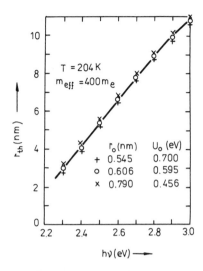

FIGURE 6.22. Zero-field ($\mathcal{E} = 0$) $r_{th} = r_{th}(h\nu)$ dependences in Pc calculated for different values of the initial separation distance r_0 from which thermalization starts.[344] The chosen r_0 values are equal to the nearest neighboring distances in the **ab** plane of the Pc crystal, U_0 is the corresponding potential energy value.

Figure 6.22 shows the $r_{th}(h\nu)$ dependences calculated for different r_o values, equal to the nearest-neighboring distances in the Pc crystal. As may be seen, the spectral $r_{th}(h\nu)$ curve is practically independent of these variations in r_o. Consequently, a constant value of $r_o = 0.79$ nm may be used in calculations as an averaged magnitude for the initial radial separation distance (see Section 6.3.2). The independence of $r_{th}(h\nu)$ for the given range of r_o may have the following origin. The value of r_o determines the Coulombic potential $U_o(r_o) = -e^2/4\pi\epsilon\epsilon_0 r_o$; thus a decrease of r_o causes, according to Eq. (6.51), a corresponding increase in kinetic energy $E_k(r_o)$, and, as a result, r_{th} remains practically invariant. As the hot electron loses its energy, its probability distribution asymptotically approaches a Maxwellian one for the given temperature $\langle E_k \rangle \rightarrow 3/2kT$. It was convenient to fix the thermalization time operationally in a similar way as in the SM approach,[306] namely

$$\langle E_k \rangle (\tau_{th}) = \alpha.3/2(kT),$$

where $\alpha = 1.05$. This means that at the given τ_{th} the value of $\langle E_k \rangle$ approaches the thermalization equilibrium value within a 5% limit. This figure also characterizes the accuracy of numerical calculations.

The calculated thermalization time τ_{th} correlates with r_{th} and also increases with photon energy $h\nu$, having, e.g., in Pc crystals a typical value of 0.8×10^{-12} s for $h\nu = 2.9$ eV at $T = 204$ K. τ_{th} increases with temperature in a similar fashion as the relaxation time τ_r, presumably due to less effective inelastic energy scattering with rising temperature (see Section 6.3.4). Thus, the τ_{th} value for $h\nu = 2.9$ eV at

$T = 293$ K equals 1.7×10^{-12} s and is greater by a factor of 1.6 than the corresponding τ_r value, i.e., $\tau_r = 1.08 \times 10^{-12}$ s (see Section 6.3.2). Unfortunately, there are no reliable data reported on τ_{th} values in Ac-type crystals, except for rough estimates which are of the same order of magnitude (see Ref. 350).

6.3.3.B. Dissociation Probability $\Omega(h\nu, T, \mathcal{E})$ of Thermalized CP States

For adjustment of the parameters of the model the calculated total dissociation probability dependence $\Omega(h\nu, T, \mathcal{E})$ of thermalized CP states were compared with the experimental photogeneration quantum efficiency $\eta(h\nu, T, \mathcal{E})$ curves at the given T and $\mathcal{E}$ values. The total $\Omega(h\nu, T, \mathcal{E})$ dependence for the given set of angular distributions of the mean thermalization distance values r_{th} (see Fig. 6.29) were obtained by summation of Ω, over all ith CP states:

$$\Omega(h\nu, T, \mathcal{E}) = \sum_i \Omega_i[r_{th}(h\nu, T, \mathcal{E}), T, \mathcal{E}], \qquad (6.53)$$

where Ω_i is calculated according to the Onsager formula (6.7) for the given r_{th} values as dependent on $h\nu$, T, and $\mathcal{E}$ for the given ith CP state.

It has been shown (Section 6.2.3) that the photogeneration quantum efficiency curves $\eta(h\nu, T, \mathcal{E})$ in the threshold spectral region can be described by the CP-state dissociation efficiency $\Omega(h\nu, T, \mathcal{E})$ at $T, \mathcal{E} = \text{const}$, using a single numerical coefficient b [see formula (6.14)], where b is an empirical parameter which can be expressed as a product of photogenerated geminate pair yield Φ_0 and carrier transition coefficient G [see formulas (6.15) and (6.16)].

In Eq. (6.15) $G(\mathcal{E})$, in case of blocking, noninjecting electrodes, can be determined according to the range-limited theory, i.e., from Hecht's law (6.16) and Φ_0 directly from approximation data (see for details Ref. 349).

It has been shown (see Section 6.2.3 and Ref. 349) that the parameter b determines the vertical shift of the $\Omega(h\nu)$ curve in semilogarithmic scale and thus, when the final adjustment of the variable parameters of the model has been reached, the experimental $\eta(h\nu)/b$ curve should coincide with the calculated $\Omega(h\nu)$ curve at $T, \mathcal{E} = \text{const}$. Figure 6.23 shows the calculated $\Omega(h\nu)$ curves for different $\mathcal{E}$ values at $T = 293$ K and the corresponding $\eta(h\nu)/b$ dependences (experimental points) for Pc crystals. The details of the experimental techniques of preparing thin-layer Pc samples of oriented crystallites by vacuum evaporation and photoconductivity measurement technique are described in Section 6.2.1.

As can be seen from Fig. 6.23, the final adjusted values of the parameters, namely, $m^* = 1550 m_e$, $\beta_m = 1.3 \times 10^{-16}$ s^{-1} kg, and $E_G^{Ad} = 2.43$ eV, agree well with experimental data and properly simulate the observed electrical field dependence.

The crucial test for the simulation model was the adjustment of the temperature dependence of the parameters. The final results of this adjustment are demonstrated

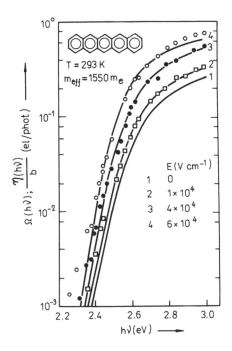

FIGURE 6.23. Dissociation efficiency $\Omega(h\nu)$ curves of thermalized CP states, as calculated according to Eq. (6.53), and corresponding experimental photogeneration quantum efficiency $\eta(h\nu)/b$ dependences (experimental points) in Pc for different electric field $\mathcal{E}$ values at $T = 293$ K.[344]

in Figs. 6.24a and 6.24b, which show the calculated $\Omega(h\nu)$ curves at $T = 293$ K (a) and $T = 204$ K (b) and the corresponding $\eta(h\nu)/b$ dependences at $\mathcal{E} = 1\times10^4$ V cm^{-1} for three different Pc-crystal samples in each case.

As can be seen from Fig. 6.24, the empirical expressions proposed for the temperature dependence of the variable parameters (6.46), (6.48), and (6.49) provide satisfactory agreement with experimental data and support the validity of the model.

The temperature dependence of the shapes of the calculated $\Omega(h\nu)$ curves are illustrated in Fig. 6.25 (the corresponding experimental values of η/b in this case are shown only for $h\nu = 2.9$ eV). It should be emphasized here that the experimentally observed rise of the slope of $\eta(h\nu)$ curves with increasing temperature[349] could only be simulated by a T-dependent effective electron mass according to Eq. (6.46) (see also Figs. 6.24a and b). Increasing m_{eff} causes a decreasing initial velocity v_0 of the hot electron [see Eq. (6.45)]. As a result the electron is more effectively retarded by the Coulombic potential at small separation distances r_{th} in the threshold spectral region. Consequently, the slope of the $\Omega(h\nu)$ curve increases with temperature and thus gives a good fit to experimental $\eta(h\nu)$ dependences.

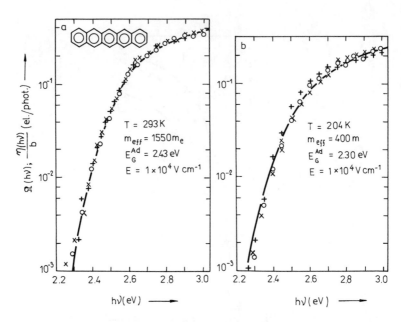

FIGURE 6.24. $\Omega(h\nu)$ curves, as calculated according to Eq. (6.53) and corresponding experimental $\eta(h\nu)/b$ dependences (experimental points of three different Pc crystal samples) at $\mathcal{E} = 1 \times 10^4$ V/cm and $T = 293$ K (a) and $T = 204$ K (b). Approximation of these $\eta(h\nu)/b$ dependences has been used for final adjustment of the parameters m_{eff} and E_G^{Ad}.[344]

This effect is supposed to be responsible also for different T-dependent $r_{\text{th}}(h\nu)$ curves in the threshold spectral region (see Figs. 6.21a and b).

6.3.3.C. Diffusive motion and mobility of the electron

Figure 6.26 shows the calculated Gaussian distribution curves versus diffusive motion of the electron during thermalization. These distribution curves visualize the evolution of the thermalization process in Pc crystals. Although the dispersion parameter σ_r increases, as can be seen, with increasing separation distance r, nevertheless the principal requirement for the validity of the model (see Ref. 306), namely, $\sigma_r(r)^2 \ll \langle r \rangle^2$ is fulfilled within the entire thermalization range (see also Fig. 6.27). It also shows that substituting the Gaussian distribution by the δ-function distribution of thermalization distances, often used in practice (see, e.g., Refs. 73 and 349), may be regarded only as a zero-order approximation.

Figure 6.27 demonstrates the variation of the mean separation distance $\langle r \rangle$ and mean radial velocity $\langle v \rangle$ with time t, as well as the variation of the corresponding dispersion parameters σ_r and σ_v. As can be seen, the mean radial velocity decreases in an approximately exponential fashion, and, as the electron reaches the thermalization area ($t \rightarrow \tau_{\text{th}}, \langle r \rangle \rightarrow r_{\text{th}}$), the value of σ_v already exceeds the magnitude of $\langle v \rangle$, and thus the "memory" of the electron of radial motion is completely lost.

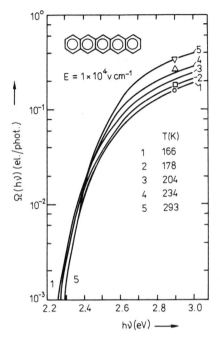

FIGURE 6.25. $\Omega(h\nu)$ curves, as calculated according to Eq. (6.53) for different values of temperature T in Pc. Corresponding experimental values of η/b are shown only at $h\nu = 2.9$ eV, for the following values of temperature T: $T = 166$ (○), 178 (□), 234 (△), and 293 K (▽).[344]

It is also possible to evaluate the temperature dependence of the *mean* electron mobility $\bar{\mu}$ during thermalization at zero field ($\mathcal{E} = 0$) according to the following formula:

$$\bar{\mu} = (e/kT)(\langle r_{th}^2 \rangle / \tau_{th}). \tag{6.54}$$

As can be seen from Fig. 6.28, curve 1, the value of $\bar{\mu}$ decreases with increasing temperature from $\bar{\mu} = 113$ cm^{2}V^{-1}s^{-1} at 165 K to $\bar{\mu} = 25.4$ cm^{2}V^{-1}s^{-1} at 293 K, presumably due to the dependence of m_{eff} on T.

Although the model has been initially developed only for the description of photogeneration processes, and all adjustments of the variable parameters are based on data of thermalization and dissociation stages of charge-carrier photogeneration, it may be shown that the simulation results can be extended for the description of transport properties of already thermalized electrons.

The effective mass $m_{eff}(T)$ and the microscopic mobility $\mu_0(T)$ of a thermalized charge carrier are related by the following formula:[348,355]

$$m_{eff}(T) = \frac{3e^2 l_o^2}{[\mu_0(T)]^2 \cdot kT}. \tag{6.55}$$

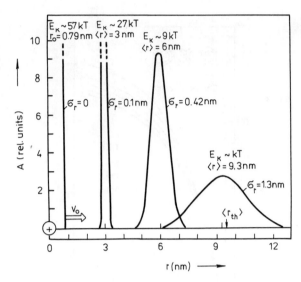

FIGURE 6.26. Evolution of the thermalization process in a Pc crystal, taking into account diffusive motion of the electron. Distribution curves and corresponding $\langle r \rangle$ and σ_r values have been calculated at $T = 204$ K, $\mathcal{E} = 0$, $m_{eff} = 400\ m_e$, and $E_G^{Ad} = 2.30$ eV.[344]

Hence,

$$\mu_0(T) = \frac{\sqrt{3}\,e l_o}{[m_{eff}(T)kT]^{1/2}}. \qquad (6.56)$$

The results of the evaluation of the $\mu_0(T)$ dependence in Pc, using the values of the $m_{eff}(T)$, according to Fig. 6.19, under assumption of constant mean free path l_o of the carrier, equal to the lattice constant $l_o = a_o = 0.79$ nm, are presented in Fig. 6.28, curve 2.

As can be seen from Fig. 6.28, we obtain an appropriate $\mu_0(T)$ dependence, namely, μ_0 decreases with temperature and has a value of 3.1 cm^{2}V^{-1}s^{-1} at 166 K and 0.91 cm^{2}V^{-1}s^{-1} at 293 K, i.e., it is of the same order as the reported microscopic mobility values in Ac-type crystals (see Ref. 288, p. 338, and Ref. 154).

Further, it may be shown that the deduced $\mu_0(T)$ dependence can be approximated as $\mu_0 \sim T^{-n}$ with $n = 2.1$ (see Fig. 6.28, curve 2′). This result is also in agreement with experimentally observed typical $\mu_0(T)$ dependences within the **ab** plane of Ac-type crystals. However, $\mu_0(T)$ dependences have not, as far as we know, been experimentally estimated for Pc single crystals.

The $\mu_0(T)$ dependence, obtained as a "by-product" from simulation data, may be regarded as the most crucial test of the validity of the model and demonstrates its applicability even outside the initially anticipated scope. These results may be regarded as a challenge to experimental physicists to confirm the predicted $\mu_0(T)$ dependence in Pc crystals. The physical consequences of apparent inconsistency

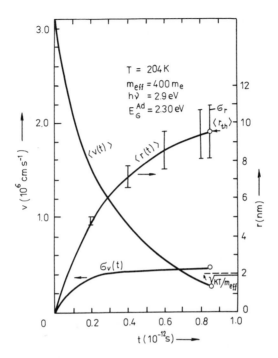

FIGURE 6.27. Variation of mean separation distance $\langle r \rangle$, mean radial velocity $\langle v \rangle$, and corresponding dispersion parameters σ_r and σ_v with time t during thermalization of the electron in Pc.[344]

between a band-like type mobility T dependence ($\mu_0 \sim T^{-n}$) and the concept of incoherent hopping ($l_o \approx a$) according to the model will be discussed later (see Section 6.3.4).

6.3.3.D. Field-induced electron drift during thermalization

It has been demonstrated experimentally (see Refs. 349 and 350 and Section 6.2.7) that at higher electric fields ($\mathcal{E} \geqslant 10^4$ V cm^{-1}) and sufficiently large values of r_{th} ($r_{th} \geqslant 4$ nm) the influence of $\mathcal{E}$ on r_{th} due to the hot electron drift during thermalization should be taken into account. The experimentally observed electron drift in Pc crystals has been successfully simulated using the adjusted parameters of the thermalization model. Figure 6.29 shows calculated trajectories of the mean $\langle r \rangle$ values of the electron during thermalization in the electric field $\mathcal{E} = 1.2 \times 10^5$ V cm^{-1}. In Fig. 6.29 the electron trajectories in the **ab** plane of the Pc crystal are shown for different values of the angle θ between the initial radial velocity v_0 and the external electric field $\mathcal{E}$ between $\theta = 0$ and $\theta = \pi$.

These simulation data concur with the experimentally observed electron drift during thermalization at the field strength $\mathcal{E} \geqslant 10^4$ V cm^{-1} (see Section 6.2.7). The

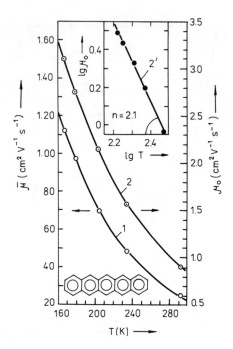

FIGURE 6.28. (1) Temperature dependence of the mean electron mobility $\bar{\mu}(T)$ during thermalization in Pc, evaluated according to Eq. (6.54). (2) Temperature dependence of stationary microscopic mobility $\mu_0(T)$ of a thermalized electron in Pc, evaluated according to Eq. (6.56) (2') $\mu_0(T)$ dependence in a log-log plot.[344]

isochrons for $\tau = $ const in Fig. 6.29 visualize the dynamics of the thermalization process in the presence of a strong external field $\mathcal{E}$.

Figure 6.30 shows calculated maximum thermalization length dependences along the direction of the external field (see Fig. 6.29), and the effective thermalization length $r_{eff} = f(\mathcal{E})$ evaluated from experimental current-voltage characteristics by the Onsager formula for different values of $h\nu$ in Pc crystals (see Section 6.2.7). Although the values r_{max} and r_{eff} are not, strictly speaking, physical equivalent ($r_{eff} < r_{max}$), nevertheless Fig. 6.30 demonstrates semiquantitatively the symbistic character of both dependences on electric field value. As may be seen, at strong electric fields ($\mathcal{E} \geqslant 10^4$ V cm^{-1}) and sufficiently high values of r_{th} ($r_{th} \geqslant 5$ nm) the thermalization time is long enough to provide a noticeable electron drift during thermalization. The symbistic behavior of both calculated $r_{max}(\mathcal{E})$ and experimental $r_{eff}(\mathcal{E})$ dependences may be regarded as additional evidence supporting the dynamic thermalization model.

It will be shown in Chapter 7 that at still higher electric fields ($\mathcal{E} \geqslant 10^5$ V cm^{-1}) (or at lower temperatures) it is possible to create hot charge carriers which move along the field with the mean drift velocity v_d greater than the mean thermal velocity v_{th} ($v_d > v_{th}$) and do not thermalize at all.

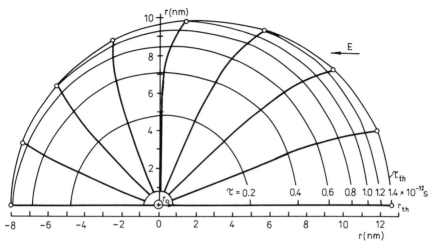

FIGURE 6.29. Calculated mean $\langle r \rangle$ values of electron trajectories in the **ab** plane of a Pc crystal for different values of angle θ from $\theta = 0$ to $\theta = \pi$ between initial radial velocity v_0 and external electrical field $\mathcal{E}$ at $\mathcal{E} = 1.2 \times 10^5$ V/cm. The isochrons for $\tau = $ const visualize the dynamics of the thermalization process. The calculation was performed with the following parameter values: $T = 204$ K; $m_{eff} = 400 \, m_e$; $E_G^{Ad} = 2.30$ eV; $\beta = 2.19 \times 10^{12}$ s^{-1}; $r_o = 0.79$ nm.[344]

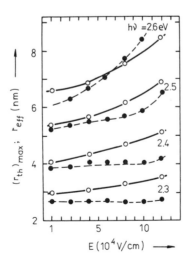

FIGURE 6.30. Calculated maximum thermalization length dependences $\langle r_{max} \rangle = f(\mathcal{E})$ along the direction of external field $\mathcal{E}$ at $T = 204$ K (filled circles) (other parameters as in Fig. 6.29) and effective thermalization length dependences $r_{eff} = f(\mathcal{E})$ (open circles) as evaluated from experimental current-voltage characteristics by the Onsager formula at $T = 205$ K for different values of $h\nu$ in a Pc crystal.[344]

6.3.4. Discussion of the Simulation Results

It can be shown that the most important result of the simulation studies, namely, the necessity of introducing a heavy quasiparticle (instead of a "bare" electron) with effective mass m_{eff} of the order of $(10^2-10^3)m_e$, is consistent with the extended polarization model of ionic conductivity states in organic molecular crystals (see Ref. 332, p. 94, and Ref. 351).

According to this phenomenological model (see Chapter 4) there exist two purely electronic states for electron and hole in the crystal, namely, E_P^+ and E_P^- (see Figs. 4.8–4.10). These states in Ac-type crystals are actually many-electron interaction levels since their location in the energy diagram of the crystal is determined by electronic polarization energies P_{eff}^- and P_{eff}^+ of the lattice by electron and hole, respectively. Consequently, these levels are, in fact, electron polaron states separated by the gap E_G, later identified (see Ref. 318) as optical energy gap E_G^{Opt}.

Due to the pronounced localization of thermal or epithermal charge carriers in Ac-type crystals with characteristic residence time on the lattice site $\tau_h = (2-5)\times10^{-14}$ s (see Chapter 3, Fig. 3.9), which is almost by an order of magnitude greater than the characteristic vibronic relaxation time $\tau_v = \hbar/E = (1\times10^{-14})-(2\times10^{-15})$ s for typical vibronic energies $E = 480-3000$ cm^{-1}, a strong coupling of the charge carrier with intramolecular vibrational modes should take place, resulting in the formation of MP states. Corresponding ionic state conductivity levels M_P^- and M_P^+ are separated by the adiabatic energy gap E_G^{Ad}, where $E_G^{Ad} = E_G^{Opt}-2E_b$ (see Ref. 307 and Figs. 4.8–4.10 in Chapter 4). Here the value of $2E_b$ is equal to the energy gains as a result of formation of positive and negative molecular polarons within the crystal, i.e., E_b, in first approximation, is equal to the relaxation energy of the neutral molecule with the localized charge carrier transforming into a genuine molecular ion. The relaxation energy E_b per ion is estimated to be $\approx 0.1-0.15$ eV for polyacene molecules[307] (see Table 3.8).

Thus, the detected heavy quasiparticles in a Pc crystal may be identified as molecular ion states, and the corresponding conductivity levels M_P^- and M_P^+, in a more general conceptional approach, as MP states (see Chapters 3 and 4).

It may be shown that the extrapolated spectral threshold of intrinsic photogeneration $E_{th} = 2.25\pm0.05$ eV in Pc (Fig. 4.1 and Ref. 349) gives the lower limit of the E_G^{Ad} value. A different method for evaluation of E_G^{Ad}, proposed by Sebastian et al.[317,318] (see also Chapter 4, Fig. 4.2), is based on graphical analysis of photon energy versus a r_{th}^{-1} plot. Asymptotic extrapolation of $h\nu(r_{th}^{-1})$ dependence at a slope of $-e^2/4\pi\epsilon\epsilon_0$, using the reported values of r_{th},[349] yields E_G^{Ad} equal to 2.47 eV for Pc (see Fig. 4.10).

It may be seen from Fig. 6.19 that the E_G^{Ad} values, as fitted to experimental $\eta(h\nu)$ dependences (see Fig. 6.24) in the simulation procedure, lie within the limits of both independent estimates.

On the other hand, electric-field modulated absorption spectra of polycrystalline Pc layers, yield, according to Sebastian et al.,[317,318] (see also Chapter 4, Fig. 4.7), a value of $E_G^{Opt} = 2.8$ eV for the optical energy gap.

These data confirm that the total relaxation energy of electron and hole to the corresponding MP states, namely, M_P^- and M_P^+, in Pc equals $2E_b = 2E_G^{Opt} - E_G^{Ad} = 0.3-0.35$ eV (see Ref. 307 and Chapter 3, Table 3.8). This value is of the same order as the independent estimates discussed above.

Thus, available experimental, as well as simulation data, support the view that photogenerated low-energy ($E \lesssim 1$ eV) charge carriers in Pc crystals are not simply quasifree electrons (or holes) but actually emerge as molecular ions, or, in more generalized approach, as adiabatic MPs (see Chapter 3, Fig. 3.10).

Consequently, the description of thermalization in terms of the AI model (see Fig. 6.1) is based on the assumption that the ejected electron, after the first scattering event at $r_o = a$, is transformed into a heavy quasiparticle, obtains the corresponding effective mass m_{eff} of a molecular polaron, due to strong coupling with intramolecular vibrations, and starts thermalization with initial velocity v_0 according to Eq. (6.45). The alternative approach, namely, the model of direct optical CT transition (see Fig. 6.2 and Ref. 351) gives an even more feasible physical picture. According to the Franck-Condon principle, formation of ions in a direct optical CT transition occurs at fixed nuclear coordinates. After the charge transition the nuclear skeleton responds to the new electronic configuration and relaxes to the new equilibrium state of the ion (see Chapter 3). According to Sebastian et al.[318] the nuclear relaxation energy, gained as a result of formation of a molecular ion, can be delivered as kinetic energy to the electron. As indicated by the authors, a similar process takes place, e.g., in case of photoemission. Thus, energy distribution curves of photoemitted electrons demonstrate the fact that nuclear relaxation energy of a molecule during ionization, emerging as a difference between vertical and adiabatic ionization potentials, can be effectively delivered to the electron without significant losses via other dissipative channels.

According to this approach, the electron-hole pair, created by optical CT transition, can use the excess energy gained from nuclear relaxation for further separation against the coulombic field. This process, as illustrated in Fig. 6.31, can be eventually regarded as thermalization which starts at $r_{CT} = r_o = 0.79$ nm in case of a CT transition from a molecule at (0, 0, 0) to a molecule at (1, 0, 0). The kinetic energy E_k obtained in this case by the electron, or more precisely, by the MP, equals

$$E_k(r_o) = h\nu_{CT} - E_G^{Ad} + |U_{CP}(r_o)|.$$ (6.57)

The calculated thermalization length r_{th} for this lowest CT transition with $h\nu_{CT} = 2.35$ eV is rather small: ≈ 2 nm (see Fig. 6.31).

However, the optical CT transition can also take place to higher excited states of the ion, to the energy level CT*, as shown in Fig. 6.31. In this case the initial kinetic energy E_k equals

$$E_k^*(r_o) = h\nu_{CT}^* - h\nu_{CT} + 2E_b = h\nu_{CT}^* - E_G^{Ad} + |U_{CP}(r_o)|.$$ (6.58)

As can be seen from Fig. 6.31, in such a case for $r_{CT} = r_o = 0.79$ nm and $h\nu_{CT} = 2.7$ eV the thermalization distance is considerable, i.e., ≈ 6.7 nm.

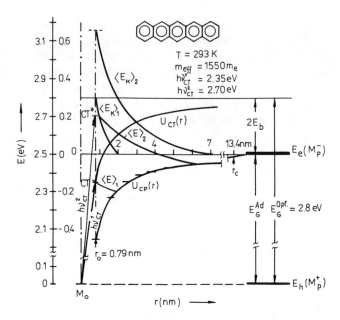

FIGURE 6.31. Energy diagram representing the thermalization process in terms of the CT transition model. According to this model the electron-hole pair, created by optical CT transition, can use the excess energy gained from nuclear relaxation for further separation against the Coulombic field. $\langle E(r) \rangle$ and $\langle E_k(r) \rangle$ curves show the variation of the average total and kinetic energies as functions of the separation distance r; $U_{CT}(r)$ and $U_{CP}(r)$ are potential energy curves for unrelaxed and relaxed ionic states, respectively; E_G^{Opt} and E_G^{Ad} are the corresponding optical and adiabatic energy gap.[344]

Comparison of Eqs. (6.51) and (6.58), and corresponding graphical representations of the thermalization process in Figs. 6.20 and 6.31 shows that both approaches give similar results and that their difference may be more semantic than physical. At least, both models, in our opinion, should not be regarded as mutually exclusive, but rather as complementary CP-state population channels (see Section 6.2.4) the contribution of each of them depending on the spectral range. Independently of the pathway by which the electron reaches the molecule at r_o, the further process of thermalization may be regarded as equivalent for both approaches. However, the CT-transition model seems physically more realistic, since it postulates the formation of an ionic state at the very start of thermalization at $r_{CT} = r_o$ and thus better satisfies the initial requirements of the simulation procedure than the AI model.

Finally, let us consider the possible physical nature of the MP discussed above (see also Chapter 3, Section 3.6, and Chapter 4, Sections 4.1–4.3) which ought to have unusual properties. On the one hand, it should be, according to the simulation model, a heavy quasiparticle, moving from one crystal site to another by incoherent hopping. On the other hand, ir should manifest a bandlike temperature-dependent mobility, namely, $\mu_0 \sim T^{-n}$, without Arrhenius activation energy.

The genesis of the polaron concepts and their physical nature have been already discussed in Chapter 3 (see Section 3.6.1). A considerable part of Chapter 4 has been devoted to the energetic aspects of the formation of conductivity states of MP in Ac-type crystal. Here, and in Chapter 7, we shall mainly discuss the transport properties of molecular polarons.

As previously mentioned, a simple model of a MP was proposed by Holstein[136] in 1959 (see also Ref. 136a). However, Holstein's model of a small-radius localized polaron involves thermally activated motion and seems not to be applicable in our case. Therefore one should search for an appropriate prototype in lattice polaron theories which have been elaborated for both large and small polaron approaches (see Refs. 12, 247, 285, and 288). In the early stages of the development of the small lattice polaron theory, an expression of the effective mass dependence on temperature, equivalent with Eq. (6.46), was proposed by Yamashita and Kurosawa[416] and Sewell.[324] However, this correspondence may serve only as a formal basis for the description of MP behavior. Later Eagles[97] developed a theory of an adiabatic nearly small lattice polaron, the effective mass of which may range from about 10–1000 times the bare electron mass. This approach, accordingly, provides a model prototype for a nearly small MP.

However, one should notice that Eagles' model is an intermediate between the large and the small polarons and, thus, obviously, connected with a lot of purely mathematical problems. Perhaps, owing to this, the nearly small polaron approach has not gained a high popularity in polaron theory studies (see Ref. 247).

It has been suggested[288a] that taking into account the intermediate status of the nearly small MP it may be called a mesopolaron.

Concerning the possible transport mechanism of a nearly small polaron one should mention that according to Firsov (see Ref. 285) a specific polaron transport mechanism can take place in a wide intermediate temperature range, namely, hopping by tunneling, which is different from both thermally activated hopping and bandlike polaron transport. Such an approach might rationalize some apparently contradictory simulation results, particularly the temperature dependence of mobility, namely, $\mu_0 \sim T^{-n}$. In the case of the intermediate nearly small polaron model one should not expect a sharp band-to-hopping transition, as predicted by Holstein's theory (see Ref. 288, p. 357), but rather smooth transfer over the intermediate T range where the polaron moves by incoherent hopping, simultaneously retaining some features of bandlike motion, such as the temperature dependence of mobility.

Although the term "hopping by tunneling" is now quite common, Martin Pope has suggested[288a] the use of the physically more adequate term "stepping by tunneling" since "hopping" is often associated with processes that require thermal activation energy. By the way, the term "stepwise tunneling" is often used by quantum chemists (see, e.g., Refs. 209a and 209b).

A molecular adiabatic nearly small polaron may be visualized physically as a molecular ion polarizing dipole-active intramolecular vibrational modes of nearest neighbor molecules via long-range Coulombic interactions (see Chapter 3, Section

3.6.1). As a result, it would give an additional contribution to the effective value of E_b and enhance interaction energy J_{nm} with neighboring molecules, thus presumably increasing the hopping probability of the polaron (see Section 3.6.2).

A similar approach, based on a generalized Huang-Rhys model, has been proposed by Salaneck et al.[305] in order to interpret temperature-dependent photoemission linewidth in molecular solids.

Microscopic theories of nearly small polaron (mesopolaron) formation and transport phenomena have been discussed in some detail in Sections 1.B.8 and 3.6.2, and in Chapter 5.

There still remains a controversy concerning the applicability of exponential dependence of the effective mass m_{eff} of the carrier on temperature according to the empirical equation (6.46). As already mentioned, this $m_{eff}(T)$ relation has been introduced in Ref. 344 as an ad hoc assumption giving the best fit to simulation results (see Fig. 6.19) and somehow based on its resemblance to a similar equation proposed in early polaron theories.[324,416] However, as indicated in Ref. 356, the exponential $m_{eff} = f(T)$ dependence, proposed by Yamashita and Kurosawa[416] and Sewell[324] may be assumed to be valid for a polaron band approach description of charge carrier transport at low temperatures. Similar exponential $m_{eff} = f(T)$ dependence has been theoretically derived for a MP [see Eq. (1B.8.18)].

Actually, in our case, a band model description of charge carrier transport may be physically reasonable only at the low temperature range when the mean free path l_o of the carrier is of the order of at least several lattice constants ($l_o > a_o$) (see Chapter 7, Figs. 7.21 and 7.28).

As the temperature increases, the effective width δE_P of the polaron band drastically decreases, inversely to the exponentially growing effective mass of the carrier, namely, $\delta E_P(T) = 1/m_{eff}(T)$. Consequently, the charge carrier becomes strongly localized, and the band model approach ought to be replaced by a classical corpuscular representation.[356] Also, one should keep in mind that the SM model, used for simulation, is derived in terms of a corpuscular representation. Additionally, in Eq. (6.55), in relating the effective mass $m_{eff}(T)$ dependence with that of microscopic mobility $\mu_0(T)$, one actually incorporates a power law rather than exponential T dependence.

It has been recently shown in Ref. 356 that in the high T region the exponential $m_{eff}(T)$ dependence (6.46) should be replaced by a power-law dependence $m_{eff}(T) \sim T^{2n-1}$, where n is the slope of the $\mu_0(T) \sim T^{-n}$ relation [see Chapter 7, formula (7.19)].

In terms of this approach it may be shown that the exponential curve, namely $m_{eff} = 18.7 m_e \exp(T/66.4)$, giving the best fit for simulation results (see Fig. 6.19), can be replaced by a power-law curve $m_{eff}(T) \sim T^{2n-1} = T^{2 \times 2.12 - 1} = T^{3.24}$. As can be seen from Fig. 6.32, both curves practically coincide in the temperature range from 166 to 293. It will be shown in Chapter 7 (see Section 7.2.3) that the exponential $m_{eff}(T)$ dependence is actually valid in the low T range, while in the high T region it should be approximated by a power-law dependence, namely,

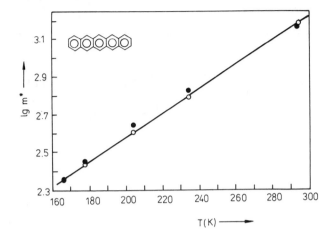

FIGURE 6.32. Temperature dependence of the relative effective mass $m^*(T)$ in Pc crystals in semilog plot. Open circles, approximated according to the exponential law (6.46) with m_0^* = 18.7 and T_0 = 66.4 K (see Fig. 6.19); filled circles, approximated according to the power-law [see Eq. (7.19)] with $2|n| - 1 = 3.24$.

$m_{\text{eff}}(T) \sim T^{2n-1}$ having a physical sense for the high temperature region, where $l_o \approx a_o$.

The results presented above may serve as a phenomenological and conceptual basis for the development of a comprehensive and self-consistent nearly small MP theory. Such a theory should be capable of describing all observed and predicted peculiarities of MP behavior, also taking into account the characteristic features of an organic crystal. This seems to be a real challenge to theorists working in the polaron field.

One of the most difficult tasks, in our opinion, would be to describe the initial, most controversial stage of photogeneration, namely, AI and/or optical CT transition which must be represented in terms of quantum mechanics and subsequently "joined" to a quasiclassical treatment used for describing thermalization, dissociation, and mobility processes (see Section 5.15).

CHAPTER 7

Charge Carrier Transport Processes in OMC

A courageous scientific imagination was needed to realize that not the behavior of bodies, but the behavior of something between them, that is, the field, may be essential for ordering and understanding events.

ALBERT EINSTEIN

In atomic theory we have fields and we have particles. There are two different ways of describing the same thing — two different points of view. We use one or the other according to convenience.

PAUL A. M. DIRAC

7.1. EXPERIMENTAL DETERMINATION OF CHARGE CARRIER TRANSPORT CHARACTERISTICS AND PARAMETERS

The last chapter of this book is devoted to the transport phenomena of charge carriers in OMC. First, we discuss in some detail recent experimental studies of charge carrier transport characteristics in the main model compounds of OMC–Nph, Ac, and α-perylene.

In the second part the results of computer simulation of experimental characteristics are presented, based on the application of a modified SM model (see Section 6.3) and MP approach (Chapter 3). Figuratively speaking, the leitmotif of this chapter will be the MP in action.

Finally, we briefly discuss the charge carrier transport phenomena in real crystals containing local trapping states.

7.1.1. Some General Problems and Specific Features

The charge carrier transport phenomena belong to the most complicated electronic processes in organic crystals and, at the same time, highly controversial with regard to their theoretical interpretation.

We wish to reiterate that, in the first place, this is due to the polaronic nature of the charge carriers. As it has been already discussed in considerable detail in Chapter 3, a charge carrier in OMC, owing to strong localization, may interact with the electronic and nuclear subsystems of the crystal, which leads to the formation of electronic, molecular, and in some cases also, lattice polarons.[288,332,351] Actually, it is the problem of the many-particle interaction that should be treated in terms of quantum electrodynamics. However, due to the very complicated nature of these processes, at present the problem has been dealt with mainly in phenomenological approaches, in terms of a model description of charge carrier transport phenomena.[344,348,354-356,406]

The second problem is connected with the fact that the polaron is a charged quasiparticle. In case of a neutral quasiparticle, like an exciton, its motion may be primarily characterized by diffusion parameters (see Chapter 5). No wonder therefore that the present transport theories are developed mainly for exciton motion in OMC.[288,351] In the case of the charge carriers the situation is more complicated. The motion of a polaron as a charged quasi-particle is influenced both by the intrinsic Coulombic and the external electric fields as well as by stochastic forces causing diffusive random walk. Thus, the motion of a polaron should be, as the first approximation, described in terms of a generalized Fokker-Planck equation [see Eq. (6.41)] the solution of which may yield the distribution function of position and velocity of the particle under investigation.

However, it is difficult to find such a solution in an explicit analytical form. We shall use in this chapter the extended SM approach[344,351] based on the Fokker-Planck equation (see Section 6.3.1) which allows, in a framework of a model description, to find, in simulation procedure, the mean values of polaron positions and velocities, their distribution and evolution in time (see Section 6.3.3). In addition there are also a number of purely experimental difficulties.

In the case of exciton transport there are available a great variety of direct optical and different indirect methods for the determination of diffusion parameters of an exciton.

In the case of charge carriers the situation is different. Here, primarily, we are interested in determining such transport parameters as the microscopic (intrinsic) mobility μ_0 and corresponding drift velocity v_d in an electric field $\mathcal{E}$, as well as the dependences of these parameters on $\mathcal{E}$ and T, namely, $\mu_0(T, \mathcal{E})$ and $v_d(T, \mathcal{E})$ in a wide range of temperature and electric field. The diffusion parameters of the carrier, its mean thermal velocity, etc., are less important for understanding the carrier transport mechanisms.

For the determination of the $\mu_0(T, \mathcal{E})$ and $v_d(T, \mathcal{E})$ dependences in OMC there actually is only one single direct experimental method available, namely, measurements of transient pulse photoconductivity characteristics.

The pulse photoconductivity method, known as the carriers time-of-flight (TOF) technique, actually is a macroscopic, integral method of measurement. Non-equilibrium charge carriers, generated by a short light pulse should be collected during their lifetime, on the electrodes of a macroscopic sample. Then, analyzing the shape of the photoconductivity pulse, it is possible to determine the transit time of the charge carriers and to obtain the $\mu_0(T, \mathcal{E})$ and $v_d(T, \mathcal{E})$ dependences for given T and $\mathcal{E}$ values.

The main experimental problem is to measure the genuine, intrinsic microscopic mobility μ_0, and then to do it down to low temperatures. Such measurements may be realized only in ultrapure, carefully grown, and specially treated practically perfect single crystals.[155–158,406,407]

Since charge carrier mobilities are very sensitive even to small concentrations of impurities (of the order of $\leqslant 1$ ppm, see Ref. 154, p. 178) only ultrapure perfect crystals permit measurements of trap-uninfluenced microscopic mobilities. The presence of shallow trapping states, either of impurity origin (e.g., Tc in the Ac crystal) or of structural origin (see Refs. 288 and 332) sets the low-temperature limit below which the microscopic mobility measurements become impossible. Below this temperature one actually measures the multiple-shallow-trapping controlled drift mobility μ_d, which may be considerably smaller than the intrinsic microscopic mobility μ_0, namely, $\mu_d < \mu_0$[154,156,157,158] (see Section 7.4).

Another technical problem is connected with the fact, that the mobility in anisotropic Ac-type crystals is actually a tensor μ_{ij}. It means that, to obtain the components of the mobility tensor, one should perform the measurements along definite crystallographic directions of the crystal.

That means that the crystal should be carefully cut along definite planes to prepare a sample, and the corresponding crystal orientations must be strictly controlled. In addition, one needs a good laser technique providing nanosecond and subnanosecond light pulses in a definite spectral range, and a sensitive electric pulse measurement technique.

Owing to the above-described specific requirements and experimental difficulties it is no wonder that correct and highly plausible $\mu_0(T, \mathcal{E})$ measurements down to very low temperatures have been performed only recently by Karl and coworkers.[154–157,158,406,407]

Historically charge carrier mobilities μ_0 in OMC were first measured in the early sixties by LeBlanc[212] and Kepler[179,180] who introduced the time-of-flight (TOF) technique in the field of organic materials.

Kepler measured the parameter μ_{ij} in three crystallographic directions of an anthracene crystal in the temperature range from 200 to 300 K (see also Ref. 154, p. 162). He also studied the mobility (μ_0) dependence on pressure p up to 3 kbar along different crystallographic axes of Ac.

There followed a number of sporadic μ_0 measurements by a number of authors in different organic crystals. However, these studies were not systematic and usually covered a narrow temperature range. These, and a number of later μ_0 measurement data have been compiled in an earlier[153] and also in a more recent review[154] by Karl (see also Ref. 288, p. 338).

The first successful μ_0 measurements in Ac down to liquid N_2 temperature were reported by Karl and Probst.[153,292]

Mobilities that strongly increased with decreasing temperature, reaching, for holes, $\mu_0^+ = 55$ cm^2/Vs at 120 K, were observed in durene crystals by Burstein and Williams.[52] Still higher values of electron mobility μ_0^- were reported by Stehle and Karl in perylene in 1981, reaching a maximum value of $\mu_0^- = 100$ cm^2/Vs at 30 K (see Ref. 406). In the late seventies Schein and coauthors performed a series of mobility $\mu_0(T)$ measurements in Nph crystals, also down to 30 K.[312,313,315,316]

The authors of Refs. 313 and 315 obtained almost temperature-independent electron mobility $\mu_{c'c'}^-$ in the c'-direction of a Nph crystal in the temperature range from 300 to 100 K, which resembled the $\mu_{c'c}^-(T)$ behavior in Ac (see Ref. 117). However, the $\mu_{c'c'}^-(T)$ dependence started to rise exponentially below 100 K. Such behavior of the $\mu_{c'c'}^-(T)$ curve was interpreted by the authors of Refs. 313 and 315 as a sign of possible "hopping to band" transitions (see also Ref. 288, p. 356 and Ref. 332, p. 36).

It will be shown later (see Section 7.2.3) that the cause of this rise of $\mu_{c'c'}^-(T)$ characteristics at low temperatures may be of quite different nature. Schein and McGhie[314] anticipated possible electric field effect on charge carrier mobilities, which might shed some light on the underlying transport mechanism. They performed mobility $\mu^-(\mathcal{E})$ measurements for electrons in a Nph crystal in the low-mobility c'-direction and actually observed no effect up to the field strength $\mathcal{E} \leqslant 1.7 \times 10^5$ V/cm at 160 K.

The first successful and reliable measurements of pronounced field dependences of carrier mobilities $\mu = \mu(\mathcal{E})$ have been obtained by Karl and coworkers.[155,406,407] In contrast to the previous attempts, authors confined the search for field dependence of $\mu = \mu(\mathcal{E})$ at lower temperatures, providing higher mobility values.

Further development of improved ultrapurification and crystal growth and treatment techniques[155,156,158,356] allowed Karl and coworkers to reach final success. They could obtain highly reliable and reproducible charge carrier mobility characteristics $\mu_0(T, \mathcal{E})$ for Nph and perylene crystals down to the helium temperatures[154,155,406,407] and for Ac down to 20 K.[158] As a result of highly meticulous studies, Karl and coauthors have made, in our opinion, two fundamental discoveries in the field of transport phenomena in OMC. First, they demonstrated that in hydrocarbon crystals the mobility T dependence $\mu_0 \sim T^{-n}$ holds down to very low temperatures and, as a result, the μ_0 value drastically increases at low T range exceeding 400 cm^2/Vs in Nph and 100 cm^2/Vs in perylene crystals.[154,155,406,407]

Second, the authors were the first to observe a prominent electric field dependence of the carrier mobilities at low temperatures.[154,155,356,406,407] These $\mu(\mathcal{E}, T)$ dependences provided overwhelming evidence that, at the given electric field and temperature range, *hot* charge carriers are created, which move in the electric field at a constant drift velocity $v_d = $ const, greater than the mean thermal velocity v_{th} of the carrier at the given temperature.

Bitterling and Willig[33,34] have recently demonstrated that it is possible to create, at high electric fields $\mathcal{E} \geqslant 5 \times 10^5$ V/cm, hot holes in an Ac crystal also at room temperature.

7.1.2. Methodology of Experimental Measurements

The microscopic drift mobility of a charge carrier μ_0 in anisotropic organic crystals of orthorhombic, monoclinic, or triclinic symmetry (see Ref. 154, p. 110) is actually a tensor and should be denoted as μ_{ij}. μ_{ij} represents charge carrier drift mobility tensor components, relating the drift velocity v (with components v_i) of holes and electrons, respectively, to the applied electric field $\mathcal{E}$ with components $\mathcal{E}_j$, according to the equation:[154]

$$v_i = \mu_{ij}\, \mathcal{E}_j. \tag{7.1}$$

As mentioned before the experimental technique used for the determination of the drift mobility is based on the TOF.[153,179,212] The quantity which is measured in a TOF for a general crystal orientation is related to the drift velocity component parallel to the applied electric field, $v_\parallel$. The ratio of cause, $\mathcal{E}$, and response $v_\parallel$, the "mobility in the given direction," $\mu_\parallel$, is in general a certain linear combination of all six mobility tensor components. Therefore, in a general situation, at least six measurements in six independent directions are required for an evaluation of the six independent tensor elements of the mobility tensor (see Ref. 154, p. 110). The tensor components are usually given in the respective orthogonalized crystallographic axes system $\mathbf{a}/a$, $\mathbf{b}/b$, $\mathbf{c}/c$ (for orthorhombic), $\mathbf{a}/a$, $\mathbf{b}/b$, $\mathbf{c}^*/c^*$ (for monoclinic), and $\mathbf{a}'/a'$, $\mathbf{b}^*/b^*$, $\mathbf{c}/c$ (for triclinic crystal symmetry), where $\mathbf{a}'/a' = (\mathbf{b}^*/b^*) \times (\mathbf{c}/c)$, $\mathbf{b}^*/b^* = (\mathbf{c}/c) \times (\mathbf{a}'/a')$. In matrix notation

$$\mu_{ij} = \begin{pmatrix} \mu_{aa} \mu_{ab} \mu_{ac} \\ \mu_{ab} \mu_{bb} \mu_{bc} \\ \mu_{ac} \mu_{bc} \mu_{cc} \end{pmatrix}, \tag{7.2}$$

where the star has to be added to the subscript c in monoclinic symmetry and the subscripts a and b have to be replaced by a' and b^* in triclinic symmetry.

From symmetry requirements $\mu_{ab} = \mu_{bc} = 0$ in monoclinic crystals, and all mixed indiced components $\mu_{ab} = \mu_{bc} = \mu_{ac} = 0$ for orthorhombic and higher-symmetry crystals.

The principal axes mobility components are designated μ_{11}, μ_{22}, and μ_{33}; in matrix notation

$$\mu'_{ij} = \begin{pmatrix} \mu_{11} & 0 & 0 \\ 0 & \mu_{22} & 0 \\ 0 & 0 & \mu_{33} \end{pmatrix} \tag{7.3}$$

with the directions 1,2,3 chosen as close as possible to the orthogonalized crystallographic directions $\mathbf{a}^{(')}$, $\mathbf{b}^{(*)}$, $\mathbf{c}^{(*)}$, respectively. For orthorhombic (and higher) symmetry $\mu_{11} = \mu_{aa}$, $\mu_{22} = \mu_{bb}$, $\mu_{33} = \mu_{cc}$. For monoclinic symmetry $\mu_{22} = \mu_{bb}$; the other two principal directions are generally oblique with respect to the crystallographic axes $\mathbf{a}$ and $\mathbf{c}^*$, forming an angle θ between $\mathbf{a}$ and the principal direction **1**. This angle θ is counted in the same direction as the monoclinic angle β

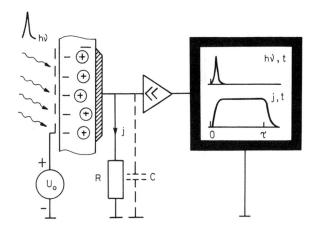

FIGURE 7.1. Schematic representation of the setup used for measuring electron and hole mobilities by the time-of-flight method according to Ref. 406.

$(\beta > 90°)$, defined as $\beta = (360/2\pi)$. arc $\sin(\mathbf{a} \times \mathbf{c}/a.c)$. Since the angle θ is not fixed by symmetry, the tensor is allowed to rotate (e.g., with temperature) about the crystallographic **b** axis in monoclinic crystals, whereas it is completely free to rotate about any axis in triclinic crystals.[154]

Some of the numerical tensor data may be represented by approximate analytical expressions

$$\mu_{ij}(T) = \mu_{ij}(300 \text{ K})\{T[\text{K}]/300\}^{n_{ij}} \qquad (7.4)$$

$$\theta(T) = \theta(300 \text{ K}) + m_\theta(T[\text{K}] - 300), \qquad (7.5)$$

where n_{ij} designates the temperature exponent of the tensor component μ_{ij}.

The essential features of the time-of-flight (TOF) method is illustrated in Fig. 7.1.[406] The crystal slices of typical dimensions ~ 50 mm$^2 \times (0.2...1)$ mm are sandwiched between a metal rear electrode and a semitransparent front electrode. Charge carriers of both signs are optically generated close to the front surface of the crystal by strongly absorbed laser pulses of nanosecond duration. Thus, the following experimental conditions should be ensured. First, the mean penetration depth of light $1/\alpha$ (where α is the absorption coefficient in cm^{-1}) should be considerably smaller than the thickness L of the crystal, namely, $1/\alpha \ll L$. Second, excitation should take place in the spectral region of intrinsic photogeneration (see Chapter 6). Under such conditions charge carriers of both signs are generated close to the front electrode (Fig. 7.1). A potential difference U_0 applied across the sample generates an electric field parallel to the normal to the platelet. If positive potential is applied (Fig. 7.1) to the front electrode, the electrons are collected by the electrode and the holes are drawn through the bulk of the crystal. While the holes are moving with constant drift velocity v_d, a constant current $j(t)$ flows in the low-

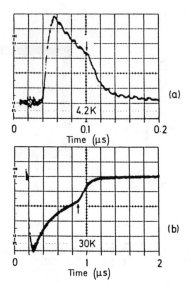

FIGURE 7.2. Typical hole (a) and electron (b) drift current pulses obtained by the time-of-flight method.[406] The experimental parameters are (a) $T = 4.2$ K, $\mathcal{E} = 12$ kV/cm, $L = 1.01$ mm; ordinate scale 13.2 μA (per major) division, equivalent to ≈ 40 μA/cm^2 per division. (b) $T = 30$ K, $\mathcal{E} = 10$ kV/cm, $L = 1.01$ mm; ordinate scale 6.6 μA (per major) division, equivalent to ≈ 20 μA/cm^2 per division. The arrival time kinks are marked by an arrow.

resistance external circuit. As soon as the "hole sheet" of thickness $d \approx 1/\alpha$ reaches the rear electrode, the photocurrent $j(t)$ pulse amplitude drops under ideal conditions to zero. However, in real situation the rise and decay time of current pulse are dependent, among other things, on the stray capacitance C.

Typical hole (a) and electron (b) TOF drift current pulses in a Nph crystal are illustrated, according to Warta and Karl,[406] in Fig. 7.2. The kink on the current pulse indicates the arrival time of the "current sheet", i.e. the time-of-flight $t = \tau_{tr}$ of the carriers. For crystal thickness L, the carrier drift velocity v_d can be determined as

$$v_d = L/\tau_{tr}. \qquad (7.6)$$

The corresponding carrier mobility μ equals

$$\mu = v_d/\mathcal{E} = L/\tau_{tr}\mathcal{E} = L^2/\tau_{tr}U_0. \qquad (7.7)$$

The main advantage of the TOF method is that the v_d and μ values can be measured independently for electrons and holes by simply changing the field polarity in the sample (see Fig. 7.2).

To avoid instrumental distortions of the transit current pulse the external circuit (see Fig. 7.1) should be of low resistance ($RC \ll \tau_{tr}$) and the duration τ_e of the exciting laser pulse must be short compared with the carrier transit time τ_{tr} to be measured ($\tau_e \ll \tau_{tr}$). However, the main factor, which causes the distortion of current pulse form, is the carrier trapping phenomenon. It sets up the most drastic condition for TOF measurements, namely, the mean life time of the carrier for trapping in deep traps τ_D should be greater than (or at least of the same order as) the carrier's transit time τ_{tr}

$$\tau_D \geqslant \tau_{tr}. \tag{7.8}$$

The condition (7.8) actually determines the low-temperature limit of reliable TOF measurements since τ_D is temperature dependent: $\tau_D = f(T)$. The τ_D value decreases with decreasing temperature due to the shift of the demarcation level separating shallow and deep traps relative to the carriers transit time τ_{tr}. At low T shallow traps, due to this shift, turn into deep ones. As the result τ_D decreases and as it becomes smaller than τ_{tr} the amplitude of the current pulse $j(t)$ due to trapping is severally distorted, and a distinct "kink" (see Fig. 7.2) disappears, making it impossible to determine the correct value of TOF τ_{tr}.

We will illustrate the situation, briefly describing the experimental conditions of low temperature $\mu_0(T)$ and $v_d(T)$ measurements in Nph and α-perylene ($\alpha-$Pl) crystals, recently carried out by Karl and coworkers.[155,158,406,407]

The experiments were performed using ultrapure, carefully treated and prepared Nph and α-Pl single crystals of thickness $L = (0.1-1) \times 10^{-1}$ cm, L being by several orders of magnitude greater than the penetration depth of strongly absorbed exciting light ($d = 1/\alpha \approx 10^{-5}$ cm). The authors of Refs. 406 and 407 used a KrF excimer laser, emitting single $\lambda = 248$-nm pulses of 8 ns halfwidth for Nph crystals, and a N_2 laser, emitting $\lambda = 337$-nm pulses of 0.8 ns halfwidth for Pl crystals. To avoid possible instrumental distortion of the current pulse form the measurements were performed at conditions $RC \ll \tau_{tr}$, where RC is the time constant of the circuit used. The magnitude of R was from 50 Ohms to 20 kOhms. The TOF τ_{tr} measurements in Nph crystals showed[406] that at $T = 300$ K the trapping of carriers of both signs in deep traps practically does not affect the form of the carrier pulse, since the estimated mean life time τ_D for trapping in deep traps equals $\tau_D \geqslant 100$ μs, i.e., $\tau_D \gg \tau_{tr}$. In case of holes τ_D sharply decreases with decreasing temperature down to 500 ns at $T = 70$, however, retains the value of 70 ns even at $T = 4.2$ K.

Under these conditions a distinct kink on the $j(t)$ dependence emerges at $T = 4.2$ K (shown by an arrow on Fig. 7.2a) indicating that the τ_{tr} value equals 0.1 μs $= 100$ ns. This allowed the authors of Ref. 406 to measure the most important transport characteristics $\mu^+(T)$ and $\mu^+(\mathcal{E})$ in Nph crystals over a wide temperature range from 300 down to 4.2 K (see Fig. 7.3).

On the other hand, the mean lifetime for deep trapping of electrons τ_D equals 500 ns already at 25 K and is smaller than the τ_{tr} value (see Fig. 7.2b). Owing to this circumstance it was not possible to measure $\mu^-(T)$ dependence for electrons below 25 K (see Fig. 7.3).

These data illustrate the importance of ultrahigh purity and structural perfection of the crystals in order to obtain reliable $\mu_0(T)$ dependences at low temperatures.[158,406,407]

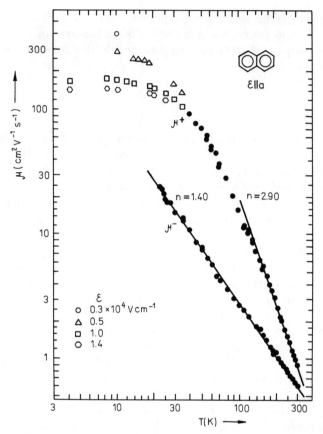

FIGURE 7.3. Electron μ^- and hole μ^+ mobilities versus temperature T in log-log plot in the crystallographic **a**-direction of ultrapure naphthalene according to Ref. 406. In the low-T region the hole mobility μ^+ becomes field $\mathcal{E}$ dependent. The solid lines indicate a $\mu \sim T^{-n}$ (7.9) power-law with exponents n as indicated in the figure.

7.1.3. Experimental Charge Carrier Transport Characteristics. Field-Generated Hot Carriers

Figures 7.3–7.11 show some typical temperature (T) and electric field ($\mathcal{E}$) dependences of two main charge carrier transport characteristics — the carrier mobility $\mu_{ij} = \mu_{ij}(T, \mathcal{E})$ and drift velocity $v_d = v_d(T, \mathcal{E})$ curves obtained in recent years by Karl and coworkers in Nph and α-perylene (α-Pl) crystals.[154–156,158,406,407]

These experimental dependences have been used as a data base for computer simulation and model description of charge carrier transport processes in OMC[348,355,356] (see Section 7.2).

Typical experimental mobility $\mu(T)$ dependences in log μ versus log T plot in the **a**-direction of an Nph crystal are shown in Fig. 7.3.[406]

As can be seen from Fig. 7.3, in the high-temperature region field-independent mobility regime of power-law dependence, represented by solid lines, is observed both for electrons (μ^-) and holes (μ^+). Such power-law dependence of $\mu(T)$ is a characteristic feature for a number of aromatic hydrocarbon as well as of hetero-cyclic compounds (see Refs. 154 and 288) and may be presented in a general way as

$$\mu_{ij}(T) = a T^{-n_{ij}}, \qquad (7.9)$$

where the slope value n_{ij} of the log μ versus log T graph is taken, for convenience sake, as $n_{ij} > 0$ (see Fig. 7.3).

Figure 7.3 demonstrates that the exponent n_{aa}^+ for holes exceeds the exponent n_{aa}^- for electrons by a factor of 2 ($n_{aa}^+ \approx 2 n_{aa}^-$). As a result, the value of μ_{aa}^+ is growing faster than that of μ_{aa}^- with decreasing T and at $T = 30$ K, μ_{aa}^+ is already by an order of magnitude greater than μ_{aa}^-.

In the region of low temperatures (and high mobilities) Karl and coworkers discovered a remarkable new physical effect[155,406,407] which has not been reported reliably for any organic material before:[407] the mobilities become electric field-dependent, the μ value decreases with rising field, namely, $\mu = \mu(\mathcal{E})$. This extraordinary behavior for hole mobility μ_{aa}^+ is demonstrated in Fig. 7.3. Below $T \leq 30$ K μ_{aa}^+ considerably deviates from the power-law (7.9) dependence and becomes strongly field-dependent. This effect is caused by deviation from Ohmic behavior which is most clearly demonstrated in drift velocity $v_d^+(\mathcal{E})$ versus $\mathcal{E}$ plot (Fig. 7.4).[406] Already at $T = 31$ K the observed drift velocities fell well below the straight line through the origin, representing μ_0 value according to Ohm's law. Thus, the $v_d(\mathcal{E})$ dependence becomes sublinear with tendency to saturation $v_d(\mathcal{E}) \rightarrow (v_d)_s =$ const. at higher field strength.

One should emphasize that, in this case, the saturation value of the drift velocity $(v_d)_s$ is greater than the mean thermal velocity of the charge carrier v_{th} at given temperature. This means that the carrier moves in the electric field at constant mean velocity $(v_d)_s > v_{th}$ and does not thermalize at all, viz. it emerges as a hot charge carrier. Thus, Karl and coworkers were the first who observed in 1984–1985[155,406,407] the electric field-induced hot carrier generation in organic crystals at low temperatures.

The $\mu_0(\mathcal{E})$ dependence at sufficiently high electric fields and the corresponding saturation effect of the mean drift velocity $v_d(\mathcal{E}) = \mu_0(\mathcal{E}) \cdot \mathcal{E} = (v_d)_s =$ const of the hot charge carriers have been explained in terms of field-induced increase of charge carrier scattering on acoustic and optical lattice phonons.[158,406] These effects are discussed later in some detail. Without this increased carrier scattering effects the ohmic mobility μ_0^+ of holes at 4.2–10 K would be about 2000 cm^2/Vs (see Fig. 7.4). However, in a real situation, due to this scattering effect, the charge carrier mobility does not exceed 300 cm^2/Vs.

It should be mentioned that, in case of electron mobility $\mu_{aa}^-(T)$ along the **a**-direction, the hot carrier generation regime could not be experimentally reached due to increased trapping below 25 K (see Fig. 7.3 and Section 7.1.2). The situation has been more favorable in case of charge carrier transport for an electric field $\mathcal{E}$

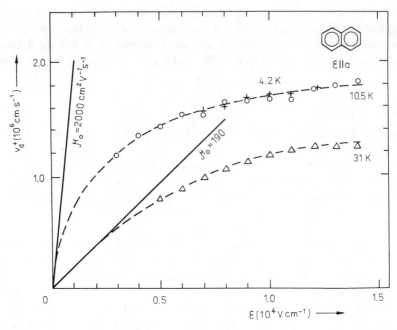

FIGURE 7.4. Hole drift velocity v_d^+ ($\mathcal{E}$) as dependent on electric field $\mathcal{E}$ in the **a** — direction of naphthalene crystal according to Ref. 406. Low-field ohmic mobilities μ_0 are indicated by tangential lines.

parallel to the crystallographic $\mathbf{c'}$ axis of an Nph crystal (Fig. 7.5). In this case the $\mu(T)$ dependences have been obtained for charge carriers of both signs down to 4 K.[154] As a result, on these $\mu(T)$ characteristics the T regions of thermalized charge carriers $\mu(\mathcal{E})$ = const emerge where the power-law (7.9) holds, as well as the T regions of hot carrier generation, where $\mu = \mu(\mathcal{E})$.

It is interesting to notice that the $\mu^-(T)$ dependence demonstrates an anomalous behavior: in the relatively wide T range from 100 to 300 K the electron mobility μ^- does not depend on temperature and practically remains constant, namely, $\mu^-(\mathcal{E})$ = const (Fig. 7.5). This anomaly of electron mobility was studied earlier by Schein et al.[312–316] These authors suggested that the transition from the T region ($T < 100$ K), where $\mu^-(T)$ increases with decreasing temperature, to the region $T > 100$ K where $\mu(T)$ = const, corresponds to the anticipated "band to hopping regime" transition. However, this interpretation was controversial since none of the proposed hopping transport models could reasonably describe a constant, T-independent carrier mobility over such a wide temperature range. In Section 7.2 we will describe an alternative approach in the framework of which the mobility $\mu(T)$ = const region can be described phenomenologically as a superposition of two independent carrier transport mechanisms. Thus, the complicated S-shaped $\mu^-(T)$ dependence (Fig. 7.5) in the $\mathbf{c'}$-direction of an Nph crystal indicates the presence of

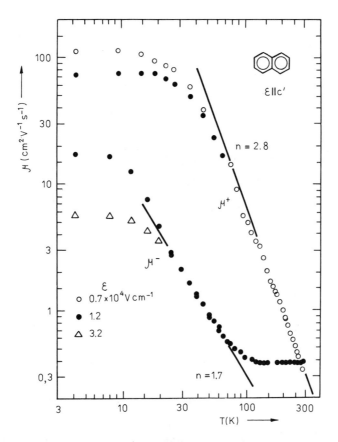

FIGURE 7.5. Electron μ^- and hole μ^+ mobilities versus temperature T in a log-log plot in the c'-direction of naphthalene crystal according to Ref. 154. In the low-T region both hole μ^+ and electron μ^- mobilities become field-dependent.

at least three transport mechanisms, which manifest themselves in different temperature regions.

As discussed in Section 7.1.2, the mobility of charge carriers in highly aniso-tropic crystals of monoclinic or triclinic symmetry group, to which the polyacenes belong, is actually a tensor μ_{ij} [see Eq. (7.2)].

As an illustration, a schematic picture of cross sections of an ellipsoid of mobility tensor μ_{ij} of a negative charge carrier (electron) in **ac**-plane of an Nph crystal at different temperatures is shown in Fig. 7.6.[154,155] In case of a Nph crystal belonging to the monoclinic group of symmetry, only the main direction **2** of the mobility tensor corresponds to the **b**-axis of the crystal, namely, $\mu_{22} = \mu_{bb}$ [see Eq. (7.3)]. The other two main directions of the tensor are declined relative to the crystallo-graphic axes **a** and **c**′, forming an angle θ between the **a**-axis and the main direction **1** (see Fig. 7.6).

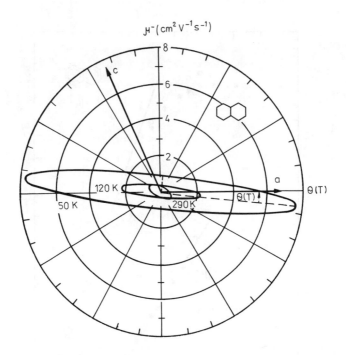

FIGURE 7.6. Schematic picture of cross sections of the ellipsoid of mobility tensor μ_{ij} of a negative charge carrier (electron) in **ac** plane of a naphthalene crystal at different temperatures according to Ref. 155.

Since the angle θ is not fixed by the crystal symmetry, the mobility tensor may rotate with changing temperature, according to Eq. (7.5). Thus, in case of a monoclinic crystal the rotation takes place around the **b**-axis (see Fig. 7.6).

It is interesting to mention that the value of angle θ and its temperature dependence $\theta(T)$ is strongly dependent on the sign of the charge carrier.[154] Thus, for electrons in Nph crystals strongly pronounced $\theta(T)$ dependence is observed: at $T = 293$ K $\theta(T) = -27°$; at $T = 200$ K $\theta(T) = -13°$ and at $T = 100$ K $\theta(T) = -6°$ (see Fig. 7.6).

In contrast to this, in the case of holes in an Nph crystal the angle equals $-2°$ and is not dependent on temperature, i.e., the mobility tensor is fixed, and its main axes practically correspond to the orthogonal axes of the crystal.

A similar situation is also in the case of an Ac crystal: for electrons the parameter m_θ in formula (7.5) equals $m_\theta = -0.033°/$K, while for holes $m_\theta = 0$ in a wide range of temperature from 100 to 300 K.

Figure 7.7 shows the experimental $\mu_{22}^-(T) = \mu_{bb}^-(T)$ dependence for electrons in an Nph crystal and corresponding $\mu_{ij}^-(T)$ dependences for two main mobility tensor components, namely, $\mu_{11}^-(T) = \mu_{33}^-(T)$.

If the mobility $\mu_{11}^-(T)$ and $\mu_{33}^-(T)$ dependences are compared with the respective mobility dependences $\mu_{aa}^-(T)$ (Fig. 7.3) and $\mu_{c'c'}^-(T)$ (Fig. 7.5), along the crystal

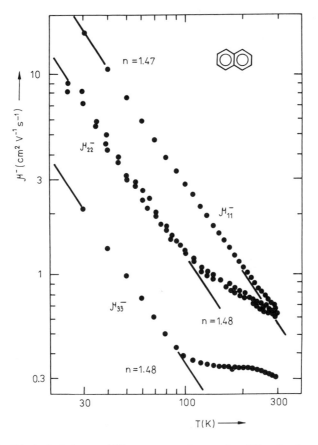

FIGURE 7.7. Principal electron mobilities μ_{ii} versus temperature T in a log-log plot in a naphthalene crystal according to Ref. 155. The $\mu_{11}^-(T)$ and $\mu_{33}^-(T)$ dependences were obtained from 11 series of measurements in different directions in the **ac** plane of the crystal whereas μ_{22}^- represents experimental points from four samples with $\mathcal{E} \| b$.

axes **a** and **c'** one may see how strongly the value of the exponent n_{ij}, as well as the temperature range in which the power-law (7.9) holds, depends on the crystallographic direction of the movement of charge carrier. It is also interesting to compare the $\mu_{ij}^-(T)$ dependences in the high-T region where the power-law is not observed. Thus, if in the **c'**-direction of the crystal $\mu_{c'c'}^-(T) = $ const (Fig. 7.5) in the main direction of the mobility tensor **3**, the $\mu_{33}^-(T)$ dependence is more complicated and of slowly decreasing shape (Fig. 7.7). For the $\mu_{22}^-(T)$ dependence [which equals the $\mu_{bb}^-(T)$ dependence] one observes considerable decrease in the slope n_{22} value at $T > 120$ K, declining from the power-law with $n_{22} = 1.48$.

In Tables 7.1 and 7.2, the values of μ_{ij} and n_{ij} are shown for charge carriers of both signs, moving in different crystallographic directions of Nph and Ac.

It is interesting to notice that the value of mobility μ_{ij} in the **c'**-direction ($\mu_{c'c'}$),

TABLE 7.1. Mobility of the Positive Charge Carrier (Hole) along the Main Crystallographic Axes ($\mu_{aa}, \mu_{bb}, \mu_{c'c'}$) and the Principal Axes of the Mobility Tensor ($\mu_{11}, \mu_{22}, \mu_{33}$) [See Eq. (7.3)] in Nph and Ac crystals, and the Corresponding Exponents n_{ij} of the Mobility Temperature Dependences $\mu_{ij} \sim T^{-n}$ [Eq. (7.9)][a]

Crystal	Hole mobility μ_{ij} cm²/Vs [b]						Exponents n_{ij} of the mobility temperature dependences $\mu_{ij} \sim T^{-n}$ [c]		
	μ_{aa}	μ_{bb}	$\mu_{c'c'}$	μ_{11}	$\mu_{22}=\mu_{bb}$	μ_{33}	n_{aa}	n_{bb}	$n_{c'c'}$
Naphthalene ($C_{10}H_8$)	0.94 (293)	1.48 (293)	0.32 (293)	0.95 (290)	1.50 (290)	0.32 (290)	2.9 ($100 \leqslant T \leqslant 300$) (see Fig. 7.3)	2.5 ($78 \leqslant T \leqslant 300$)	2.8 ($78 \leqslant T \leqslant 300$) (see Fig. 7.5)
Anthracene ($C_{14}H_{10}$)	1.13 (300)	2.07 (300)	0.73 (300)	1.28 (300)	2.07 (300)	0.57 (300)	1.5 ($90 \leqslant T \leqslant 300$)	1.26 ($110 \leqslant T \leqslant 300$)	1.43 ($10 \leqslant T \leqslant 300$)

[a]Karl N.: Organic Semiconductors, in Landolt-Börnstein, Group III, Vol. 17, Semiconductors, Springer, Berlin, Heidelberg, 1985, pp. 106–218; Warta W. and Karl N.: Phys. Rev. B **32**, 1172 (1985); Warta W., Stehle R., and Karl N.: Appl. Phys. A **36**, 163 (1985); Karl N., Marktanner, Stehle R., and Warta W.: Synth. Metals **41-43**, 2473 (1991).

[b]The temperature of determination is given in parentheses.

[c]The T interval in the limits of which the dependence holds, is given in parentheses.

TABLE 7.2. Mobility of Negative Charge Carrier (Electron) Along the Main Crystallographic Axes ($\mu_{aa}, \mu_{bb}, \mu_{c'c'}$) and the Principal Axes of the Mobility Tensor ($\mu_{11}, \mu_{22}, \mu_{33}$) in Nph, Ac, and Perylene Crystals, and the Corresponding Exponents n_{ij} of the Mobility Temperature Dependences $\mu_{ij} \sim T^{-n}$ [Eq. (7.3)] in Nph, Ac, and Perylene Crystals, and the Corresponding Exponents n_{ij} of the Mobility Temperature Dependences $\mu_{ij} \sim T^{-n}$ [Eq. (7.9)] [a]

Crystal	Electron mobility μ_{ij} cm²/Vs [b]						Exponents n_{ij} of the mobility temperature dependences $\mu_{ij} \sim T^{-n}$ [c]		
	μ_{aa}	μ_{bb}	$\mu_{c'c'}$	μ_{11}	$\mu_{22}=\mu_{bb}$	μ_{33}	n_{aa}	n_{bb}	$n_{c'c'}$
Naphthalene ($C_{10}H_8$)	0.62 (293)	0.64 (293)	0.44 (293)	—	—	—	1.40 ($20 \leq T \leq 300$) (see Fig. 7.3)	1.50 ($25 \leq T \leq 80$) 0.55 ($120 \leq T \leq 300$) (see Fig. 7.7)	1.7 ($25 \leq T \leq 80$) −0.04 ($120 \leq T \leq 300$) (see Fig. 7.5)
Anthracene ($C_{14}H_{10}$)	1.73 (300)	1.05 (300)	0.39 (300)	1.74 (300)	1.05 (300)	0.38 (300)	1.26 ($40 \leq T \leq 300$)	0.84 ($150 \leq T \leq 300$)	−0.16 ($150 \leq T \leq 300$)
Perylene ($C_{20}H_{12}$)	2.37 (300)	5.53 (300)	0.78 (300)	2.40 (300)	5.53 (300)	0.75 (300)	1.78 ($40 \leq T \leq 300$) at $\mathcal{E} \leq 5\times10^3$ V/cm (see Fig. 7.9)	1.72 ($60 \leq T \leq 300$)	2.16 ($40 \leq T \leq 300$) (see Fig. 7.10)

[a]Karl N.: Organic Semiconductors, in Landolt-Börnstein, Group III, Vol. 17, Semiconductors, Springer, Berlin, Heidelberg, 1985, pp. 106–218; Warta W. and Karl N.: Phys. Rev. B **32**, 1172 (1985); Warta W., Stehle R., and Karl N.: Appl. Phys. A **36**, 163 (1985); Karl N., Marktanner, Stehle R., and Warta W.: Synth. Metals **41-43**, 2473 (1991).
[b]The temperature of determination is given in parentheses.
[c]The T interval in the limits of which the dependence holds, is given in parentheses.

for carriers of both signs, is only a little lower than μ_{ij} in the **ab**-plane, namely, μ_{aa} and μ_{bb}.

These data are in contradiction with theoretical predictions. According to the calculations of Glaeser and Berry,[117] the charge carrier transfer integral J_{ij} in the **c**-direction of an Ac crystal is by two orders of magnitude smaller than in the **a** or **b** directions (see Chapter 3). Therefore one should expect a strong anisotropy of charge carrier mobility analogous to the anisotropy of singlet and triplet exciton migration in Ac-type crystals.[288] This problem was discussed in Chapter 3, Section 3.2.3. To explain the lack of expected anisotropy, Duke and Schein[95] have suggested that the carrier transfer rate in the **c'**-direction may be enhanced by rotational (libronic) movements of the molecules of the lattice (see Fig. 3.12), i.e., stimulated by lattice phonons of definite vibrational mode. This effect has been described in terms of a model Hamiltonian in Chapter 3 [see Eq. (3.6.10b)]. Further, as may be seen from Tables 7.1 and 7.2, the exponent n_{ij} of the $\mu(T)$ dependence (7.9) does not show a specific or characteristic value, but rather a wide spectrum of values depending on the nature of the crystal, crystallographic direction of the transport, and the sign of the charge carrier (see, e.g., Fig. 7.3). Schein[312] tried to analyze the observed $\mu_{ij}(T)$ dependences and the corresponding exponent n_{ij} values in terms of wide ($\delta E > kT$) and narrow ($\delta E < kT$) band approaches. For a wide band one should expect the value of $n_{ij} \approx 3/2$, and for a narrow one $n_{ij} = 2$ or 3. It is easy to see from Tables 7.1 and 7.2 that the observed n_{ij} values do not, on the whole, support any of these expectations.

After all, it has been shown in Section 3.2 that this kind of approach is not correct. One should not confuse the concepts of band structure of the charge carrier energy $E(\mathbf{k})$ and the band transport of carriers. The criterion for the applicability of the last concept is that the mean free path of the carrier $\bar{l}_o$ should be considerably greater than the lattice constant a_o, namely, $\bar{l}_o \gg a_o$ [see formula (3.2.10)]. Since, for the high-T region, where the $\mu(T)$ dependence (7.9) holds, $\bar{l}_o \approx a_o$ (see Section 7.3 and Fig. 7.15), the band transport paradigm is not valid.

It is shown in Section 7.3 that the charge carrier transport in OMC may be described in the framework of the nearly small MP model, as was introduced in Sections 3.6.1 and 4.1–4.3 (for detail, see Refs. 344, 348, 351, 355, and 356).

Now let us consider the possible physical mechanisms causing the sublinear decline from Ohm's law at higher electric fields $\mathcal{E}$, finally leading to saturation of the drift velocities $v_d(\mathcal{E}) \rightarrow (v_d)_s = \text{const.}$ As mentioned earlier, Karl and coworkers[158,406] have interpreted these effects in terms of field induced specific scattering mechanisms of the charge carriers on acoustic and optical phonons of the lattice. For thermalized charge carriers, in the state of thermal equilibrium, the carrier's relaxation time τ_r is assumed to be isotropic and constant (i.e., momentum- and energy-independent). In this case mobility μ may be simply described by equation $\mu = (e/m_{eff})\tau_r$, i.e., the mobility μ is also constant, field-independent at the given temperature.

However, at higher electric fields, the field-induced charge carrier velocity component becomes comparable to the thermal velocity. As a consequence the mean scattering time τ_r becomes markedly reduced due to acceleration of the carrier in the electric field $[\tau_r \to \tau_r(\mathcal{E})]$ resulting in faster approach to the scattering sites. This effect is the principal cause for field-induced non-linearities of carrier transport observable as field-dependent decrease in mobility.[158]

If elastic (or pseudoelastic) scattering is sufficiently effective to directionally randomize the field-induced momentum increase of the charge carriers before inelastic scattering sets in, leading to energy relaxation, an elevated effective temperature may be ascribed to the charge carrier ensemble.[158] In this case the above equation should be slightly modified replacing the scattering time τ_r, pertaining to lattice temperature, with a reduced scattering time $\tau_r(\mathcal{E})$, corresponding to the elevated electron temperature

$$\mu(\mathcal{E}) = (e/m_{\text{eff}})\tau_r(\mathcal{E}).$$

Shockley has used this "temperature model" to describe the field-induced increase of the scattering rate $\beta(\mathcal{E})$ [and corresponding decrease of the scattering time $\tau_r(\mathcal{E})$ and, hence, decrease of the mobility $\mu_{ij}(\mathcal{E})$] by acoustic phonons.[327] He derived an analytical expression in the framework of acoustic deformation potential scattering, which is shown to be valid for the initial field-induced deviations from Ohm's law:

$$\beta(\mathcal{E}) = 1/\tau_r(\mathcal{E}) = \beta_0\{1/2 + [1/4 + 3\pi/32(\mu_0 \mathcal{E}/u_L)^2]^{1/2}\}^{1/2}. \quad (7.10)$$

The corresponding carrier's drift velocity $v_d(\mathcal{E})$ equals:[406]

$$v_d(\mathcal{E}) = \mu_0 \mathcal{E} \sqrt{2}\{1 + [1 + 3\pi/8(\mu_0 \mathcal{E}/u_L)^2]^{1/2}\}^{-1/2}. \quad (7.11)$$

The formula of Eq. (7.11) contains only two parameters: the low-field Ohmic mobility μ_0 and longitudinal sound velocity u_L.

Applied to organic crystals (Nph and perylene),[154,155,158,406,407] by taking the values for μ_0 and u_L from experiment, formula (7.11) leads to a surprisingly close fit for the initial deviations from Ohm's law. This may be illustrated by the $v_d^+(\mathcal{E})$ curves in the a-direction of a Nph crystal (Fig. 7.8).[406] The solid lines show a fit to experimental points by Eq. (7.11), describing acoustic-phonon scattering of hot charge carriers in a deformation-potential approximation. The average u_L value of an Nph crystal $u_L = 4 \times 10^5$ cm/s at 6 K was used; the slopes of the dashed tangents represent the low-field Ohmic mobilities used for approximation. Similar satisfactory fit has been obtained also for sublinear electron drift velocities in α-Pl crystals (see Refs. 155, 158, and 406).

It is important to emphasize that the acoustic phonon scattering model [Eq. (7.11)] is actually applicable only at the very initial phase of field-induced deviations from Ohm's law. With further increase of electric field the observed $v_d(\mathcal{E})$ curves markedly decline from the parabolic $v_d(\mathcal{E}) \sim \sqrt{\mathcal{E}}$ dependence and show a definite tendency to saturation $v_d(\mathcal{E}) \to (v_d)_s = $ const. That means that at higher electric fields above some threshold value $\mathcal{E} \geqslant \mathcal{E}_s$, a new, very efficient scattering channel appears, causing inelastic energy loss of field-accelerated charge carriers. It

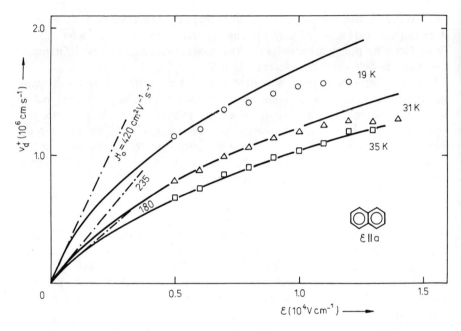

FIGURE 7.8. Hole drift velocity $v_d^+(\mathcal{E})$ as dependent on electric field $\mathcal{E}$ in the **a**-direction of naphthalene crystal according to Ref. 406. The solid lines represent a fit to the experimental points by Eq. (7.11) describing acoustic phonon scattering of hot holes in a deformation-potential approximation. The slopes of the dashed tangents represent the low-field Ohmic mobilities μ_0 used in the model.

has been shown by Warta and Karl[406] that an average field-independent (saturated) drift velocity of the carrier $(v)_s = \text{const}$ is caused by cyclic repetition of carrier acceleration in the electric field, ending in an emission of an optical lattice phonon $h\nu_{ph}$ at every cycle, when the kinetic energy of the accelerated carrier $(m_{eff}v_{max}^2)/2 = h\nu_{ph}$, which yields

$$(v_d)_s = \left[\frac{h\nu_{ph}}{2m_{eff}}\right]^{1/2}, \qquad (7.12)$$

where $(v_d)_s = v_{max}/2$.

This allows one to estimate the $m_{eff}(T)$ dependence in the low T-region directly from experimental $(v_d)_s(T)$ dependences, taking into account the possible range of optical phonon energies for the given crystal (see Section 7.2.3).

This generative optical phonon scattering loss mechanism will be discussed in some detail in Section 7.2.3 in the framework of computer simulation approach.[355,356] Thus, two critical field values may be assigned, the first one $\mathcal{E}_c$, at which $v_d(\mathcal{E}) \approx v_{th}$, and the sublinear decline deviation from the Ohm's law starts manifesting the beginning of field-induced hot carrier generation in the acoustic scattering range. The second threshold value of $\mathcal{E}$, namely, $\mathcal{E} = \mathcal{E}_s$, can be

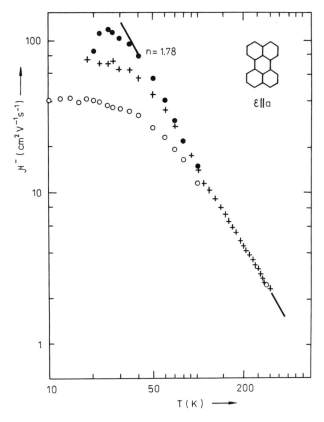

FIGURE 7.9. Electron μ^- mobilities versus temperature T in a log-log plot in **a**-direction of a perylene crystal according to Refs. 154 and 407. The crystal thickness $L = 252$ μm.

assigned to the onset of dominant scattering on optical phonons from which the saturation region of the $v_d(\mathcal{E})$ dependence starts.

Another important conclusion may be derived from the observed $v_d(\mathcal{E})$ dependences of hot charge carrier transport. At sufficiently low temperature and high electric field the charge carriers reach drift velocities much higher than both the mean thermal velocity of the carrier $(v_d(\mathcal{E}) > v_{th})$ and the longitudinal sound velocity u_L the crystal $(v_d(\mathcal{E}) > u_L)$. This fact excludes all those lattice polaron model concepts where a charge carrier must be preceded by lattice distortions (formation of acoustic phonon packet) for preparing resonant sites.[158] However, it is demonstrated later (see Sections 7.2.3 and 7.3) that both thermalized and hot carrier transport and the dependence of their effective masses on temperature $(m_{eff} = f(T))$ may be described in the framework of the nearly small MP model.

For α-perylene crystals, as it can be seen from Figs. 7.9–7.11, the $\mu_{ij}(T)$ dependence (7.9) emerges only in case of negative charge carrier transport.[154,155,158,407]

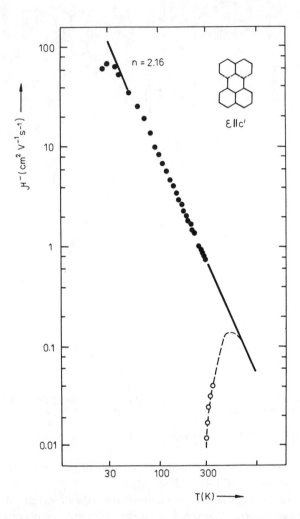

FIGURE 7.10. Electron μ^- mobilities versus temperature T (filled circles) in a log-log plot in **c′**-direction of a perylene crystal according to Refs. 155 and 407. The crystal thickness $L = 306$ μm. The open circles in the lower right corner represent previously available mobility data. The dashed line is a Hoesterey-Leton-type shallow trapping model fit with trap parameters: $E_t = 0.27$ eV and $N_t/N_0 = 1.7 \times 10^{-3}$ mol/mol, revealing that the purity of the crystals used in these older measurements was insufficient for obtaining the true microscopic carrier mobility.

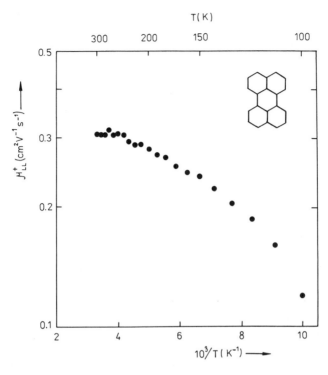

FIGURE 7.11. Hole mobility μ^+ plotted logarithmically versus reciprocal temperature $1/T$ in an oblique LL-direction of a perylene crystal according to Ref. 155.

In this case, too, like in polyacene crystals, at the low-T region ($T < 80$) in the fields of the order $\mathcal{E} \geqslant 5 \times 10^4$ V/cm one observes hot carrier generation.

The corresponding values of μ_{ij}, as may be seen from Table 7.2, are of the same order as in Nph and Ac crystals. However, the anisotropy of the mobility is more pronounced: the μ_{bb} value is almost by an order of magnitude greater than that of $\mu_{c'c'}$.

The mobility tensor μ_{ij} for electrons in an α-Pl crystal rotates around the **b**-axis of the crystal and the value of angle θ ($\theta > 0$), in contrary to Nph (see Fig. 7.7), increases with decreasing temperature. Thus, e.g., at $T = 300$ K the value of $\theta = 8.2°$, while at $T = 60$ K $\theta = 24°$.[154] The spectrum of n_{ij} values of α-Pl, similar as in the case of polyacenes, does not support any specific kind of band transport model. On the contrary, in case of α-Pl the criterion of nonapplicability of band transport $\bar{l}_o \approx a_o$, holds down to ~ 80 K (see Fig. 7.28). Figure 7.10 also illustrates the dramatic influence of impurity centers on the $\mu_{c'c'}$ value and its T-dependence. This example demonstrates the importance of working with ultrapure crystals in order to obtain reliable $\mu_{ij}(T)$ measurements.[156,157,407]

Figure 7.11 demonstrates a $\mu_{ij}(T)$ dependence, not typical for aromatic hydrocarbon crystals. As can be seen, in case of positive charge carriers the mobility

$\mu_{LL}^+(T)$ increases with increasing temperature, i.e., the charge carrier transport process is actually a thermally activated one. However, the plot of μ_{LL}^+ versus $1/T$ graph shows that the thermally activated mobility $\mu_{LL}^+(T)$ cannot be described by a simple Arrhenius formula with constant activation energy, namely, $\mu_{ij} \sim \exp[-E_a/kT]$, as one should expect in case of a small lattice polaron. The effective E_a magnitude decreases, as can be seen from Fig. 7.11, with growing temperature and reaches zero value near room temperature. It should be mentioned that in this case the absolute value of the mobility μ_{LL}^+ is small, of the order of 0.1–0.2 cm^2/Vs. It indicates the presence of marked localization of the holes as compared with the mobilities of electrons (Figs. 7.9–7.10). For such pronounced localization the formation of a small-lattice polaron or a more complicated polaron-type quasiparticle is actually possible (see Section 3.6.1).

It is interesting to mention that very similar thermally activated mobility $\mu_{ij}(T)$ dependence has been found for principal tensor components of hole mobility $\mu_1^+(T)$ and $\mu_2^+(T)$ in the 2,3-dimethylnaphthalene crystal with static dipolar disorder and for electron mobility in the c'-direction of a pyrine crystal.[155]

Recently Karl and coworkers[158] reported the observation of hot electrons in the low-T region (20–40 K) in the **a**-direction of an Ac crystal. The saturation region of $v_d(\mathcal{E}) \to (v_d)_s = $ const starts at $\sim 2.5 \times 10^4$ V/cm.

Earlier Bitterling and Willig[33,34] have demonstrated that it is also possible to create *hot* positive charge carriers in ultrapure Ac crystals at room temperature (300 K), however, at much higher field strength $\mathcal{E} > 4 \times 10^5$ V/cm. These authors attribute the carrier drift velocity saturation $v_d(\mathcal{E}) \to (v_d)_s = $ const along the **c**′-direction in Ac crystals as due to cyclic excitation of low-energy intramolecular vibrations with $h\nu = 395$ cm^{-1}. They also estimated the free path l_o of the hole to be equal to the lattice constant, namely, $l_o \approx c_o = 9.2$ Å.

Thus, at present a rich experimental data bank of charge carrier transport characteristics and parameters are available that have been used for computer simulation of transport processes in Nph, Ac, and α-Pl crystals.[344,348,354–356] In the next sections we shall present analyses of the simulation results and their interpretation in terms of polaron models.

7.2. COMPUTER SIMULATION OF CHARGE CARRIER TRANSPORT CHARACTERISTICS

7.2.1. Some General Considerations

Analyzing their experimental data, Karl and coworkers[158,406] concluded that the observed charge carrier transport characteristics may be interpreted in terms of a classical band-type transport model only in the low T region where the mean free path l_o of the carriers is of the order of at least several lattice constants. However, at high T when l_o is of the order of the lattice constant a_o, ($l_o \approx a_o$), the band transport picture is hardly consistent (see the criterion of the band type transport approach validity, Section 3.2.3). According to Karl et al.,[406] their data provide a new challenge for developing a general transport theory that is able to explain both low and high T regimes and link them in a unified description.

A number of other authors have also questioned the validity of band theory approaches for the description of carrier transport at such extremely small free path values. Thus, in our opinion it has been conclusively proved by Glaeser and Berry[117] that in case of carrier scattering at every lattice site ($l_o \approx a_o$) the band model approach is not valid and should be substituted by some kind of hopping model (see also a detailed analysis of the limits of band theory applicability for Ac-type crystals in the monograph Ref. 332, p. 29).

The authors of Ref. 34 have also come to the conclusion that their experimental results "give strong support to the notion that charge carrier transport in Ac-type crystals represents an intermediate case to which neither conventional band theory nor the concept of a small polaron apply."

This mobility puzzle creates a number of controversial questions. On one hand, the mean free path l_o of the carrier in polyacene crystals, from room temperature down to $T = 150$ K, and in perylene crystals even down to ~ 80 K as concluded both from theoretical and experimental estimates (see Figs. 7.15 and 7.28), is actually of the order of a lattice constant and strongly suggests a hopping model. On the other hand, the typical $\mu(T)$ dependences in polyacene and α-Pl crystals $\mu \sim T^{-n}$ (see Section 7.1.3) is often supposed to favor some form of band-type carrier transport.

In this case, a conventional small-lattice polaron approach is not applicable, since for a hopping motion of a small polaron temperature-activated behavior is required (see Ref. 228, p. 356), which contradicts the experimentally observed $\mu(T)$ dependences.

In any case, the carrier hopping motion would have to be thermally nonactivated, e.g., of the tunneling type.

Recently Movaghar has presented a comprehensive review of existing carrier transport theories for molecular solids comprising alternative band, polaron and hopping mechanisms of conduction (see Refs. 250 and 251 and references therein).

However, current transport theories are not, in general, suitable for quantitative description of the experimental transport characteristics in Ac-type crystals and has been mainly used for qualitative interpretation of T- and $\mathcal{E}$ -dependent behavior of the carrier mobility.

The controversial character of this problem is vividly reflected in the above-mentioned review by Movaghar.[250,251] The author emphasizes that the band-type transport approach may be valid only at low temperatures. However, the typical mobility dependence $\mu(T) \sim T^{-n}$ at higher temperatures "cannot be said to be completely understood" and "a unique and general description for the behaviour... is at present lacking."[251]

From this point of view a model description of transport phenomena provides certain advantages over generalized theories. It has been shown that, in terms of a phenomenological model, it is possible to perform computer simulation of experimental transport characteristics and provide a quantitative, numerical description of these characteristics.[344,348,354-356]

Thus, computer simulation studies of separation mechanisms of photogenerated charge pairs in Pc crystals[344] have demonstrated (see Section 6.3.3) that the temperature dependence of the carrier mobility $\mu(T) = aT^{-n}$ may be phenomenologically interpreted in terms of an adiabatic nearly small MP model[334,348,351] (see Sections 3.6.1 and 6.3.4). We should reiterate here that, in this approach, the carrier is considered as a heavy polaron-type quasiparticle formed as the result of carrier interaction with intramolecular vibrations of the local lattice environment. It has been shown in Section 6.3.3 that in the case of Pc crystals the effective mass m_{eff} of the MP increases exponentially with increasing T according to Eq. (6.46) and actually determines the observed $\mu(T)$ dependence $\mu(T) \sim T^{-n}$. However, in the studies of Ref. 344 computer simulations were performed only in the 165–300 K temperature range for which experimental data for Pc crystals were available (see Section 6.2).

In this chapter we present comparative simulation studies of experimental charge carrier transport characteristics in Nph, Ac, and Pl crystals reported in a number of recent papers and reviews by Karl and coworkers[154-156,158,406,407] and discussed in some detail in the previous section.

The present simulation studies demonstrate once more the plausibility of the MP model[334,344,348,351] for the description, at least in a phenomenological way, of charge carrier separation and transport processes over a wide range of temperature (4.2–300 K) and electric field ($\mathcal{E} = 10^3-10^6$ V/cm) in a framework of a unified theoretical approach. The MP approach is also consistent with the data of ionized state energy levels in polyacene crystals, since it has been shown that the conductivity levels of thermalized carriers in these crystals are not free electron but relaxed MP states (see Section 4.3).

7.2.2. Simulation Procedure and Model Considerations

In the simulation studies the modified version[344] of the initial SM model[306] has been used. According to this approach[344] the charge carrier is considered as an adiabatic nearly-small molecular polaron (MP) produced in a photoionization process with effective mass m_{eff} and excess kinetic energy $E_k = (m_{eff}v_o^2)/2$, where v_o is the radial velocity at an initial position r_o with respect to the parent ion at zero time ($t = 0$).

We have used the set of equations (6.42–6.43) of the modified SM model (see Section 6.3.1) and similar simulation procedure.

As shown in Section 6.3.1, there are two basic input quantities of the model that may be formally regarded as adjustable parameters, namely, effective mass of the carrier m_{eff} and carrier relaxation rate constant β. In Ref. 344 the $m_{eff}(T)$ and $\beta(T)$ dependences were evaluated in the simulation process by using the experimental thermalization length $r_{th} = r_{th}(\mathcal{E}T)$ and the quantum yield dependences

$\eta = \eta(h\nu, \mathcal{E}, T)$ of the photoconductivity action spectra. In the present simulation the $m_{eff}(T)$ values were evaluated using two independent approaches:[355,356]

1. From the saturation values of the drift velocity $(v_d)_s(T)$-dependences of hot carriers in the low-T region (4–100 K) reported by Karl *et al.*[154,155,158,406,407]

2. From the reported temperature dependences of the mobilities $\mu(T)$ of thermalized carriers in the high-T region (100–300 K).[154,155,158,406,407]

Equation (7.12) allows one to estimate the $m_{eff}(T)$ dependence of the low-T region directly from reported experimental $(v_d)_s(T)$ dependences, taking into account the possible range of optical phonon energies for the given crystal

$$m_{eff}(T) = \frac{h\nu_{ph}}{2|(v_d)_s(T)|^2} \, . \qquad (7.13)$$

The relaxation rate constant $\beta_s(\mathcal{E})$ in the saturation $(v_d)_s = $ const region can be obtained analytically from Eq. (6.42b). For $(v_d)_s = $ const, $\langle v_i \rangle = (v_d)_s$; $d\langle v_i \rangle/dt = 0$; in addition, the first (Coulombic) term of Eq. (6.43) is also equal to zero (no counter charge close by). Thus

$$(v_d)_s = \frac{e\,\mathcal{E}}{m_{eff}\,\beta_s(\mathcal{E})} = \text{const.} \qquad (7.14)$$

Hence, the relaxation rate constant in the saturation region β_s can be evaluated, for the given m_{eff} and $\mathcal{E}$ values, according to the following formula[355]

$$\beta_s(\mathcal{E}) = \frac{e\,\mathcal{E}}{m_{eff}(v_d)_s} \, . \qquad (7.15)$$

Formula (7.15) may also be derived from the expression of maximum kinetic energy $(m_{eff}v_{max}^2)/2$ of the carrier, gained in the electric field $\mathcal{E}$ over the mean free path l_o at the end of every acceleration cycle, namely, $(m_{eff}v_{max}^2)/2 = e\,\mathcal{E}\,l_s = e\,\mathcal{E}\cdot 2(v_d)_s\tau_s(\mathcal{E})$, where $\tau_s(\mathcal{E})$ is the mean relaxation time in the saturation region $\tau_s(\mathcal{E}) = 1/\beta_s(\mathcal{E})$ and $v_{max} = 2(v_d)_s$ (see Ref. 356 and Fig. 7.18).

In the sublinear region of the $v_d(\mathcal{E})$ dependence, hot charge carriers accelerated in the electric field $\mathcal{E}$ are presumably scattered by acoustic lattice phonons.[355,356,406] In this case the field-dependent relaxation rate constant $\beta_a(\mathcal{E})$ may be calculated according to Eq. (7.10) obtained in terms of the acoustic deformation potential model (see Ref. 406) which gives a parabolic type of $v_d(\mathcal{E}) \sim \sqrt{\mathcal{E}}$ dependence. However, $\beta_a(\mathcal{E})$ values, as estimated according to Eq. (7.10), yield satisfactory agreement between calculated and experimental curves mainly at the initial part of the $v_d(\mathcal{E})$ curve, but deviate considerably on approaching the saturation region, when both acoustic and optical scattering mechanisms become competitive.

It has been shown[355] that an effective scattering rate constant $\beta_a(\mathcal{E})$ in the sublinear $v_d(\mathcal{E})$ region may be estimated according to the following empirical formula, analogous to Eq. (7.15):

$$\beta_a(\mathcal{E}) = \frac{e\mathcal{E}}{m_{\text{eff}}v_d(\mathcal{E})} . \tag{7.16}$$

$\beta_a(\mathcal{E})$ values, calculated according to Eq. (7.16), and used as the input parameters in the simulation procedure provide a good agreement with the experimental $v_d(\mathcal{E})$ curves over the whole sublinear range up to the very saturation limit $v_d(\mathcal{E}) \to (v_d)_s$ (see, e.g., Fig. 7.12). Since $v_d(\mathcal{E}) < (v_d)_s$, the $\beta_a(\mathcal{E}) > \beta_s(\mathcal{E}_s)$, where $\mathcal{E}_s$ is the threshold field strength above which generation of optical phonons becomes dominant and the sublinear $v_d(\mathcal{E})$ dependence passes over to the saturation region of $(v_d)_s = $ const.

In the high-temperature regime (100–300 K) of thermalized carriers the $m_{\text{eff}}(T)$ values were estimated from reported experimental $\mu(T)$ dependences[154,155,158,406,407] according to the following formula:[348,355]

$$m_{\text{eff}}(T) = \frac{3e^2 l_o^2}{[\mu(T)]^2 kT} , \tag{7.17}$$

which results from relating the expressions of the mean thermal velocities: $v_{\text{th}} = \mu(T)kT/el_o$ and $v_{\text{th}} = \sqrt{(3kT/m_{\text{eff}})}$, respectively.

In general, formula (7.17) is an isotropic one. However, if the experimental main components of the mobility tensor are used, one obtains the corresponding main m_{eff} tensor components.

The $\beta(T)$ values were estimated according to the following formula:[348,355]

$$\beta(T) = \frac{\mu(T)kT}{3el_o^2} . \tag{7.18}$$

These estimates in the high-T regime were carried out under the assumption of a constant mean free path l_o of the carrier, equal to the lattice constant in the corresponding direction of carrier transport in the crystal.

As is shown in Ref. 355, independent m_{eff} estimates for the low (10–35 K) and high (150–300 K) temperature regimes confirm the applicability of the exponential $m_{\text{eff}}(T)$ dependence in agreement with formula (6.46) (at least in the case of positive carrier transport along **a** and **c**'-axis in Nph crystals) (see Figs. 7.15 and 7.16). However, in the intermediate temperature region equation (6.46) yields apparently underestimated m_{eff} values and shows that the assumption of $l_o = $ const is not valid. Therefore, for this temperature region (50–150 K), the m_{eff} values in Ref. 355 were determined by interpolation according to Eq. (6.46). This approach allowed the authors to estimate also the temperature dependence of l_o in this temperature range using the reduced m_{eff} value in the Eq. (6.46) (see Figs. 7.15 and 7.16).

However, the general applicability of the exponential $m_{\text{eff}}(T)$ dependence (6.46) over the whole T range (4–300 K) was later questioned by Silinsh, Shlihta, and Jurgis[356] (see also Section 6.3.4).

On one hand, phenomenologically introduced exponential $m_{\text{eff}}(T)$ dependence (6.46)[344,348,355] has demonstrated in a number of cases its validity for approximation of experimental data in the low, as well as in the high-T regimes (see Figs. 6.19, 7.15, and 7.16). The exponential $m_{\text{eff}}(T)$ dependence is also in accordance with theoretical predictions of the MP theory [see Chapter 1, Section 1B.8, Eq. (1B.8.18)].

On the other hand, it is possible to demonstrate that in the high-T region of thermalized charge carriers one should expect a power-law dependence of $m_{\text{eff}}(T)$.[356]

Indeed, for the high-T region, where the mobility $\mu \sim T^{-n}$ formula (7.9) holds, it is possible to derive the $m_{\text{eff}}(T)$ dependence from Eq. (7.17). Thus, inserting formula (7.9) in (7.17), one obtains the following equation:[356]

$$m_{\text{eff}}(T) = \frac{3e^2 l_o^2}{ka^2} \cdot T^{2|n|-1}. \tag{7.19}$$

As shown in Section 6.3.4 and as will be demonstrated later, at a definite value of slope n the $m_{\text{eff}}(T)$ dependence calculated either by exponential (6.46) or power-law (7.19) formulas, practically coincide in the high-T regime. The choice of the appropriate approximation formula in this case depends on physical or model considerations. This problem will be discussed below (see Section 7.3).

Further, in the framework of the modified SM model [see Eqs. 6.42 and 6.43], mean charge carrier trajectories $r(t)$ and corresponding velocities $v(t)$ are calculated as a function of temperature T, electric field $\mathcal{E}$, and angle θ between the initial radial velocity vector $\mathbf{v}_o$ of the carrier and the external electric field $\mathcal{E}$. Also the dispersion parameters of r and v, namely, σ_r and σ_v, are evaluated; these are caused by stochastic force effects on carrier motion. The initial charge carrier-parent ion separation length r_o, from which thermalization starts, has been taken equal to the nearest-neighbor distance along the corresponding crystal axis, e.g., $r_o = a_o$ or $r_o = c_o'$.

7.2.3. Simulation Results and Their Interpretation

7.2.3.A. Positive Charge Carrier (MP$^+$) Transport in the a-Direction of a Naphthalene Crystal

Figure 7.12 shows the computer simulation results of drift velocity $v_d^+(\mathcal{E})$ dependences of the positive charge carrier in the hot carrier transport regime at low T in a Nph crystal, as reported in Refs. 155 and 406. The input parameters $m_{\text{eff}}(T)$ and $\beta(T)$ of the simulation model were determined according to Section 7.2.2.

The possible range of m_{eff} in the low T region ($T < 35$ K) was estimated by formula (7.13) using reported experimental $(v_d)_s(T)$ results (Fig. 7.12) and limiting

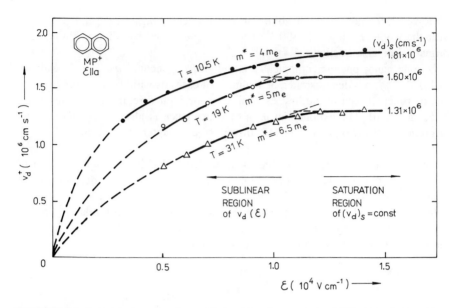

FIGURE 7.12. Calculated mean values of the positive charge carrier (MP$^+$) drift velocity $v_d^+(\mathcal{E})$ in the **a**-direction of naphthalene crystal as a function of the electric field $\mathcal{E}$ (solid lines).[355] The experimental points of $v_d^+(\mathcal{E})$ are those of Refs. 155 and 406. m^* denotes the relative mass of the carrier; $(v_d)_s$ the saturation value of the drift velocity.

values of optical phonon energies $h\nu_{ph}$, which lie between 46 and 125 cm^{-1} for Nph crystals (see Ref. 332, p. 19). The corresponding m_{eff} ranges are shown by vertical bars in Figure 7.15. From the obtained range of m_{eff} values those were used for the calculation which provided the best fit for an exponential $m_{eff}(T)$ dependence, Eq. (6.46), over the whole T range, i.e., those that lie on a straight line in the $m_{eff}(T)$ semilog plot, Fig. 7.15. The corresponding values of the parameter $\beta_s(\mathcal{E})$ were determined according to formula (7.15).

In the sublinear region of $v_d(\mathcal{E})$ the corresponding parameter $\beta_a(\mathcal{E})$ was evaluated according to formula (7.16).

As can be seen from Fig. 7.12, simulation provides a good agreement of the calculated $v_d(\mathcal{E})$ curves with experimental data[155,406] both in the saturation and in the sublinear region.

The computer-simulated experiment also allows one to visualize the conditions for hot carrier generation, as well as to predict the evolution of carrier mean trajectories and velocities under the influence of Coulombic and external electric fields. Figure 7.13 shows a typical picture of the mean carrier trajectories in the **ab**-plane of a Nph crystal, as a function of angle θ between the initial velocity vector $\mathbf{v}_o$ and the external field $\mathcal{E}$. The simulation has been performed using the following input parameters: $T = 31$ K, $m^* = 6.5m_e$ (see Fig. 7.12), $v_o = 2.7\times10^7$ cm/s, $\beta_s = 8.1\times10^{12}$ s^{-1}, and $\mathcal{E} = 4\times10^4$ V/cm ($\mathcal{E} \parallel a$). (In Fig. 7.13, for sake of conve-

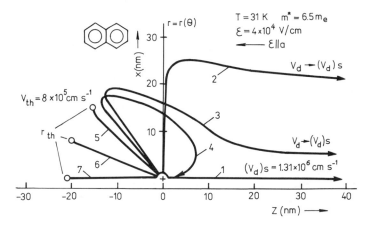

FIGURE 7.13. Calculated mean $\langle r \rangle$ values of charge carrier trajectories $r(\theta) = \{x(t), z(t)\}$ in the **ab**-plane of the naphthalene crystal for different values of the angle θ between the initial radial velocity vector v_o of the carrier and the external electric field $\mathcal{E} = -E_z$. 1, $\theta = 0°$; 2, 90°; 3, 130°; 4, 135°; 5, 140°; 6, 160°; 7, 180°.[355]

nience, the motion of a negative charge carrier in the field of a positive parent ion is shown; only one-half of the angular range, $0°-180°$ is depicted.)

Figure 7.13 demonstrates that as a function of the angle θ there appear three types of mean carrier trajectories. In the angular range between $\theta = 0°$ and $\theta \approx 130°$ hot carriers are created. As may be seen in Fig. 7.13, carriers moving along the direction of the field $\mathcal{E}$ at $\theta = 0°$ (trajectory 1) do not thermalize, but leave the Coulombic capture radius r_c and obtain a constant mean drift velocity, $(v_d)_s = 1.31 \times 10^6$ cm/s, as reported in Refs. 155 and 406, which is greater than mean thermal velocity $(v_{th} = 8 \times 10^5$ cm/s), or sound velocity $(u_L = 4 \times 10^5$ cm/s). Carriers ejected from the parent ion at angles $0° < \theta \leq 130°$ are deflected by the external field and also obtain the reported $(v_d)_s$ value for the given parameters T, $\mathcal{E}$, and m^* (trajectories 2,3).

On the other hand carriers, ejected at angles $180° \geq \theta \geq 140°$ thermalize within the Coulombic radius r_c forming bound charge pair (CP) states (trajectories 5–7). In a relatively narrow intermediate range of θ ($130° < \theta < 140°$) a peculiar type of mean trajectories emerges. After the deflection by the external field $\mathcal{E}$ the carrier is "trapped" by the Coulombic field of the parent ion and either recombines (see trajectory 4), or passes near the ion and, after making a loop, finally thermalizes at negative z values.

Figure 7.14a illustrates the temporal evolution of mean carrier drift velocities $v_d(t)$ as a function of the angle θ for the trajectories 1, 3, and 7 shown in Fig. 7.13. As may be seen, carriers moving along the field ($\theta = 0$) lose their excess energy, due to fast scattering, on a picosecond time scale, however, they do not thermalize, but continue moving along the field with a velocity $(v_d)_s > v_{th}$. On the other hand, carriers moving against the field, effectively thermalize. Their initial velocity v_o

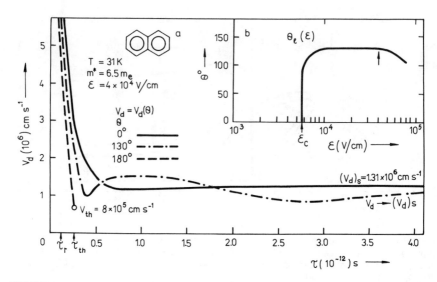

FIGURE 7.14. (a) Calculated mean values of charge carrier drift velocities $v_d = v_d(t)$ in the naphthalene crystal as a function of time for different values of the angle θ; τ_{th}, thermalization time; $\tau_r - 1/\beta$, relaxation time; other parameters as in Fig. 7.13. (b) Calculated limiting angle $\theta_l(\mathcal{E})$ of hot carrier generation as a function of electric field $\mathcal{E}$; $\mathcal{E}_c$ denotes the critical field of hot carrier generation.[355]

decreases practically exponentially with a relaxation time $\tau_r = 0.12 \times 10^{-12}$ s and a final thermalization time $\tau_{th} = 0.26 \times 10^{-12}$ s (see Fig. 7.14a).

More complicated, as may be seen, is the $v(t)$ dependence for a carrier, ejected at $\theta = 130°$. The initial velocity in this case also decreases rapidly, however, it does not reach the v_{th} value. The velocity increases again, apparently due to the field of the parent ion, decreases, and finally asymptotically approaches a constant value of $(v_d)_s = 1.31 \times 10^6$ cm/s.

Figure 7.14b shows the limiting angle θ_1, determining the sector, within the boundaries of which a hot carrier can be created at given T, as a function of the electric field $\mathcal{E}$. As can be seen, for $T = 31$ K, at critical field value $\mathcal{E}_c = 0.575 \times 10^4$ V/cm, a wide sector from $\theta = 0°$ up to $\theta = 90°$ is abruptly formed, within the boundaries of which hot carriers are generated. With growing $\mathcal{E}$, the boundary of the hot carrier sector also increases and saturates at $\theta = 130°$, for $\mathcal{E} > 10^4$ V/cm. At further increase of $\mathcal{E}$ above 4×10^4 V/cm the $\theta_1(\mathcal{E})$ dependence begin to decrease.

This means that at the critical field value $\mathcal{E}_c$ one half of all photogenerated carriers emerge as free hot carriers, and within the limits of $1 \times 10^4 \leqslant \mathcal{E} \leqslant 4 \times 10^4$ V/cm their contribution reaches the value of 3/4.

Figure 7.15 shows in a logarithmic plot the temperature dependence of the effective carrier mass m^* (in m_e units) discussed above. The possible range of m^* in the low-T region (H) of the hot carrier regime (10–35 K) was determined according to formula (7.13) but in the high-T region (T) of thermalized carriers (150–300 K)

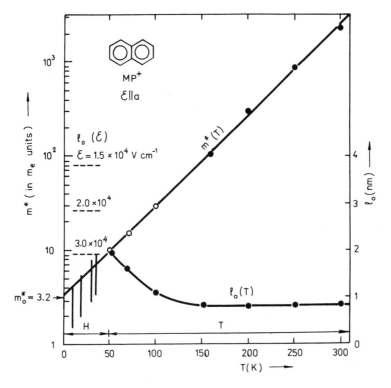

FIGURE 7.15. Estimated values of the relative effective mass m^* of a positive charge carrier (MP$^+$) as a function of temperature and mean value of the free path l_o as a function of temperature $l_o(T)$ and electric field $l_o(\mathcal{E})$ in the **a**-direction of the naphthalene crystal [only three discrete values of $l_o(\mathcal{E})$ are shown]. H, temperature range of hot carriers; T, temperature range of thermalized carriers.[355]

according to formula (7.17), under the assumption of a constant mean free path of the carrier of $l_o = a_o = 0.82$ nm. As can be seen, in the low-T region the best fit to the exponential $m_{\text{eff}}(T)$ formula (6.46) is provided by the upper values of m^*, corresponding to the highest energy of optical phonons in a Nph crystal, i.e., 125 cm^{-1}.

The values of m^*, adjusted by interpolation in the intermediate T-region (50–150 K), allow us to evaluate, according to Eq. (7.17), the T-dependence of $l_o = l_o(T)$. As can be seen from Fig. 7.15, the increase of $l_o(T)$ begins at ca.150 K and reaches the value of $l_o \approx 2$ nm at 50 K. Such an increase in l_o is reasonable and confirms theoretical and empirical prediction (see, e.g., Ref. 288, p. 362). It is also not surprising that in the low-T region the m^* values are relatively small, only 4–7 electron masses; the extrapolated m_o value at $T = 0$ is $3.2m_e$. These values agree with Warta and Karl's estimates.[406]

On the other hand, at room temperature a MP becomes a rather heavy quasiparticle having m^* of the order of $10^3 m_e$. A similar value of m^* at room temperature was obtained by independent, more rigorous simulation procedure in Ref. 344 for Pc crystals (see Section 6.3.3 and Fig. 6.19). The l_o value for the hot carrier in the low-T (10–15 K) range is electric field-dependent, i.e., $l_o = l_o(\mathcal{E})$ and can be evaluated by the following equation:[348]

$$l_o(\mathcal{E}) = \frac{h\nu_{\text{ph}}}{e\,\mathcal{E}}. \qquad (7.20)$$

Thus, the $l_o(\mathcal{E})$ value, in case of hot carrier scattering on optical phonons, decreases with increasing electric field $\mathcal{E}$ and lies in the range between 4 and 2 nm for $\mathcal{E} = (1.5–3.0)\times10^4$ V/cm (see Fig. 7.15).

7.2.3.B. Positive Charge Carrier (MP⁺) Transport in the c′-Direction of a Nph Crystal

Figure 7.16 demonstrates that the $m^*(T)$ and $l_o(T)$ dependences for the transport of a positive charge carrier in the c′-direction of a Nph crystal are similar to those of the a-direction (see Fig. 7.15). The $m^*(T)$ dependence can be approximated by an exponential function according to Eq. (6.46), which in the low-T hot carrier region (H) (10–60 K) gives the best fit for the upper values of m^*, corresponding to $h\nu_{\text{ph}} = 125$ cm^{-1}. The $l_o(T)$ dependence in this case also begins to increase below $T < 150$ K (Fig. 7.16). However, the values of effective masses are greater and lie between 17.8 and 47.9 m_e in the low-T region (10–50 K), reaching $m^* \approx 10^4 m_e$ at $T = 300$ K. Apparently, due to this, the saturation drift velocities $(v_d)_s$ are smaller in the c′-direction (see Figs. 7.12 and 7.17).

It is interesting to notice that for the positive charge carriers in the a and c′-directions of Nph crystals the exponential $m_{\text{eff}}(T)$ dependences [Eq. (6.46)] holds for the whole temperature range from 10 K up to 300 K (see Figs. 7.15 and 7.16). However, one should expect that in the high-T region the $m_{\text{eff}}(T)$ dependence would be of a power-law type [Eq. (7.19)].

The cause of this apparent contradiction is the following.[356] It may be shown that in the high-T region ($150 < T \leq 300$ K) these $m_{\text{eff}}(T)$ dependences can be also linearized in log m_{eff} versus log T plot according to Eq. (7.19). In this case the slope n of the dependence $\mu(T) \sim T^{-n}$ is large, namely, $n_{aa} = 2.9$ and $n_{c'c'} = 2.8$.[154] As a result, the corresponding exponents $(2n-1)$ of the power-type $m_{\text{eff}}(T)$ dependence (7.19) is given by $(2n_{aa}-1) = 4.8$ and $(2n_{c'c'}-1) = 4.6$, respectively. It may be shown that for these $(2n-1)$ values the power-law type (7.19) and exponential (6.46) dependences practically agree in the $150 < T \leq 300$ K range.

As can be seen from Fig. 7.17, the calculated $v_d(\mathcal{E})$ curves are in good agreement with experimental data both in the saturation and sublinear regions. Also, in this case, the parameter $\beta_a(\mathcal{E})$ in the transition region of $v_d(\mathcal{E}) \to (v_d)_s$ was determined by the empirical formula (7.16).

It is interesting to mention that the extrapolated value of $m_o^* = 14.5 m_e$ at $T = 0$ in the c′-direction of Nph is very near to that of the reported effective mass of holes in anthracene $m^* = 11 m_e$, as determined experimentally by Burland and Konzel-

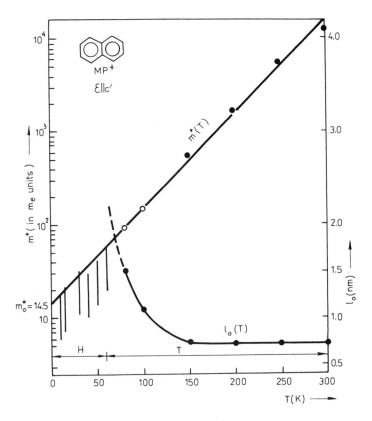

FIGURE 7.16. Estimated values of the relative effective mass m^* of a positive charge carrier (MP^+) as a function of temperature and mean value of free path $l_o(T)$ in the $\mathbf{c}'$-direction of the naphthalene crystal.[355]

mann at $T = 2$ K.[51] Figure 7.18a shows a schematic picture of cyclic acceleration of a charge carrier in an electric field $\mathcal{E}$, resulting in optical phonon emission at the end of every cycle, and subsequent return to zero velocity. Figure 7.18a depicts the transport of a positive charge carrier in the $\mathbf{c}'$-direction of the Nph crystal in the hot carrier region at $T = 10$ and $\mathcal{E} = 3\times10^4$ V/cm which yields, under these conditions, the reported mean $(v_d)_s$ value $(v_d)_s = 0.9\times10^6$ cm/s (see Fig. 7.17). In this case the carrier reaches the maximum velocity $v_{max} = 2(v_d)_s = 1.8\times10^6$ cm/s during the acceleration time $\tau_s = 3.1\times10^{-13}$ s. This provides the kinetic energy E_k necessary for optical phonon excitation $E_k = m_{eff} v_{max}^2/2 = h\nu_{ph} = 0.015$ eV $= 125$ cm^{-1}. However, the simulation results demonstrate that the schematic picture of cyclic velocity changes, from $v = 0$ to v_{max} and abruptly back to $v = 0$, as displayed in Fig. 7.17a, is oversimplified. Figure 7.17b shows the actual statistical distribution of velocities due to stochastic forces, as taken into account in our simulation model (see Ref. 344). As may be seen, the velocity distribution curve

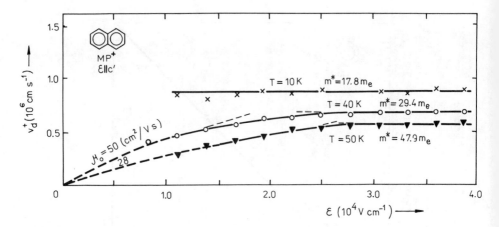

FIGURE 7.17. Calculated mean values of the positive charge carrier (MP$^+$) drift velocity $v_d^+(\mathcal{E})$ in the **c′**-direction of the naphthalene crystal as a function of the electric field $\mathcal{E}$ (solid lines),[355] and experimental points of $v_d^+(\mathcal{E})$ according to Ref. 406. μ_0, low-field Ohmic mobility.

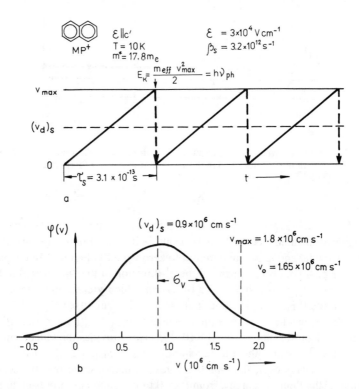

FIGURE 7.18. (a) Schematic diagram of the cyclic inelastic scattering mechanism of a hot charge carrier (MP) under creation of optical lattice phonons in the naphthalene crystal; (b) Corresponding distribution curve $\varphi(v)$ of carrier velocities v; σ_v denotes the dispersion parameter of velocities; v_o is the initial velocity of the carrier.[355]

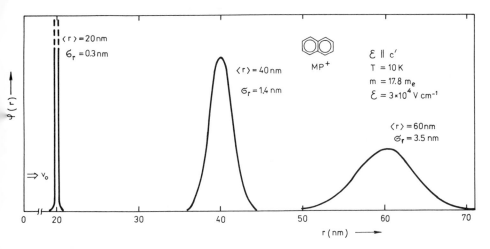

FIGURE 7.19. Distribution curve $\varphi(r)$ of the carrier coordinates r moving in the c'-direction of a naphthalene crystal; $\langle r \rangle$, mean distance from the initial position $r = 0$; σ_r denotes the corresponding dispersion parameter at distance r.[355]

$\varphi(v)$ is pretty wide, characterized by a dispersion parameter $\sigma_v \approx 0.5 \times 10^6$ cm/s. The peak of the Gaussian curve corresponds to the reported mean drift velocity $(v_d)_s = 0.9 \times 10^6$ cm/s, but the right-side distribution tail reaches values greater than v_{max} while the left-side tail reaches even negative values of v. However, it is important to emphasize that the velocity dispersion parameter σ_v remains constant and does not change during the hot carrier transport, apparently due to the cyclic nature of the velocity change.

A quite different behavior is observed in the distribution function of charge carrier position. As can be seen from Fig. 7.19, the dispersion parameter σ_r of the carrier coordinates r increases monotonically with increasing distance from the origin $r = 0$. In other words, the stochastic deviations of the coordinates of the carriers accumulate during their transport which thus exhibits a Gaussian distribution, however, with $\sigma_r^2 \ll r^2$ (see Fig. 7.19).

Figures 7.20 and 7.21 illustrate the physical conditions which determine the transition from the scattering by optical phonons [saturation region of $v_d = (v_d)_s = \text{const}$] to scattering by acoustic phonons [sublinear region of $v_d(\mathcal{E})$ dependence] (see Fig. 7.12).

Figure 7.20 shows the calculated field dependences of relaxation rate constants for scattering on optical $\beta_s(\mathcal{E})$ and acoustic $\beta_a(\mathcal{E})$ phonons, as well as the corresponding mean scattering times $\tau_s(\mathcal{E}) = 1/\beta_s(\mathcal{E})$ and $\tau_a(\mathcal{E}) = 1/\beta_a(\mathcal{E})$. These dependences have been obtained for positive charge carier transport in the c'-direction of the Nph crystal at $T = 40$ K and $m^* = 29.5m_e$ (see Fig. 7.17). As may be seen from Fig. 7.20, the $\beta_s(\mathcal{E})$ decreases and $\tau_s(\mathcal{E})$ increases with decreasing electric field $\mathcal{E}$. At $\mathcal{E} < 2 \times 10^4$ V/cm $\beta_s(\mathcal{E})$ becomes smaller than $\beta_a(\mathcal{E})$ and, accordingly, $\tau_s(\mathcal{E})$ larger than $\tau_a(\mathcal{E})$. This means that below

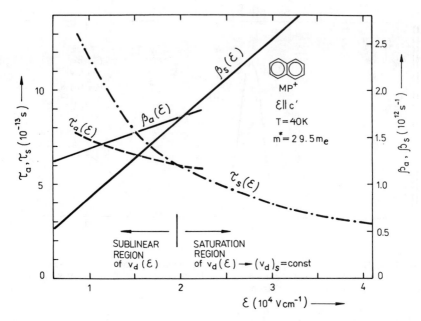

FIGURE 7.20. Calculated curves of carrier relaxation rate constants $\beta_a(\mathcal{E})$ and $\beta_s(\mathcal{E})$ in case of scattering at acoustic and optical lattice phonons, respectively, as a function of the electric field $\mathcal{E}$, as well as the corresponding mean relaxation time dependences, $\tau_a(\mathcal{E})$ and $\tau_s(\mathcal{E})$, for carriers moving in **c'**-direction of the naphthalene crystal.[355]

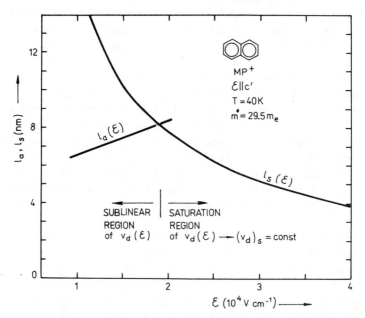

FIGURE 7.21. Calculated curves of the carrier mean free path for the case of scattering at acoustic and optical lattice phonons $l_a(\mathcal{E})$ and $l_s(\mathcal{E})$, respectively, as a function of the electric field $\mathcal{E}$, for carriers moving in **c'**-direction of the naphthalene crystal.[355]

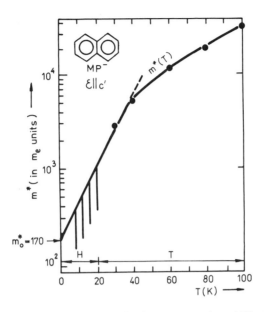

FIGURE 7.22. Estimated values of relative effective masses $m^* = m^*(T)$ of negative charge carrier for the **c′**-direction of the naphthalene crystal.[355]

$\mathcal{E} = 2 \times 10^4$ V/cm scattering at acoustic phonons becomes dominant, and the saturation region of $(v_d)_s$ = const passes over to the sublinear region of the $v_d(\mathcal{E})$ dependence (see Fig. 7.17). Figure 7.21 illustrates the corresponding field dependences of carrier mean free path for scattering at optical $l_s(\mathcal{E})$ and acoustic $l_a(\mathcal{E})$ phonons.

Also, in this case, the inequality $l_a(\mathcal{E}) < l_s(\mathcal{E})$ below $\mathcal{E} = 2 \times 10^4$ V/cm determines the transition from carrier scattering at optical to scattering at acoustic phonons, corresponding to the transition from the saturation to the sublinear region of the $v_d(\mathcal{E})$ dependence (see Fig. 7.17).

7.2.3.C. Negative Charge Carrier Transport in the **c′**-Direction of a Naphthalene Crystal

The T-dependence of negative charge carrier mobility $\mu(T)$ in the **c′**-direction of a Nph crystal shows strange anomalies. In this case, the $\mu(T)$ dependence $\mu(T) = aT^{-n}$, typical for polyacene crystals, appears only in a relatively narrow T-range below ca.80 K. Also, the n value is smaller ($n = 1.48$) as compared with the positive carrier $\mu(T)$ dependence ($n = 2.8$). However, the principal anomaly is the fact that in the higher T-range (100–300 K) the carrier mobility becomes nearly temperature-independent, having a practically constant value $\mu \approx 0.4$ cm^2/Vs.[154] Karl et al.[155] observed also hot negative charge carriers at $T < 20$ K at $\mathcal{E} > 1 \times 10^4$ V/cm (see Fig. 7.23).

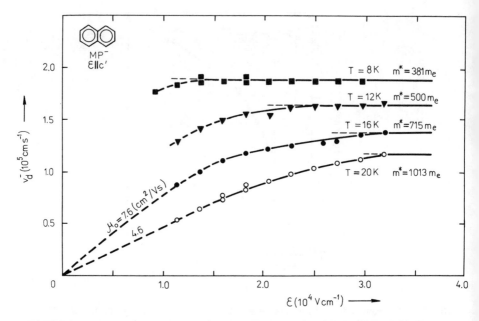

FIGURE 7.23. Calculated mean values of the drift velocity of negative charge carrier, $v_d^-(\mathcal{E})$, in the **c'**-direction of the naphthalene crystal as a function of the electric field $\mathcal{E}$ (solid lines)[355] and experimental points of $v_d^-(\mathcal{E})$ according to Ref. 355.

The above-mentioned anomalies manifest themselves also in the $m^*(T)$ dependence (Fig. 7.22). In the hot carrier regime (H) at low-T (8–20 K) and above up to 50 K, the $m^*(T)$ dependence follows the exponential law of $m_{eff}(T)$ increase (6.46) in a similar way as in the case of positive charge carriers (see Figs. 7.15 and 7.16). However, above $T = 30$ K the $m^*(T)$ curve declines from the exponential dependence toward smaller m^* values. One should also notice that the values of m^* of negative carriers are considerably greater than those of positive carriers. Even in the low-T region of hot carriers for T from 8 to 20 K the corresponding m^* values lie between 381 and $1013 m_e$ (see Fig. 7.22 and 7.23). The extrapolated m_o^* value at $T = 0$ K is also considerable, viz. $m_o = 170 m_e$, while m^* at higher temperatures $(T > 40$ K) reaches a magnitude of $10^4 m_e$ (see Fig. 7.22). Such large effective masses apparently are the cause of the low drift velocities in the hot carrier regime, which range about $(1–2) \times 10^5$ cm/s (see Fig. 7.23). On the whole, the simulation results of the $v_d^-(\mathcal{E})$ dependences are in good agreement with reported experimental data also in this case (see Fig. 7.23).

A number of authors have discussed the possible physical reasons causing the phenomenon of a nearly T-independent carrier mobility over such a wide temperature range (100–300 K) (for references and details see Ref. 332, p. 37 and Ref. 288, p. 360).

Karl[155] has proposed a new empirical approach, according to which the effective nearly T-independent mobility μ_{eff} in this region can be presented as a superposition

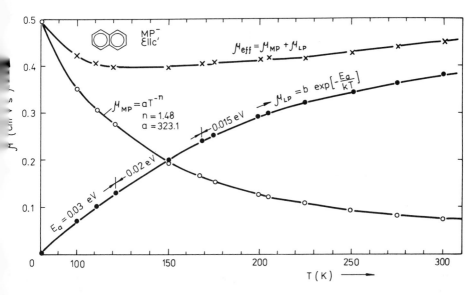

FIGURE 7.24. Deconvolution of the temperature dependences of the experimental negative charge carrier mobilities $\mu_{\text{eff}}(T)$ in the c'-direction of naphthalene single crystal: $\mu_{\text{MP}}(T)$ and $\mu_{\text{LP}}(T)$ are the evaluated T-dependences of nearly small molecular polaron (MP) and small lattice polaron (LP) mobility contributions according to Karl's experimental data.[155] E_a is the activation energy underlying the small lattice polaron hopping contribution.

of two different $\mu(T)$ dependences: $\mu_1 = aT^{-1.48}$ and $\mu_2 = b \exp[-E_a/kT]$.

Silinsh et al. have suggested[355] that these two mobility $\mu(T)$ components may be regarded as a manifestation of two different, coexisting polaron type transport mechanisms: a nearly small MP, as described above, which moves by stepping via tunneling without activation energy displaying the typical $\mu(T)$ dependence $\mu_{\text{MP}} = aT^{-n}$ (see Refs. 334, 344, 348, and 351), and a small-lattice polaron which moves by thermally activated hopping and thus exhibits a typical exponential mobility T-dependence: $\mu_{\text{LP}} = b \exp[-E_a/kT]$ (see Ref. 288, p. 353).

In this case, the effective carrier mobility $\mu_{\text{eff}}(T)$ is given by the sum of these two contributions:

$$\mu_{\text{eff}}(T) = \mu_{\text{MP}} + \mu_{\text{LP}} = aT^{-n} + b \exp[-E_a/kT], \qquad (7.21)$$

where E_a is the activation energy of lattice polaron hopping motion.

Figure 7.24 shows the $\mu(T)$ dependences of both mobility components: $\mu_{\text{MP}}(T)$ and $\mu_{\text{LP}}(T)$ as well as the resulting $\mu_{\text{eff}}(T)$ curve which is in a satisfactory agreement with experimental data.[155] The $\mu_{\text{MP}}(T)$ and $\mu_{\text{LP}}(T)$ deconvolution of the experimental data into the contributions was obtained from Karl.[155]

It should be mentioned that the empirical $\mu_{\text{LP}}(T)$ dependence is not a single exponential curve. Actually, the E_a value is slightly different for different T-ranges

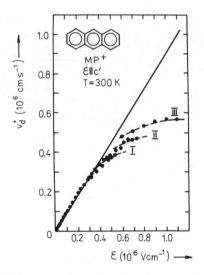

FIGURE 7.25. Calculated mean values of positive charge carrier drift velocity $v_d^+(\mathcal{E})$ in the c'-direction of an anthracene single crystal at $T = 300$ K as the function of the electric field (broken lines I, II, III for the three possible saturation values).[355] Experimental points of $v_d^+(\mathcal{E})$ according to Refs. 33 and 34.

and decreases with increasing temperature (as indicated by arrows on Fig. 7.24) (see Ref. 155).

Equation (7.21) is in accord with some earlier theoretical predictions (see e.g., Ref. 257). It has been shown that there may exist a relatively wide temperature region in which the two modes of quasiparticle motion can coexist, namely, a "coherent" (in this case tunneling) and an "incoherent" (hopping) type of motion, a superposition of which yields the nearly T-independent mobility μ (or diffusion coefficient D) values. A lattice polaron formation is not characteristic for polyacene crystals.[288,332,351] Only in case of the described negative charge carrier motion in Nph crystals, when at $T \geq 100$ K the effective mass $m^* > 10^4 m_e$ of the carrier causes strong localization, its residence time on a lattice site may become comparable with or greater than the relaxation time necessary for lattice polaron formation ($\tau_h \geq \tau_l$, see Ref. 351, p. 166), and such formation emerges as a physical reality.

Thus, it is reasonable to assume that the nearly T-independent carrier mobility μ_{eff} in the high-T region (100–300 K) is actually caused by formation of small-lattice polarons.

7.2.3.D. Hot Charge Carriers at Room Temperature in Ac Crystals

Figure 7.25 shows the simulation results of $v_d(\mathcal{E})$ dependences of hot positive charge carrier transport in the c'-direction of Ac crystals at $T = 300$ K and $\mathcal{E} > 3 \times 10^5$ V/cm, as reported in Refs. 33 and 34.

As may be seen from Fig. 7.25, the $v_d(\mathcal{E})$ shows a sublinear electric field dependence at $\mathcal{E} > 3\times10^5$ V/cm, and three rather distinct regions of v_d saturation. According to Willig,[409] "these steps may or may not be real. Their existence should be checked in new careful measurements." Willig has kindly presented a $v_d(\mathcal{E})$ plot of earlier unpublished results, obtained in 1978 under different experimental conditions, in which a similar, rather distinct step between 6 and 7×10^5 V/cm appears "too large for assignment to uncertainty." However, this step occurs at different field strength than any of the steps in Fig. 7.25 which may be regarded as "strong indication for a systematic error as origin of these steps."

Since at $T = 300$ K carrier scattering takes place at every lattice site (see Figs. 7.15 and 7.16), the mean free path in this case has a discrete value, equal to the intermolecular distance in the $\mathbf{c}'$-direction of the crystal, namely, $l_o = c_0'$ $= 0.92$ nm. This means that at high electric field strengths the carrier can accumulate more kinetic energy over this distance than necessary for excitation of a single phonon. Thus, for the three possible $v_d(\mathcal{E})$ saturation regions I, II, and III (see Fig. 7.25) the carrier obtains over the distance $l_o = 0.92$ nm kinetic energies equal to 0.0478, 0.067, and 0.092 eV, respectively, in accordance with formula (7.20). These saturation region kinetic energies correspond approximately to three, four, and six optical phonons of energy $h\nu_{ph} = 125$ cm^{-1} (0.0155 eV). Thus, simulation results speak in favor of multiphonon scattering of hot carriers at high electric fields at room temperature. In our opinion, the probability of multiphonon scattering of hot carriers should be considerably greater than that of excitation of intramolecular vibrations, as suggested in Ref. 34. This is due to the much greater cross section of a nearly-small MP as compared with that of localized intramolecular vibrational modes.[344] It should be mentioned that, according to Mott and Davis,[247] the process of multiphonon emission should have a greater probability than that of single phonon emission.

The scattering time τ_s of the carrier becomes relatively small at very high fields (e.g., $\tau_s = 8.10^{-14}$ s at $\mathcal{E} = 1.10^6$ V/cm in the saturation III region, see Fig. 7.25) and thus may be comparable with the characteristic time necessary for MP buildup[351] (see Table 3.7).

7.2.3.E. Negative Charge Carrier (MP⁻) Transport in the **a**-direction of an α-perylene Crystal

Figure 7.26 shows the computer simulation results of drift velocity $v_d^-(\mathcal{E})$ dependences of the negative charge carrier in the hot carrier transport regime at low T ($T < 100$ K) in an α-Pl crystal as reported in Refs. 154, 155, and 407. The input parameters $m_{eff}(T)$ and $\beta(T)$ of the simulation model were determined as in Section 7.2.2. As may be seen from Fig. 7.26, the saturation regime of $v_d(\mathcal{E}) \rightarrow (v_d)_s =$ const, at which the scattering at optical lattice phonons prevails, can be reached only at considerably higher field strength $\mathcal{E} \geqslant 6.10^4$ V/cm than in Nph crystals (where saturation begins at $\mathcal{E} \geqslant 1.10^4$ V/cm for both types of carriers) (see Fig. 7.12).

In the sublinear region of $v_d(\mathcal{E})$ the corresponding parameter $\beta_a(\mathcal{E})$ was evalu-

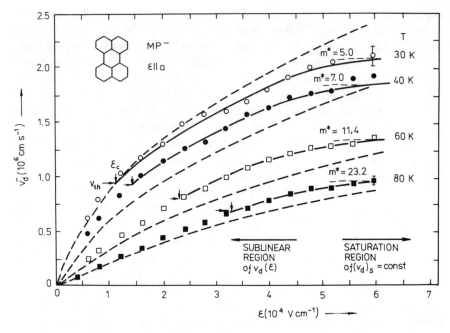

FIGURE 7.26. Calculated mean values of negative charge carrier (MP⁻) drift velocities $\bar{v}_d(\mathcal{E})$ in the **a**-direction of an α-perylene crystal as a function of electric field $\mathcal{E}$: solid lines, computer simulation data[356]; dashed lines, approximation by the parabolic-type $v_d(\mathcal{E}) \sim \sqrt{\mathcal{E}}$ formula (7.11) according to Refs. 154 and 155. The experimental points of $\bar{v}_d(\mathcal{E})$ are those of Refs. 154 and 155. m^* denotes the relative effective mass of the carrier. The arrows show the critical field $\mathcal{E}_c$ strength values where $v_d \approx v_{th}$ and below which the carrier thermalization sets in (where v_{th} is the mean thermal velocity of the carrier).

ated according to formula (7.16). As can be seen from Fig. 7.26, the simulation curves provide good agreement with the experimental points of the $v_d(\mathcal{E})$ dependences. On the other hand, the analytically derived parabolic-type $v_d(\mathcal{E}) \sim \sqrt{\mathcal{E}}$ formula (7.11) gives less satisfactory agreement with experimental data.

The arrows in Fig. 7.26 show the critical field $\mathcal{E}_c$ at which thermalization of charge carriers takes place. As may be seen from Fig. 7.26, the $\mathcal{E}_c$ values, at which $v_d \approx v_{th}$, is temperature-dependent. It may be shown that the temperature dependence of the critical field $\mathcal{E}_c(T)$, above which hot carriers can be created, follows an empirical formula, proposed earlier in Ref. 348:

$$\mathcal{E}_c(T) \sim \exp(aT^{1/2}). \tag{7.22}$$

Figure 7.27 shows the $\mathcal{E}_c(T)$ dependence, indicated by the arrows on Fig. 7.26 in a log $\mathcal{E}_c$ versus $T^{1/2}$ plot. As can be seen from Fig. 7.27, the proposed empirical formula (7.22) may be regarded as valid for the description of the $\mathcal{E}_c(T)$ dependence.

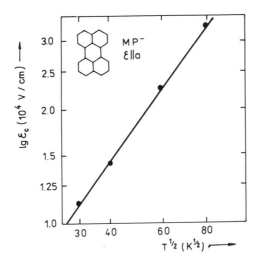

FIGURE 7.27. Estimated temperature-dependent critical electrical field $\mathcal{E}_c(T)$ values at which hot, nonthermalized charge carriers can be created plotted as log $\mathcal{E}_c$ versus $T^{1/2}$. The $\mathcal{E}_c$ values, indicated by the arrows in Fig. 7.26, have been used.[356]

On the whole, the $v_d(\mathcal{E})$ dependences in Fig. 7.26 demonstrate that in the electric field range from $\mathcal{E}_c$ up to $\mathcal{E} = 6.10^4$ V/cm the hot carrier scattering on the acoustic lattice phonons prevails.

The possible range of $m_{\text{eff}}(T)$ values in the low-T region ($T \leqslant 100$ K) was estimated by formula (7.13) using experimental $(v_d)_s(T)$ data at $\mathcal{E} = 6.10^4$ V/cm (see Fig. 7.26 and reported characteristic optical phonon frequences $h\nu_{\text{ph}}$ of α-Pl crystals taking as limiting values optical phonon energies $h\nu_{\text{ph}}$ which lie between 200 and 104 cm^{-1},[194] see also Ref. 332, p. 41).

This excludes a number of lower-frequency modes. However, simulation results demonstrate that in the saturation region the inelastic scattering at higher-energy optical phonons (see Figs. 7.15, 7.16, 7.28, and 7.32) prevails. The corresponding $m_{\text{eff}}(T)$ ranges in the low-T hot carrier region ($T \leqslant 100$ K) are shown by vertical bars in Fig. 7.28.

In the high-temperature region (see Fig. 7.28) the $m_{\text{eff}}(T)$ values were estimated from reported experimental $\mu(T)$ dependences[154,155,407] according to formula (7.17) under the assumption of a constant mean free path l_o of the carrier, equal to the lattice constant in the direction of the **a**-axis, namely, $l_o = 1.14$ nm.

As may be seen from Fig. 7.28, the calculated m_{eff} values of a thermalized carrier at $T = 100$ K lie on a straight line in the semilog plot which coincides with the upper end of the vertical bars of the possible $m_{\text{eff}}(T)$ range of hot carriers. First, these results show that in this T range ($30 \leqslant T \leqslant 100$ K) the $m_{\text{eff}}(T)$ dependence follows the exponential law (6.46). Second, the coincidence of the line with the upper end of the vertical bars indicates that scattering most probably occurs at the

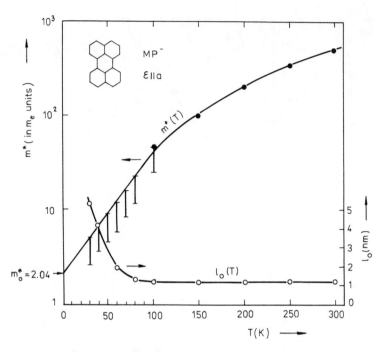

FIGURE 7.28. Estimated values of relative effective mass $m^*(T)$ of a negative charge carrier (MP$^-$) and mean value of carrier free path $l_o(T)$ in the **a**-direction of α-perylene as a function of temperature T.[356] The vertical bars indicate the possible m^* range in the hot carrier regime [formula (7.13)]; the filled circles indicate the m^* values in the temperature range of thermalized carriers, evaluated according to formula (7.17).

high-energy optical phonons around $h\nu_{ph} = 200$ cm^{-1} rather than at those at lower energies.

It should be emphasized that the agreement at $T = 100$ K of the m^* value, estimated by formula (7.17) with the upper end of the vertical bar of the m^* range, calculated according to formula (7.13), may serve as an additional indication of the validity of both approaches.

The extrapolated m_{eff} value at zero temperature m_{eff}^o is small — practically equal to twice the free electron masses $m_o^* \approx 2m_e$. The parameter T_o in the approximation formula (6.46) equals 32.3 K (~ 22 cm^{-1}).

As can be seen from Fig. 7.28, at higher temperatures, above 100 K, the $m_{eff}(T)$ curve declines from the exponential dependence (6.46) and shows a tendency toward saturation. This behavior of the $m_{eff}(T)$ is reminicent of a similar character of the $m_{eff}(T)$ curve in the case of negative charge carrier transport in the **c$'$**-direction of a Nph crystal (see Fig. 7.22).

Hence, in the high-T region one may expect a power-law dependence of $m_{eff}(T)$ [see formula (7.19)] instead of an exponential one (6.46). Indeed, it may be shown that in the temperature range $100 < T \leqslant 300$ K the $m_{eff}(T)$ dependence (see Fig.

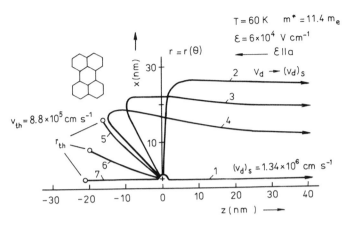

FIGURE 7.29. Calculated mean $\langle r \rangle$ values of charge carrier trajectories $r(\theta) = \{x(t), z(t)\}$ in the ab-plane of an α-perylene crystal for different values of the angle θ between the initial radial velocity v_o of the carrier and the external electric field $\mathcal{E} = -E_z$;[356] (1) $\theta = 0°$, (2) $\theta = 80°$, (3) $\theta = 120°$, (4) $\theta = 135°$, (5) $\theta = 140°$, (6) $\theta = 160°$, (7) $\theta = 180°$.

7.28) can be linearized in $\log m_{\text{eff}}$ versus $\log T$ plot with the slope value equal to $(2n_{aa} - 1) = 2.56$ for $n_{aa} = 1.78$ (see Ref. 154). However, this approximation does not fit for the low-T region at $T < 100$ K.

Figure 7.28 also demonstrates the extremely small mean free path l_o of the carrier, equal to the lattice constant in the **a**-direction, namely, $a_o = 1.14$ nm, which remains so small down to very low temperatures. Only below $T = 80$ K do the l_o values begin to increase. It ought to be remembered that in Nph crystals this increase of l_o begins already at ca. 150 K (see Figs. 7.15 and 7.16). These data confirm the feasibility of using this l_o value ($l_o = a_o$) in formula (7.17) for evaluating the effective mass temperature dependence $m^*(T)$ in the high-T region.

Figure 7.29 shows a plot of computer-simulated *mean* trajectories of negative charge carriers in the **ab**-plane of α-Pl under the influence of Coulombic and external electric fields as a function of the angle θ between the initial velocity vector $\mathbf{v}_o$ of the ejected carrier and the external field $\mathcal{E}$.[356] The simulation has been performed using the following input parameters: $T = 60$ K, $m^* = 11.4m_e$ (see Fig. 7.26), $\beta = \beta_s = 6.7 \times 10^{12}$ s^{-1}, and $\mathcal{E} = 6.10^4$ V/cm ($\mathcal{E} \| a$). The initial kinetic energy E_k of the ejected electron is assumed to be $E_k = (m_{\text{eff}} v_0^2)/2 = 1.3$ eV and the corresponding value of $v_o = 2.1 \times 10^7$ cm/s.

Figure 7.29 demonstrates that there are two main types of mean carrier trajectories as a function of angle θ. In the range between $\theta = 0°$ and $\theta = 135°$ hot carriers are created from the very beginning. Carriers moving along the direction of the field $\mathcal{E}$ at $\theta = 0°$ (trajectory 1) do not thermalize but leave the Coulombic capture radius r_c and obtain a constant mean drift velocity, $(v_d)_s = 1.34 \times 10^6$ cm/s, as reported in Refs. 155 and 356, which is greater than the mean thermal velocity $(v_{\text{th}} = 8.8 \times 10^5$ cm/s. Carriers ejected from the parent ion at angles $0° < \theta \leq 135°$ are deflected by

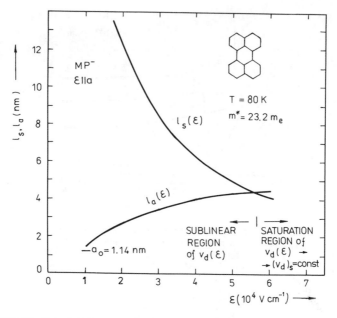

FIGURE 7.30. Calculated curves of carrier mean free path $l_a(\mathcal{E})$ and $l_s(\mathcal{E})$ for the case of scattering by acoustic and optical lattice phonons, respectively, as a function of the electric field $\mathcal{E}$ for carrier moving in the **a**-direction of an α-perylene crystal.[356]

the external field and also obtain the reported $(v_d)_s$ value for given parameters $T, \mathcal{E}$, and m^* (trajectories 2–4).

On the other hand, carriers ejected at angles $180° \geqslant \theta \geqslant 135°30'$ thermalize within the Coulombic radius r_c forming bound CP states which need additional thermal activation to become free carriers according to the Onsager theory (see Section 6.2.3),[344] and references therein) (trajectories 5–7). On the whole, the mean carrier trajectories in Fig. 7.29 are rather similar to those of a Nph crystal (see Fig. 7.13).

However, there is a minor difference. In the case of a naphthalene crystal there appears a relatively wide range of angles ($130° < \theta < 140°$) at which a rather peculiar type of mean trajectories appears. In the case of α-Pl this intermediate range of transition from hot to thermalized carriers is very narrow, of the order of $30'$ ($0.5°$); i.e., the transition occurs at $\theta = 135°15' \pm 15'$.

The calculated mean values of carrier drift velocities $v_d = v_d(t)$, the distribution curves of carrier coordinates $\varphi(r)$ and velocities $\varphi(v)$, and the dispersion parameters σ_r and σ_v are similar to those of the naphthalene crystal (cf. Figs. 7.14, 7.18, and 7.19).

Figure 7.30 shows the calculated field dependences of the carrier mean free path for scattering by optical $l_s(\mathcal{E})$ and by acoustic $l_a(\mathcal{E})$ lattice phonons. These dependences have been obtained for negative charge carrier transport in the **a**-direction of the α-Pl crystal at $T = 80$ K and with $m^* = 23.2m_e$ (see Fig. 7.26).

The $l_s(\mathcal{E})$ dependence was evaluated according to formula:[355]

$$l_s(\mathcal{E}) = \frac{h\nu_{ph}}{e\,\mathcal{E}}. \qquad (7.23)$$

Thus, the $l_s(\mathcal{E})$ value decreases as the reciprocal of field strength $\mathcal{E}$ (see Fig. 7.30).

The $l_a(\mathcal{E})$ value was evaluated in the following way. The mean scattering time at acoustic phonons $\tau_a(\mathcal{E})$ in the sublinear region of $v_d(\mathcal{E})$ is related to the relaxation rate constant as $\tau_a(\mathcal{E}) = 1/\beta_a(\mathcal{E})$. The value of $\beta_a(\mathcal{E})$ can be determined according to formula (7.16).

Consequently, the mean free path $l_a(\mathcal{E})$ equals

$$l_a(\mathcal{E}) = \tau_a(\mathcal{E}).2v_d(\mathcal{E}) = \frac{2m_{eff}[v_d(\mathcal{E})]^2}{e\,\mathcal{E}}. \qquad (7.24)$$

Thus, the $l_a(\mathcal{E})$ dependence may be obtained directly from the experimental sublinear $v_d(\mathcal{E})$ dependences (Fig. 7.26) for given $\mathcal{E}$ and T values.

As can be seen from Fig. 7.30, $l_a(\mathcal{E})$ decreases with decreasing field $\mathcal{E}$ and approaches the lattice constant value $a_o = 1.14$ nm at $\mathcal{E} \approx 1.10^4$ V/cm. Thus, at this low field limit l_a becomes equal to the mean thermal scattering length $l_o = a_o$ (see Fig. 7.28).

By the way, the l_o value may be evaluated independently from other simulation data, namely, $l_o = \tau_r \cdot v_{th}$, where τ_r is the carrier relaxation time $\tau_r = 1/\beta \approx \tau_{th}/2$ (see Fig. 7.14) and v_{th} is the mean thermal velocity of the carrier. For thermalized carriers at $T = 80$ K in α-Pl $\tau_r = 2.13 \times 10^{-13}$ s, and $v_{th} = 7.0 \times 10^5$ cm/s. Consequently, $l_o = 1.49$ nm.

As we see, this rough evaluation result is rather close to the lattice constant value $a_o = 1.14$ nm and the l_a value at $\mathcal{E} = 1.10^4$ V/cm (see Fig. 7.30).

Returning once more to Fig. 7.30, one can see that at field values above $\mathcal{E} = 6 \times 10^4$ V/cm the inequality $l_s(\mathcal{E}) < l_a(\mathcal{E})$ determines the transition from carrier scattering at acoustic to scattering at optical phonons, corresponding to the transition from the sublinear dependence of $v_d(\mathcal{E})$ to the saturation region of carrier drift velocity $(v_d)_s = $ const (see Fig. 7.26).

7.2.3.F. Negative Charge Carrier (MP⁻) Transport in the c'-direction of an α-perylene Crystal

The computer simulation results of the reported[154,155] drift velocity $v_d^-(\mathcal{E})$ dependences of the negative charge carrier in the hot carrier transport regime in the c'-direction at low temperatures ($T \leqslant 80$ K) are shown in Fig. 7.31.

As can be seen from Fig. 7.31 the calculated $v_d(\mathcal{E})$ curves are in good agreement with experimental data both in the sublinear and in the saturation region. On the other hand, approximation by a parabolic-type $v_d(\mathcal{E}) \sim \sqrt{\mathcal{E}}$ formula (7.11), according to Ref. 154 gives less satisfactory agreement with experimental data.

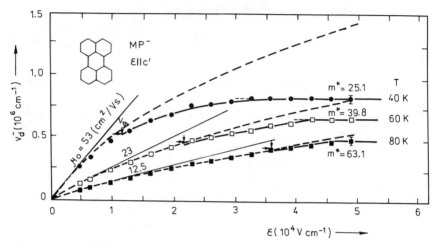

FIGURE 7.31. Calculated mean values of negative charge carrier (MP$^-$) drift velocities $v_d^-(\mathcal{E})$ in the **c**$'$-direction of an α-perylene crystal as a function of the electric field $\mathcal{E}$: the solid lines represent computer simulation data;[356] the dashed lines represent approximation by the parabolic-type $v_d(\mathcal{E}) \sim \sqrt{\mathcal{E}}$ formula (7.11) according to Refs. 154 and 155. The experimental points of $v_d(\mathcal{E})$ are those of Refs. 154 and 155. The arrows show the critical-field values $\mathcal{E}_c$ at which $v_d \approx v_{th} \cdot \mu_0$ is the low-field Ohmic mobility.

In the **c**$'$-direction the $(v_d)_s$ values are smaller than those in the **a**-direction by a factor of 2. However, the saturation region starts at lower field strength (see Figs. 7.26 and 7.31). Also, in this case the critical field values $\mathcal{E}_c$ for hot carrier creation (shown by arrows on Fig. 7.31) are strongly temperature-dependent and can again be satisfactorily approximated by formula (7.22).

The effective mass T-dependence (Fig. 7.32) was estimated according to a procedure similar to that as described in Section 7.2.3.E. As can be seen, for the negative charge carrier transport in the **c**$'$-direction, the exponential law (6.46) for $m_{eff}(T)$ holds both in the hot and partially in the thermalized carrier regions up to 200 K. Only at higher temperatures, namely, at $T > 200$ K, does the $m_{eff}(T)$ curve deviate from the exponential dependence (6.46). It is interesting to notice that in this case the straight line in the semilog plot crosses the middle part of the vertical bars of possible $m^*(T)$ range. This may indicate that the scattering of the carriers takes place at middle-range optical lattice phonons, with frequencies around 130-160 cm^{-1} (cf. Ref. 194 and Ref. 332, p. 41). The extrapolated m_0^* at $T = 0$ in this case equals 10. The parameter T_o in (6.46) is 43.1 K (30 cm^{-1}). In this case the $m_{eff}(T)$ dependence in Fig. 7.32 can also be linearized in the high-T region ($100 < T < 300$ K) in the log m_{eff} versus log T plot according to formula (7.19). Since in the **c**$'$-direction $n_{c'c'} = 2.15$, the corresponding $(2n_{c'c'}-1)$ value in the power-law (7.19) equals 3.3. This explains why the exponential $m_{eff}(T)$ dependence extends up to ~ 200 K in Fig. 7.32. It may be shown that the exponential function of $m_{eff}(T)$ (6.46) and the power-law function of $m_{eff}(T) \sim T^{2n-1}$ (7.19) practically

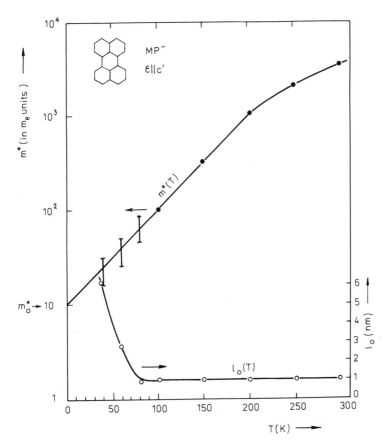

FIGURE 7.32. Estimated values of relative effective mass $m^*(T)$ of a negative charge carrier (MP$^-$) and the mean value of carrier free path $l_o(T)$ in the **c'**-direction of α-perylene as a function of temperature T.[356] The vertical bars indicate the possible range of m^* in the hot carrier regime [formula (7.13)]; the filled circles indicate m^* values of thermalized carriers [formula (7.17)].

coincide for $2n-1 \approx 3$ in the 100–150 K range.

Thus, the deviation of the exponential $m_{\text{eff}}(T)$ dependence at higher temperatures seems to be common for negative charge carriers in α-Pl (see Figs. 7.28 and 7.32) as well as in Nph crystals (see Fig. 7.22).

As can be seen from Fig. 7.32 in the **c'**-direction the carrier mean free path l_o remains constant $l_o \approx c'_o = 1.1$ nm over a wide T range down to 80 K. Only below 80 K an increase in the l_o value was observed. Thus, a small value of l_o, practically equal to the lattice constant, is a general property of polyacenes and of α-Pl over a wide temperature range. This characteristic feature should be taken into account in new trials of developing more realistic transport models of charge carriers in organic molecular crystals.

7.2.3.G. Positive Charge Carrier Transport in an α-perylene

In contrast to the activationless mobility temperature dependences of the type $\mu^- \sim T^{-n}$, characteristic for negative charge carriers, the positive carriers in α-Pl exhibit thermally activated mobility $\mu^+(T)$ dependence. Karl and coworkers[155] have found that the $\mu^+(T)$ value in an oblique direction $\mathbf{L}$ of the crystal ($\mathcal{E} \| L$) increases with temperature. (The L-direction forms the following angles with the main crystallographic axes: $\angle \mathbf{L,a} = 54.5°$, $\angle \mathbf{L,b} = 90°$, and $\angle \mathbf{L,c'} = 35.3°$). However, the $\mu^+(T)$ dependence in this case is not a simple exponential $\mu^+(T) \sim \exp[-E_a/kT]$. The activation energy E_a actually decreases with increasing temperature and practically reaches zero value at 300 K. The mobility μ^+ is small, on the order of 0.1–0.3 cm^2/Vs in the 120–300 K region[154,155] (see Section 7.1.3). Very often such a thermally activated mobility in organic crystals is caused by multiple shallow trapping due to the presence of chemical impurities or structural defects.[156,157] In such a case, instead of the real microscopic mobility, one observes a small effective trap-limited mobility. However, the α-Pl crystals, used in $\mu^+(T)$ measurements of Ref. 155, were obtained after very careful purification, and the observed mobility should be regarded as a microscopic one. Karl and coworkers[155] have also found similar thermally activated mobility dependence for the principal tensor components μ_1^+ and μ_2^+ in the case of positive charge carrier transport in the 2,3-dimethylnaphthalene and for the mobility $\mu_{c'c'}^-$ of negative carriers in the $\mathbf{c'}$-direction of a pyrene crystal (see Section 7.1.3). In all these cases the mobility value is small, below 0.4 cm^2/Vs. This means that marked localization of charge carriers prevails and, consequently, the mean residence time of a carrier on a particular molecule may be long enough to form a lattice polaron.[155,351] Since, however, the activation energy E_a of mobility changes with temperature, a simple small-lattice polaron model may not be valid and one should try to search for a modified, more universal polaron-type hopping transfer mechanism (see Ref. 251).

7.3. MODEL DESCRIPTION OF CHARGE CARRIER TRANSPORT PHENOMENA IN OMC. ADVANTAGES AND LIMITATIONS

Summarizing the main results of this chapter, we shall show that the model description of the carrier transport phenomena possesses certain advantages over analytical transport theories, but has also some inherent shortcomings.

Actually, we have used in our studies several types of phenomenological models.

First, some considerations will be given to the modified SM model,[344] successfully used for computer simulation of the experimental carrier transport characteristics in the previous section. The main advantage of this model is, in our opinion, that it provides possibility of calculating the mean trajectories and drift velocities of charge carriers and simulate in a computer experiment practically all reported charge carrier mobility data[33,34,154,155,158,406,407] in polyacene and perylene crystals. It also allows to visualize the conditions for hot carrier generation and predict the evolution of carrier *mean* trajectories and velocities under the influence of

Coulombic and external electrical fields, as well as to evaluate the statistical dispersion of these transport parameters due to stochastic forces.

Second, to give a physical interpretation of the simulation data we have introduced, in a phenomenological approach, a model of adiabatic nearly small MP.[334,344,348,355,356] The present successful simulation of reported experimental mobility data (Section 7.2) demonstrates the plausibility of the MP model approach for a phenomenological description of the charge carrier separation and transport phenomena in polyacene and perylene crystals over a wide temperature (4.2–300 K) and electric field ($\mathcal{E} = 10^3 - 10^6$ V/cm) range.

In this approach, a charge carrier emerges as a heavy quasiparticle with temperature-dependent effective mass $m_{eff}(T)$, which can be described by an exponential [Eq. (6.46)] or a power-law [Eq. (7.19)] function. It has been demonstrated that the carrier's effective mass $m_{eff}(T)$ is actually a tensor, depending on the carrier's sign and the direction of transport in the crystal.

It has been shown in Section 3.6 that the nearly small MP may be formed as a result of the interaction of a charge carrier with intramolecular vibrations of the molecule on which it is localized during the residence time, and also with polar, infrared (IR)-active modes of the nearest-neighbor molecules (see also Refs. 334, 344, 348, 355, and 356).

In Section 3.6.2, the conditions of MP formation and transport have been analyzed in the framework of a model Hamiltonian approach. It has been shown that one of the necessary conditions for the transport of a MP is the existence of a slight dispersion of the vibrational modes, i.e., a dependence of the coupling constant $g_{\lambda n}(\mathbf{k})$ and vibron energy $h\nu(\mathbf{k})$ on the wave vector $\mathbf{k}$ (see also Refs. 339 and 351).

Thus, the nearly small MP may be regarded as a slightly delocalized ionic state in a neutral crystal which moves by stepping via tunneling from one crystal site to another (see Fig. 3.10).

These model descriptions of hot charge carrier thermalization and MP formation processes has been recently theoretically substantiated by Čápek.[59a] This author has proposed a microscopic theory which considers combined effects of hot electron relaxation and transfer processes during the thermalization. The theory is based on time — convolutionless GME approach and demonstrates an interplay of polaron cloud formation, i.e., relaxation and gradual loss of coherence of carrier propagation with increasing time as the hot carrier "cools" down. (see Section 5.15). This approach extends the quasi-classical interpretation of the generalized Fokker-Planck equations in terms of the SM's model (see Section 6.3.1) to the domain of quantum mechanics.

It has been demonstrated in Section 7.2 that for the low-T region of carrier transport the best approximation is provided by the exponential $m_{eff}(T)$ dependence [Eq. (6.46)]. It is shown to be valid both for the positive and the negative charge carrier transport in Nph crystals (see Figs. 7.15, 7.16, and 7.22) and for the negative charge carriers in α-Pl crystals (see Figs. 7.28 and 7.32).

In this connection one should remember that in the early stages of the development of small-lattice polaron theories, an expression of the m_{eff} dependence on temperature, equivalent with Eq. (6.46), was proposed by Yamashita and Kurosawa[416] and Sewell[324] (see also Ref. 12) assumed to be valid for a polaron band approach description of charge carrier transport at low temperatures.

In Section 1B.8 of Chapter 1, the exponential $m_{\text{eff}}(T)$ dependence has been theoretically derived for a nearly small MP [see Eq. (1B.8.18)].

Actually, in our case, a band model description of charge carrier transport may be physically reasonable only at the lowest temperatures when the mean free paths l_o of the carrier is of the order of at least several lattice constants $l_o > a_o$ (see Figs. 7.15, 7.16, 7.22, 7.28, and 7.32). As the temperature increases, the effective width δE_p of the polaron band drastically decreases, inversely to the exponentially growing effective mass of the carrier, namely, $\delta E_p = 1/m_{\text{eff}}$.

Consequently, the charge carrier becomes strongly localized, and the band model approach ought to be replaced by a corpuscular representation of the charge carrier.

It should be emphasized once more that the most characteristics feature of both the positive and negative charge carrier mobility in polyacene crystals and negative charge carrier in α-Pl crystals is the lack of thermal activation yielding a decreasing $\mu(T)$ dependences with increasing temperature according to a typical power law $\mu(T) \sim T^{-n}$, where n may assume a wide range of values ($n = 0$–2.9), different for negative and positive charge carriers and for different directions of movement in the lattice (see Tables 7.1 and 7.2). Similar $\mu(T)$ dependences have been observed for a number of other aromatic and heterocyclic OMC (see Ref. 154 and references therein). It has sometimes been argued that such negative $\mu(T)$ dependence speaks in favor of a band-type, coherent carrier motion, similar in nature to transport in conventional inorganic semiconductors. However, such a simple assumption is in contradiction to other important transport features in OMC.

First, the mean free path l_o of carriers over a wide T regime (150–300 K in polyacenes and 80–300 K in α-Pl), where the $\mu(T) \sim T^{-n}$ dependence holds, is of the order of one lattice constant ($l_o = 0.5$–1.0 nm) (see Figs. 7.15, 7.16, 7.22, 7.28, and 7.32). This means that the carrier is scattered practically at every lattice site and the band theory transport model, according to criterion ($l_o \approx a_o$) (see 3.2.3) is not valid for description of such type of motion.

The computer simulation results in Section 7.2 unequivocally demonstrate that the apparent contradiction of the two main carrier transport characteristics, namely, $l_o \approx a_o$ and $\mu(T) \sim T^{-n}$ in the high-T region may be solved in the framework of the nearly-small MP model which predicts the carrier transport by stepping via tunneling from one crystal site to another without any activation energy.

The simulation also shows (see Section 7.2) that in the high-T regime the power-law $m_{\text{eff}}(T)$ dependence (7.19), which has been derived in terms of corpuscular charge carrier representation, is the most appropriate approximation function.

Thus, we can conclude that in the low-T region, where indeed $l_o > a_o$, the carrier may be regarded as a partially delocalized quasiparticle, and one can think in terms of a "polaron-band" model. It is appropriate to apply, in this temperature range, the exponential $m_{\text{eff}}(T)$ dependence [see Eqs. (1B.8.18) and (6.46)].

On the other hand, in the high-T region where $l_o \approx a_o$, the carrier should be regarded as a ballistic particle and described in the framework of corpuscular representation. In this temperature region the power-law $m_{eff}(T)$ dependence (7.19) seems to be physically reasonably sound since it has been derived in terms of such representation.

Thus, in our opinion, the nearly small MP approach reconcile both the low- and high-temperature transport regimes.

As shown in Chapter 4, the molecular polaron approach is consistent with the energy structure of ionized states in polyacene crystals, since it has been shown that the conductivity levels of thermalized carriers in these crystals are not those of free electrons but relaxed molecular polaron states separated by an adiabatic energy gap E_{Ad} (see also Refs. 334, 344, 351, and 307).

The results presented above may serve as a phenomenological and conceptual basis for further development of a comprehensive and self-consistent nearly small MP theory — a challenge to theoreticians working in the polaron field (see Sections 1B.8 and 3.6.2)

An interesting alternative approach for the description of charge carrier transfer in organic solids has been proposed by Larsson et al.[209a,209b] The authors try to apply time-dependent quantum chemical kinetic theories based on generalized Marcus charge transfer models (see Refs. 223a and 223b). It may be shown in terms of this approach that in a situation when the so-called Marcus reorganization energy λ [which is equivalent to the molecular ion formation energy $(E_b)_i$, see p. 130] is smaller than the term $\Delta = 2t = 2J_{ij}$, where J_{ij} is the electron transfer integral, i.e., if $\lambda = (E_b)_i < \Delta = 2J_{ij}$, the electron transfer from molecule i to molecule j takes place without an activation energy, and, vice versa, if $\lambda > 2J_{ij}$, the activation energy of the order of $\lambda/4$ is needed for electron transfer. The first criterion describes semiquantitatively the activationless electron transfer in the ab plane of Ac-type crystals, while the second criterion $\lambda > 2J_{ij}$—the electron transport conditions in the c' direction (see Sec. 7.2.3C)—may be regarded as complementary to more conventional polaron transport theories.

The specific cases of thermally activated carrier transport, discussed in Section 7.2.3.G, apparently require a novel small-lattice polaron model approach, taking into account the temperature dependence of the activation energy $E_a = E_a(T)$ for polaron hopping.

However, at this point we should indicate that the contemporary theories of the hopping transport do not necessarily yield temperature-independent activation energy E_a; other activation forms are also possible (see, e.g., Ref. 329 and Sections 5.9 and 5.13).

We also wish to discuss briefly the phenomenon of anomalous temperature-independent mobility of negative charge carriers over a wide T range (100–300 K) in the c'-direction of Nph and Ac crystals, described in Section 7.2.3.C. A number of authors have argued about the possible physical reasons for such anomalous mobility behavior (for references and details see Ref. 332, p. 37 and Ref. 288, p. 360).

Recently Kenkre *et al.*[170] have proposed, according to the authors, a new unified theory of the mobilities of photoinjected electrons in a Nph crystal, which is based on polaronic description of charge carriers, as suggested by Silbey and Munn.[329] This theory provides a quantitative description of the T-dependence of the drift mobilities μ_{ij} of electrons in three directions of the Nph crystal (including c'-direction, see Fig. 7.5) over a wide range of temperatures, as reported in Ref. 155. The main conclusion of the theory[170] is that the photoinjected electrons in a Nph naphthalene crystal move as polarons in all crystal directions over the whole temperature range.

The theory proposed by Kenkre *et al.*,[170] although developed for a specific mobility case in a Nph crystal, is, in general, in good accordance with the phenomenological polaron models developed in the present work.

On the other hand, Karl[155] has proposed a simple empirical approach, according to which the effective nearly T-independent carrier mobility in the c'-direction of Nph may be presented as a superposition of the components of two independent but differently temperature dependent $\mu(T)$ tensors: $\mu_1 = a T^{-1.48}$ and $\mu_2 = b \exp(-E_a/kT)$. We have extended this approach in terms of polaron models (see Section 7.2.3.C) assuming that these two mobility $\mu(T)$ components may be regarded as a manifestation of two different, coexisting transport mechanisms: a nearly small MP, which moves by stepping via tunneling without activation energy, displaying a typical $\mu(T)$ dependence, $\mu_{MP} = a T^{-n}$; and a small-lattice polaron which moves by thermally activated hopping and thus exhibits a typical exponential mobility dependence (see also Ref. 355). However, in this case, a more detailed and comprehensive polaron theory is required.

Finally, we should emphasize, that our studies demonstrate the plausibility of the MP approach for a description of hot carrier creation and transport phenomena (see Section 5.15). Both experimental and our simulation data show that there exists a strongly temperature-dependent critical electric field value $\mathcal{E}_c$ at which hot carriers can be created. However, the possible theoretical implications of the proposed empirical formula (7.22) of the observed $\mathcal{E}_c(T)$ dependence are not yet clear and require further studies.

7.4. IMPACT OF LOCAL TRAPPING STATES IN REAL OMC ON THE CHARGE CARRIER TRANSPORT CHARACTERISTICS

As shown in Section 7.1.2, charge carrier traps (see Sections 4.6.1 and 4.6.2) are omnipresent under normal conditions in any real organic crystal, and it is extremely difficult to get beyond impurity-limited transport of the carriers.[157]

One of the most sensitive transport characteristic of charge carriers is the TOF current pulse (see Section 7.1.2 and Figs. 7.1 and 7.2). The presence of deep charge carrier traps leads to rapid decay of the current pulse before the carriers reach the opposite electrode (see Fig. 7.1). This is the case when the trap residence time is long compared with the trap-free microscopic transit time $(\tau_{tr})_0$ and therefore the

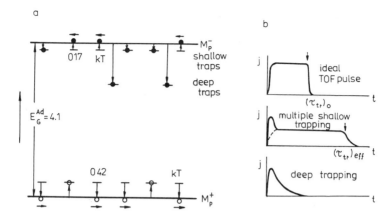

FIGURE 7.33. (a) Schematic diagram of shallow and deep charge carrier traps in Ac-type crystal (see Fig. 4.24), (b) ideal time-of-flight current pulse, multiple shallow trapping limited pulse (an initial generation of free charge carriers of both signs has been assumed for the solid curve), and pulse shape under deep trapping conditions; no arrival time kink can be identified in the latter case (according to Ref. 157).

determination of τ_{tr} is impossible (see Fig. 7.33a).

The presence of shallow traps leads to multiple trap-limited diminished current pulse amplitude and an increased thermally activated effective transit time $(\tau_{tr})_{eff} > (\tau_{tr})_0$ which are caused by a sequence of trapping and thermal release events, with only a fraction of charge carriers being momentarily free to move.[157] This situation, when the trap residence time is smaller than $(\tau_{tr})_0$ is shown schematically in Fig. 7.33b.

An ideal TOF current pulse is next to impossible to observe in practice and one can only approach it under conditions of minimal (as far as possible) trap density in the crystal (see Fig. 7.2).

Karl[157] has recently performed kinetic analysis and calculated shallow trap limited effective mobilities μ_{eff} of charge carriers in organic crystals.

As one might expect from condition (7.8) and related arguments (see Section 7.1.2) the temperature dependence of effective mobility $\mu_{eff} = \mu_{eff}(T)$ should reveal a maximum at T_m, dependent on the density N_t of traps and their trapping depth E_t.

As shown in Ref. 157, in the high-temperature region for $T > T_m$ the microscopic mobility is dominant, and $\mu(T)$ follows the usual for aromatic crystals T dependence, namely

$$\mu_{eff} = \mu_o = aT^{-n}. \tag{7.25}$$

In the low-temperature region, below T_m, the effective mobility decreases according to the expression

$$\mu_{eff} \sim T^{-n} \exp[-E_t/kT] \tag{7.26}$$

in which the exponential term of thermally activated mobility dominates and, consequently, μ_{eff} becomes smaller than the microscopic mobility μ_o, namely, $\mu_{\text{eff}} < \mu_o$. The $\mu_{\text{eff}}(T)$ curves, calculated in Ref. 157 as dependent on the parameter E_t, are shown in Fig. 7.34. Similar $\mu_{\text{eff}}(T)$ dependences have been obtained with the density N_t of the trapping states as a parameter.[157] These dependences demonstrate that with decreasing trapping depth E_t and trap density N_t the limiting temperature T_m of microscopic mobility $\mu_o \sim T^{-n}$ is steadily shifted to lower temperatures. These computed data have been confirmed experimentally (see, e.g., Fig. 7.10, or Fig. 13 in Ref. 157).

These extremely drastic conditions illustrate how difficult is it to obtain microscopic mobility dependences $\mu_o = \mu_o(T)$ down to liquid helium temperatures (see Section 7.1.3).

On the other hand, Kotani[196] has recently demonstrated, that the presence of antitraps, e.g., 5% of 2,3-dimethylnaphthalene (DMN) in the Ac host crystal only slightly lowers the mobility of electrons and holes, not changing the character of its $\mu(T)$ dependence.

Charge carrier transport in *disordered* organic solids, e.g., molecularly doped polymers and related disordered organic photoconductors, has been discussed extensively in recent review articles by Baessler[15a] and by Borsenberger, Margin, Van der Auweraer, and Schryver.[35c] It has been shown that in disordered materials the charge carrier transport occurs by hopping through the Gaussian distribution manifold of local states with superimposed positional disorder (see Sec. 4.6.1). In this case the carrier drift mobilities μ_d are extremely low and their transport exhibits *dispersive* (non-Gaussian) character. The distribution width of the hopping site manifold increases with increasing permanent dipole moments of the material and causes further decrease of the carrrier mobility.[35b]

7.5. POSSIBLE CHARGE CARRIER TRANSPORT MECHANISMS IN LB MULTILAYERS

We have discussed briefly in Section 4.7 the unusual energy structure of LB multilayers of vanadyl phthalocyanine (VOPhc) (see Fig. 4.26), consisting of a sequence of quantum wells separated by isolating aliphatic spacers.

The charge carrier transport in such a multilayer assembly presumably occurs by electron tunneling from the conduction level of one potential well directly to the neighboring one[353] (see Fig. 4.26c).

The existance of this kind of charge carrier transport mechanism in LB films has been confirmed experimentally by studies of electron tunneling via VOPhc LB monolayers.[96]

It has been shown (see Ref. 353) that at electric fields below $\mathcal{E} \leqslant 10^6$ V cm^{-1} the integral current via VOPhc LB films is actually trap-controlled (see Fig. 4.26c), and, to keep the current flowing, a conductivity activation energy E_a is needed.

However, at higher electric fields ($\mathcal{E} > 10^6$ V cm^{-1}) the authors of Ref. 353 observed an interesting reversible field effect. At a definite critical field $\mathcal{E}_c$ the E_a

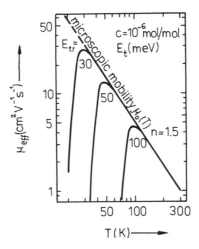

FIGURE 7.34. Calculated effective mobilities μ_{eff} as a function of temperature in a $\log(\mu)/\log(T)$ plot for a trap concentration of 10^{-6} mole/mole, and detrapping activation energy $E_{tr} = 30$, 50, and 100 meV; $\mu_0 \approx T^{-1.5}$ (according to Ref. 157).

value steeply decreases to zero, and the conductivity thus becomes quasi-"trap-free."

However, it has been shown in Ref. 353 that the effect of $E_a \to 0$ is not connected with dielectric breakdown of the film. The $E_a(\mathcal{E})$ dependences are completely reversible and reproducible, since the real breakdown fields of the film are very high and lie above $\mathcal{E} > 10^7$ V cm^{-1}.

A possible explanation of the onset of a trap-free regime at $\mathcal{E} > \mathcal{E}_c$ may be the following. It may be shown that at field strength $\mathcal{E} = 10^6$ V cm^{-1} the potential shift $\Delta\varphi$ would be approximately 0.5 eV over 50 Å. That means that a local trapping state E_t below one potential well would be at same level as the conduction level of the neighboring well, and charge carriers might be tunneling from this trapping state directly to the conduction state of the film (see Fig. 4.26c). This means that the trapping states may be depopulated by an appropriate field value thus providing an onset of a trap-free regime. This effect should not be temperature-dependent.

However, the authors of Ref. 353 observed that the $\mathcal{E}_c$ value actually is temperature-dependent and decreases with decreasing T. Therefore it was suggested that this reversible effect may serve as evidence that at $\mathcal{E} > \mathcal{E}_c$ hot, nonthermalized charge carriers are created with mean drift velocity v_d greater than the mean thermal velocity v_{th} of the carrier. It has been shown above, both experimentally and by computer simulation (see Section 7.2.2) that one of the characteristic features of hot carrier generation in organic crystals is the T-dependence of $\mathcal{E}_c$, namely, $\mathcal{E}_c \sim \exp a\sqrt{T}$. In the case of VOPhc LB films this $\mathcal{E}_c(T)$ dependence roughly holds. This may serve as indirect evidence of hot carrier high-field-induced generation in the LB films. This assumption has been proved by preliminary computer simulation.[353] As shown in Section 7.3, the computer simulation approach has been

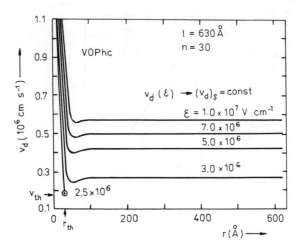

FIGURE 7.35. Calculated mean values of the charge carrier drift velocity v_d as a function of electric field $\mathcal{E}$ in an VOPhc LB multilayer assembly (l = 630 Å, n = 30); T = 300 K; E_k = 1.3 eV; m_{eff} = 1020 m_e; β = 2×10^{13} s^{-1}.[353]

successfully used for model description of charge carrier transport and hot carrier generation processes in polyacene and α-Pl crystals, based on carefully measured experimental characteristics (see Section 7.1.3). However, in case of LB films not all necessary experimental input data were available for such simulation. Nevertheless, some rough estimates could be obtained using approaches analogous to those of model description of carrier transport in molecular crystals.

In Ref. 353 the simulation was performed for VOPhc LB film with the number of layers n = 30 and thickness l = 630 Å. It was assumed that the saturation value of the drift velocity of the hot carrier in an LB film $(v_d)_s$ is approximately equal to the observed $(v_d)_s$ value in Ac crystals at room temperature, namely, $(v_d)_s$ = 6×10^5 cm s^{-1} (see Sections 7.2.3D and Fig. 7.25). It was also assumed that the inelastic scattering of hot carriers takes place mainly on the C-H stretching modes, of aliphatic chains at $\hbar\omega$ = 0.373 eV, as proved by inelastic electron tunneling spectroscopy measurements of VOPhc monolayers.[96]

Under such assumptions one can evaluate the effective mass m_{eff} of the carrier and its relaxation rate constant β at the given drift velocity v_d and field strength $\mathcal{E}$. The obtained m_{eff} and β values were the main input parameters for the simulation. Figure 7.35 illustrates the calculated $v_d(\mathcal{E})$ dependences for the VOPhc LB film.

As can be seen from Fig. 7.35, at field values $\mathcal{E}$ = $\mathcal{E}_c \geqslant$ 3×10^6 V cm^{-1}, hot charge carriers actually can be created with $v_d(\mathcal{E}) > v_{th}$. These data are in good agreement with experimental $\mathcal{E}_c$ values at which $E_a \rightarrow 0$ (see Ref. 353). On the other hand, a hot carrier with kinetic energy E_k = 1.3 eV, injected at $\mathcal{E} \leqslant$ 2.5×10^6 V cm^{-1}, thermalizes moving along the field at the mean thermalization distance r_{th} = 27 Å (see Fig. 7.35) obtaining a mean thermal velocity v_{th} = 1.8×10^5 cm s^{-1}.

The simulation approach[353] allows one to estimate also the possible time-of-flight τ_{tr} of a hot carrier across an VOPhc LB film ($n = 30$, $l = 630$ Å).

It was shown that τ_{tr} decreases with increasing electric field and obtains the value of ~ 10 ps at $\mathcal{E} = 1 \times 10^7$ V cm^{-1}. Thus, the transfer time of a carrier across a monolayer is very short indeed and lies in the subpicosecond domain, in agreement with indirect estimates (see Ref. 98). This may be of great importance for future application of LB films in molecular electronics.

It may be concluded from the above results that the feasibility of electric field induced hot charge carrier generation is a general property of both molecular crystals and LB molecular assemblies. However, it takes place only at field strength $\mathcal{E} \geqslant \mathcal{E}_c$, characteristic for a given temperature.

CHAPTER 8

Epilogue: Summing Up and Looking Ahead

I invented for them the art of numbering, the bases of all sciences,
and the art of combining letters, memory of all things,
mother of the Muses and the source of all the other arts.

AESHYLUS, PROMETHEUS BOUND

The everlasting universe of Things
Flows through the Mind, and rolls its rapid waves,
Now dark — now glittering—now reflecting gloom —
Now lending splendor, where from secret springs
The source of human thought its tribute brings.

PERCY BYSSHE SHELLY

The authors have tried, in the present work, to give a comprehensive treatment of the physical principles underlying the dynamics of excitonic and electronic processes in OMC in an attempt to summarize the present state of experimental and theoretical research in this field.

First, we should reiterate that OMC, due to weak molecular interaction, possess a number of specific features, making them essentially different from covalent or ionic crystals, as illustrated in Table 1.1. Hence, they cannot be described in the framework of traditional solid-state physics.

In the exposition of the material under discussion we have tried, as the reader will notice, to keep a balance between theoretical, analytical approaches and phenomenological, model-type descriptions.

Partly, for tutorial purposes, we have chosen the two levels of presentation intentionally, to satisfy both theorists and experimental physicists, as well as people from interdisciplinary areas. Such a structural and conceptional organization of the book reflects, to some extent, the scientific backgrounds and dispositions of the authors.

However, it should be emphasized that both approaches — rigorous analytical theory and phenomenological treatment, based on model-type descriptions — may be regarded as a natural counterparts of cognitive research, particularly in new fields of science still in the developmental stage.

The above statement can be illustrated by some historical aspects of introduction

378

of electronic polarization and electronic polaron concepts in organic solid-state physics.

The theory of electronic polaron by Toyozawa[386] and a phenomenological model of electronic states in organic crystals, taking into account the many-electron interaction phenonemon of electronic polarization, known as Lyons model,[122,220] appeared almost simultaneously in the sixties (see Chapter 3). The Lyons model was further extended and later modified by a number of authors (see monographs[288,332,351] and Chapters 3 and 4) and at present has become a conceptual cornerstone for interpretation and calculation of energy level structure of electronic states in perfect and imperfect organic crystals.

In the framework of a modified phenomenological approach microelectrostatic self-consistent polarization field methods were developed which allowed one to calculate with sufficiently high accuracy the electronic polarization energies and construct the electronic energy state diagrams in polyacene crystals. In this context, one should especially mention the refined polarization energy calculations by Munn and his group (see Sections 3.5.3–3.5.5), using Fourier transformation method and submolecular approach, including charge-quadrupole interaction terms.

These refined calculations made it possible to reevaluate the earlier polarization energy data and take into account certain asymmetry of positive and negative polaron states in the energy diagram of a crystal (see Sections 3.5.6 and 4.5).

However, the splendid achievements of the microelectrostatic treatment of polarization energies could not conceal the fact that a real, dynamic theory of the electronic polaron had remained behind the scene.

This was noticed by Sir Nevill Mott who indicated that "one cannot help feeling that there is a gulf between those who work in polaron theory, particularly large polarons, and investigators of conduction in organic crystals."[246] In Section 1B of Chapter 1 we have tried "to bridge this gulf." In Chapter 1, Section 1B we made an attempt to formulate a more generalized theory of OMC based on Hamiltonian description of all kinds of interaction of charge carriers and excitons with surrounding lattice, including formation of electronic and lattice polarons. This generalized treatment, as the reader will have noticed, has further served as a conceptual basis for analysis and interpretation of more specific cases. It is important to reiterate that in terms of such generalized approach it was possible to show that the electronic polarization is responsible for the formation both of electronic and excitonic polarons, as well as the formation of nonvalent dispersion-type molecular interaction forces. Thus, electronic polarization seems to be the dominant inherent property of organic crystals, emerging in all major interaction processes.

When speaking about specific cases of application of general theory, one should remember that the dynamic treatment of electronic polarization has been shown to be practically equivalent to advanced microelectrostatic approaches, if small quantum corrections are taken into account (see Section 3.4).

In this case the correspondence between dynamic and microelectrostatic approaches demonstrates, in our opinion, the complementary nature of analytical and phenomenological treatment of the problem. More controversial is another specific interaction problem, namely, the exciton self-trapping phenomenon. It has been traditionally treated in terms of the adiabatic approximation. However, recently the validity of the adiabatic approximation has been seriously questioned (see Section 2.2). It has been argued that one should, at least at the present stage of knowledge, be critical about any theory of the exciton-phonon interaction dynamics which is formulated in the adiabatic representation and is connected with the adiabatic approximation. Due to this argument, we have preferred to use the crude (rigid) approximation throughout practically the whole book. Obviously, this problem requires further extended theoretical studies.

Summing up the main results of the energy structure studies of electronic states in an ideal organic crystal, we should like first to remind the reader that the most important breakthrough in the field was connected with the experimental determination of the optical E_G^{Opt} and adiabatic E_G^{Ad} energy gaps in polyacene crystals (see Section 4.1-4.3). These results unequivocally demonstrate that the conduction levels in OMC are the self-energy states of vibrationally relaxed electronic polaron, i.e., the states of a molecular polaron. In terms of a model description, a nearly small MP may be envisaged as slightly delocalized ionic state in a neutral crystal which is formed as a result of interaction of charge carrier with intramolecular vibrational modes of the molecule, on which it is localized during the residence time, as well as with polar IR-active modes of the nearest-neighbor molecules (see Section 3.6.1 and Fig. 3.10).

The MP model has been confirmed by computer simulation data (see Sections 6.3; 7.2, and 7.3) and a generalized lattice polaron theory based on Hamiltonian representation (see Chapter 1, Section 1B.8). Particular conditions of nearly small polaron formation and activationless transport have also been analyzed in the framework of a model Hamiltonian approach (see Section 3.6.2).

Thus, the plausibility of the existence of molecular polarons in Ac-type crystal has been unambiguously confirmed, both experimentally and in terms of phenomenological models and analytical polaron theories. However, the development of a more general and comprehensive nearly small MP theory is still required.

We hope that such fruitful results obtained for polyacene crystals, will stimulate future studies of other classes of organic crystals.

Considerable advancement has also been achieved in deeper understanding of the photogeneration micromechanisms in OMC (see Chapter 6). It has been shown that charge carrier photogeneration and separation phenomena in OMC are highly complicated and proceed via multistep processes.

At the near-threshold spectral region of intrinsic photogeneration there emerge two competitive photogeneration mechanisms, namely, AI of excited molecular states and direct optical charge transfer on nearest-neighbor molecules. At higher excitation energies the AI mechanism becomes dominant and appears in two possible modes, i.e., as direct ionization of photoexcited neutral molecular states, and autoionization from fixed electronic levels after intramolecular relaxation.

Experimental and computer-simulated studies confirm the viability of the modified ballistic SM model and an extended Onsager's approach for description of photogenerated charge carrier thermalization and separation processes. They also confirm the plausibility of the MP model for interpretation of these phenomena. However, there still remains some uncertainty in interpretation of the very first step of photogeneration, namely, the act of AI. Here, it might be viable to involve a "hybrid" approach of quantum and classical representations. Some recent achievements in this field are briefly covered in Section 5.15.

The charge carrier transport phenomena belong to the most complicated electronic processes in organic crystals and, at the same time, are highly controversial with regard to their interpretation.

We have discussed in detail the recent achievements in the field of transport theories, based on the generalized stochastic Liouville equation and GMEs, as well as improvements and refinements of approaches and their relations to some earlier transport theories (see Chapter 5).

These contemporary generalized approaches allow one to describe both the coherent (bandlike) and diffusive motion of excitons and charge carriers in a framework of a single unified model, as well as time- and temperature-dependent dynamics of transitions from one mode of transport to the other one. However in this field there remain a number of unsolved or only partially solved problems, as formulated in Chapter 5.

Some principal controversies have been connected with charge carrier transport phenomena in Ac-type crystals (see Chapter 7).

First, there is the so-called charge carrier "mobility puzzle." This puzzle arises from the empirical fact that, on one hand, the mean free path l_o of the carrier in Ac-type crystals from room temperature down to $T \approx 150$ K is actually of the order of lattice constant a_o ($l_o \approx a_o$) and strongly suggests a hopping model approach. On the other hand, the typical $\mu(T)$ dependences in polyacene and α-Pl crystals $\mu \sim T^{-n}$ are often supposed to speak in favor of some band-type carrier transport.

The other problem is connected with the polaron nature of charge carriers in OMC, the fact, that they are charged quasiparticles. The present transport theories have been developed mainly for neutral quasiparticles, namely, excitons. In case of charge carriers the situation is more complicated. The motion of a polaron as a charged quasiparticle is influenced by the intrinsic Coulombic and the external electric fields, as well as by stochastic forces causing diffusive random walk. Therefore the motion of a polaron should be, as the first approximation, described in terms of a generalized Fokker-Planck equation the solution of which may yield the distribution function of position and velocity of the particle under investigation.

However, as already mentioned, it is difficult to find such a solution in explicit analytical form.

Instead, we have used a modified SM approach, based on the Fokker-Planck equations, which allows, in the framework of a model description, to find, in simulation procedure, the mean values of polaron positions and velocities, as well as their statistical distribution and evaluation in time (see Section 7.2). From this point of view, a model description of charge carrier transport phenomena provides certain advantages over generalized analytical theories (see Section 7.3). We have demonstrated that in terms of such a phenomenological model it is possible to perform computer simulation of reported experimental charge carrier transport characteristics and provide a quantitative, numerical description of these characteristics (see Section 7.2). In addition, it allows us to visualize the mean carrier trajectories and velocities and their evolution. Thus, computer simulation gives a deeper insight in the physics of the process under investigation.

Finally, it should be stressed that these simulation studies demonstrate once more the plausibility of the MP model for the description of charge carrier separation and transport processes over a wide range of temperature (4.2–300 K) and electric field ($\mathcal{E} = 10^3$–10^6 V/cm) in a framework of a unified theoretical approach. It also allows us to describe the conditions of hot, nonthermalized charge carrier creation, as dependent on electric field and temperature.

The above discussion raises an apparently philosophical question about the advantages and limitations of an analytical and a phenomenological, model descriptions and their respective cognitive power in solving physical problems.

These seemingly opposite cognitive approaches may be formulated as deductive logic and axiomatic thinking versus inductive intuitive physical imagination and "visual" thinking.

In this connection we should like to remind the reader of the epigraph by Louis de Broglie, introducing Chapter 2: "May it not be universally true that the concepts produced by human mind, when formulated in slightly vague form, are roughly valid for reality, but that when extreme precision is aimed at, they become ideal forms whose content tends to vanish away."

This warning by de Broglie is apparently aimed against hyperformalization in scientific approaches, particularly analytical ones. According to the Louis de Broglie credo, as cited in his famous book "Sur les sentiers de la science," "One cannot say that rigorous axiomatic theories are fruitless; however, in general, they rarely promote a real breakthrough in science. The main reason is that axiomatic methods actually try to avoid inductive intuition — the sole method which allows to get over the frontiers of already known; axiomatic method may be good for classification or tutorial purposes, however, it is not a method of discovery."

Another main "architect" of the modern quantum physics, Niels Bohr, is often named as a brilliant example of intuitive, visual thinking. Bohr's favorite definitions of "a great truth" as "a truth whose opposite is also a great truth" and "No progress without a paradox," obviously helped him in "jumping out" of the axiomatic systems of the classical physics and creation of the new paradigms of quantum physics. On the other hand, overestimated reliance on pure intuition is also dangerous since it sometimes reduces to the declaration "I believe."

There is still another aspect that should be kept in mind. To describe a real physical situation or a set of experimental data, more than one model may be proposed. In such a case one should find additional criteria to choose the most appropriate model. Our experience in this book demonstrates that the two approaches, analytical and phenomenological, should not be regarded as mutually exclusive, but complementary.

In the spirit of the above discussion let us conclude this Epilogue with an inscription from Antoine Pevsner's "Construction in the Third and Fourth Dimension" devoted to Niels Bohr "Author of the Principle of Complementarity": "Contraria Sunt Complementa."

References

1. Abbi S. C. and Hanson D. M.: J. Chem. Phys. *60*, 319 (1974).
2. Abe S.: J. Phys. Soc. Jpn. *57*, 4029 (1988).
3. Abe S.: J. Phys. Soc. Jpn. *57*, 4036 (1988).
4. Abe S.: J. Phys. Soc. Jpn. *59*, 1496 (1990).
5. Abele I. H., Ducmanis D. Z., Muzikante I. J., Taure L. F., and Silinsh E. A. Izv. Acad. Nauk Latv. SSR, Ser. Fiz. Tekh. Nauk *4*, 15 (1989).
6. Abrahams S. C., Robertson J. M., and White J. G.: Acta Crystallogr. *2*, 233 (1949).
7. Agranovich V. M.: Exciton Theory (in Russian). Nauka, Moscow 1968.
8. Agranovich V. M. and Galanin M. D.: Energy Transfer of Electronic Exitation in Condensed Media (in Russian). Nauka, Moscow 1978.
9. Aladekomo J. B., Arnold S., and Pope M.: Phys. Stat. Solidi *B80*, 333 (1977).
10. Aleksandrov S. B., Freimanis J. F., Grishin E. P., Yurel' S. P., and Buhbinder Y. E. : Izv. Akad. Nauk Latv. SSR, Ser. Fiz. Tekh. Nauk *1*, 23 (1978).
11. Aleksandrov V. V., Belkind A. I., Kaulach I. S., Muzikante I. J., Silinsh E. A., and Taure L. F., In: Organic Semiconductors (in Russian), M. V. Kurik, ed. Kiev, 1976, pp. 20–24.
12. Appel J.: Solid State Phys. *25*, 193 (1968).
13. Argyres P. N. and Kelley P. L.: Phys. Rev. *134*, A98 (1964).
14. Arnold S., Fave J. L., and Schott M.: Chem. Phys. Lett. *28*, 412 (1974).
15. Baessler H.: Phys. Stat. Solidi (b) *107*, 9 (1981).
15a. Baessler H.: Phys. Stat. Solidi (b) *175*, 15 (1993).
16. Baessler H. and Killesreiter H.: Phys. Status Solidi (b) *53*, 183 (1972).
17. Baessler H. and Killesreiter H.: Mol. Cryst. Liq. Cryst. *24*, 21 (1973).
18. Baessler H. and Vaubel G.: Phys. Rev. Lett. *21*, 615 (1968).
19. Balescu R.: Physica *38*, 98 (1968).
20. Balode D. R., Murashow A. A., Silinsh E. A., and Taure L. F.: Izv. Acad. Nauk Latv. SSR Ser. Phys. Tekh. Nauk *4*, 51 (1982).
21. Balode D. R., Silinsh E. A., and Taure L. F.: Izv. Acad. Nauk Latv. SSR, Ser. Phys. Tekh. Nauk *1*, 35 (1978).
22. Barvík I. and Čápek V.: Phys. Rev. *B40*, 9973 (1989).
23. Batley M., Johnston L. J., and Lyons L. E.: Aust. J. Chem. *23*, 2397 (1970).
24. Batt R. H., Braun C. L., and Hornig J. F.: J. Chem. Phys. *49*, 167 (1968); Appl. Optics *3*, 20 (1969).
25. Battaglia M. R., Buckingham A. D., and Williams J. H.: Chem. Phys. Lett. *78*, 421 (1981).
26. Belkind A. I. and Grechov V. V.: Phys. Status Solidi (a) *26*, 377 (1974).
27. Belkind A. I. and Kalendarev R. I.: Phys. Status Solidi (a) *14*, 681 (1972).
28. Benderskii V. A.: Zhur. Strukt. Chimii *4*, 415 (1963).
29. Berkina B. I. and Prihotyko A. F., eds: Cryocrystals. Naukova Dumka, Kiev (in Russian), 1983.
30. Bertalanffy L. von: General Systems Theory: Foundations, Development, Applications. Braziller, New York, 1968.
31. Birks J. B.: Photophysics of Aromatic Molecules. New York, London, 1970.
32. Birks J. B. Photophysics of Aromatic Molecules. In: Organic Molecular Photophysics. Wiley, London, etc., 2, 409, 1975.
33. Bitterling K. and Willing F. In: 11th Molecular Crystal Symposium, Lugano, Switzerland, Sept. 30–Oct. 8, 1985, p. 23.
34. Bitterling K. and Willig F.: Phys. Rev. *B35*, 7973 (1987).
35. Blinov L. M. and Kirichenko N. A.: Soviet Solid State Physics *12*, 1574 (1970); *14*, 2490 (1972).

384

35a. Blinov L. M., Palto S. P., and Udal'yev A. A.: Mol. Mat. *1*, 65 (1992).
35b. Borsenberger P. M. and Fitzgerald J. J.: J. Phys. Chem. *97*, 4815 (1993).
35c. Borsenberger P. M., Margin E. H., Van der Auweraer M., and De Schryver F. C.: Phys. Stat. Solidi (a) *140*, 9 (1993).
36. Bounds P. J. and Munn R. W.: Chem. Phys. *44*, 103 (1979).
37. Bounds P. J. and Munn R. W.: Chem. Phys. *59*, 41 (1981).
38. Bounds P. J. and Munn R. W.: Chem. Phys. *59*, 47 (1981).
39. Bounds P. I. Petelenz P., and Siebrand W.: Chem. Phys. *63*, 303 (1981).
40. Bounds P. I. and Siebrand W.: Chem. Phys. Lett. *75*, 414 (1980).
41. Bounds P. J. and Siebrand W.: Chem. Phys. Lett. *85*, 496 (1982).
42. Bounds P. I., Siebrand W., Eisenstein I., Munn R. W., and Petelenz P.: Chem. Phys. *95*, 197 (1985).
43. Braun C. L.: J. Chem. Phys. *80*, 4157 (1984).
44. Bree A. and Lyons L. E.: J. Chem. Soc. 1956, 2662.
45. Brodin M. S., Marisova S. V., and Prihodko A. F.: In: Exitations in Molecular Crystals (in Russian), M. S. Brodin, ed. Naukova Dumka, Kiev, 1973, pp. 50–83.
46. Brown C. J.: J. Chem. Soc. A. 1968, 2488.
47. Brown C. J.: J. Chem. Soc. A. 1968, 2494.
48. Bucholz J. S. and Somorjai G. A.: J. Chem. Phys. *66*, 573 (1977).
49. Buckingham R. A.: Proc. R. Soc. A *168*, 264 (1938).
50. Buckingham R. A. and Corner J.: Proc. R. Soc. A *189*, 118 (1947).
51. Burland D. M. and Konzelmann U., J. Chem. Phys., *67*, 319 (1977).
52. Burstein Z. and Williams D. F.: Phys. Rev. *B15*, 5759 (1977).
53. Camerman A. and Trotter J.: Acta Crystallogr. *18*, 636 (1965).
54. Campillo A. J., Hyer R. C., Shapiro S. L., and Swenberg C. E.: Chem. Phys. Lett. *48*, 495 (1977).
54a. Capra F.: The Tao of Physics. Collins Publishing Group, Glasgow 1983.
55. Čápek V.: Czech. J. Phys. *B28*, 567 (1978).
56. Čápek V.: Czech. J. Phys. *B28*, 773 (1978).
57. Čápek V.: Czech. J. Phys. *B29*, 439 (1979).
58. Čápek V.: Czech. J. Phys. *B34*, 130 (1984).
59. Čápek V.: Z. Phýsik *B60*, 101 (1985).
59a. Čápek V.: Z. Physik B *92*, 523 (1993).
60. Čápek V. and Barvík I.: J. Phys. C: Solid St. Phys. *18*, 6149 (1985).
61. Čápek V. and Barvík I.: Czech. J. Phys. *B37*, 997 (1987).
62. Čápek V. and Barvík I.: J. Phys. C: Solid St. Phys. *20*, 1459 (1987).
62a. Čápek V. and Barvík I.: Phys. Rev. A *46*, 7431 (1992).
63. Čápek V. and Krausová D.: Czech. J. Phys. B *37*, 1201 (1987).
64. Čápek V. and Munn R. W.: Phys. Stat. Sol. (b) *108*, 521 (1981).
65. Čápek V. and Munn R. W.: J. Chem. Phys. *76*, 4674 (1982).
66. Čápek V. and Munn R. W.: Phys. Stat. Sol. (b) *109*, 245 (1982).
67. Čápek V. and Spivcka V.: Czech. J. Phys. B *34*, 115 (1984).
68. Čápek V. and Szöcs V.: Phys. Stat. Sol. (b) *125*, K137 (1984).
69. Čápek V. and Szöcs V.: Phys. Stat. Sol. (b) *131*, 667 (1985).
70. Čápek V. and Szöcs V.: Czech. J. Phys. *B36*, 1182 (1986).
71. Chablo A., Crickshank D. W. J., Hincliffe A., and Munn R. W.: Chem. Phys. Lett. *78*, 424 (1981)
72. Chance R. R. and Braun C. L.: J. Chem. Phys. *59*, 2269 (1973).
73. Chance R. and Braun C. L.: J. Chem. Phys. *64*, 3573 (1976).
74. Choi S. I., Jortner J., Rice S., and Sillbey R.: J. Chem. Phys. *41*, 3294 (1964).
75. Craig D. P. and Walmsley S. H.: Exitons in Molecular Crystals. Theory and Applications. Benjamin, New York, 1968.
76. Cruickshank D. W.: Acta Crystallogr. *9*, 915 (1956).
77. Cruickshank D. W.: Acta Crystallogr. *10*, 470 (1957).
78. Cruickshank D. W.: Acta Crystallogr. *10*, 504 (1957).
79. Cruickshank D. W.: Rev. Mod. Phys. *30*, 163 (1958).
80. Cruickshank D. W.: Tetrahedron *17*, 155 (1962).
81. Cruickshank D. W. and Sparks R. A.: Proc. R. Soc. A *258*, 270 (1960).
82. Davydov A. S.: Theory of Light Absorption in Molecular Crystals (in Russian). AN USSR, Kiev 1951
83. Davydov A. S.: Phys. Stat. Sol. *27*, 51 (1968).

84. Davydov A. S. Theory of Molecular Excitons (in Russian). Nauka, Moscow 1968; English trans. Plenum Press, New York, 1971.
85. Davydov A. S.: Solid State Theory (in Russian). Nauka, Moscow 1976.
86. Davydov A. S. Solitons in Molecular Systems (in Russian). Naukova dumka, Kiev, 1988.
87. Davydov A. S. and Sheka E. F.: Phys. Status Solidi *11*, 877 (1965).
88. Dewar M. J. S.: The Molecular Orbital Theory of Organic Chemistry. McGraw-Hill, New York, 1969.
89. Dewar M. J. S., Hashmall J. A., and Venier C. G.: J. Am. Chem. Soc. *90*, 1953 (1968).
90. Dexter D. L.: J. Chem. Phys. *21*, 836 (1953).
91. Donaldson D. M., Robertson J. M., and White J. G.: Proc. R. Soc. A *220*, 311 (1953).
92. Dow J. D. and Redfield D.: Phys. Rev. B *5*, 594 (1972).
93. Dows D. A., Hsu L., Mitra S. S., Brafman D., Hayek M., Daniels W. B., and Crawford R. K.: Chem. Phys. Lett. *22*, 595 (1973).
94. Druger S. D. and Knox R. S.: J. Chem. Phys. *50*, 3143 (1969).
95. Duke C. B. and Schein L. B.: Physics Today, Feb. (1980).
96. Durandin A. D., Rutkis M. A., and Silinsh E. A. J. Mol. Electron. *7*, 179 (1991).
97. Eagles D. M.: Phys. Rev. *145*, 645 (1968); *181*, 1278 (1969).
98. Eichenberger R., Williy F., and Stovek W.: Mol. Cryst. Liq. Cryst. *175*, 19 (1989).
99. Eiermann R., Parkinson G. M., and Baessler H.: J. Phys. Chem. *87*, 544 (1983).
100. Eisenstein I. and Munn R. W.: Chem. Phys. *77*, 47 (1983).
101. Eisenstein I., Munn R. W., and Bounds P. J.: Chem. Phys. *74*, 307 (1983).
102. Eley D. D., Inokuchi H., and Willis M. R.: Discuss. Faraday Soc. *28*, 54 (1959).
103. Ern V.: Phys, Rev. Lett. *22*, 8 (1969).
104. Ern V., Suma A., Tomkiewicz Y., Avakian P., and Groff R. P.: Phys. Rev. B *5*, 3222 (1972).
105. Fano U.: Phys. Rev. *124*, 1866 (1961).
106. Firsov Yu. A. Small-radius polarons. Transport phenomena (in Russian). In: Polaroni, Yu. A. Firsov, ed. Nauka, Moscow, 1975, pp. 207–423
107. Fischer S. F. In: Organic Molecular Aggregates, P. Reineker, H. Haken, and H. C. Wolf, eds. Springer-Verlag, Berlin-Heidelberg-New York-Tokyo, 1983, pp. 107–119,
107a. Fleming G. R. and Wolynes P. G.: Physics Today, May 1990, p. 36.
108. Förster T.: Ann. Physik (Leipzig) *2*, 55 (1948).
109. Fowler W. B.: Phys. Rev. *151*, 657 (1966).
110. Frenkel J. I.: Collection of Selected Works (in Russian). AN USSR, Moscow, 1958.
111. Fröhlich, H.: Adv. Phys. *3*, 325 (1954).
112. Frolova F. K. and Kurik M. V.: Phys. Stat. Sol. *33*, K99 (1969).
113. Furikawa M., Mizuno K., Matsui A., Tamai N., and Yamazaki I.: Chem. Phys. *138*, 423 (1989).
114. Gailis A. K., Durandin A. D. and Silinsh E. A.: Izv. Acad. Nauk Latv. SSR, Ser. Fiz. Tekh. Nauk *1*, 21 (1989)
115. Geacintov N. and Pope M.: J. Chem. Phys. *47*, 1194 (1967).
116. Geacintov N. and Pope M.: J. Chem. Phys. *50*, 814 (1969).
117. Glaeser R. M. and Berry R. S.: J. Chem. Phys. *44*, 3797 (1966).
118. Gosar P. and Choi S.: Phys. Rev. *150*, 529 (1966).
119. Grabert H. In: Springer Tracts in Modern Physics Vol. 95. G. Hohler, ed. Springer, Berlin-Heidelberg-New York, 1982, p. 46–68
120. Griffith C. H., Walker M. S., and Goldstein P.: Mol. Cryst. Liq. Cryst. *33*, 149 (1976).
121. Grover M. and Silbey R.: J. Chem. Phys. *54*, 4843 (1971).
122. Gutman F. and Lyons L. E.: Organic Semiconductors, Wiley, New York, 1967.
123. Hadni A., Wyncke B., Horlot G., and Gerrbaux X.: J. Chem. Phys. *51*, 3514 (1969).
124. Hager. J., Smith M. A., and Wallace S. C.: J. Chem. Phys. *83*, 4820 (1985); *84*, 6771 (1986).
125. Haken H.: Advances in Synenergetics, 3rd ed. Berlin, Heidelberg, etc., Springer Verlag, 1983.
126. Haken H., Reineker P.: Z. Physik, *249*, 253 (1972).
127. Haken H. and Schottky W.: Ztschr. Phys. Chem. *16*, 218 (1958).
128. Haken H. and Strobl G.: Proc. Intern. Symp., American University, Beirut, Lebanon, 1967. A. B. Zahlan, ed. University Press, Cambridge, UK, 1967, pp. 311–214
129. Harison W. A.: Solid State Theory. McGraw-Hill, New York, 1970.
130. Hayashi S., Okada I., Ide N., and Kojima K. J. Phys. Condens. Matter *4*, 379 (1992).
131. Heisel F., Mishe J. A., Schott M., and Sipp B.: Mol. Cryst. Liq. Cryst. *41*, 251 (1978).
132. Hesse R. and Baessler H.: Phys. Stat. Sol. (b), *101*, 485 (1980).

133. Higasi K., Baba H., and Rembaum A.: Quantum Organic Chemistry. Wiley Interscience, New York, London 1965.
134. Hiraya A., Achiba Y., Mikami N., and Kimura K.: J. Chem. Phys. *82*, 1810 (1985).
135. Holstein T.: Ann. Phys. (New York) *8*, 325 (1959).
136. Holstein T.: Ann. Phys. *8*, 343 (1959).
136a. Holstein T. and Turkevich L.: Phys. Rev. B *38*, 1901 (1988); *ibid.* *38*, 1923 (1988).
136b. Hoyland J. R. and Goodman L.: J. Chem. Phys. *36*, 12 (1962).
137. Hubavc I. and Svarvcek M.: Int. J. Quant. Chem. *23*, 403 (1988).
138. Hug G. L.: Ionic States in Molecular Crystals, PhD thesis, University of Chicago, Chicago, 1970.
139. Hug G. and Berry R. S.: J. Chem. Phys. *55*, 2516 (1971).
140. Hughes R. C. Geminate Recombination of Charge Carrier in Organic Solids. In: Annual Report, Conference on Electronic Insulation and Dielectric Phenomena. National Academy of Sciences, Washington, D.C., 1971, pp. 8–16.
141. Hughes R. C.: J. Chem. Phys. *55*, 5442 (1971).
142. Ioselevitch A. S. and Rashba E. I.: Soviet JETP *88*, 1873 (1985).
143. Ioselevich A. S. and Rashba E. I.: Solid State Commun. *55*, 705 (1985).
144. Iyechika Y., Yakushi K., Ikemoto I., and Kuroda H.: Acta Crystallogr. *38*, 766 (1982).
145. Jamashita J. and Kurosawa T.: Phys. Chem. Solids *5*, 34 (1958).
146. Jankowiak R., Kalinowski J., Konys M., and Bucherf J.: Chem. Phys. Lett. *65*, 549 (1979).
147. Jones W., Ramdas S., and Thomas J. M.: Chem. Phys. Lett. *54*, 490 (1978).
148. Jortner J.: Phys. Rev. Lett. *20*, 244 (1968).
149. Jurgis A. J. and Silinsh E. A.: Phys. Stat. Sol. (b) *53*, 735 (1972).
150. Kadashchuk A. K., Ostapenko N. I., Skrishevskii Y. A., and Sugakov V. J.: Soviet JETP Lett. *46*, 207 (1987).
150a. Kadashchuk A. K., Ostapenko N. I., Skrishevskii Y. A., Shpak M. T., Silinsh E. A., and Shlihta G. A.: Soviet Solid State Physics *32*, 1312 (1990).
151. Kadashchuk A. K., Ostapenko N. I., and Sugakov V. J. : Mol. Cryst. Liq. Cryst. *201*, 167 (1991).
151a. Kalinowski J., Godlewski J., and Mondalski P.: Mol. Cryst. Liq. Cryst. *175*, 67 (1989).
151b. Kalinowski J., Godlewski J., and Mondalski P.: In: Defect Control in Semiconductors, K. Sumino, ed. North-Holland, Amsterdam, New York, Oxford, Tokyo, 1990, Vol. 2, p. 1705.
152. Kao K. C. and Hwang W., Electrical Transport in Solids. With Particular Reference to Organic Semiconductors. Pergamon Press, Oxford-New York-Toronto-Sydney-Paris-Frankfurt, 1981.
153. Karl, N. Organic Semiconductors, in: Festkörperprobleme. Adv. Solid State Phys., *14*, 161 (1974).
154. Karl N. Organic Semiconductors. In: Landolt-Börnstein, Group III, Vol. 17, Semiconductors, Springer, Berlin, 1985, pp. 106–218.
155. Karl N.: In: 11th Molecular Crystal Symposium, Lugano, Switzerland, Sept. 30–Oct. 8, 1985, p. 135.
156. Karl N.: J. Cryst. Growth *99*, 1009 (1990).
157. Karl N.: In: Defect Control in Semiconductors, Vol. 2, K. Sumino, ed. Elsevier, North-Holland, Amsterdam, New York, Oxford, Tokyo, 1990.
158. Karl N., Marktanner J., Stehle R., and Warta W.: Synthetic Metals **41–43**, 2473 (1991).
159. Karl N., Rohrbacher H., and Siebert D.: Phys. Status Solidi (a) *4*, 105 (1971).
160. Kato K., and Braun C. L.: J. Chem. Phys. *72*, 172 (1980).
161. Katoh R. and Kotani M.: Chem. Phys. Letters *166*, 258 (1990).
162. Katoh R. and Kotani M.: Mol. Cryst. Liq. Cryst. *183*, 447 (1990).
163. Katz. J. L., Rice S. A., Choi S. I., and Jortner J.: J. Chem. Phys. *39*, 1683 (1963).
164. Kaulach I. S. and Silinsh E. A.: Phys. Status Solidi (b) *75*, 247 (1976).
165. Kenkre V. M.: Phys. Rev. *B11*, 1741 (1975).
166. Kenkre V. M.: Phys. Rev. *B12*, 2150 (1975).
167. Kenkre V. M.: Phys. Rev. *B22*, 2089 (1980).
168. Kenkre V. M.: Chem. Phys. Lett. *82*, 301 (1981).
169. Kenkre V. M. In: Exciton Dynamics in Molecular Crystals and Aggregates. Springer Tracts in Modern Physics 94, G. Hohler, ed. Springer, Berlin-Heidelberg-New York, 1982, pp. 1–109.
170. Kenkre V. M., Andersen J. P., Dunlap D. M., and Duke C. B. Phys. Rev. Lett., *62*, 1165 (1989).
171. Kenkre V. M. and Knox R. S.: Phys. Rev. *B9*, 5279 (1974).
172. Kenkre V. M. and Knox R. S.: J. Luminescence *12/13*, 187 (1976).
173. Kenkre V. M., Montroll E. W., and Shlessinger M. F.: J. Stat. Phys. *9*, 45 (1973).
174. Kenkre V. M. and Paris P. E.: Phys. Rev. *B27*, 3221 (1983).
175. Kenkre V. M. and Rahman S.: Phys. Lett. *A50*, 170 (1974).

176. Kenkre V. M. and Schmid D.: Chem. Phys. Lett. *94*, 603 (1983).
177. Kenkre V. M. and Wong Y. M.: Phys. Rev. *B22*, 5716 (1980).
178. Kenkre V. M. and Wong Y. M.: Phys. Rev. *B23*, 3748 (1981).
179. Kepler R. G.: Phys. Rev. *119*, 1226 (1960).
180. Kepler R. G. In: Organic Semiconductors, Brothy J. J., Buttery J. W., eds. Macmillan, New York, 1962, p. 1.
181. Killesreiter H. and Baesler H.: Phys. Status Solidi (b) *51*, 657 (1972).
182. Kitaigorodsky A. I.: Tetrahedron *14*, 230 (1961).
183. Kitaigorodsky A. I.: J. Chim. Phys. *63*, 9 (1966).
184. Kitaigorodsky A. I.: Molecular Crystals and Molecules. Academic Press, New York, 1973.
185. Kittel Ch.: Introduction to Solid State Physics, 2nd ed. Wiley, New York, London, 1956.
185a. Klimkans A. and Larsson S.: Chem. Phys. (1994). Accepted for publication.
186. Knight J. C. and Davis E. A.: J. Phys. Chem. Solids *35*, 543 (1974).
186a. Knox R. S.: Theory of Excitons. Academic Press, New York, 1963.
187. Kobayashi T. in: Crystals, Growth, Properties and Application 13, Organic Crystals I: Characterization, Springer-Verlag, Berlin, Heidelberg, New York, 1991, p. 6.
188. Kobayashi T., Fujiyishi Y., Iwatsu F., and Uyeda N.: Acta Crystallogr. *37*, 692 (1981).
189. Kobayashi T., Fujiyoshi Y., and Uyeda N.: Acta Crystallogr. *38*, 356 (1982).
190. Kobayashi T., Yamamoto-Asaka N., Maeda T., and Kawase N. in: Defect Control in Semiconductors, K. Sumino, ed. North-Holland, Amsterdam, New York, Oxford, Tokyo, 1990, Vol. 2, p. 1679.
191. Kobayashi T., Yase K., and Uyeda N.: Acta Crystallogr. B *40*, 263 (1984).
191a. Kojima K.: Progr. Crystal Growth Charact. *23*, 369 (1991).
192. Köngeter A. and Wagner M.: J. Chem. Phys. *92*, 4003 (1990).
193. Koopmans T.: Physica Bd. *1*, 104 (1934).
194. Kosic T., Schosser C., and Dlott D. D.: Chem. Phys. Lett. *96*, 57 (1983).
195. Kotani M.: Mat. Sci. *13*, 121 (1987).
196. Kotani M.: In: Defect Control in Semiconductors, Vol. 2, K. Sumino, ed. Elsevier, North-Holland, Amsterdam, New York, Oxford, Tokyo, 1990, p. 1697.
197. Kotani M., Morikawa E., and Katoh R.: Mol. Cryst. Liq. Cryst. *171*, 305 (1989).
198. Krivenko T. A., Dementjev V. A., Bokhenkov E. L., Kolesnikov A. I., and Sheka E. F.: Mol. Cryst. Liq. Cryst. *104*, 207 (1984).
199. Krotchuk A. S., Kurik M. V., and Smishko G.: Optika Spektrosk. *30*, 442 (1971).
200. Kühne R. and Reineker P.: Sol. St. Comm. *29*, 279 (1979).
201. Kurik M. V.: Phys. Stat. Sol. (a) *8*, 9 (1971).
202. Kurik M. V., Mangira V. S., and Tsikora L. I.: Soviet Fizika Tverd. Tela *13*, 716 (1971).
203. Kurik M. V. and Piryatinski Y. P. In: 11th Molecular Crystal Symposium, Sept. 30–Oct. 4, 1985, Lugano, Switzerland, 1985, pp. 174–177.
204. Kurik M. V. and Silinsh E. A.: Izv. Akad. Nauk. Latv. SSR, Ser. Fiz. Tekh. Nauk *3*, 47 (1984).
205. Kurik M. V., Spendiarov N. N., Savtchinskii P. I., and Frolova E. K.: Soviet Fizika Tverd. Tela *12*, 1443 (1970).
206. Kurik M. V. and Tsikora L. I.: Phys. Stat. Solidi *66*, 695 (1974).
207. London F.: Z. Phys. Chem. *11*, 222 (1930).
208. Lang I. G. and Firsov Yu. A.: Soviet JETP *43*, 1843 (1962).
209. Lang I. G. and Firsov Yu. A.: Soviet JETP *45*, 378 (1963).
209a. Larsson S. Intern. J. Quant. Chem. *30*, 31 (1986).
209b. Larsson S. and Braga M.: Chem. Phys. *176*, 367 (1993).
210. Lax M.: J. Chem. Phys. *20*, 1753 (1952).
211. Lax M.: Rev. Mod. Phys. *32*, 25 (1960).
212. LeBlanc O. H., Jr.: J. Chem. Phys. *33*, 626 (1960).
213. LeBlanc O. H., Jr.: J. Chem. Phys. *36*, 1082 (1962).
214. LeFevre C. G. and LeFevre R. J. W.: Rev. Pure Appl. Chem. *5*, 261 (1955).
215. LeFevre R. J. W., Radom L., and Ritchie G. L. D.: J. Phys. Chem. Soc. Sect. *B*, 775 (1968).
216. LeFevre R. J. W. and Sundaram K.: J. Chem. Soc. *1963*, 4442.
217. Lennard-Jones J. E.: Cambridge Philos. Soc. *1963*, 469 (1931).
218. Lin S. H., Boeglin A., and Lin S. M.: J. Photochem. *39*, 173 (1987).
219. Lochner K., Reimer B., and Baessler H.: Phys. Stat. Sol. (b) *76*, 533 (1976).
220. Lyons L. E.: J. Chem. Soc. 1957, 5001.
221. Lyons L. E. and Mackie J. C.: Proc. Chem. Soc. *71*, 3 (1962).
222. Lyons L. E. and Milne K. A.: J. Chem. Phys. *65*, 1474 (1974).

223. Mackie J. C.: Ph.D. thesis, University of Sydney, Sydney, 1963.
223a. Marcus R. A.: J. Chem. Phys. *24*, 966 (1956); *ibid. 24*, 979 (1956); *ibid. 43*, 679 (1965).
223b. Marcus R. A. and Sutin N.: Biochim. Biophys. Acta *811*, 265 (1985).
224. Martienssen W.: J. Phys. Chem. Sol. *2*, 257 (1957).
225. Mathieson A., Robertson J. M., and Sinclair V. C.: Acta Crystallogr. *3*, 245 (1950).
226. Matsui A.: J. Phys. Soc. Jpn. *21*, 22212 (1966).
227. Matsui A.: J. Opt. Soc. Am. *B* 1990, 1615.
228. Matsui A., Iemura N., and Nishmura: J. Luminiscence *24/25*. 445 (1981).
229. Matsui A. and Mizuno K.: Exciton dynamics in organic molecular crystals. In: 5th Intern. Conf. on Excited States in Solids, Lyon, July 1–4, 1985, Lyon, 1985.
230. Matsui A. and Mizuno K.: Proc. Intern. Soc. Opt. Eng. *1054*, 224 (1989).
231. Matsui A., Mizuno K., Tamai N., and Yamasaki I.: J. Luminescence *40, 41*, 455 (1988).
232. Matsui A., Mizuno K., Tamai N., and Yamasaki I.: Proc. Intern. Soc. Opt. Eng. *910*, 131 (1988).
233. Matsui A., Ohno T., Mizuno K., Yokoyama T., and Kobayashi M.: Chem. Phys. *111*, 121 (1987).
234. Melz P.: J. Chem. Phys. *57*, 1694 (1972).
235. Merrifield R. E.: J. Chem. Phys. *28*, 647 (1958).
236. Miyazaki H. and Hanamura E.: J. Phys. Soc. Jpn. *50*, 1310 (1981).
237. Miyazaki H. and Hanamura E.: J. Phys. Soc. Jpn. *51*, 818 (1982).
238. Miyazaki H. and Hanamura E.: J. Phys. Soc. Jpn. *51*, 828 (1982).
239. Mizuno K. and Matsui A.: J. Phys. Soc. Jpn. *55*, 2427 (1986).
240. Mizuno K. and Matsui A.: J. Luminescence *38*, 323 (1987).
241. Mizuno K. and Matsui A.: In: 5th Int. Conf. Molecular Systems, Oct. 13–17, 1991, Okazaki, Japan, 1991.
242. Mizuno K., Matsui A., and Sloan G. J.: J. Phys. Soc. Jpn., *53*, 2799 (1984).
243. Mokichev N. N. and Pachomov L. G.: Soviet Sol. State Phys. *24*, 3389 (1982).
243a. Moler C. and Van Loan C.: SIAM Rev. *20*, 801 (1978).
244. Montroll E. W. and Shlesinger M. E. In: Studies in Statistical Mechanics, Vol. XI, J. L. Lebowitz and E. W. Montroll, eds. North-Holland, Amsterdam-Oxford-New York-Tokyo, 1984, pp. 1–121.
245. Mori H.: Progr. Theor. Phys. *33*, 423 (1965).
246. Mott N.: Contemporary Phys. *22*, (1981).
247. Mott N. F. and Davis E. A.: Electron Processes in Non-Crystalline Materials, Clarendon Press, Oxford, 1979.
248. Mott N. F. and Gurney R. W.: Electronic Processes in Ionic Crystals, 2nd ed., University Press, London, Oxford, 1940; Clarendon, Oxford 1948, reprinted Dover, New York, 1964.
249. Mott N. F. and Littleton M. J.: Trans. Faraday Soc. *34*, 485 (1938).
250. Movaghar B.: J. Molec. Electron. *3*, 183 (1987).
251. Movaghar B.: J. Molec. Electron. *4*, 79 (1988).
251.a Mozumder A., Pimblott S. M., Clifford P., and Green N. J. B.: Chem. Phys. Lett. *142*, 385 (1987).
252. Mulliken: J. Am. Soc. *74*, 811 (1952).
253. Munn R. W.: Chem. Phys. *84*, 413 (1984).
254. Munn R. W.: Second Intern. Symposium on Molecular Electronics and Biocomputers, Moscow, Sept. 11–18, 1989.
255. Munn R. W., Nicholson J. R., Schwob H. P., and Williams D. F.: J. Chem. Phys. *58*, 3928 (1973).
255a. Munn R. W.: Mol. Cryst. Liq. Cryst. *288*, 23 (1993).
256. Munn R. W. and Siebrand W.: J. Chem. Phys. *52*, 47 (1970).
257. Munn R. W. and Silbey R.: J. Chem. Phys. *68*, 2439 (1978).
258. Munn R. W. and Silbey R.: J. Chem. Phys. *83*, 1843 (1985).
259. Munn R. W. and Silbey R.: J. Chem. Phys. *83*, 1854 (1985).
260. Munn R. W. and Williams D. F.: J. Chem. Phys. *59*, 1742 (1973).
261. Mason R. and Williams G.: J. Chem. Soc. Dalton Trans. 1979, 676.
262. Muzikante I. J., Taure L. F., Rampans A., and Silinsh E. A.: Izv. Acad. Nauk Latv. SSR, Ser. Fiz. Tekh. Nauk *4*, 18 (1983); *1*, 76 (1986) and *6*, 31 (1987).
263. Nagy P.: J. Phys. C: Sol. St. Phys. *20*, 5527 (1987).
264. Nakada J.: J. Phys. Soc. Jpn. *20*, 246 (1965).
265. Nakajima S.: Progr. Theor. Phys. *20*, 948 (1958).
266. Nasu K.: J. Luminiscence *38*, 90 (1987).
267. Nasu K. and Toyozawa Y.: J. Phys. Soc. Jpn. *50*, 235 (1981).
268. Nedbal L.: Phys. Stat. Sol. (b) *109*, 109 (1982).
269. Nedbal L. and Szöcs V.: J. Theor. Biol. *120*, 411 (1986).

270. Nešpurek S., Zmeškal O. and Schauer F.: a. Phys. Status Solidi (a) 75, 591 (1983); b. Phys. Status Solidi (a) 85, 619 (1984); c. J. Phys. Sol. State Phys. 19, 7231 (1986).
271. Nešpurek S. and Koropecki I.: Mat. Sci. (Poland) 123, 181 (1987).
272. Nešpurek S. and Silinsh E. A.: Phys. Status Solidi (a) 34, 747 (1976).
273. Nešpurek S. and Sworakowski J.: Phys. Status Solidi (a) 46, 273 (1978).
274. Noolandi J. and Hong. K. M.: J. Chem. Phys. 70, 3230 (1979).
275. Okada I., Ide N., and Kojima K. In: Defect Control in Semiconductors, Vol. 2, K. Sumino, ed. North-Holland, Amsterdam, New York, Oxford, Tokyo, 1990, p. 1667.
276. Onsager L.: J. Chem. Phys. 2, 599 (1934).
277. Onsager L.: Phys. Rev. 54, 554 (1938).
278. Ostapenko N. and Sugakov V. in: Defect Control in Semiconductors, Vol. 2, K. Sumino, ed. North-Holland, Amsterdam, New York, Oxford, Tokyo, 1990, p. 1701.
279. Pai D. M. and Enck R. C.: Phys. Rev. B11, 5163 (1975).
280. Palacin S., Ruaudel-Teixier A., and Barraud A.: Mol. Cryst. Liq. Cryst. 156, 331 (1988).
281. Pauli W. In: Probleme der moderne Physik, Festschrift zum 60. Geburtstag A. Sommerfelds. P. Debye and S. Hirzel, ed. Leipzig, 1928, p. 30–45
282. Peier W.: Physica 57, 565 (1972),
283. Persin A. J. and Kitaigorodsky A. T.: The Atom-Atom Potential Method. Springer, Berlin, 1987.
284. Petelenz P.: Phys. Stat. Sol. (b) 73, 293 (1976).
284a. Pimblott S. M., Mozumder A., and Green N. J. B.: J. Chem. Phys. 90, 6595 (1989).
285. Polarons, Y. A. Firsov, ed. (in Russian). Nauka, Moscow, 1975.
286. Pope M. and Burgos J.: Mol. Cryst. 3, 215 (1967).
287. Pope M. and Kallmann, H.: Discus. Faraday Soc. 51, 7 (1971).
288. Pope M. and Swenberg C. E.: Electronic Processes in Organic Crystals. Clarendon Press, Oxford University Press, Oxford, New York, 1982.
288a. Pope M.: Private communication (1993).
289. Popovic Z. D. and Sharp J. H.: J. Chem Phys. 66, 5076 (1977).
290. Powell C. and Socs Z. G.: J. Lumin. 11, 1 (1975).
291. Primas H.: Helv. Phys. Acta 34, 36 (1961).
292. Probst K. H. and Karl N.: Phys.. Stat. Solidi A27, 499 (1975).
292a. Rackowski S. and Scher H.: J. Chem. Phys. 89, 7242 (1988).
292b. Rackowski S.: Chem. Phys. Lett. 178, 19 (1991).
293. Ramdas S., Thomas J. M., and Goringe N. J.: J. Chem. Soc. Faraday Trans. II 73, 551 (1977).
294. Rashba E. I.: Self-trapping of excitons in solids. In: Defects in Insulating Crystals. V. M. Tuchkevich and K. K. Shvarts, ed. Zinatue Riga and Springer, Berlin-Heidelberg-New York, 1981, pp. 255–266
295. Rashba E. L.: Autolocalization of excitons. In: Excitons (in Russian). Nauka, Moscow, 1985.
296. Reineker P. In: Exciton Dynamics in Molecular Crystals and Aggregates. Springer Tracts in Modern Physics 94. G. Hohler, ed. Springer, Berlin-Heidelberg-New York, 1982, pp. 111–226.
297. Reineker P., Keiser B., and Jayannavar A. M.: Phys. Rev. A39, 1469 (1989).
298. Reineker P. and Kühne R.: Phys. Rev. B21, 2448 (1980).
299. Ries B., Schönherr G., Baessler H., and Silver M.:, Phil. Mag. B49, 597 (1984).
299a. Risken H.: The Fokker-Planck Equation. Methods of Solution and Application, 2nd ed. Springer-Verlag, Heidelberg, 1989.
300. Rittner E. S., Hunter R. A., and De Pre F. K.: J. Chem. Phys. 17, 198 (1949).
301. Robertson J. M.: Proc. R. Soc. A. 142, 659 (1933).
302. Robertson J. M.: Rev. Mod. Phys. 30, 155 (1958).
303. Robertson J. M., Sinclair V. C., and Trotter J.: Acta Crystallogr. 14, 697 (1961).
304. Salaneck W. R.: Phys. Rev. Lett. 40, 60 (1978).
305. Salaneck W. R., Duke C. B., Eberhardt W., Plummer E. W., and Freund H. J.: Phys. Rev. Lett. 45, 280 (1980).
306. Sano H. and Mozumder A.: J. Chem Phys. 66, 689 (1977).
307. Sato N., Inokuchi H., and Silinsh E. A.: Chem. Phys. 115, 269 (1987).
308. Sato H., Kawasaki M., Toya K., and Kimura K.: J. Phys. Chem. 91, 751 (1987).
309. Sato N., Seki K., and Inokuchi H.: J. Chem. Soc. Faraday Trans. II 77, 1621 (1981).
310. Sato N., Seki K., Inokuchi H., Harada Y., and Takahashi T.: Solid State Commun. 41, 759 (1982).
311. Scheidt W. R. and Dow W.: J. Am. Chem. Soc. 99, 1101 (1977).
312. Schein L. B.: Phys. Rev. B15, 1024 (1979).
313. Schein L. B., Duke C. B., and McGhie A. R.: Phys. Rev. Lett. 40, 197 (1978).

314. Schein L. B. and McGhie A. R.: Chem. Phys. Lett. *62*, 356 (1979).
315. Schein L. B. and MGhie A. R.: Phys. Rev. *B20*, 1631 (1979).
316. Schein L. B. Warta W., McGhie A. R., and Karl N.: Chem. Phys. Lett. *75*, 267 (1980).
316a. Scher H.: Mol. Cryst. Liq. Cryst. *288*, 41 (1993).
317. Sebastian L., Weiser G., and Baessler H.: Chem. Phys. *61*, 125 (1981).
318. Sebastian L., Weiser G., Peter G., and Baessler H.: Chem. Phys. *75*, 103 (1983).
319. Seiferheld U., Ries B., and Baessler H.: J. Phys. C. *16*, 5189 (1983).
320. Seki. K., Harada Y., Ohno K., and Inokuchi H.: Bull. Chem. Soc. Jpn. *47*, 1608 (1974).
321. Selsby R. G. and Grimison A.: Int. J. Quant. Chem. *72*, 3677 (1968).
322. Selsby R. G. and Grimson A.: Int. J. Quant. Chem. *12*, 527 (1977).
323. Sevian H. M. and Skinner J. L.: J. Chem. Phys. *91*, 1775 (1989).
324. Sewell G. L.: Phil. Mag. *3*, 1361 (1958).
325. Sherman A. V.: Phys. Stat. Sol. (b) *141*, 151 (1987).
326. Sherman A. V.: Phys. Stat. Sol. (b) *145*, 319 (1988).
327. Shockley W., Bell System Techn. J. *30*, 990 (1951).
327a. Siebrand W.: J. Chem. Phys. *41*, 3574 (1964).
328. Silbey R., Jortner J., Rice S. A., and Vala M. T., Jun.: J. Chem. Phys. *42*, 733 (1965); *ibid.* *43*, 2925 (1965).
329. Silbey R. and Munn R. W.: J. Chem. Phys. *72*, 2763 (1980).
330. Silinsh E. A.: Phys. Status Solidi (a) *3*, 817 (1970).
331. Silinsh E. A., Izv. Nauk Latv. SSR, Ser. Phys. Tekn. Nauk, *2*, 26 (1977).
332. Silinsh E. A.: Organic Molecular Crystals. Their Electronic States. Springer-Verlag, Berlin, Heidelberg, New York, 1980.
333. Silinsh E. A. Local electronic states of structural origin in organic molecular crystals, In: Defects in Insulating Crystals, V. M. Tuchkevich and K. K. Shvarts, eds. Zinatne, Riga; Springer, Heidelberg, 1981, pp. 107–134.
334. Silinsh E. A. XIth Molecular Crystal Symposium, Lugano, Switzerland, 1985, p. 277.
335. Silinsh E. A. Local trapping states of structural origin in organic semiconductors: nature, distribution characteristics, methods of control, in: Defect Control in Semiconductors, Vol. 2, K. Sumino, ed. North-Holland, Amsterdam, New York, Oxford, Tokyo, 1990, p. 1679.
336. Silinsh E. A., Bouvet M., and Simon J.: Molecular Materials, Section C of Mol. Cryst. Liq. Cryst. (1994). (Submitted for publication.)
337. Silinsh E. A., Belkind A. I., Balode D. R., Biseniece A., Grehov V. V., Taure L. F., Kurik M. V., Vertsimaha Y. J., and Bok J.: Phys. Stat. Solidi (a) *25*, 339 (1974).
338. Silinsh E. A. and Čápek V.: Izv. Akad. Nauk Latv. SSR, Ser. Fiz Tekh. Nauk *6*, 15 (1987).
339. Silinsh E. A. and Čápek V., Izv. Akad. Nauk Latv. SSR, Ser. Fiz. Tekh. Nauk, *2*, 32 (1988).
340. Silinsh E. A., Čápek V., and Nedbal L.: Phys. Stat. Sol. (b) *102*, K149 (1980).
341. Silinsh E. A. and Inokuchi H.: Chem. Phys. *149*, 373 (1991).
342. Silinsh E. A. and Jurgis A. J.: Izv. Akad. Nauk Latv. SSR, Ser. Fiz. Tekh. Nauk *1*, 73 (1977).
343. Silinsh E. A. and Jurgis A. J.: Izv. Akad. Nauk Latv. SSR, Ser. Fiz. Tekh. Nauk *2*, 21 (1977).
344. Silinsh E. A. and Jurgis A. J.: Chem. Phys. *94*, 77 (1985).
345. Silinsh E. A. and Jurgis A. J.: Mater. Sci. *13*, 247 (1987).
346. Silinsh E. A., Jurgis A. J., Gailis A. K., and Taure L. F.: Energy spectra of ionized states for perfect and defective molecular crystals. In: Electric Properties of Organic Solids. Karpacz, Poland, Sept. 1–7, 1974, Wroclaw, 1974, pp. 40–68
347. Silinsh E. A., Jurgis A. J., Kolesnikov V. A., and Muzikante I. J.: In: 10th Molecular Crystal Symposium, St. Jovite, Quebec, Canada, St. Jovite, 1982, p. 264.
348. Silinsh E. A., Jurgis A. J., and Shlihta G. A.: J. Mol. Electron. *3*, 123 (1987).
349. Silinsh E. A., Kolesnikov V. A., Muzikante I. J., and Balode D. R.: Phys. Stat. Solidi (B) *113*, 379 (1982).
350. Silinsh E. A., Kolesnikov V. A., Muzikante I. J., Balode D. R., and Gailis A. K. On charge carrier photogeneration mechanism in organic molecular crystals, preprint LAFI-34; submitted to Intern. Conf. on Defects in Insulating Crystals, Riga, May 18–23, 1981 Inst. of Physics, Latvian SSR Academy of Sciences, Salaspils, Latvia, 1981.
351. Silinsh E. A., Kurik M. V., and Čápek V.: Electronic Processes in Organic Molecular Crystals. Localization and Polarization Phenomena (in Russian). Zinatne, Riga, Latvia, 1988.
352. Silinsh E. A., Muzikante I. J., Rampans A., and Taure L. F.: Chem. Phys. Lett. *105*, 617 (1984).
353. Silinsh E. A., Muzikante I. J., Taure L. F., and Shlihta G. A.: J. Mol. Electron. *7*, 127 (1991).
354. Silinsh E. A., Shlihta G. A., and Jurgis A. J.:, In: Proceed. of 6th Intern. Conference on Energy and

Electron Transfer, Vol. 2. Prague, Czechoslovakia, Aug. 14–18, 1989, p. 191.

355. Silinsh E. A., Shlihta G. A., and Jurgis A. J.: Chem. Phys. *138*, 347 (1989).
356. Silinsh E. A., Shlihta G. A., and Jurgis A. J.: Chem. Phys. *155*, 389 (1991).
357. Simon J. and Andre J. J. Molecular Semiconductors, Springer-Verlag, Berlin, Heidelberg, New York, Tokyo, 1985.
358. Sinclair V. C., Robertson J. M., and Mathieson A. M.: Acta Crystallogr. *3*, 251 (1950).
359. Skála L.: Int. J. Quant. Chem. *23*, 497 (1983).
360. Skála L. and Bílek O.: Acta Univ. Carolinas-Math. Phys. *23*, 61 (1982); Phys. Stat. Sol. (b) *114*, K51 (1982).
361. Skála L. and Kenkre V. M.: Phys. Lett. *114*, 395 (1986).
362. Skála L. and Kenkre V. M.: Z. Physik *B63*, 259 (1986).
363. Skettrup T.: Phys. Rev. *B18*, 2622 (1978).
364. Sokolov F. F.: Proc. Estonian Acad. Sci. *25*, 422 (1976).
365. Sokolov F. F. and Hizhnakov V. V.: Phys. Stat. Sol. (b) *75*, 669 (1976).
366. Sondermann U., Kutoglu A., and Baessler H.: J. Phys. Chem. *89*, 1735 (1985).
367. Stevens B.: Spectrochim. Acta, *18*, 439 (1962).
368. Sumi H.: J. Phys. Soc. Jpn. *36*, 770 (1974).
369. Sumi H.: J. Phys. Soc. Jpn. *38*, 825 (1975).
370. Sumi H.: J. Chem. Phys. *67*, 2943 (1977).
371. Sumi H.: J. Chem. Phys. *70*, 3775 (1979).
372. Sumi H. and Sumi A.: J. Soc. Jpn. *56*, 2211 (1987).
373. Suzuki M., Yokoyama T., and Ito M.: Spectrochim. Acta A *24*, 1091 (1968).
374. Sworakowski J.: Mol. Cryst. Liq. Cryst. *11*, 1 (1970).
375. Sworakowski J.: In: Electric Properties of Organic Crystals, Karpacz, Poland, Sept. 1–7, 1974, Wroclaw 1974, p. 191.
376. Sworakowski J.: Proceedings of 7th Intern. Symposium on Electrets, Berlin, Sept. 25–27, 1991, p. 45.
376a. Sworakowski J.: Abstracts of the 6th Symposium on Unconventional Photoactive Solids, Leuven, Belgium, August 15–19, 1993, p. 159.
377. Szöcs V. and Barvík I.: J. Theor. Biol. *122*, 179 (1986).
378. Szöcs V. and Barvík I.: J. Phys. C: Solid St. Phys. *21*, 1533 (1986).
379. Takase T. and Kotani M.: Private communication (1987).
380. Tanaka J.: Bull. Chem. Soc. Jpn. *36*, 1237 (1963).
381. Taure L., Abele I., Muzikante I, and Silinsh E. A.: Dunkel- and Photoleitfähigkeit of Vanadyl-Phthalozyan in Langmuir-Blodgett Schichten, 26. Tagung Organishe Festkorper, June 27–28, 1989, Chemnitz, Germany.
382. Thaxton G. D., Jarningen R. C., and Silver M.: J. Chem. Phys. *66*, 2461 (1962).
383. Tokuyama M. and Mori H.: Prog. Theor. Phys. *55*, 411 (1976).
384. Tomkiewitz Y., Groff R., and Avakian P.: J. Chem. Phys. *54*, 4504 (1971).
385. Tomura M. and Takahasi Y.: J. Phys. Soc. Jpn. *25*, 647 (1968).
386. Toyozawa Y.: Progr. Theor. Phys. (Kyoto) *12*, 421 (1954).
387. Toyozawa Y.: Progr. Theor. Phys. *26*, 29 (1961).
388. Toyozawa Y.: ISSP Tech. Rep. A 119 (1964).
389. Toyozawa Y.: Localization and Delocalization of an Exciton in in the Phonon Field. In: Organic Molecular Aggregates, P. Reineker, H. Haken, and H. C. Wolf, eds. Springer-Verlag, Berlin-Heidelberg-New York-Tokyo, 1983, pp. 90–106
390. Toyozawa Y.: In: Highlights of Condensed Matter Theory, Tipog. Compositori, Bologne, Italy, 1985, pp. 798–830.
390a. Treutlein H., Schulten K., Brünger A. T., Karplus M., Deisenhofer J., and Michel H.: Proc. Natl. Acad. Sci. USA *89*, 75 (1992).
391. Trlifaj M.: Czech. J. Phys. *6*, 533 (1956).
392. Troitsky V. i., Bannikov V. S., and Berzina T. S.: J. Mol. Electron. *5*, 147 (1989).
393. Trotter J.: Acta Crystallogr. *15*, 289 (1962).
394. Turro N. J.: Modern Molecular Photochemistry. Benjamin, New York, 1978.
395. Ukei K.: Acta Crystallogr. B *29*, 2290 (1973).
396. Urbach F.: Phys. Rev. *92*, 1324 (1953).
397. Uyeda N., Kobayashi T., Ishizuka K., Fujiyoshi Y.: Chem. Scripta, *14*, 47 (1978–1979).
398. Venzl G. and Fischer S. F.: Phys. Rev. B *32*, 6437 (1985).
399. Vilfan J.: Phys. Status Solidi (b) *59*, 351 (1973).

400. Vilken A. J., Grechov V. V., and Murashov A. A.: Izv. Akad. Nauk Latv. SSR, Ser, Fiz. Tekh. Nauk 6, 25 (1989).
401. Wagner M.: J. Chem. Phys. 82, 3207 (1985).
402. Wagner M: Non-radiative transitions: dangers and deficiencies of the adiabatic potential sheets. In: Proc. XIIth Int. Conf. on Defects in Semiconductors, L. E. Kimerling and J. M. Parsey, eds. Metallurg. Soc. AIME, Warrandale, PA, 1985, pp. 229–234.
403. Wagner M. and Koengeter A.: J. Chem. Phys. 88, 7550 (1988).
404. Wagner M. and Köngeter A.: J. Chem. Phys. 91, 3036 (1989).
405. Wang S., Mahutte C. K., and Matsura M.: Phys. Stat. Sol. (b) 51, 11 (1972).
406. Warta W. and Karl N.: Phys. Rev. B32, 1172 (1985).
407. Warta W., Stehle R., and Karl N.: Appl. Phys. A36, 163 (1985).
408. Wheeler J. A.: Niels Bohr, the man. Physics Today, Oct. 1985, p. 66.
409. Willig P.: Private communication (1989).
410. Zagrubskii A. A., Petrov V. V., and Lovcjus V. A.: Izv. Akad. Nauk Latv. SSR, Ser. Fiz. Tekh. Nauk 5, 29 (1981).
410a. Zhao Y., Brown D. W., and Lindenberg K.: J. Chem. Phys. 100, 2335 (1994).
411. Zhukov V. D., Kurik M. V., Piryatinski Yu. P., and Tsikora L. N.: Soviet Sol. State Phys. 25, 1030 (1983).
412. Ziolo R. F., Griffiths C. H., and Traup J. M.: J. Chem. Soc., Dalton Trans. 2300 (1980).
413. Zwanzig R.: J. Chem. Phys. 33, 1338 (1960)
414. Zwanzig R. W. In: Lecture in Theor. Phys. Vol. 3, Boulder, CO, 1960. W. E. Brittin, B. W. Downs, and J. Downs, eds. Interscience, New York, 1961, pp. 106–141
415. Zwanzig R.: Physica 30, 1109 (1964).
416. Yamashita J. and Kurosawa T.: J. Phys. Chem. Sol. 5, 34 (1958).
417. Yarkony D. and Silbey R.: J. Chem. Phys. 67, 5818 (1977).

Subject Index

A

Acoustic (lattice) phonons in OMC, 22, 334, 354, 364

Acridine
  as guest molecule in Ac host crystal, 191

Activation energy
  of CP state thermal dissociation (carrier geminate escape), 257
  of dark conductivity, 193
    in SCLC regime, 193
  of photogeneration, 257, 280

Ac-type crystals (*see also* Organic molecular crystals, Polyacenes)
  interaction time scale in, 126, 128
  interaction types in, 126, 128
  typical interaction energies in, 126, 128
  typical localization (residence) time in, 126

Adiabatic approximation, 30, 66
  traditional treatment, 66

Adiabatic (dynamic) representation, 6, 30, 66
  versus crude representation, 66

Anthracene
  crystal structure, 11
  electron affinity, 121, 161
  intramolecular vibrational modes, 20
  ionization energy, 121, 161
  molecular structure, 10
  electric quadrupole moments, 117

Anthracene crystal
  absorption spectrum, 276
  autoionization quantum yield in, 287
  band structure for carriers in, 92
    for excitons in, 64
  charge carrier mobility in, 332, 333

charge carrier traps in, 184, 186, 187, 191

charge pair (CP) state energies in, 172, 174

charge transfer (CT) state energies in, 154, 158, 176

Davydov splitting in, 62, 63

dislocation defects in, 187, 188

effective mass of carriers in, 350

electron affinity in, 161

electronic polarization energy in, 110, 113, 115, 118, 122, 123, 161

energy gap in, 123, 130, 148, 149, 150, 156, 158, 161
  adiabatic, 123, 130, 148, 149, 150, 158, 161
  optical, 123, 130, 148, 156, 158, 161

energy level diagram of, 161

energy levels of neutral states in, 161

energy levels of polaron states in, 161

energy spectra of local states in, 182, 187, 188

exciton diffusion length in, 65

exciton dispersion-type interaction energy, 62, 63

exciton resonance interaction energy, 62

exciton self-trapping parameters in, 71

exciton traps in, 190

guest molecules as trapping states in, 191

hopping (residence) time of carriers in, 126

hot charge carriers in, 358

intrinsic photoconductivity threshold in, 147, 158, 265

intrinsic carrier generation in, 265

ionization energy of, 121, 161